Vijai Kumar Gupta, Susanne Zeilinger, Edivaldo Ximenes Ferreira Filho, María-del-Carmen Durán-Domínguez-de-Bazúa, Diane Purchase (Eds.)
Microbial Applications

Vijai Kumar Gupta, Susanne Zeilinger,
Edivaldo Ximenes Ferreira Filho,
María-del-Carmen Durán-Domínguez-de-Bazúa,
Diane Purchase (Eds.)

Microbial Applications

Recent Advancements and Future Developments

DE GRUYTER

Editors

Dr. Vijai Kumar Gupta, Ph. D.
National University of Ireland Galway
School of Natural Sciences,
Discipline of Biochemistry,
Glaway, Irland
E-mail: vijaifzd@gmail.com
Prof. Dr. Susanne Zeilinger
University of Innsbruck
Institute of Microbiology
Technikerstrasse 25, 6020 Innsbruck, Austria
E-mail: Susanne.Zeilinger@uibk.ac.at
Prof. Dr. Edivaldo X. Ferreira Filho
Laboratory of Enzymology,
Department of Cellular Biology,
University of Brasília,
Brasília, DF, CEP 70910-900, Brazil
E-mail: eximenes@unb.br
Prof. Dr.
María-del-Carmen Durán-Domínguez-de-Bazúa
Universidad Nacional Autónoma de México
Departamento Ingeniería Química
04510 Mexico, D.F., Mexico
E-mail: mcduran@unam.mx
Dr. Diane Purchase
Middlesex University London
Faculty of Science & Technology,
Department of Natural Sciences
The Burroughs, London NW4 4BT, Great Britain
E-mail: D.Purchase@mdx.ac.uk

The book has 47 figures and 32 tables.

ISBN 978-3-11-041220-8
e-ISBN (PDF) 978-3-11-041278-9
e-ISBN (ePUB) 978-3-11-041282-6

Library of Congress Cataloging-in-Publication Data
A CIP catalog record for this book has been applied for at the Library of Congress.

Bibliographic information published by the Deutsche Nationalbibliothek
The Deutsche Nationalbibliothek lists this publication in the Deutsche Nationalbibliografie; detailed bibliographic data are available in the Internet at http://dnb.dnb.de.

Cover image: kasto80/iStock/Thinkstock
Typesetting: Satzstudio Borngräber, Dessau-Roßlau
Printing and Binding: CPI books GmbH, Leck
♾ Printed on acid-free paper
Printed in Germany

www.degruyter.com

Preface

An enormously diverse group of microbes contribute to our health, food, agriculture, and environment. Over the last century, microorganisms have been exploited for the production of numerous enzymes, food ingredients, pharmaceuticals, solvents and chemical feedstocks etc.

Since very ancient times man have been utilizing microbial population for the production of priceless products via various fermentation processes which includes traditional fermented foods and beverages such as bread, beer, wine, vinegar, yoghurt and cheese, as well as fermented fish, meat and vegetables. Currently, more than 3500 traditionally fermented foods exist in the world. Fermentation is both transformation and preservation techniques for food and also improves its nutritional and organoleptic qualities. Additionally, a well conducted fermentation process favours useful microbial flora, to the detriment of undesirable flora in order to prevent spoilage and promote taste and texture.

Enzymes are being known to mankind since the ancient human civilization. Microbial enzymes are currently acquiring much attention as they are preferred due to their economic feasibility, high yields, consistency, ease of product modification and optimization, regular supply, rapid growth of microbes on inexpensive media, stability, and greater catalytic activity. Current applications are focused on many different markets including pulp and paper, leather, detergents and textiles, pharmaceuticals, chemical, food and beverages, biofuels, animal feed and personal care, among others. Also, they play a major role in the diagnosis, treatment, biochemical investigation, and monitoring of various dreaded diseases.

Bioenergy is major stake holder in meeting global future energy needs and is the only renewable source that can replace fossil fuels in all energy markets. The recent period has seen a renewed intense focus on biomass to bioenergy conversion technologies and processes, with the aim of developing economical and sustainable solutions at commercial scale. First generation bioenergy used starch-based agricultural products as substrates and second-generation bioenergy uses the lignocellulose present in biomass, agricultural residues and wastes. Hydrolytic enzymes generally derived from microbial sources is the preferred option as they selectively convert carbohydrate-rich biopolymers in biomass to fermentable sugars, without formation of by-products that inhibit downstream bioenergy and biorefinery conversion processes. Utilization of each fraction in biomass provides an effective way to minimize environmental pollution, address food security problems and improve agricultural waste management approaches.

Currently, the bacterial resistance, especially to most commonly used antibiotics has proved to be a severe therapeutic problem. Therefore, we are forced to develop an alternative or supportive treatment for successful cure of life-threatening infections. Endolysins are enzymes used by bacteriophages at the end of their replication cycle to degrade the peptidoglycan of the bacterial host from within, resulting in cell lysis and release of progeny virions. Endolysins when applied externally can make them interesting antimicrobial candidates, particularly in light of increasing bacterial drug resistance. The development of molecular biology and novel generation methods of sequencing has opened up new possibilities in the design of engineered phages and recombinant phage-derived proteins in combating multidrug resistant infections.

The new found awareness in human safety and environmental conservation has kindled fresh enthusiasm for natural sources of colors. Among microbes, bacteria have immense potential to produce pigments. Bacterial pigments due to their better biodegradability and higher compatibility with the environment, offer promising avenues for various applications in food, pharmaceuticals, cosmetics and textiles. Production of pigments by fermentation has a number of advantages: cheaper production, easier extraction, higher yields, and abundant raw materials. Optimization of fermentation process and the medium components are reported as key strategies for economic recovery of pigments.

Antimicrobial peptides encompass a wide variety of structural motifs which confer their antimicrobial activity. Peptaibols and peptaibiotics show interesting physico-chemical and biological properties including the formation of pores in bilayer lipid membranes, as well as antibacterial, antifungal, occasionally antiviral activities, and may elicit plant resistance. They have unusual amino acid content which are the result of non-ribosomal biosynthesis.

Throughout history the metabolic activities of fungi have been harnessed for various industrial applications. Fungi are indispensable biotechnological tools central to the commercial manufacture of fermented food, pharmaceuticals, organic acids, enzymes. Currently, fungal derived enzymes that degrade plant derived biomass are being utilized for the development of bioprocesses for biofuel and renewable chemical production. The metabolic and enzymatic diversity of fungi will continue to be developed for several industrial products.

Cyanobacteria have been identified as a rich source of biologically active compounds with antiviral, antibacterial, antifungal and anticancer activities and secondary metabolites including exopolysaccharides, vitamins, toxins, enzymes pharmaceuticals and polyhydroxyalkanoates. Cyanobacterial hydrogen has been considered as a very promising source of alternative energy. Additionally, they are also used in wastewater treatment, food, fertilizers and aquaculture.

Recent advances in genomics, metagenomics, proteomics, efficient expression systems, metabolic engineering and emerging recombinant DNA techniques have facilitated the discovery of new microbial applications. Metagenomic libraries should be constructed to discover new functional genes in microorganisms that are involved in the biosynthesis of biotechnological relevant compounds.

Most recent research, current technologies and advancements on applications of microorganisms in various areas are extensively covered in the book. The merits and drawbacks of each microbial application is described in the book. All the chapters in the book are derived from international scientific experts in their respective research areas. Chapters are also designed to facilitate early stage researchers, and enables them to easily grasp the concepts, methodologies and applications of microbial technologies.

Professor Pramod W. Ramteke
Jacob School of Biotechnology and Bioengineering,
Sam Higginbottom Institute of Agriculture, Technology and Sciences,
Allahabad- 211007
INDIA

List of Authors

Chapter 1
*Mary Anne Amalaradjou, Indu Upadhyaya, and Kumar Venkitanarayanan**
Department of Animal Science, University of Connecticut, Storrs, CT, USA
*Corresponding author email: kumar.venkitanarayanan@uconn.edu

Chapter 2
Gibson Nyanhongo, Enrique Herrero Acero, Georg Guebitz
Institute for Environmental Biotechnology
University of Natural Resources and Life Sciences, Vienna, Austria

*Maria Daniela Dela Justina Matuchaki, Martinho Rau, Jürgen Andreaus**
Department of Chemistry, Universidade Regional de Blumenau – FURB, Blumenau, Brazil
*Corresponding author email: jandr@furb.br

Chapter 3
Parisa Rahimi Tamandegani
Department of Plant Protection, College of Agriculture, Bu Ali Sina
University of Hamedan, Ahmadi Roshan Str. 6517838695, Hamedan, Iran

Sedigheh Karimi Dorcheh
Institute of General Microbiology and Microbe Genetics,
Friedrich-Schiller-University Jena, Neugasse 24, 07743 Jena, Germany

*Khabat Vahabi**
Institute of General Botany and Plant Physiology,
Friedrich-Schiller-University Jena, Dornburger Str. 159, 07743 Jena, Germany
*Corresponding author email: khabat.v@gmail.com

Chapter 4
Swapnil Ganesh Sanmukh and Sérgio Luis Felisbino
Laboratory of Extracellular Matrix, Departament of Morphology; Institute of Biosciences;
Sao Paulo State University (UNESP) – Botucatu, SP, Brazil. Zip Code 18618–689
*Corresponding author: felisbin@ibb.unesp.br

Krishna Suresh Khairnar and Waman Narayan Paunikar
Environmental Virology Cell, National Environmental Engineering Research Institute (CSIR-NEERI),
Nehru Marg, Nagpur-440020, Maharashtra (India)
*Corresponding author: k_khairnar@neeri.res.in

Chapter 5
Avnish Kumar, Monika Asthana, Krishn Gopal Jain*
Department of Biotechnology, School of Life Sciences, Dr. Bhim Rao Ambedkar University,
Agra-282004, Uttar Pradesh, India

Vinod Singh
Department of Microbiology, Barkatullah University, Hoshangabad Road Habibganj,
Bhopal 462026, Madhya Pradesh, India
*Corresponding author email: avnishkumar81@gmail.com

Chapter 6

Marcela Claudia Pagano, Luiz Henrique Rosa
Federal University of Minas Gerais, Belo Horizonte, Minas Gerais, Brazil

Natalia V. Fernández
Universidad Nacional del Comahue, INIBIOMA – CONICET, Argentina
*Corresponding autor email: marpagano@gmail.com

Chapter 7

Shadia M. Abdel-Aziz, M. Fadel*
Microbial Chemistry Department, National Research Centre, 33 El Bohouth st. (former El Tahrir st.) – Dokki – Giza – Egypt – P.O.12622

Vijai K. Gupta
MGBG, Discipline of Biochemistry,
School of Natural Sciences, National University of Ireland Galway, Ireland

D Sukmawati
Department of Biology, Universitas, Negeri Jakarta, Indonesia
*Corresponding author email: abdelaziz.sm@gmail.com

Chapter 8

Nádia S. Parachin
Grupo Engenharia de Biocatalisadores,
Instituto de Ciências Biológicas, Campus Darcy Ribeiro,
Universidade de Brasília, Brasil.

*Grace F. Ghesti**
LabCCerva, Instituto de Química,
Campus Darcy Ribeiro, Universidade de Brasília, Brasil.
*Corresponding autor email: grace@unb.br

Chapter 9

*Massimiliano Marvasi**
Natural Science Department, School of Science and Technology,
Middlesex University, London, The Burroughs, NW4 4BT, London, UK
*Corresponding author email: M.Marvasi@mdx.ac.uk

George A. O'Connor, Max Teplitski
Soil and Water Science Department, University of Florida,
Gainesville, FL, USA

Chapter 10

*Mattia Pia Arena, Pasquale Russo, Giuseppe Spano**
Department of Agriculture, Food and Environment Sciences
Via Napoli 25, 71122, University of Foggia, Foggia, Italy
*Corresponding author email: giuseppe.spano@unifg.it

Daniela Fiocco
Department of Clinical and Experimental Medicine,
University of Foggia, Foggia, Italy

Chapter 11
*Leonora Rios de Souza Moreira†, Caio de Oliveira Gorgulho Silva†, Barbara Calheiros Neumann, Edivaldo Ximenes Ferreira Filho**
Laboratory of Enzymology, Department of Cellular Biology, University of Brasília, Brasília, DF, CEP 70910-900, Brazil.
† These authors contributed equally to this work
*Corresponding author email: eximenes@unb.br

Chapter 12
*Shafiquzzaman Siddiquee**
Biotechnology Research Institute, Universiti Malaysia Sabah, Jln UMS,
88400 Kota Kinabalu, Sabah, Malaysia
*Corresponding author email: shafiqpab@ums.edu.my/shafiqpab@yahoo.com

Chapter 13
Karla Victoria Jiménez-Guevara, Leonel E. Amábilis-Sosa, Landy Irene Ramírez-Burgos, Irina Salgado-Bernal,
and María-del-Carmen Durán-Domínguez-de-Bazúa
Departamento Ingenieria Quimica, Universidad Nacional Autónoma de México, Conjunto E, Laboratorios 301–303
(Laboratorios de Ingeniería Química Ambiental y de Química Ambiental)
Mexico
*Corresponding author email: mcduran@unam.mx

Chapter 14
Abdollah Ghasemian and Zahra Moradpour
Department of Pharmaceutical Biotechnology,
Faculty of Pharmacy, Urmia University of Medical Sciences,
P.O. Box 57157-1441, Urmia, Iran
email: ghasemianabdollah@umsu.ac.ir, moradpourzahra@umsu.ac.ir

Chapter 15
*Vedavyas R. Niveditha and Kandikere R. Sridhar**
Department of Biosciences, Mangalore University, Mangalagangotri,
Mangalore 574 199, Karnataka, India
*Corresponding author email: kandikere@gmail.com

Contents

Mary Anne Amalaradjou, Indu Upadhyaya,
and Kumar Venkitanarayanan

1 Microbial applications in the food industry

1.1 Abstract

Microorganisms including bacteria, yeast and mold have been used for millennia in the production of fermented foods, the first and foremost microbial application in the food industry. In addition to production of varied food products, these microbes have been employed in the production of food ingredients, including but not limited to organic acids, food enzymes, amino acids, food color, sugar alcohols and antioxidants. The production of food and food ingredients primarily involves the use of starter cultures that are involved in the technological or fermentation process. The advancement of system biology with the availability of 'omics' based technologies that incorporate experimental, bioinformatics, computational and mathematical approaches has revolutionized bioprocessing in the food industry. Use of these modern cutting edge tools have enabled the elucidation of the genomics, transcriptomics and metabolomics of food microorganisms. In addition, a better understanding of microbial biology has facilitated the knowledge-based manipulation of bacteria for food and food ingredient production, metabolic engineering for production of nutraceuticals and molecular mining of unknown activities. This chapter discusses the various applications of microorganisms in food production, food protection, nutraceutical and ingredient manufacture, biology of the microbes involved in food production and bioengineering in the food industry.

1.2 Introduction

Microorganisms, including bacteria, yeast and fungi play a prominent role in the production of diverse foods and food ingredients. Microbes have been employed in the production of fermented foods and beverages, organic acids, amino acids, flavoring agents, food colors, vitamins and single cell protein [1]. Fermentation, particularly with reference to the production of alcoholic beverages and fermented foods (dairy, vegetable and meat) is the most common application of microorganisms in the food industry. The production of fermented foods is one of the oldest food processing technologies known to man. The earliest use of microorganisms in foods dates to earlier than 7000 BC, when sugars were converted to alcohols by yeast to make beer in Sumeria and Babylonia. Although fermentation initially originated as an unplanned or inevitable outcome that resulted when raw material was left unpreserved, earliest records of organized mass production of fermented foods such as bread date back to the Romans around 3000 BC [2].

The manufacturing of fermented foods was originally artisanal in nature, with the methodologies and knowledge associated with the manufacturing being transferred from generation to generation. These food producers manufactured relatively small quantities of products that were distributed locally [3]. However, the advent of the industrial revolution in the seventeenth century had a significant impact on food fermentation practices and on our understanding of the underlying process. Industrialization helped in the establishment of large scale production of fermented foods to cater to large masses in the towns and cities. Additionally, the developments in microbiology in the latter half of the 19th century led to our understanding of the biological basis of fermentation. Comprehension of the role of microorganisms such as bacteria, yeast and mold in fermentation eventually fueled the development of more controlled and efficient fermentation processes.

The increased and large-scale industrial production of fermented foods led to a need for the production of foods with consistent quality and safety. This was aided by the microbiologists' efforts to identify and cultivate microorganisms capable of performing fermentations. This eventually resulted in the identification of starter cultures, their characterization and production on a large scale to meet the food industry needs. The commercial development of starter cultures had a major impact on the fermentation processes used, thereby ensuring product consistency and reliability of fermentation. With the advent of the 21st century, fermentation processes and food production were further revolutionized through the use of biotechnology in generating microbial strains that promote highly controlled, efficient, economical and large-scale fermented food production. The availability of tools for genetic manipulation and strain characterization of microbial cultures associated with food fermentations transformed the industrial process. Although the generation of genetically modified (GM) microbial cultures for food processes is tempered due to concerns over regulatory issues and public perceptions, GM cultures are currently being used in the production of microbial enzymes and other food ingredients, including monosodium glutamate and amino acids [5].

Current starter cultures used in food applications have two main roles in food processing. The first role is in the actual production process or technology involved in product manufacturing such as fermented foods or organic acids in case of fermentation, whereas the second application in foods is as a functional agent, providing health benefits to the host beyond their nutritive function. Often microbial cultures employed in the technological processes do not possess functional attributes; hence, adjunct cultures are added to microbially processed foods to enhance their health benefits.

Presently, more than 2000 different fermented food products are consumed worldwide. Many of these products are ethnic and produced to meet local demands. However with increased globalization, cultural mixing, introduction of exotic foods and consumer's interest in natural and health foods, future consumption of fermented foods will increase significantly worldwide. The increasing body of scientific evidence on the health benefits associated with the consumption of fermented foods such as live cultures in yogurt on gastrointestinal health have further increased the demand and consumption of fermented foods. The timely advancement of biotechnological platforms and genetic information on microorganisms involved in food fermentation has facilitated the custom production of fermented foods not only possessing specific flavor and functional characteristics, but also imparting nutritional benefits to the consumer [5].

1.3 Biology of microorganisms used in food industry

The production of fermented foods and other food ingredients through microbial fermentation relies on the growth and metabolism of specific cultures present in the starting material such as milk for cheese or molasses for sugar alcohol production. These cultures that are primarily responsible for product manufacture are referred to as starter cultures. The starter cultures can naturally be present or are inoculated into food materials to bring about desired changes in the final product. These changes may include enhanced preservation, improved food safety, enhanced nutritional and functional values and increased sensory attributes and economic value [5]. Similar to the assorted variety of fermented foods produced, a wide diversity also exists in the starter cultures involved in the production process. The commonly used starter cultures include bacteria, yeast and molds.

The bacterial starter cultures predominantly include lactic acid bacteria (LAB), which consists of Gram positive cocci and rods that metabolize sugar via the homofermentative or heterofermentative pathway [6]. Although LAB include 10 genera, most starter cultures belong to *Lactobacillus*, *Lactococcus*, *Leuconostoc* and *Streptococcus*. The choice of the LAB used depends on their fermentative capacity and the desired attributes in the final product. The LAB used in the manufacture of fermented dairy products are selected to perform four major functions, including lactose fermentation, acidification of milk, generation of flavor and flavor precursors and modification of product texture characteristics [3, 7, 8]. Although dairy LAB produce multiple effects on the product, some starter cultures are selected to perform one primary function. For example, application of *Pediococcus acidilactici* as a sausage starter culture is primarily designed to reduce meat pH to a level inhibitory to undesirable competitors and pathogens [5, 9]. Similarly, organisms such as *Oenococcus oeni* (wine fermentation culture) are used to increase wine pH for de-acidifying overly acidic wines [5, 10]. Other non-LAB culture frequently used in food fermentations include *Priopionibacterium freudenrrichii* subsp. *Shermanii* (Emmenthaler and Swiss cheese), *Brevibacterium linensis* (Limburger, Meunster, Brick cheese), Micrococcus spp. (dry fermented sausage) and *Acetobacter aceti* (Vinegar production).

Yeast starter cultures are most commonly used in bread manufacture. Among the different yeast starter cultures, *Saccharomyces cerevisae* (baker's yeast) is the predominant culture used in bread making. *S. cerevisae* strains used in bread making are specifically selected for their ability to rapidly produce large amounts of carbon dioxide. On the other hand, yeast cultures used in the manufacture of wine, beer and other alcoholic beverages are selected for their ability to produce large amounts of ethanol. In addition to these attributes, starter cultures used in the beverage industry are selected for their ability to flocculate, grow at high sugar concentrations and contribute to flavor [10, 5, 57]. Some of the commonly used yeast cultures in fermentation include *S. cerevisae*, *S. pastorianus*, *S. bayanus* and *S. carlsbergensis*. On the other hand, mold starter cultures are commonly employed in the production of several cheese types. These include *Penicillium roqueforti* (blue mold cheese) and *P. camemberti* (white mold cheese). Besides cheese, fungal cultures are widely used in the manufacture of several oriental fermented products, including tempeh (*Rhizopus microsporus* subsp. *Oligosporus*) and miso and soy sauce (*Aspergillus oryzae* and *A. soyae*).

Although starter cultures are selected for attributes specific to the product, there are other pre-requisites for selecting them [7]. These include i) safety – it is critical that starter cultures are screened for the absence of virulence factors and thus are free of pathogens and toxins, ii) technological effectiveness – ability to dominate over naturally occurring microorganisms, and free of bacteriophages and microorganisms that can inhibit the fermentation process, iii) economic viability – the propagation and production of starter cultures must be economically feasible. The starter cultures must be robust to withstand freeze drying, and stable under defined storage conditions for several months with no reduction in activity [58]. In addition to screening natural isolates for culture improvement, recent technological advancements in systems biology have not only enabled rapid screening, but also the rational development of engineered starter strains with modified physiological, biochemical and genetic properties.

1.4 Role of microorganisms in the food industry

1.4.1 Fermented foods

Food fermentation has been practiced for millennia resulting in a diverse variety of products, including milk, meat, vegetables and fruits. In most cases, fermentation involves the oxidation of carbohydrates to generate a range of products, including organic acids such as acetic and lactic acids, alcohol and carbon dioxide [59]. Fermentation provides for a preservative effect by restricting the growth of spoilage and pathogenic organisms. Further, fermentation results in other desirable qualities, including the production of flavor compounds (diacetyl and acetaldehyde), production of bioactive compounds, and the associated nutritional and health benefits [59].

1.4.1.1 Dairy products

Among the various fermented food products, dairy products are popular as convenient, stable and nutritious health foods. As mentioned earlier, LAB are the most common starter cultures used in dairy fermentations. However, in several milk-based products, additional bacteria referred to as secondary or adjunct cultures are often used to influence product flavor and texture [58, 60]. For example, *P. freudenreichii* is incorporated in Swiss cheese for "eye formation" through the production of carbon dioxide [3]. Besides defined cultures, other types of adjunct cultures used include a varied mixture of undefined yeasts, molds and bacteria. These microorganisms are either added to milk or smeared on cheese surface to impart unique flavor [61].

Yogurt is made by fermenting milk using a starter culture comprised of *Streptococcus thermophilus* and *Lactobacillus bulgaricus* in a 1:1 ratio. [3, 58, 62]. *S. thermophilus* is primarily added to help reduce milk pH, while *L. bulgaricus* is responsible for yogurt flavor development. Both cultures also produce extracellular polymers that contributes to yogurt texture and viscosity.

Cheese is produced by concentrating milk casein, fat and minerals through dehydration. The basic steps in cheese making include acidification, coagulation, dehydration and salting. In case of cheese, the starter cultures are primarily used for acid production. The most common starter cultures used in cheese are *Lactococcus lactis* subsp. *Lactis*, *Lc. lactis* subsp. *Cremoris*, *S. thermophilus*, *L. helveticus* and *L. delbrueckii* subsp. *Bulgaricus*. Lactic acid produced by these LAB is responsible for the acidic flavor of unripened cheese and milk coagulation. Further, starter cultures also aid in flavor development, production of proteolytic and lipolytic enzymes that are critical to ripening and suppression of spoilage and pathogenic organisms [63, 64].

Kefir is a fermented milk beverage that is prepared by the addition of kefir grains that contain acid-producing *L. delbrueckii* subsp. *bulgaricus* and alcohol-producing *Torula* spp. [3]. Additionally, *L. delbrueckii* is used in the manufacture of Bulgarian buttermilk, while *L. helveticus* is employed in the making of cultured butter milk. Another fermented dairy product is cultured butter that is made from milk fat using LAB that primarily produce the flavor compound diacetyl. These cultures include *Lc. lactis*, *Leuconostoc citrovurum* and *Leu. Dextranicum* [60, Tab. 1.1].

Tab. 1.1: Starter and adjunct cultures employed in the production of fermented foods and beverages.

Starter/adjunct cultures	Application	Reference
Dairy/Dairy products:		
Lactobacillus kefiri, Lactobacillus kefiranofaciens	Kefir	11, 12
Lactobacillus acidophilus, Kluyveromyces spp.	Koumiss	13, 14
Propionobacterium	cheese with eyes	15
Penicillium roqueforti	Blue mold cheese	16
Penicillium camemberti	External mold cheese	17
Brevibacterium linens	Surface ripened cheese	18
Bread and Beverages:		
Lactobacillus alimentarius, Lactobacillus brevis, Lactobacillus casei, Lactobacilllus curvatus, Lactobacillus delbrueckii, Lactobacillus hilgardii, Lactobacillus pentosus, Lactobacillus plantarum, Lactobacillus pontis, Lactobacillus panis, Lactobacillus reuteri, Lactobacillus sanfranciscensis, Lactobacillus bavaricus Pediococcus acidilactici	Sour dough	19, 20, 21, 22, 23, 24, 25, 26
Lactobacillus fermentum	Swedish sour dough	27
Sachharomyces cerevisiae	Bread, beer, wine	28, 29
Saccharomyces bayanus subsp *uvarum, S. pastorianus (S. carlsbergensis)*	Beer	30, 31
Candida milleri, Candida humilis	Bread	32, 33
Issatchenkia oreintalis, Oenococcus oeni	Wine	34, 35
Lactobacillus sakei	Rice wine	36
Meat/Meat products:		
Pediococcus acidilactici	Dry sausage	37
Lactobacillus planatarum	Salami, dry sausage	38
Lactobacillus sakei, Lactobacillus curvatus	Dry and fermented sausage	39
Enterococcus faecium	Spanish style dry sausage	40
Staphylococcus xylosus, Lactobacillus bavaricus	Dry sausage	41

Tab. 1.1: (continued)

Starter/adjunct cultures	Application	Reference
Lactococcus lactis	Merguez sausage	42
Enterococcus casseliflavus	Cacciatore sausage	43
Lactobacillus rhamnosus	North European-type dry sausage	44
Fresh produce:		
Leuconostoc mesenteroides	Sauerkraut, Pickles, Kimchi, Olives	45
Leuconostoc fallax	Sauerkraut, olives	46
Lactobacillus plantarum, Lactobacillus brevis, Pediococcus pentosaceus	Sauerkraut, Kimchi, Pickles, Olives	47
Ethnic/traditional products *		Substrate
Rhizopus oligosporus	(a) Bongkrek (Indonesia) (b) Oncom (Indonesia) (c) Tempeh (Indonesia)	(a) Coconut press cake (b) Peanut press cake (c) Soybeans
Corynebacterium manihot, Lactobacillus plantarum, Streptococcus spp.	Gari (West Africa)	Cassava root
Leuconostoc mesenteroides, Enterococcus faecalis, Torulopsis, Candida, Trichosporon pullulans	Idli (Southern India)	Rice and black gram dal
Lactobacillus delbrueckii	Mahewu (South Africa)	Maize
Cephalosporium spp., Fusarium spp.	Ogi (Nigeria, West Africa)	Maize
Aspergillus spp.	(a) Ogi (Nigeria, West Africa) (b) Soy sauce (Orient, Japan, China, Phillipines)	(a) Maize (b) Soybeans and wheat
Penicillium spp.	Ogi (Nigeria, West Africa)	Maize
Neurospora intermedia	Oncom (Indonesia)	Peanut press cake
Zygosaccharomyces rouxii	Soy sauce (Orient, Japan, China, Phillipines)	Soybeans and wheat
Saccharomyces cerevisiae	(a) Nan (India) (b) Busa (Turkistan, Egypt)	(a) White wheat flour (b) Rice, millet
Lactobacillus spp.	Busa (Turkistan, Egypt)	Rice, millet
Lactobacillus spp.	Pito (Nigeria)	Guinea corn/maize

Tab. 1.1: (continued)

Starter/adjunct cultures	Application	Reference
Lactobacillus spp.	Sorghum beer (South Africa)	Sorghum/Maize
Zymomonas mobilis, Saccharomyces spp.	Pulque (Mexico)	Agave
Leuconostoc kimchi, Leuconostoc gelidum Leuconostoc inhae, Leuconostoc citreum Lactococcus lactis, Weissella kimchii	Kimchi (Korean)	48, 49, 50, 51, 52, 53

* Compiled from Caplice and Fitzgerald (1999) [3], Jay (1996) [54], Beuchat (1997) [55] and Lucke (1998) [56]

1.4.1.2 Meat products

Inoculation of meats during processing to ascertain controlled microbial activity results in the development of fermented meat products with unique sensory characteristics [65]. The largest category of fermented meat products include dry and semi-dry sausages. These fermented meats are produced by lactic fermentation of a mixture of chopped or ground meat, fat, salt, curing agents, sugar and spices [56]. The predominant species involved in lactic fermentation are *L. sake*, *L. curvatus*, *S. carnosus* and *Micrococcus varians*. Besides bacteria, yeast and mold mixture consisting of *Debaryomyces hansenii*, *Candida famata*, *Penicillium nalgiovense* and *P. chrysogenum* are used as starter cultures in fermented meats [66]. The use of starter cultures in meat ensures consistent flavor, texture, shelf stability and food safety of the final product [37, 39, Tab. 1.1].

1.4.1.3 Vegetable and Fruit products

Vegetables and fruits naturally have a high microbial load that are critical for the fermentation processes. Hence most plant fermentations occur as a consequence of providing optimum growing conditions that selectively favor LAB over the predominant background population [67]. However commercial production of fermented plant products has led to use of defined starter cultures. For example, commercial production of cucumber pickles employs a starter culture consisting of *L. mesenteroides*, *Enterococcus fecalis*, *Pediococcus cerevisiae*, *L. brevis* and *L. plantarum*. Similarly, the predominant culture in the production of sauerkraut (product of lactic acid fermentation of cabbage) is *L. plantarum*. The commercial production of vinegar involves the alcoholic fermentation of sugar-containing plant extracts by *Acetobacter* spp. [58]. In case of fermented fruit juices, the common LAB used include *L. plantarum*, *L. casei*, *L. acidophilus*, *Lc. lactis* and *Leu. mesenteroides* [3]. One of the most popular fermented fruit products is wine, which is manufactured by the fermentation of the crushed fruit material by yeast and bacteria. The main yeast used in wine fermentation is *S. cere-*

visae that metabolizes simple sugars present in grapes to produce alcohol. Based on the amount of residual sugar, wine is either termed dry (0.1 to 0.2 % residual sugar) or sweet (1–20 % residual sugar, [2]). In addition to the above mentioned products, there are several indigenous products with complex sensory, biochemical and nutritional characteristics that are produced with undefined cultures. These are often produced locally by different ethnic groups and may not be available commercially [55, 68, Tab. 1.1].

1.4.2 Food safety and spoilage

In light of consumer demand for foods with improved shelf-life and safety with minimal chemical preservatives, the application of probiotics against foodborne pathogens has become a common practice in the food industry. It is well known that starter probiotics can produce a wide range of antimicrobial metabolites and proteins, thereby inhibiting undesirable flora in food products. The inhibition of spoilage and pathogenic bacteria by LAB has been well studied and characterized over the past 50 years [69, 70, 71]. Lactic acid bacteria exert a preservative effect on food systems by limiting the growth of spoilage and/or pathogenic flora in food products. The metabolic products produced by LAB include organic acids such as lactic, acetic and propionic acids, provide an acidic environment that is unfavorable to the growth of pathogenic and spoilage microorganisms. Besides the antimicrobial metabolites, LAB produce antimicrobial peptides called as bacteriocins which possess anti-pathogenic effects in different food systems. Bacteriocins are produced by a variety of Gram-positive and Gram-negative species, however, those produced by LAB are of particular interest to the food industry [72, 73] since they have a GRAS (generally been regarded as safe) status.

Bacteriocins have a broad spectrum of action against an array of Gram-positive bacteria, including foodborne pathogens such as *Listeria monocytogenes* and *Staphylococcus aureus*, and food spoilage microorganisms such as *Clostridium tyrobutyricum* [73, 74, 75]. The antibacterial properties of bacteriocins such as lacticin 3147 and the more commonly used Nisin, suggest that they have the potential to be used in improving food safety and prophylaxis and treatment of animal and human infections. Employing such bacteriocin-producing cultures can result in an enhanced effect on product safety. The production of different bacteriocins by LAB has been studied in several complex, multicomponent food systems comprising a variety of related micro-environments.

Listeria monocytogenes is a significant foodborne pathogen in many parts of the world. Since *L. monocytogenes* can survive refrigeration temperatures and high salt concentrations, it is extremely resilient and ubiquitous in the environment. In recent years, a majority of the *L. monocytogenes* outbreaks have been caused due to ingestion of contaminated ready to eat meat, fresh produce and cheese. Furthermore, other foodborne pathogens, including many Gram-negative pathogens such as *E. coli* O157, *Campylobacter* and *Salmonella* also [76] result in severe illnesses every year. Although the nature of the Gram-negative cell wall restricts the activity of LAB bacteriocins, these antimicrobial proteins may be used in combination with other treatments, such as high hydrostatic pressure (HHP) to increase their effectiveness. Thus, bacteriocins may be best applied in combination with other antimicrobial interventions to provide multiple hurdles to prevent the growth of pathogenic and spoilage bacteria, especially where contamination could occur post-production.

1.4.2.1 Meat products

Being an excellent source of nutrients, meat products serve as substrates for the growth of microorganisms [77]. During processing, fermented meat products are subjected to a short drying process resulting in products with relatively high water activity. The processing of these products also produces insufficient acid that fails to inhibit the growth of pathogens such as *L. monocytogenes*. For example, a study conducted by Foegeding and co-workers (1992) [37] demonstrated the ability of *L. monocytogenes* to multiply in comminuted cured pork due to quick ripening with elevated temperatures and slow acid production. When inoculated with a strain of *Lactobacillus sake* that produces an anti-listerial bacteriocin, a reduction in *L. monocytogenes* counts by about 1 log in comminuted cured pork was observed. A similar anti-listeria effect was also observed with different fermented meat products [37, 39].

Refrigeration and vacuum packaging are two of the most commonly employed post-processing steps in the meat industry to reduce the population of pathogenic and spoilage bacteria. However, these operations cannot negate the selection of psychotropic spoilage bacteria, mainly *Enterobacteriaceae*, *Pseudomonas* spp. and *Brochothrix thermosphacta* [78]. In addition, some mesophilic species such as *Salmonella* and pathogenic *E. coli* can grow and survive in slightly temperature-abused refrigerated foods, thereby compromising meat safety. Although many studies have selected and characterized an array of LAB strains against pathogens *in vitro*, Maragkoudakis *et al.* (2009) [79] were the first group to use live protective LAB on chicken meat. Two strains (*Enterococcus faecium* PCD71 and *L. fermentum*) were applied to raw chicken meat, causing reduced growth of *L. monocytogenes* and *S.* Enteritidis. Additionally, no LAB-induced spoilage or reduction of the nutritional value of the meat product was observed. Similarly another study demonstrated the role of protective cultures on sliced beef primarily against spoilage bacteria [77]. *L. sakei* and *L. curvatus* of meat origin are the most common strains used in meat to prevent spoilage [80]. Specifically, *L. sakei* CETC 4808, a known bacteriocin producer was found to significantly reduce spoilage bacteria on the surface of vacuum-packaged sliced beef without adversely affecting the chemical and sensory qualities [78]. Likewise, another strain, *L. curvatus* CRL705 was able to control the growth of spoilage microorganisms naturally present on the meat [80]. Protective cultures have also been used for extending shelf-life of cooked meat products such as ham. For example, *L. sakei* 10A, isolated from turkey meat, demonstrated antagonistic activity against *Leu. mesenteroides* and *B. thermosphacta* [81].

One of the most commonly used bacteriocins in meat systems is nisin produced by *Lc. lactis*. Nisin has been mainly used in combination with nitrates to prevent clostridial growth in meat [82, 83]. Similarly, nisin could be used in meat under specific systems such as sausage to control spoilage caused by LAB. Davies *et al.* 1999 [84] examined the influence of fat content and phosphate emulsifier on the effectiveness of nisin in sausage, and found that lower fat contents correlate with high nisin activity. Other studies have used nisin in combination with lactic acid to demonstrate the potential synergistic effect of the preservatives to inhibit Gram negative organisms [85, 86]. In addition to the above-mentioned pathogens, nisin is also effective at inhibiting *B. thermosphacta* when incorporated in a cold meat-binding system [87]. However, a few studies have questioned the efficacy of nisin in meat products. For example, Rose *et al.* in 1999 [88], observed that nisin is inactivated by glutathione due to glutathione S-transferase activity. Glutathione is found abundantly in raw meat, and this reaction potentially diminishes the activity of nisin. In light of potential difficulties in using nisin in raw meat, other bacteriocins such as Leucocin A, enterocins, sakacins and the carnobactericins A and B to maintain the shelf life of fresh meat have been examined (Tab. 1.2).

Tab. 1.2: Microbes employed in promoting food safety and preventing food spoilage.

Bacteria/Bacteriocin	Food Product	Target organism/s	Reference
Streptococcus sp CNCM I-841	Probiotic supplement	*Clostridium sp. Listeria monocytogenes, Enterococcus fecalis*	89
Lactobacillus delbrueckii	Bulgarian yellow cheese	*L. monocytogenes, S. aureus, Ent. faecalis, E. coli, Yersinia enterocolitica, Y. pseudotuberculosis*	90
Enterococcus mundtii	Fresh produce	*L. monocytogenes, C. botulinum*	91
Lactococcus lactis supsp. *cremoris* R	Radish	*Clostridium, Staphylococcus, Listeria*, and *Leuconostoc* spp.	92
Lactococcus lactis subsp. *lactis* (NisZ)	Bean-sprouts	*L. monocytogenes* Scott A	93
Enterococcus faecalis 226	Whey	*L. monocytogenes*	94
Leuconostoc mesenteroides Y105	Goat's milk	*L. monocytogenes*	95

Tab. 1.2: (continued)

Bacteria/Bacteriocin	Food Product	Target organism/s	Reference
Leuconostoc carnosum Ta11A (LeuA)	Meat	*L. monocytogenes*	96
Lactococcus lactis subsp. lactis (Nis)	Sauerkraut	*L. monocytogenes*	95
Lactobacillus plantarum UG1	Meat	*L. monocytogenes, Bacillus cereus, C. perfringens, C. sporogenes*	97
Lactococcus lactis DPC3147	Irish kefir grain	*Clostridium, Enterococcus, Listeria, Leuconostoc* spp.	74
Lactobacillus plantarum BFE905	"Waldorf" salad	*L. monocytogenes*	98
Carnobacterium piscicola CP5	French mold-ripened soft cheese	*Carnobacterium, Listeria,* and *Enterococcus* spp.	99
Lactobacillus plantarum WHE92 (PedAcH)	Munster cheese	*L. monocytogenes*	100
Carnobacterium piscicola JG126	Spoiled ham	*L. monocytogenes*	101
Enterococcus faecalis EFS2	Traditional French cheese	*L. inocua*	102
Lactobacillus plantarum SA6	Fermented sausage	*Lactobacillus* spp.	103
Brevibacterium lines M18	Red smear cheese	*Listeria* and *Corynebacterium* spp.	104
Lactobacillus bavaricus (bavA)	Sour doughs	*L. monocytogenes*	26
E. faecium PCD71	Chicken meat	*S.* Enteriditis, *L. monocytogenes*	79
Lactobacillus fermentum ACA-DC179	Chicken meat	*S.* Enteriditis s, *L. monocytogenes*	79
Lactobacillus sakei CETC 4808	Beef meat	*Enterobacteriaceae, Pseudomonas spp., B. thermosphacta*	78
Lactobacillus curvatus CRL705	Beef meat	Spoilage LAB, B. thermos-phacta, Listeria spp.	80

Tab. 1.2: (continued)

Bacteria/Bacteriocin	Food Product	Target organism/s	Reference
Lactobacillus sakei 10A	Ham	*Spoilage bacteria*	81
Pseudomonas putida LTH 5878	Iceberg lettuce	*S.* Enteriditis Typhimurium, *S. aureus, L. innocua.*	105
L. mesenteroides CM135, CM160, PM249	Iceberg lettuce Golden delicious apples	*S.* Enteriditis Typhimurium, *E. coli, L. monocytogenes*	106
Lactobacillus curvatus LR55	Non-fermented pickles	*L. monocytogenes*	107
Carnobacterium divergens V41	Cold-smoked salmon	*L. monocytogenes*	108
Lactobacillus casei T3	Cold-smoked salmon	*L. innocua*	109
Lactobacillus plantarum Pe2,	Cold-smoked salmon	*Listeria innocua*	109
Carnobacterium piscicola Sal3	Cold-smoked salmon	*Listeria innocua*	109
Leuconostoc gelidum EU2247	Cooked and fresh peeled shrimp	*Vibrio spp., Salmonella spp.*	110
Lactococcus piscium EU2441	Cooked and fresh peeled shrimpgog	*C. botulinum, S. aureus; Shewanella putrefaciens, Photobacterium phosphoreum, Aeromonas spp., Pseudomonas spp.*	110

1.4.2.2 Vegetables, fruits and fresh produce

Many outbreaks in the past decade have been associated with foodborne pathogens such as *Salmonella*, *Listeria* and *E. coli* O157:H7 in minimally processed vegetables and fruits, including pre-washed and pre-cut salads, cantaloupes, prepared fruit salads and other fresh produce [26, 98, 105] This is mainly due to the fact that raw fruits and vegetables with their high microbial load cannot be pasteurized or cooked without compromising quality, thus being sold in a ready- to-use form devoid of preservatives [3, 106]. Moreover, the high humidity, increased cut surface, with a resultant release of nutrients can promote microbial growth. Additionally, the antimicrobial effect of industrial washing of fresh produce with chlorine or ozone has been found to be minimal [106]. In this regard, a strain of *Pseudomonas putida* was found to possess potential antibacterial property against foodborne pathogens [105]. This strain possessed no risk potential and was proposed as a post-harvest or process step application in the production line prior to the final washing. However, LAB constitute the majority of the strains applied as protective cultures in fruits and vegetables. Furthermore, three *Leu. mesenteroides* strains, isolated from fresh fruit and vegetables have been used as bioprotective cultures against *S.* Typhimurium, *E. coli* and *L. monocytogenes* in wounded apples and lettuce leaf without any adverse effects on the sensory or visual characteristics of the product [106, Tab. 1.2].

1.4.2.3 Seafoods

The consumption of seafoods has vastly increased in recent years [111], with renewed efforts targeted at biopreservation to ensure safety and quality of minimally processed food products. In order to reduce the water activity (a_w) and prevent spoilage, salt or sugar along with a mild processing technique, such as cold smoking, is frequently applied. However spoilage or pathogenic microorganisms, namely *Clostridium botulinum* type E and *L. monocytogenes* frequently contaminate seafoods. Although *C. botulinum* type E can be controlled by combining salt and low temperatures, *L. monocytogenes* has the ability to grow at refrigeration temperatures and tolerate low a_w. Protective cultures containing LAB have been investigated to improve the microbial safety and control spoilage of seafoods [112]. Since some LAB strains can secrete active bacteriocins at high salt concentration and low temperatures under aerobic and anaerobic atmospheres [113], they are excellent candidates for large-scale application in seafood safety. A common probitotic, *Carnobacterium divergens* V41 was applied to sterile cold-smoked salmon co-inoculated with a mixture of *L. monocytogenes* strains [108]. In samples containing high numbers of natural microbiota (>10^4 to 10^5 CFU/g), the artificially inoculated LAB quickly became dominant over spoilage and foodborne bacteria. Similarly, the potential antibacterial activity of three LAB strains was evaluated on cold-smoked salmon artificially contaminated with *L. innocua* and stored under vacuum at 4°C [109]. The combination of *L. casei* T3 and *L. plantarum* PE2 was the most effective culture due to a competition against the pathogen. Additionally, ready-to-eat seafoods such as cooked and peeled shrimps are very susceptible to the colonization of pathogens and spoilage bacteria [110]. The growth of these microorganisms can be controlled by psychotropic LAB, which are capable of delaying the sensory spoilage of the products, besides inhibiting the growth of *L. monocytogenes* and *S. aureus* (Tab. 1.2).

1.4.3 Organic acids

Organic acids, including citric acid, lactic acid, fumaric acid, propionic acid, malic acid, gluconic acid, acetic acid and kojic acid are produced by microbial fermentation. Besides the food industry, these organic acids are widely used in pharmaceutical and chemical industries. Although most organic acids can be produced chemically, these procedures require harsh conditions with several intermediary steps that make large scale production impractical. On the other hand, microbial fermentation is a simpler process that requires less energy input and is cost effective. Further, microbial fermentation also produces organic acids in a pure form for use in industrial applications [114].

Citric acid is one of the world's major fermentation products with an estimated annual production of over 550,000 tons [1, 114]. It is primarily used as an acidulant or pH regulator, flavor enhancer and preservative in the food and beverage industry besides its application as a leavening agent in bakery products. Traditionally citric acid was produced from citrus fruits, but now it is mostly produced by microbial fermentation using *Aspergillus niger*, *Candida* species, *Penicillium luteum*, *Mucor piriformis* and *Saccharomycopsis lipolytica* [1, 114]. The commonly used starting materials for citric acid fermentation include starch from potatoes, starch hydrolysates, glucose syrup, sucrose, sugarcane molasses and sugar beet molasses [115]. During citric acid biosynthesis, glucose is metabolized to pyruvate during glycolysis, which is further decarboxylated to acetyl CoA. Citrate synthase subsequently converts acetyl CoA to citrate in the tricarboxylic acid cycle [TCA; 114, 116].

Another widely used organic acid in the food industry is lactic acid. Strains of *Lactobacillus delbruckii*, *L. casei*, *L. helveticus* and *L. acidophilus* are commonly employed in the commercial fermentation of sugar to lactic acid [117]. Lactic acid is used as a preservative, acidulant, buffering and pickling agent in foods, including meat. Semi refined sugar, molasses or whey are used as the common carbon sources in the industrial fermentation of lactic acid. Lactic acid is produced as an end product of anaerobic glycolysis during fermentation through the use of the enzyme lactate dehydrogenase that is produced by the lactic cultures [118, 119].

Kojic acid is a by-product in the fermentation of malting rice used in the production of sake. This organic acid is produced by several *Aspergillus* species, *Bacterium xylinoides* and *Glycinium roseum* [116]. Kojic acid is a mild inhibitor of pigment formation in plant and animal tissues and hence is used on cut fruit surfaces and seafood to prevent color changes. Additionally, it is used as a precursor to flavor enhancers in foods [121]. However, its main use is in the cosmetic industry as a skin whitener [122, Tab. 1.3].

Tab. 1.3: Microorganisms employed in the production of food ingredients and nutraceuticals.

Organism	Application/Product	Reference
Organic acids		
Aspergillus niger, Candida species, Penicillium luteum, Mucor piriformis, Saccharomycopsis lipolytica	Citric acid	1, 114
Lactobacillus delbruckii, L. casei, L. helveticus, L. acidophilus	Lactic acid	117
Aspergillus species, *Bacterium xylinoides, Glycinium roseum*	Kojic acid	116, 120, 121, 122
Food enzymes		
Aspergillus oryzae	Takamylase	123, 124
Aspergillus species	Glucose oxidase	125
Hansenula polymorpha	Hexose oxidase	125
A. niger, A. oryzae, Bacillus amyloliquefaciens, B. stearothermophilus, Mucor miehei, M. pusillus	Protease	125
Candida rugosa, C. antartica, Pseudomonas alcaligenes, P. mendocina, Burkholderia cepacia	Lipase	126, 127
Rhizomucor meihei	Aspartic protease	128
Amino acids		
Corynebacterium glutamicum	Glutamate, lysine, threonine and phenylalanine	125
Bacillus methanolicus	Lysine	129
Nutraceuticals and food color		
Dunalliela salina, Xanthophyllomyces dendrorhus, Haematococcus pluvialis, Blakeslea trispora	β-carotene	130–133
Dunaliella bardwil	Lutein	134, 135
Flavobacterium multivorum	Zeaxanthin	134, 135
Xanthophyllomyces dendrorhous Brevibacterium spp., *Mycobacterium lacticola*	Astaxanthin	136, 137
Sugar alcohol		
Zygosaccharomyces, Debaromyces, Hanensula Pichia	Erythritol	138
Candida spp.	Xylitol	139–141
Zymomonas mobilis	sorbitol	142

1.4.4 Food enzymes

Enzymes have been used in food processing since ancient times in the production of cheese, beer, wine and vinegar [143]. Although enzymes are produced by plants, fungi, bacteria and yeast, microbial enzymes offer more advantages in food applications. The use of microbial enzymes enables low production costs, development of controlled and standardized processes, large-scale production in industrial fermenters, rapid culture development and genetic manipulation along with the use of non-burdensome methods. In addition, microbial enzymes possess a wide range of physical and chemical characteristics and prevent the effects brought about by seasonality associated with the use of non-microbial enzymes [144, 145]. Therefore, the above mentioned attributes make microbial enzymes ideal biocatalysts for various food applications.

Microbial enzymes are used in the food industry for baking, and the production of juices, wine and sugar. Amylases are the most widely used enzymes in the bakery industry. The predominant amylase used in baking is called takamylase, which is obtained from *A. oryzae*. The enzyme is used to break down starch in wheat flour to reduce dough viscosity and increase bread volume and crumb firmness [123, 124]. Besides amylases, xylanases from *Aspergillus* species, glucose oxidase from *Aspergillus* and hexose oxidase from *Hansenula polymorpha* are used to improve bread volume and quality [125]. Pectinases produced by *Aspergillus* species, are a heterogenous group of enzymes that degrade pectin, destabilize cell wall and help in improving extraction, color and aroma of fruits juices and wine [146].

Enzymes used in sugar production are primarily involved in the conversion of plant materials to sugar during the bioethanol production process. These include amylase and invertase produced by *Aspergillus* and glucose isomerase produced by *Streptomyces* species. Amylases are used to convert starch-like polysaccharides to sugar syrups. This class of enzymes are among the first to be produced in an industrial fermentation process [147]. Invertase converts sucrose to fructose and glucose, where the resulting inverted sugar syrup is sweeter than the original product. Glucose isomerase (converts glucose to fructose) is also important in the food industry due to its application in the production of high-fructose corn syrup [124, 145].

A wide variety of other enzymes, including protease, lipase, lactase and milk-clotting enzymes are used in the production of dairy products. Proteases are produced by several organisms, including *A. niger*, *A. oryzae*, *Bacillus amyloliquefaciens*, *B. stearothermophilus*, *Mucor miehei* and *M. pusillus* [125, Tab. 1.3]. These enzymes catalyze the cleavage of peptide bonds in proteins and are used in cheese manufacturing. Similarly lipases from *Candida rugosa*, *C. antartica*, *Pseudomonas alcaligenes*, *P. mendocina* and *Burkholderia cepacia* are employed in the digestion of milk fat in cheese production [126, 127] Another group of enzymes popularly used in cheese manufacture are the milk clotting enzymes (microbial rennets), which represent the single largest group of enzymes used in food production. The most common microbial rennet is aspartic protease from *Rhizomucor meihei*. This enzyme has a high specificity for milk kappa-casein and helps in curd formation and cheese production [128]. Besides aspartic protease, alternative enzymes from *Endothia parasitica* are also employed in some cheese applications.

1.4.5 Amino Acids

Amino acids are extensively used in the food industry as feed additives, and as starting materials for the synthesis of other compounds for use in medicine, cosmetics and chemical industry. Among the different amino acids that are produced, L-glutamate is used as a flavor enhancer, glycine as a sweetener, aspartate and phenylalanine as the low-calorie sweetener aspartame, and lysine and methionine as food and feed additives [1, 148]. Amino acids can be produced by microbial fermentation, extraction from animal and plant protein hydrolysates, chemical synthesis and enzymatic transformation. As with other products, microbial fermentation allows for the large scale production of amino acids, thereby making the process highly cost-effective and cheaper [1].

Corynebacterium glutamicum is a versatile bacterium that is commercially used in the production of glutamate, lysine, threonine and phenylalanine [125]. Besides *Corynebacterium*, *Escherichia*, *Serratia*, *Bacillus*, *Hansenula*, *Candida* and *Saccharomyces* are also used in amino acid production. Bacteria do not normally accumulate large amounts of amino acids due to regulatory control of their synthesis. Hence, genetic engineering is employed to generate mutants that can produce and store large quantities of amino acids [148, 149, 150]. Lysine is an essential amino acid that is not synthesized by humans, and hence has to be obtained from the diet. Since vegetables and cereals naturally have a low lysine content, these are supplemented with lysine to increase the nutritive value [114]. Microbial fermentation to produce lysine primarily employs *C. glutamicum* and *Bacillus methanolicus* [129, Tab. 1.3]. The lysine producing *C. glutamicum* strain was developed by UV radiation induced mutation to enhance its lysine production and accumulation attributes [151].

1.4.6 Nutraceuticals and food color

Carotenoids, flavonoids and terpenoids are natural compounds that are currently incorporated in foods as food ingredients, natural food colorants, antioxidants and nutraceuticals. These compounds are also being investigated for the prevention and treatment of chronic diseases such as diabetes and cancer [152, 153, 154]. Carotenoids in particular function as light-quenching pigments and precursors of Vitamin A [155]. The demand for these compounds as nutraceuticals has increased the research into the development of commercially viable production processes. As with above mentioned compounds, these colorants were primarily chemically extracted from plant sources. However, since this process is challenging and costly, the industry has moved towards the production of natural nutraceuticals by microbial fermentation. In addition to being cost effective, microbial production is environmentally friendly and can meet increasing consumer demands [156].

Carotenogenic microbes comprise a wide variety of organisms, including algae, yeast and fungi. Among these, the most commonly used microbes for large scale β-carotene production include *Dunalliela salina*, *Xanthophyllomyces dendrorhus*, *Haematococcus pluvialis* and *Blakeslea trispora* (155; 157, 158, 159). Several of these strains have also been genetically modified to enhance carotenoid production [160]. Besides β-carotene, other commercially significant carotenoids include xanthophylls such as lutein, zeaxanthin and astaxanthin. Astaxanthin is produced by fermentation using *Xanthophyllomyces dendrorhous*, *Brevibacterium* and *Mycobacterium lacticola* [136, 137, Tab. 1.3]. Lutein and zeaxanthin are xanthophylls that lack the provitamin A activity, but are primarily used as anti-oxidants. The consumption of carotenoids, especially lutein has been shown to help maintain skin health by preventing UV-induced erythema [161, 162]. Of the microbial sources of these pigments, *Dunaliella bardwil* produces lutein and *Flavobacterium multivorum* produces zeaxanthin [134, 135]

1.4.7 Sugar alcohols

The most common carbohydrate in cellular metabolism is the monomeric form of sugars, including D-glucose, D-fructose, D-mannose, D-galactose, D-xylose and L-arabinose [163]. All of these sugars can be reduced to sugar alcohols. Some of these sugar alcohols have been identified to be sweeter than the parent compound. Thus, the identification of their sweet taste and their low caloric content prompted the development of sugar alcohol containing artificial sweeteners for use in low-caloric food products. In addition, since insulin is not involved in their metabolism, these sugar alcohols can be consumed by diabetics. Further, owing to their poor digestibility, most sugar alcohols also serve as laxatives [163]. Of the different sugar alcohols, erythritol, xylitotl, sorbitol and mannitol are used as sweeteners for human consumption.

All sugar alcohols are intermediary metabolites of the pentose phosphate pathway. The parent sugar molecule that is generated during the pathway is reduced to form the sugar alcohol. Erythritol is the only sugar alcohol that is produced by fermentation. Yeasts such as *Zygosaccharomyces*, *Debaromyces*, *Hanensula* and *Pichia* produce erythritol [138], whereas the most efficient producers of xylitol are members of the *Candida* species. Besides its use as a sweetner, erythritol exhibits antioxidant properties and prevents dental carries by inhibiting *Streptococcus mutans* [163]. In case of xylitol, it has been reported to increase the levels of retinol-binding proteins and prevent middle ear infections in children, adrenocortical suppression during steroid therapy and experimental osteoporosis [139, 140, 141]. The sugar alcohol sorbitol is naturally produced by *Zymomonas mobilis* as a metabolite during sugar metabolism [142, Tab. 1.3]. In order to meet the growing consumer needs, *L. plantarum* has been genetically engineered for large scale commercial sorbitol production [165].

1.4.8 Microalgae

Microalgae and cyanobacteria, especially of marine origin have the potential to meet the increasing demands for the development of new nutraceuticals and pharmaceuticals. These phototrophic microorganisms have the ability to biosynthesize secondary metabolites, including coenzyme Q, α-tocopherol, sulfated polysaccharides, phycobilliproteins, antioxidants, antivirals, antibiotics and anti-inflammatory agents [166–174]. Microalgea and cyanobacteria are also the primary producers of polyunsaturated fatty acids such as arachidonic acid, eicosapentanoic acid and docosahexanoic acid [175, 176]. Of the known phototrophic organisms, *Dunaliella*, *Chlorella* and *Arthrospira* (Spirulina) are primarily used for the production of biomass for functional foods. Additionally, *Arthrospira* is employed in the production of vitamins such as vitamin A and $B_{12,}$ protein supplements and pycobilliproteins that are used as food colors in gummy bears, dairy products, popsicles, soft drinks and chewing gums, [177–180].

1.5 Bioengineering in the food industry

Microorganisms have been used by humans for the production of food and food ingredients for thousands of years. Empirical strain selection for enhanced food production is the earliest form of intentional bioengineering that humans have employed. However the advent of molecular biology, genetic engineering, genomics and systems biology has provided us with a range of modern cutting edge technologies to speed up the selection process and improve on the existing microbial strains [181]. System biology differs from the traditional reductionist approach to understanding biology by the incorporation and interpretation of diverse data sets to understand the biological system [182], which generally involves the use of 'omics'-based technologies incorporating experimental, bioinformatics, computational and mathematical approaches to predict a biological system. Further the holistic nature of 'omics' technologies facilitates rational engineering of strains and development of the production process. Additionally, availability of next-generation sequencing platforms has enabled the rapid sequencing and assembly of genomes of industrial organisms [183]. The availability of genome sequences has facilitated a greater understanding of the industrially relevant microbial strains and elucidation of the genetic basis for differences between high- and low-producing strains [181]. In fact, several of these technologies have been applied towards the development of efficient, cost-effective and large scale processes for the production of food products, food ingredients, enzymes and nutraceuticals.

1.5.1 Engineered starter cultures for fermentation

Since the modern fermented food manufacturers run large scale and mechanized operations that depend on time and scheduling demands, starter culture technology in the twentieth century has progressed towards the selection of strains based on the time required for fermentation. Further, knowledge of the sugar being fermented and the end products produced are also important factors that influence culture performance and the final product [2]. The use of molecular biology and genetic engineering tools revealed that these phenotypic traits necessary for lactic acid production, culture growth and activity, casein hydrolysis and bacteriocin production were encoded by plasmid DNA. This knowledge together with gene transfer and gene exchange techniques supported the development of engineered strains with modified physiological, biochemical and genetic properties [2, 184–187]. Another concern for the dairy industry is the susceptibility of starter cultures to bacteriophages. Therefore for ensuring the survivability of the starter cultures, LAB have been engineered to express phage resistance, thus generating robust cultures. These improved starter cultures are used in food fermentation not only to improve the manufacturing process but also to enhance product quality.

1.5.2 Engineered microorganisms for food ingredient production

The application of systems biology to the microbial production of food ingredients has enhanced cost effectiveness, speed and productivity of the food industry. In this regard, systems biology has been critical to understanding the growth, physiology and metabolism of *C. glutamicum* that is commonly used in amino acid production [188]. Additionally, traditional mutagenesis in combination with molecular biology-based metabolic engineering, flux analysis, metabolism and metabolomics have been employed to enhance the production of amino acids and metabolites by *Corynebacterium* [189, 149]. These studies illustrate that a holistic approach based on knowledge from genomics, transcriptomics and metabolomics will lead to rational strain improvement.

Similar approaches have also been applied to the development of microbial enzymes for use in the food industry. For example, metagenomics has been employed to screen for catalytic activity in different microbial habitats for identifying novel enzymes [190]. For example, next-generation sequencing and systems biology techniques helped to identify the genetic factors responsible for the lipolytic activity of *P. freudenrichii* in swiss cheese ripening [191]. This study identified that *P. freudenrichii* contained 12 genes putatively encoding lipases and esterases which may be responsible for the release of volatile esters that contribute to the unique flavor of swiss cheese. The abundance of lipases in strains other than lactic acid bacteria may offer potential for biotechnological manipulation outside of their normal roles [181]. It is expected that use of systems biology will eventually lead to the generation of 'superhost' organisms, expressing heterologous biosynthetic pathways that will negate the need for multiple strains [192–194]. Further, the standardization of these techniques will facilitate rapid assembly of synthetic constructs for the expression of customized and desired biological product [195].

1.6 Conclusion

Scientific advancements, especially the advent of the post genomic era with access to systems biology, molecular biology and genome sequencing tools have enabled the elucidation of the genomics, transcriptomics and metabolomics of food microorganisms. A better understanding of microbial biology has facilitated the knowledge-based manipulation of bacteria for food and food ingredient production, metabolic engineering for production of nutraceuticals and molecular mining of unknown activities. However, the current challenge lies in the consolidation and interpretation of this wealth of information in ways that will improve culture performance, increase product production to meet consumer demands without compromising food safety and quality.

1.7 References

[1] Pai JS. Applications of Microorganisms in Food Biotechnology. Ind J Biotech 2003, 2, 382–386.
[2] Hutkins RW. Microbiology and Technology of Fermented Foods. 1st ed. Ames, IA, USA, Blackwell Publishing, 2006.
[3] Caplice E, Fitzgerald GF. Food fermentations: role of microorganisms in food production and preservation. Int J Food Microbiol 1999, 50, 131–149.
[4] Current status and options for biotechnologies in food processing and in food safety in developing countries. Guadalajara, Mexico. Food and Agricultural Organization of the United Nations, 2010 (Accessed, April 23, 2015 at http://www.fao.org/docrep/meeting/019/k6993e.pdf).
[5] Hutkins RW. Bread Fermentation. In: Hutkins RW., ed. Microbiology and Technology of Fermented Foods. 1st ed. Ames, IA, USA, Blackwell Publishing, 2006, 261–299.
[6] Axelsson L. Lactic acid bacteria: Classification and Physiology. In: Salminen S, von Wright A, Ouwehand A., eds. Lactic acid bacteria: Microbiology and functional aspects. 1st ed. New York, NY, USA, Marcel Dekker Incorporated, 2004, 1–66.
[7] Buckenhüskes HJ. Selection criteria for lactic acid bacteria to be used as starter cultures for various food commodities. FEMS Microbiol Rev 1993, 12, 253–271.

[8] Ross RP, Stanton C, Hill C, Fitzgerald GF, Coffey A. Novel cultures for cheese improvement. Trend Food
Sci Tech 2000, 11, 96–104.

[9] Hammes WP, Hertel, C. New developments in meat starter cultures. Meat Sci 1998, 49, S125–S138.

[10] Fugelsang KC. Fermentation and post-fermentation processing. In: Fugelsang KC., ED. Wine Microbiology. 1st ed. New York, NY, USA, Springer US, 1997, 132–142.

[11] Beshkova DM, Simova ED, Frengova GI, Simov ZI, Dimitrov ZP. Production of volatile aroma compounds by kefir starter cultures. Int Dairy J 2003, 3, 529–535.

[12] Güzel-Seydim ZB, Seydim AC, Greene AK, Bodine AB. Determination of organic acids and volatile flavor substances in kefir during fermentation. J Food Comp Anal 2000, 13, 35–43.

[13] Wu R, Wang L, Wang J et al. Isolation and preliminary probiotic selection of Lactobacilli from koumiss in Inner Mongolia. J Basic Microbiol 2009, 49, 318–326.

[14] Danova S, Petrov K, Pavlov P, Petrova P. Isolation and characterization of *Lactobacillus* strains involved in koumiss fermentation. Int J Dairy Tech 2005, 58, 100–105.

[15] Thierry A, Maillard MB. Production of cheese flavour compounds derived from amino acid catabolism by *Propionibacterium freudenreichii*. Le Lait 2002, 82, 17–32.

[16] Scott PM, Kennedy BP. Analysis of blue cheese for roquefortine and other alkaloids from *Penicillium roqueforti*. J Agri Food Chem 1976. 24, 865–868.

[17] Gripon JC. Mould-ripened cheeses. In: Fox PA., ed. Cheese: chemistry, physics and microbiology. 2nd ed. New York, NY, USA, Springer US, 1999, 111–136.

[18] Rattray FP, Fox PF. Recently identified enzymes of *Brevibacterium linens* ATCC 9174-A review. J Food Biochem 1998, 22, 353–373.

[19] Corsetti A, Lavermicocca P, Morea M, Baruzzi F, Tosti N, Gobbetti M. Phenotypic and molecular identification and clustering of lactic acid bacteria and yeasts from wheat (species *Triticum durum* and *Triticum aestivum*) sourdoughs of Southern Italy. Int J Food Microbiol 2001, 64, 95–104.

[20] Kazanskaya LN, Afanasyeva OV, Patt VA. Microflora of rye sours and some specific features of its accumulation in bread baking plants of the USSR. In: Holas J, Kratochvil F, eds. Developments in food science. Progress in cereal chemistry and technology. 5th ed. London, UK, Elsevier, 1983, 759–763.

[21] Corsetti A, Settanni L. Lactobacilli in sourdough fermentation. Food Res Int 2007, 40, 539–558.

[22] Reale A, Mannina L, Tremonte P et al. Phytate degradation by lactic acid bacteria and yeasts during the wholemeal dough fermentation: a 31P NMR study. J Agri Food Chem 2004, 52, 6300–6305.

[23] Corsetti A, De Angelis M, Dellaglio F et al. Characterization of sourdough lactic acid bacteria based on genotypic and cell wall protein analyses. J Appl Microbiol 2003, 94, 641–654.

[24] Settanni L, Van Sinderen D, Rossi J, Corsetti A. Rapid differentiation and in situ detection of 16 sourdough *Lactobacillus* species by multiplex PCR. Appl Environ Microbiol 2005, 71, 3049–3059.

[25] Corsetti A, Settanni L, Van Sinderen D. Characterization of bacteriocin like inhibitory substances (BLIS) from sourdough lactic acid bacteria and evaluation of their *in vitro* and *in situ* activity. J Appl Microbiol 2004, 96, 521–534.

[26] Larsen AG, Vogensen FK, Josephsen J. Antimicrobial activity of lactic acid bacteria isolated from sour doughs: purification and characterization of bavaricin A, a bacteriocin produced by *Lactobacillus bavaricus* MI401. J Appl Microbiol 1993, 75, 113–122.

[27] Spicher G, Lonner C. The microflora of sourdough. 21. The *Lactobacillus* species of sourdough from swedish bakeries. Zeitschrift fur Lebensmittel-Untersuchung und-Forschung 1985, 181, 9–13.

[28] Thiele C, Gänzle MG, Vogel RF. Contribution of sourdough Lactobacilli, yeast, and cereal enzymes to the generation of amino acids in dough relevant for bread flavor. Cereal Chem 2002, 79, 45–51.

[29] Legras JL, Merdinoglu D, Cornuet J, Karst F. Bread, beer and wine: *Saccharomyces cerevisiae* diversity reflects human history. Mol Ecol 2007, 16, 2091–2102.

[30] Naumov GI, James SA, Naumova ES, Louis EJ, Roberts IN. Three new species in the *Saccharomyces* sensu stricto complex: *Saccharomyces cariocanus, Saccharomyces kudriavzevii* and *Saccharomyces mikatae*. Int J Sys Evol Microbiol 2000, 50, 1931–1942.

[31] Martini AV, Kurtzman CP. Deoxyribonucleic acid relatedness among species of the genus *Saccharomyces* sensu stricto. Int J Sys Evol Microbiol 1985, 35, 508–511.

[32] Gänzle MG, Ehmann M, Hammes WP. Modeling of growth of *Lactobacillus sanfranciscensis* and *Candida milleri* in response to process parameters of sourdough fermentation. Appl Environ Microbiol 1998, 64, 2616–2623.

[33] Gullo M, Romano AD, Pulvirenti A, Giudici P. *Candida humilis*—dominant species in sourdoughs for the production of durum wheat bran flour bread. Int J Food Microbiol 2003, 80, 55–59.

[34] Zambonelli C, Romano P, Suzzi, G. Microorganisms of wine. In: Cantarelli C, Lanzarini G., eds. Biotechnology Applications in Beverage Production. 1st ed. Springer Netherlands. 1989, 17–30.

[35] Nielsen JC, Richelieu M. Control of flavor development in wine during and after malolactic fermentation by *Oenococcus oeni*. Appl Environ Microbiol 1999, 65, 740–745.

[36] Leroy F, De Vuyst L. Lactic acid bacteria as functional starter cultures for the food fermentation industry. Trend Food Sci Technol 2004, 15, 67–78.

[37] Foegeding PM, Thomas AB, Pilkington DH, Klaenhammer TR. Enhanced control of *Listeria monocytogenes* by *in situ*-produced pediocin during dry fermented sausage production. Appl Environ Microbiol 1992, 58, 884–890.

[38] Campanini M, Pedrazzoni I, Barbuti S, Baldini, P. Behaviour of *Listeria monocytogenes* during the maturation of naturally and artificially contaminated salami: effect of lactic-acid bacteria starter cultures. Int J Food Microbiol 1993, 20, 169–175.

[39] Hugas M, Garriga M, Aymerich MT, Monfort JM. Inhibition of *Listeria* in dry fermented sausages by the bacteriocinogenic *Lactobacillus sake* CTC494. J Appl Microbiol 1995, 79, 322–330.

[40] Callewaert R, Hugas M, De Vuyst L. Competitiveness and bacteriocin production of Enterococci in the production of Spanish-style dry fermented sausages. Int J Food Microbiol 2000, 57, 33–42.

[41] Lahti E, Johansson T, Honkanen-Buzalski T, Hill, Nurmi, E. Survival and detection of *Escherichia coli* O157: H7 and *Listeria monocytogenes* during the manufacture of dry sausage using two different starter cultures. Food Microbiol 2001, 18, 75–85.

[42] Benkerroum N, Daoudi A, Kamal, M. Behaviour of *Listeria monocytogenes* in raw sausages (merguez) in presence of a bacteriocin-producing lactococcal strain as a protective culture. Meat Sci 2003 63, 479–484.

[43] Sabia C, De Niederhäusern S, Messi P, Manicardi G, Bondi M. Bacteriocin-producing *Enterococcus casseliflavus* IM 416K1, a natural antagonist for control of *Listeria monocytogenes* in Italian sausages ("cacciatore"). Int J Food Microbiol 2003, 87, 173–179.

[44] Työppönen S, Markkula A, Petäjä E, Suihko ML, Mattila-Sandholm T. Survival of *Listeria monocytogenes* in North European type dry sausages fermented by bioprotective meat starter cultures. Food Con 2003, 14, 181–185.

[45] Fleming HP, McFeeters RF, Daeschel MA. The lactobacilli, pediococci, and leuconostocs: vegetable products. In:Gilliland SE., ed. Bacterial starter cultures for foods. 1st ed. Florida, USA, CRC Press Incorporated, 1985, 97–118.

[46] Barrangou R, Yoon SS, Breidt Jr F, Fleming HP, Klaenhammer TR. Identification and characterization of *Leuconostoc fallax* strains isolated from an industrial sauerkraut fermentation. Appl Environ Microbiol 2002, 68, 2877–2884.
[47] Breidt Jr F, McFeeters RF, Díaz-Muñiz I. Fermented vegetables. In: Doyle M, Beuchat L., eds. Food microbiology: Fundamentals and Frontiers. 3rd ed. Washington DC, VA, USA, ASM press, 2007, 783–793.
[48] Kim J, Chun J, Han HU. *Leuconostoc kimchii* sp. nov., a new species from kimchi. Int J Sys Evol Microbiol 2000, 50, 1915–1919.
[49] Kim B, Lee H, Park S, Kim J, Han H. Identification and characterization of *Leuconostoc gelidum*, isolated from kimchi, a fermented cabbage product. *J Microbiol* 2000, 38, 132–136.
[50] Kim B, Lee J, Jang J, Kim J, Han H. *Leuconostoc inhae* sp. nov., a lactic acid bacterium isolated from kimchi. Int J Sys Evol Microbiol 2003, 53, 1123–1126.
[51] Kang YJ, Kang JW, Choi JH et al. A feasibility study of application of laccase-based time-temperature indicator to kimchi quality control on fermentation process. *J Korean Societ Appl Biologic Chem* 2014, 57, 819–825.
[52] Cheigh CI, Choi HJ, Park H et al. Influence of growth conditions on the production of a nisin-like bacteriocin by *Lactococcus lactis subsp. lactis* A164 isolated from kimchi. J Biotech 2002, 95, 225–235.
[53] Choi HJ, Cheigh CI, Kim SB, et al. *Weissella kimchii* sp. nov., a novel lactic acid bacterium from kimchi. Int J Sys Evol Microbiol 2002, 52, 507–511.
[54] Jay JM. Fresh and fermented fruit and vegetable products. In: Jay JM., ed. Modern Food Microbiology. 3rd ed. New York, NY, USA, Springer US, 1996, 149–176.
[55] Beuchat LR. Traditional fermented foods. In: Doyle M, Beuchat L, Montville TJ., eds. Food Microbiology: Fundamentals and Frontiers. 1st ed. Washington, DC, VA, USA, ASM Press, 1997, 629–648.
[56] Lucke, FK, Fermented sausages. In: Wood B J B., ed. Microbiology of food fermentation. Volume 1. London, UK, Elsevier applied science publishers, 1998, 441–483.
[57] Campbell I. Wild yeasts in brewing and distilling. In: Priest FG, Campell I., eds. Brewing microbiology. 2nd ed. London, UK, Chapman & Hall, 2003, 247–266.
[58] Doyle MP, Meng J. Bacteria in food and beverage production. In: Dworkin M, Falkow S, Rosenberg E, Schleifer K, Stackebrandt E., eds. The prokaryotes. Volume 1. New York, NY, USA, Springer US, 2006, 797–811.
[59] Ross RP, Morgan S, & Hill C. Preservation and fermentation: past, present and future. Int J Food Microbiol 2002 79, 3–16.
[60] Banks JM. Cheese. In: Early R., ed. The technology of dairy products. 2nd ed. London, UK, Blackie Academic & Professional, 1998, 446.
[61] Ray, B. Characteristics and Applications of Pediocin (S) of *Pediococcus acidilactici*: Pediocin PA-1/AcH. In: Bozoglu TF, Ray B., eds. Lactic Acid Bacteria. Volume 98. Berlin, Germany, Springer Berlin Heidelberg, 1996, 155–203.
[62] Chandan RC, Shahani KM. In: Hui YH., ed. Dairy Science and Technology Handbook. Vol. 2. New York, NY, USA, VCH Publishers Incorporated , 1993, 1–56.
[63] Steele, JL. Contribution of lactic acid bacteria to cheese ripening. In: Malin EL, Tunick MH., eds. Chemistry of structure-function relationships in cheese. Vol.367. New York, NY, USA, Springer US, 1995, 209–220.
[64] Lewus, CB, Kaiser A., & Montville TJ. Inhibition of food-borne bacterial pathogens by bacteriocins from lactic acid bacteria isolated from meat. Appl Environ Microbiol 1991, 57, 1683–1688.
[65] Luche F. Fermented meat products. Food Res. Intern 1994, 27, 299–308.

[66] Jessen B. Starter cultures for meat fermentations. In: Campbell-Platt G, Cook PE., eds. Fermented meats. New York, NY, USA, Springer US, 1995, 130–159.
[67] Daeschel MA, Fleming HP. Selection of lactic acid bacteria for use in vegetable fermentations. Food Microbiol 1984, 1, 303–313.
[68] Steinkraus, KH. Lactic acid fermentation in the production of foods from vegetables, cereals and legumes. Antonie van Leeuwenhoek 1983. 49, 337–348.
[69] Price RJ, Lee JS. Inhibition of *Pseudomonas* species by hydrogen peroxide producing lactobacilli. J Milk Food Tech 1970, 33, 13–18.
[70] Klaenhammer TR. Bacteriocins of lactic acid bacteria. Biochimie 1988, 70, 337–349.
[71] Holzapfel WH, Geisen R, Schillinger U. Biological preservation of foods with reference to protective cultures, bacteriocins and food-grade enzymes. Int J Food Microbiol 1995, 24, 343–362.
[72] Nettles CG, Barefoot SF. Biochemical and genetic characteristics of bacteriocins of food-associated lactic acid bacteria. J Food Prot 1993, 56, 338–356.
[73] Deegan LH, Cotter PD, Hill C, Ross P. Bacteriocins: biological tools for bio-preservation and shelf-life extension. Int Dairy J 2006, 16, 1058–1071.
[74] Ryan MP, Rea MC, Hill C, Ross RP. An application in cheddar cheese manufacture for a strain of *Lactococcus lactis* producing a novel broad-spectrum bacteriocin, lacticin 3147. Appl Environ Microbiol 1996, 62, 612–619.
[75] Galvin M, Hill C, Ross, RP. Lacticin 3147 displays activity in buffer against Gram positive bacterial pathogens which appear insensitive in standard plate assays. Lett Appl Microbiol 1999, 28, 355–358.
[76] Adak GK, Long SM., O'Brien SJ. Trends in indigenous foodborne disease and deaths, England and Wales: 1992 to 2000. Gut 2002, 51, 832–841.
[77] Gálvez A, Abriouel H, Benomar N. Lucas R. Microbial antagonists to food-borne pathogens and biocontrol. Curr Opinion Biotech 2010, 21, 142–148.
[78] Katikou P, Ambrosiadis I, Georgantelis D, Koidis P, Georgakis SA. Effect of *Lactobacillus*-protective cultures with bacteriocin-like inhibitory substances' producing ability on microbiological, chemical and sensory changes during storage of refrigerated vacuum-packaged sliced beef. J Appl Microbiol 2005, 99, 1303–1313.
[79] Maragkoudakis PA, Mountzouris, KC, Psyrras D et al. Functional properties of novel protective lactic acid bacteria and application in raw chicken meat against *Listeria monocytogenes* and *Salmonella* Enteritidis. Int J Food Microbiol 2009, 130, 219–226.
[80] Castellano P, González C, Carduza F, Vignolo G. Protective action of *Lactobacillus curvatus* CRL705 on vacuum-packaged raw beef. Effect on sensory and structural characteristics. Meat Sci 2010, 85, 394–401.
[81] Vermeiren L, Devlieghere F, Vandekinderen I, Debevere J. The interaction of the non-bacteriocinogenic *Lactobacillus sakei* 10A and lactocin S *producing Lactobacillus sakei* 148 towards *Listeria monocytogenes* on a model cooked ham. Food Microbiol 2006, 23, 511–518.
[82] Rayman MK, Aris B, Hurst A. Nisin: a possible alternative or adjunct to nitrite in the preservation of meats. Appl Environ Microbiol 1981, 41, 375–380.
[83] Rayman KH, Ali L, Malik N, Hurst A. Failure of nisin to inhibit outgrowth of *Clostridium botulinum* in a model cured meat system. Appl Environ Microbiol 1983, 46, 1450–1452.
[84] Davies EA, Milne CF, Bevis HE et al. Effective use of nisin to control lactic acid bacterial spoilage in vacuum-packed bologna-type sausage. J Food Prot 1999, 62, 1004–1010.
[85] Ariyapitipun T, Mustapha A, Clarke AD. Microbial shelf life determination of vacuum-packaged fresh beef treated with polylactic acid, lactic acid, and nisin solutions. J Food Prot 1999, 62, 913–920.

[86] Ariyapitipun T, Mustapha A, Clarke AD. Survival of *Listeria monocytogenes* Scott A on vacuum-packaged raw beef treated with polylactic acid, lactic acid, and nisin. J Food Prot 2000, 63, 131–136.
[87] Cutter CN, Siragusa, GR. Incorporation of nisin into a meat binding system to inhibit bacteria on beef surfaces. Lett Appl Microbiol 1998, 27, 19–23.
[88] Rose NL, Sporns P, Stiles ME, McMullen LM. Inactivation of nisin by glutathione in fresh meat. J Food Sci 1999, 64, 759–762.
[89] Gomez S, Cosson C, Deschamps AM. Evidence for a bacteriocin-like substance produced by a new strain of *Streptococcus* sp., inhibitory to Gram-positive food-borne pathogens. Res Microbiol 1997, 148, 757–766.
[90] Miteva V, Ivanova I, Budakov I et al. Detection and characterization of a novel antibacterial substance produced by a *Lactobacillus delbrueckii* strain 1043. J Appl Microbiol 1998, 85, 603–614.
[91] Bennik MH, Vanloo B, Brasseur R, Gorris LG, Smid EJ. A novel bacteriocin with a YGNGV motif from vegetable-associated *Enterococcus mundtii*: full characterization and interaction with target organisms. Biochimica et Biophysica Acta 1998, 1373, 47–58.
[92] Yildirim Z, Johnson MG. Characterization and antimicrobial spectrum of bifidocin B, a bacteriocin produced by *Bifidobacterium bifidum* NCFB 1454. J Food Prot 1998, 61, 47–51.
[93] Cai Y, Ng LK, Farber JM. Isolation and characterization of nisin-producing *Lactococcus lactis* subsp. *lactis* from bean-sprouts. J Appl Microbiol 1997, 83, 499–507.
[94] Villani F, Salzano G, Sorrentino E, Pepe O, Marino P, Coppola S. Enterocin 226NWC, a bacteriocin produced by *Enterococcus faecalis* 226, active against *Listeria monocytogenes*. J Appl Bacteriol 1993, 74, 380–387.
[95] Héchard Y, Dérijard B, Letellier F, Cenatiempo Y. Characterization and purification of mesentericin Y105, an anti-*Listeria* bacteriocin from *Leuconostoc mesenteroides*. J Gen Microbiol 1992, 138, 2725–2731.
[96] Felix JV, Papathanasopoulos MA, Smith AA, Von Holy A, Hastings JW. Characterization of leucocin B-Ta11a: A bacteriocin from *Leuconostoc carnosum* Ta11a isolated from meat. Curr Microbiol 1994, 29, 207–212.
[97] Enan G, El-Essawy AA, Uyttendaele M, Debevere J. Antibacterial activity of *Lactobacillus plantarum* UG1 isolated from dry sausage: characterization, production and bactericidal action of plantaricin UG1. Int J Food Microbiol 1996, 30, 189–215.
[98] Franz CMAP, Du-Toit MN, Olasupo A, Schillinger U, Holzapfel WH. Plantaricin D, a bacteriocin produced by *Lactobacillus plantarum* BFE 905 from ready to eat salad. Lett Appl Microbiol 1998, 26, 231–235.
[99] Herbin S, Mathieu F, Brulé F, Branlant C, Lefebvre G, Lebrihi A. Characteristics and genetic determinants of bacteriocin activities produced by *Carnobacterium piscicola* CP5 isolated from cheese. Curr Microbiol 1997, 35, 319–326.
[100] Ennahar S, Aoude-Werner D, Sorokine O, et al. Production of pediocin AcH by *Lactobacillus plantarum* WHE 92 isolated from cheese. Appl Environ Microbiol, 1996 62, 4381–4387.
[101] Jack, RW, Wan J, Gordon J et al. Characterization of the chemical and antimicrobial properties of piscicolin 126, a bacteriocin produced by *Carnobacterium piscicola* JG126. Appl Environ Microbiol 1996, 62, 2897–2903.
[102] Maisnier-Patin S, Forni E, Richard J. Purification, partial characterisation and mode of action of enterococcin EFS2, an antilisterial bacteriocin produced by a strain of *Enterococcus faecalis* isolated from a cheese. Int J Food Microbiol 1996, 30, 255–270.
[103] Rekhif N, Atrih A, Lefebvrexy G. Activity of plantaricin SA6, a bacteriocin produced by *Lactobacillus plantarum* SA6 isolated from fermented sausage. J Appl Bacteriol 1995, 78, 349–358.

[104] Valdés-Stauber N, Scherer S. Isolation and characterization of Linocin M18, a bacteriocin produced by *Brevibacterium linens*. Appl Environ Microbiol 1994, 60, 3809–3814.
[105] Wei HL, Zhang LQ. Quorum-sensing system influences root colonization and biological control ability in *Pseudomonas fluorescens* 2P24. Antonie van Leeuwenhoek 2006, 89, 267–280.
[106] Trias R, Badosa E, Montesinos E, Bañeras L. Bioprotective Leuconostoc strains against *Listeria monocytogenes* in fresh fruits and vegetables. Int J Food Microbiol 2008, 127, 91–98.
[107] Reina LD, Breidt F, Fleming HP, Kathariou S, Isolation and selection of lactic acid bacteria as biocontrol agents for nonacidified, refrigerated pickles. J Food Sci 2005, 70, 7–11.
[108] Brillet A, Pilet MF, Prevost H, Cardinal M, Leroi F. Effect of inoculation of Carnobacterium divergens V41, a biopreservative strain against *Listeria monocytogenes* risk, on the microbiological, chemical and sensory quality of cold-smoked salmon. Int J Food Microbiol 2005, 104, 309–324.
[109] Vescovo M, Scolari G, Zacconi C. Inhibition of *Listeria innocua* growth by antimicrobial-producing lactic acid cultures in vacuum-packed cold-smoked salmon. Food Microbiol 2006, 23, 689–693.
[110] Matamoros S, Pilet MF, Gigout F, Prevost H, Leroi F. Selection and evaluation of seafood-borne psychrotrophic lactic acid bacteria as inhibitors of pathogenic and spoilage bacteria. Food Microbiol 2009, 26, 638–644.
[111] Calo-Mata P, Arlindo S, Boehme K, de Miguel T, Pascoal A, Barros-Velazquez J. Current applications and future trends of lactic acid bacteria and their bacteriocins for the biopreservation of aquatic food products. Food Bioprocess Technol 2008, 1, 43–63.
[112] Pilet MF, Leroi F. Applications of protective cultures, bacteriocins and bacteriophages in fresh seafood and seafood products. In: Lacroix C, ed. Protective cultures, antimicrobial metabolites and bacteriophages for food and beverage biopreservation. Cambridge, England, Woodhead Publishing Limited, 2010, 324–347.
[113] Tomé E, Pereira VL, Lopes CI, Gibbs PA, Teixeira PC. In vitro tests of suitability of bacteriocin-producing lactic acid bacteria, as potential biopreservation cultures in vacuum-packaged cold-smoked salmon. Food Cont 2008, 19, 535–543.
[114] Bhalla TC, Sharma NN, Sharma M. Production of metabolites, industrial enzymes, amino acids, organic acids, antibiotics, vitamins and single cell proteins. Shimla, India, National Science Digital Library, 2007.
[115] Rohr M, Kubicek CP, Kominek J. Citric acid. In: Rehm HJ, Reed G, eds. Biotechnology. 2nd ed. New York, NY, Verlag Chemie Weinheim, 1996, 307–345.
[116] Sauer M, Porro D, Mattanovich D, Branduardi P. Microbial production of organic acids: expanding the markets. Trend Biotech 2008, 26, 100–108.
[117] Kascak JS, Kominek J, Roehr M. Lactic acid. In Rehm HJ, Reed G, Puhler A, Stadler P, eds. Biotechnology. Vol. 6. Weinheim, Germany, VCH Press, 1996, 293–306.
[118] Kulozik U, Wilde J. Rapid lactic acid production at high cell concentrations in whey ultrafiltrate by *Lactobacillus helveticus*. Enz Microbial Technol, 1999, 24, 297–302.
[119] Kwon S, Yoo IK, Lee WG, Chang HN, Chang YK. High-rate continuous production of lactic acid by *Lactobacillus rhamnosus* in a two-stage membrane cell-recycle bioreactor. Biotech Bioengineer 2001, 73, 25–34.
[120] Wilson BJ, Miscellaneous *Aspergillus* toxins. In Ciegler A., Kadis, S. and Ajl. S.J. (eds.) Microbial Toxins, vol. VI, 1971. Pp 208–298. Academic Press, New York
[121] Le Blanc DT, Akers HA. Maltol and ethyl maltol from larch tree to successful food additives. Food Technol., 1989. 26: 78–87.
[122] Ohyama Y, Mishima Y. Melanosis-inhibitory effect of kojic acid and its action mechanism. Frag J 1990, 6, 53–58.

[123] Giménez A, Varela P, Salvador A, Ares G, Fiszman S, Garitta L. Shelf life estimation of brown pan bread: A consumer approach. Food Quality Preference 2007, 18, 196–204.

[124] Ananingsih VK, Gao J, Zhou W. Impact of green tea extract and fungal alpha-amylase on dough proofing and steaming. Food Bioprocess Technol 2013, 6, 3400–3411.

[125] McNeil B, Archer D, Giavasis I, Harvey L. Microbial Production of Food Ingredients, Enzymes and Nutraceuticals. 1st ed. Philadelphia, PA, USA, Elsevier Woodhead publishing, 2013.

[126] Jaeger KE, Reetz MT. Microbial lipases form versatile tools for biotechnology. Trends Biotechnol 1998 16, 396–403.

[127] Fox PF, McSweeney PLH. Proteolysis in cheese during ripening. Food Rev Intl 1996, 12, 457–509.

[128] Takahashi K. Aspartic Proteinases. New York, NY, USA, Plenum Press, 1998.

[129] Flickinger MC. Brief overview of the conversion of methanol to amino acids by *Bacillus methanolicus* at 50ºC. In: Bhalla TC, Sharma NN, Sharma M., eds. Food and Industrial Micorbiology. Proceed. Anaheim, CA, USA, Society for Industrial Microbiology (SIM) Annual Meeting, 2004, S87.

[130] Frengova GI, Beshkova DM. Carotenoids from *Rhodotorula* and *Phaffia*: yeasts of biotechnological importance. J Industrial Microbiol Biotech 2009, 36, 163–180.

[131] Mehta BJ, Obraztsova IN, Cerdá-Olmedo E. Mutants and intersexual heterokaryons of *Blakeslea trispora* for production of β-carotene and lycopene. Appl Environ Microbiol 2003, 69, 4043–4048.

[132] Olaizola M. Commercial production of astaxanthin from *Haematococcus pluvialis* using 25,000-liter outdoor photobioreactors. J Appl Phycol 2000, 12, 499–506.

[133] Raja R, Hemaiswarya S, Rengasamy R. Exploitation of *Dunaliella* for β-carotene production. Appl Microbiol Biotech 2007, 74, 517–523.

[134] Del Campo JA, García-González M, Guerrero MG. Outdoor cultivation of microalgae for carotenoid production: current state and perspectives. Appl Microbiol Biotech 2007, 74, 1163–1174.

[135] Masetto A, Flores-Cotera LB, Diaz C, Langley E, Sanchez S. Application of a Complete Factorial Design for the Production of Zeaxanthin by *Flavobacterium sp.* J Biosci Bioeng 2001, 92, 55–58.

[136] Rodríguez-Sáiz M, de la Fuente JL, Barredo JL. *Xanthophyllomyces dendrorhous* for the industrial production of astaxanthin. Appl Microbiol Biotech 2010, 88, 645–658.

[137] Nelis H, De Leenheer AP. Microbial sources of carotenoid pigments used in foods and feeds. J Appl Bacteriol, 1991, 70, 181–191.

[138] Moon HJ, Jeya M, Kim IW, Lee JK. Biotechnological production of ery thritol and its applications. Appl Microbiol Biotechnol, (2010). *86*(4), 1017–1025.

[139] Mäkinen M, Kontiokari T, Niemelä M. A novel use of xylitol sugar in preventing acute otitis media. Pediatrics 1998, 102, 879–884.

[140] Mäkinen KK, Isotupa KP, Mäkinen PL et al. Six-month polyol chewing-gum programme in kindergarten age children: a feasibility study focusing on mutans streptococci and dental plaque. Int Dental J, 2005, 55, 81–88.

[141] Georgieff M, Moldawer LL, Bistrian BR, Blackburn GL. TPN-induced liver damage. In: Holm E, Kasper H., eds. Metabolism and Nutrition in Liver Disease. Proceed 41. Hingham, MA, USA, MTP Press Ltd, 1985, 287–301.

[142] Viikari L. Formation of sorbitol by Zymomonas mobilis. Appl Microbiol Biotechnol, (1984). *20*(2), 118–123.

[143] Kirk, O, Borchert TV, Fuglsang CC. Industrial enzyme applications. Curr Opinion Biotech 2002, 13, 345–351.

[144] Hasan F, Shah AA, Hameed A. Industrial applications of microbial lipases. Enz Microbial Technol 2006, 39, 235–251.

[145] Soares I, Barcelos RP, Baroni S, Távora Z. Microorganism-Produced Enzymes in the Food Industry. In: Valdez B., ed. Scientific, Health and social aspects of the food industry. Rijeka, Croatia, InTech, 2012, 83–94.

[146] Arnous A, Meyer AS. Discriminated release of phenolic substances from red wine grape skins (*Vitis vinifera* L.) by multicomponent enzymes treatment. Biochem Engineer J 2010 49, 68–77.

[147] Hjort CM, Pedersen H. U.S. Patent No. 7,172,887. Washington, DC: U.S. Patent and Trademark Office, 2007.

[148] Leuchtenberger W. Amino acids—technical production and use. In: Roehr M., ed. Biotechnology, Products of primary metabolism. 2nd ed, Weinheim, Germany, VCH, 1996, 465–502.

[149] Eggeling L, Sahm H. L-Glutamate and L-lysine: traditional products with impetuous developments. Appl Microbiol Biotechnol, (1999). *52*(2), 146–153.

[150] Asakura Y, Kimura E, Usuda Y, Kawahara Y, Matsui K, Osumi T, Nakamatsu T. Altered metabolic flux due to deletion of *odhA* causes L-glutamate overproduction in *Corynebacterium glutamicum*. Appl Environ Microbiol 2007, 73:1308–1319.

[151] Suzuki N, Okayama S, Nonaka, H, Tsuge Y, Inui M, Yukawa H. Large-scale engineering of the *Corynebacterium glutamicum* genome. Appl Environ Microbiol 2005, 71, 3369–3372.

[152] Allen RG, Tresini M. Oxidative stress and gene regulation. Free Radical Biol Medicine 2000, 28, 463–499.

[153] Chemler JA, Lock LT, Koffas MA, Tzanakakis ES. Standardized biosynthesis of flavan-3-ols with effects on pancreatic beta-cell insulin secretion. Appl Microbiol Biotechnol 2007, 77, 797–807.

[154] Reddy LH, Couvreur P. Squalene: A natural triterpene for use in disease management and therapy. Adv Drug Delivery Rev, 2009, 61, 1412-1426.

[155] Frengova GI, Beshkova DM. Carotenoids from *Rhodotorula* and *Phaffia*: yeasts of biotechnological importance. J Industrial Microbiol Biotech 2009, 36, 163–180.

[156] Dufossé L, Galaup P, Yaron A et al. Microorganisms and microalgae as sources of pigments for food use: a scientific oddity or an industrial reality? Trends in Food Science & Technology 2005, 16, 389–406.

[157] Mehta BJ, Obraztsova IN, & Cerdá-Olmedo, E. Mutants and intersexual heterokaryons of *Blakeslea trispora* for production of β-carotene and lycopene. Appl Environ Microbiol 2003, 69, 4043–4048.

[158] Olaizola, M. Commercial production of astaxanthin from *Haematococcus pluvialis* using 25,000-liter outdoor photobioreactors. J Appl Phycol 2000, 12, 499–506.

[159] Raja R, Hemaiswarya S, Rengasamy R. Exploitation of Dunaliella for β-carotene production. Appl Microbiol Biotechnol 2007, 74, 517–523.

[160] Schmidt-Dannert, C, Frances A. “Library cells; for use in generation of carotenoids.” U.S. Patent Application 09/733,759.

[161] Roberts RL, Green J, Lewis B. Lutein and zeaxanthin in eye and skin health. Clinics in Dermatol, 2009, 2, 195–201.

[162] Stahl W, Sies H. Carotenoids and Protection against Solar UV Radiation. Skin Pharmacol Physiol 2002,15, 291–296.

[163] Granström TB, Izumori K, Leisola M. A rare sugar xylitol. Part I: The biochemistry and biosynthesis of xylitol. Appl Microbiol Biotechnol 2007, 74, 277–281.

[164] Yokozawa T, Kim HY, Cho EJ. Erythritol attenuates the diabetic oxidative stress through modulating glucose metabolism and lipid peroxidation in streptozotocin-induced diabetic rats. J Agri Food Chem 2002, 50, 5485–5489.

[165] Nyyssölä A, Leisola M. Production of sugar alcohols by lactic acid bacteria. Recent Res Development Biotech & Bioeng 2005, 19–39.
[166] Arad SM, Friedman OD, Rotem A. Effect of nitrogen on polysaccharide production in a Porphyridium sp. Appl Environ Microbiol 1988, 54, 2411–2414.
[167] Gantt E. Properties and ultrastructure of phycoerythrin from *Porphyridium cruentum*. Plant Physiol 1969, 44, 1629–1638.
[168] Guil-Guerrero JL, Belarbi EH, Rebolloso-Fuentes MM. Eicosapentaenoic and arachidonic acids purification from the red microalga *Porphyridium cruentum*. Bioseparation 2000, 9, 299–306.
[169] Klein B, Buchholz, R. Microalgae as sources of food ingredients and nutraceuticals. In: McNeil, B, Archer D, Giavasis, I, Harvey L., eds. Microbial Production of Food Ingredients, Enzymes and Nutraceuticals. 1st ed. Philadelphia, PA, USA, Woodhead publishing, 2013, 559–570.
[170] Durmaz G, Alpaslan M. Antioxidant properties of roasted apricot (Prunus armeniaca L.) kernel. *Food Chem* 2007, 100, 1177–1181.
[171] Chu M, Diele F, Plemmons R, Ragni S. Optimality, computation, and interpretation of nonnegative matrix factorizations. SIAM J Matrix Analysis 2004, 4–8030.
[172] Guzmán M, Galve-Roperh, I, Sánchez C. Ceramide: a new second messenger of cannabinoid action. Trend Pharmacol Sci 2001, 22, 19–22.
[173] Matsukawa R, Hotta M, Masuda Y, Chihara M, Karube I. Antioxidants from carbon dioxide fixing *Chlorella sorokiniana*. J Appl Phycol 2000, 12, 263–267.
[174] Rechter S, König T, Auerochs S et al. Antiviral activity of Arthrospira-derived spirulan-like substances. Antiviral Res 2006, 72, 197–206.
[175] De Swaaf M, Pronk J, Sijtsma L. Fed-batch cultivation of the docosahexaenoic-acid-producing marine alga *Crypthecodinium* cohnii on ethanol. Appl Microbiol Biotechnol 2003, 61, 40–43.
[176] Giménez AG, González MI, Medina AR, Grima EM, Salas SG, Cerdán LE. Downstream processing and purification of eicosapentaenoic (20: 5n-3) and arachidonic acids (20: 4n-6) from the microalga *Porphyridium cruentum*. Bioseparation 1998, 7, 89–99.
[177] Lee YK. Microalgal mass culture systems and methods: their limitation and potential. J Appl Phycol 2001, 13, 307–315.
[178] Costa JAV, Colla LM, Filho PD. Spirulina platensis growth in open raceway ponds using fresh water supplemented with carbon, nitrogen and metal ions. Zeitschrift Fur Naturforschung 2003, 58, 76–80.
[179] Watanabe F. Vitamin B12 sources and bioavailability. Experimental Biol Med 2007, 232, 1266–1274
[180] Santillan C. Mass production of Spirulina. In : Mislin H, Bachofen, R., eds. New Trends in Research and Utilization of Solar Energy through Biological Systems.1st ed. Waldshut-Tiengen, Germany, Birkhäuser Basel, 1982, 46–49.
[181] Hoskisson PA. Applying systems and synthetic biology approaches to the production of food ingredients, enzymes and nutraceuticals by bacteria. In: McNeil, B, Archer D, Giavasis, I, Harvey L., eds. Microbial Production of Food Ingredients, Enzymes and Nutraceuticals. 1st ed. Philadelphia, PA, USA, Woodhead publishing, 2013, 81–96.
[182] Noble D. Systems biology and the heart. Biosystems 2006, 83, 75–80.
[183] Parkhill J, Wren BW. Bacterial epidemiology and biology—lessons from genome sequencing. Genome Biol 2011, 12, 230.
[184] Harlander SK, McKAY LL, Schachtele CF. Molecular cloning of the lactose-metabolizing genes from *Streptococcus lactis*. Appl Environ Microbiol 1984, 48, 347–351.
[185] Kempler GM, McKay LL. Characterization of plasmid deoxyribonucleic acid in Streptococcus lactis subsp. diacetylactis: evidence for plasmid-linked citrate utilization. Appl Environ Microbiol 1979, 37, 316–323.

[186] Beres SB, Carroll RK, Shea PR et al. Molecular complexity of successive bacterial epidemics deconvoluted by comparative pathogenomics. Proc Natl Acad Sci USA 2010, 107, 4371–4376
[187] McKay LL, Baldwin KA. Induction of prophage in *Streptococcus lactis* C2 by ultraviolet irradiation. Appl Micorbiol 1973, 25, 682–684.
[188] Dong X, Quinn PJ, Wang X. Metabolic engineering of Escherichia coli and *Corynebacterium glutamicum* for the production of L-threonine. Biotech Adv 2011, 29, 11–23.
[189] Kjeldsen KR, Nielsen J. In silico genome scale reconstruction and validation of the *Corynebacterium glutamicum* metabolic network. Biotech Bioeng 2009, 102, 583–597.
[190] Uchiyama T, Miyazaki K. Functional metagenomics for enzyme discovery: challenges to efficient screening. Curr Opinion Biotech 2009, 20, 616–622.
[191] Falentin H, Deutsch SM, Jan G et al. The complete genome of *Propionibacterium freudenreichii* CIRM-BIA1T, a hardy actinobacterium with food and probiotic applications. PLoS One 2010, 5, e11748.
[192] Glass JI, Assad-Garcia N, Alperovich N et al.. Essential genes of a minimal bacterium. Proc Natl Acad Sci USA 2006, 103, 425–430.
[193] Marchisio MA. In silico implementation of synthetic gene networks. In: Weber W, Fussenegger M., eds. Synthetic Gene Networks, Vol 813. New York, NY, USA, Humana Press, 2012, 3–21.
[194] Marchisio MA, Stelling J. Automatic design of digital synthetic gene circuits. PLoS Computational Biol 2011, 7, e1001083.
[195] Müller KM, Arndt KM. Standardization in synthetic biology. In: Weber W, Fussenegger M., eds. Synthetic Gene Networks, Vol 813. New York, NY, USA, Humana Press, 2012, 23–43.

Gibson Nyanhongo, Enrique Herrero Acero,
Maria Daniela Dela Justina Matuchaki, Martinho Rau,
Georg Guebitz, and Jürgen Andreaus

2 Microbial applications for fabric and textile industries

2.1 Abstract

The application of microorganisms, but especially of microbe derived enzymes in fibre and textile processing is one of the most rapidly growing field. Apart from increasing process efficiency leading to improved product quality and enabling process integration, enzymes are turning the once high energy demanding, high water demanding and highly polluting industry into a sustainable environmentally friendly productive sector. Enzymes as biocatalysts are being used to selectively target fibre acompanying impurities such as pectin, hemicellulose, lignin, cellulose, proteins, fats, starch, greasy substances, colorants and oil during fibre recovery, wet processing operations (desizing, scouring, bleaching, dyeing), textile after-care (in detergents formulations) and remove unreacted reagents and undesired side products from bleaching, dyeing and in the treatment of colored effluents. They they are also used to achieve specific fibre modifications or attribute new characteristics to textile materials in pretreatment, dyeing and finishing (polishing, biostoning) processes. Furthermore, microorganisms and enzymes can be employed for the synthesis of dyes and pigments. Thus, this chapter gives an overview on current impressive developments in the application of enzymes in fibre and textile processing and after-care.

2.2 Introduction

In the textile industry, textile fibres go through several chemical wet processes, including pre-cleaning (desizing, scouring, bleaching), dyeing and finishing. In addition previous chemical processing steps to obtain fibres from renewable or fossil feedstock are necessary. These processes consume large amounts of energy, water and resources, and generate large amounts of effluents and waste. To develop cleaner processes, the use of biotechnological processes involving microorganisms and enzymes is growing rapidly [1, 2]. Enzymes can be used as environmentally friendly alternatives in various stages of the processing of textile fibres, such as amylase for desizing, catalase for the removal of excess hydrogen peroxide after bleaching, cellulase and laccase for denim finishing, and proteases plus other enzymes in homecare laundry detergent formulations [3].

The use of enzymes contributes to less harmful and environmental friendly processes, moreover, allows to obtain textile products with enhanced comfort for use such as softer feel, reduced tendency to pilling formation and increased brightness [4] and enables to attribute new characteristics and properties through functionalization of fibres [5].

Natural fibres are bio-based and obtained from either plants or animals. They include all natural cellulosic fibres from plants (hard or soft wood, cotton, jute, sisal, bamboo, coir, kenaf, sea grass, flax, hemp, abaca, ramie, jute, etc.) and protein based fibres (wool mainly from sheep wool although it can also be obtained from alpaca, llama, goat, camel etc.) and silk from silkworm cocoons. These fibres have found applications in automobile, textiles, home furnishing, packaging and many other composite materials [6]. Before they can be used, both the cellulose and proteins fibres have to be either extracted or cleaned and then processed into the desired products e.g. spun into yarn which are then woven or knitted into fabric material [7, 8]. The extraction, cleaning and processing technologies, although some established decades ago, are undergoing extensive transformation driven by the desire to reduce energy and water consumption, reduce or completely eliminate any toxic chemicals used during processing and ultimately minimize environmental pollution. For this purpose, enzyme based processes are fast emerging as contenders. Enzymes have been continually making inroads into many industrial processes including pulp and paper, leather, detergent, fibre, textile, pharmaceutical, chemical, food and beverage, biofuel, animal feed, personal care processing etc. over the past decades [7, 9]. The total market for industrial enzymes has continued to increase reaching $3.3 billion in 2010 and it is estimated to reach 4.4 billion by 2015, with the bulk of the enzymes being used in the detergent, textile, biofuels and, pulp and paper industries [7].

Thanks to advances in biotechnology, it is now possible to design, engineer and specifically adapt enzymes for specific applications, changing enzyme specificity, improving catalytic efficiency on non-natural substrates and increasing temperature and pH-stabilities. Moreover, host organism are designed to produce (i.e. overexpress) the desired enzyme in high amounts while inactivating genes encoding enzymes that catalyze unwanted side reactions to meet the increasing enzyme demand. This development propelled the industrial uptake of enzymes making it a classical energy efficient, greenhouse gas free and pollution mitigating technology in the field of so-called white or industrial biotechnology. This chapter therefore reviews progress in the use of microorganisms and enzymes in the fibre and textile processing industry from extraction of fibres to cleaning, processing and after-care and in the synthesis of dyes and pigments.

Cellulases and hemicellulases are produced by a large number of microorganisms such as bacteria, fungi, actinomycetes and yeasts, that are not limited to the secretion of a single type of enzyme, but several different types which work in a synergistic manner [10]. Among a large number of non-pathogenic microorganisms capable of producing useful enzymes, filamentous fungi are particularly interesting due to easy availability and high production of extracellular enzymes of great industrial potential [11].

2.3 Enzymes for the recovery of cellulose fibres

The extraction of cellulose fibres from stem, leaf or inner part of the bark of lignocellulosic materials such as wood requires the use of techniques which allow the separation of cellulose fibres from intimately bound or surrounding hemicellulose, lignin, pectin, waxes, lipids or accidentally introduced impurities without damaging the cellulose fibres. Once the fibres are obtained, some form of treatment to make them suitable for dyeing or finishing processes is necessary.

2.4 Bast fibre recovery

Bast fibres are defined as those fibres obtained from the stems or stalks of dicotyledonous plants such as flax, ramie, hemp, jute and kenaf [12]. The cellulose fibres make 60–70 % of the bast fibres and occur in bundles or aggregates of 10 to 25 elementary fibres cemented together by lignin, pectin, gum, hemicellulose, lipids, waxes etc. [13]. Traditionally the cellulose fibres have been recovered through retting (rotting away cellular tissues and pectins surrounding bast-fibre bundles) [14, 15]. Dew, water, chemical and mechanical retting system are the most commonly used methods for the recovery of bast fibres as well as bamboo cellulose fibres. The dew retting process used for more than 2000 years, relies on indigenous soil fungi to colonize the stem, degrade pectin [a complex colloidal acid polysaccharide comprising a backbone of galacturonic acid residues linked by α (1–4) linkages and side chains consisting of rhamnose, arabinose, galactose and xylose] and hemicellulose (a complex polymer made of xylan, glucuronoxylan, arabinoxylan, glucomannan and xyloglucan) while the water retting relies on naturally occurring mixed bacterial populations. During the retting process, the hemicellulose, lignin, pectin, gum or waxes and many other organic materials which surround the cellulose fibres are degraded and/or removed by the combined action of microorganisms, temperature and moisture [16]. Considering the fact that natural retting systems are spontaneous processes leading to fibres of varying quality [17], studies using molecular biology techniques to identify strains with degumming capacity from various retting systems in order to develop a higher yield, reproducible microbial retting inocula are ongoing. This approach uses culture independent molecular techniques (PCR assays targeting bacterium 16S rRNA genes) to identify key microorganisms responsible for obtaining high quality cellulose fibres from flax [18], jute [19, 20] as well as bamboo fibres [21] as depicted in Figure 2.1.

Since the 1970s, several enzymes have also been screened for their ability to rapidly recover high quality cellulose fibres. The focus has been the removal of pectin

Fig. 2.1: Effect of microbial retting on bamboo.

and hemicellulose in order to access the cellulose fibres. Various enzymes formulations containing a combination of xylanases and pectinases (polymethyl galacturonases, polygalacturonases, pectin lyases and pectinesterases) have been shown to recover bast fibres [4, 22, 23], thereby making them critical in the enzyme retting formulation [24–28]. This is not surprising since pectin being a complex polysaccharide located in the cell junctions of all tissues including the fibre bundles has been known to prevent access to the bast cellulose fibres [29, 30]. Since pectin is a complex polymer, a complex mixture of pectinases is also produced by the plant pathogenic fungi and bacteria to completely degrade the pectin.

The produced pectinase enzyme mix includes protopectinases responsible for catalyzing the solubilization of protopectin, polygalacturonases which hydrolyze the polygalacturonic acid chain, lyases which catalyze the trans-eliminative cleavage of the galacturonic acid polymer and pectin esterases which liberate pectins and methanol by de-esterifying the methyl ester linkages of the pectin backbone) [25]. Pectinases were some of the first enzymes to be used commercially since the 1930s for the preparation of wines and fruit juices. Unlike acidic pectinases used in the fruit juice industries and wine making, alkaline pectinases are mainly being targeted for the degumming, retting of ramie, hemp, jute, flax and pectin rich wastewater treatment. An alkaline pectate lyase in the presence of ethylenediaminetetraacetic acid (EDTA) as a calcium chelator produced a good yield of strong fibres compared to those obtained with a mixed-enzyme retting formulation containing cellulases [27]. The presence of cellulases particularly cellobiohydrolases damages cellulose fibres by hydrolyzing them. Foulk et al [28] using a pectinase-rich mixture demonstrated that normal atmospheric conditions were satisfactory for penetration of enzyme formulation into crimped stems and enhancing the retting process. However, the pectinases and hemicellulases should be free of cellulases especially cellobiohydrolases and exoglucanases which attack the reducing and non-reducing ends to avoid damage or hydrolysis of the cellulose fibres. It should also be noted that combined chemical treatment or mechanical treatment with enzyme treatment is necessary to produce quality cellulose fibres. Other studies have demonstrated that high quality bast fibres can be obtained by including beating while treating with cellulases and pectinases [26]. Recently, Pakarinen et al. [30] noted that accessibility of fibre hemp was enhanced by first applying steam explosion or hot alkali treatment procedures before adding the pectinases and xyalanases. Although an effective enzyme based retting system has not yet been developed, from the ongoing studies, it is clear that the development of an enzyme-chemical or enzyme-mechanical process is quite feasible in the near future.

2.4.1 Wood cellulose fibre recovery

Although the main focus of the forest industry was until recently the production of paper-grade pulps (kraft pulps), as the traditional paper market continues to decline, companies are shifting towards the production of dissolving pulps [31, 32]. Dissolving pulps are high quality cellulose reactive pulps (> 90 % cellulose content) [32] with uniform molecular weight distribution, low content of hemicellulose and lignin. They are produced using the acid sulfite process, prehydrolysis kraft process [32] and to a lesser extent through organosolv pulping process. Sulfite dissolving pulps with 90–92 % cellulose-content are used to make viscose for textiles and cellophane while sulfate dissolving pulps with 96–98 % cellulose content are used to make rayon (Fig. 2.2) for industrial products such as tire cord, high-quality fabrics, various acetate

Fig. 2.2: Viscose rayon obtained from pulping wood.

and other specialty products [33]. It is estimated that approximately 77 % of dissolving pulps produced worldwide are used in the production of rayon and acetate [34]. Removal of residual hemicellulose and any other contaminating substances is critical in obtaining high quality pulps suitable for making these high value products.

This has necessitated the use of enzymes for the removal of residual hemicellulose from dissolving pulps since they are highly specific. Xylanase preparations from different microorganisms have been explored for their ability to selectively remove residual hemicellulose in pulps [34–38]. Earlier studies showed that xylanase preparations contaminated with low amounts of cellulases (especially endoglucanase and cellobiohydrolase) resulted in the destruction of the cellulose fibres. Using genetic engineering Bernier et al. [39] isolated a xylanase coding gene from *Bacillus subtilis* strain producing both cellulase and xylanase and cloned it in *Escherichia coli* strain which does not produce either cellulase or xylanase. The produced recombinant xylanase was able to remove hemicellulose from dissolving pulps without compromising beneficial properties of the pulps. Since then many xylanase genes have been isolated from different microorganisms and expressed in *E. coli* [40]. From an economic point of view, combining xylanase pretreatment with chemical treatment of sulfite pulp leads to the production of dissolving pulps with low chlorine and higher brightness [41–44]. Gehmayr et al [45] showed that it is possible to produce high quality pulps with enhanced reactivity suitable for the production of viscose, while maintaining a good quality viscose dope by using endoglucanase. Thus, although cellulases have been shown to negatively affect the pulps when present in xylanase formulations, controlled cellulase (endoglucanase) treatment of pulps has been suggested for increasing the reactivity of dissolving pulps [46–48]. Endoglucanase hydrolyses of the amorphous cellulose regions of the short chain fibres increases the amount of reactive ends, but endoglucanase preparations should be free of cellobiohydrolases and exoglucanases to avoid reducing the cellulose's degree of polymerization and degradation to cellobiose. Ibarra et al. [44], studying the behavior of different endoglucanases on the accessibility and reactivity of hardwood and softwood dissolving-grade pulps, discovered that endoglucanase with a cellulose binding domain (CBD) and an inverting hydrolysis mechanism was the most effective in increasing the reactivity of both pulps. The authors were also able to reuse the spent process water containing enzyme several times which is important for lowering the production cost. This study basically laid the foundation for the possibility of using enzyme pretreatment to reduce the carbon disulfide charge during the viscose manufacturing process.

The increasing demand for dissolving pulps [32] and high costs associated with the production of these pulps has also triggered interest in upgrading paper-grade kraft pulps into dissolving pulps by selectively removing residual hemicellulose, hydrolyzing the cellulose fibre to produce uniform size and subsequently activating the pulps. Ibara et al [44] studied the feasibility of converting different kraft pulps from wood and non-wood pulps into commercial dissolving-grade pulps while Arnoul-Jarriault et al [31] investigated the possibility of upgrading softwood bleached kraft pulp to dissolving pulps. These new efforts are basically extending the applications of enzymes in the pulping process. This is because enzyme based processes have already made great inroads in many kraft pulping process steps. The use of enzymes in the Pulp and Paper industry dates back to 1986 [49] and has over the years managed to reduce and/or replace toxic chemicals and reduce energy demand. Enzyme based strategies continue to be investigated for almost all kraft pulp production steps including debarking, beating, bleaching, pitch control, fibre processing, removal and inhibition of slime formation, modifying starch, deinking and wastewater treatment [34, 50]. Enzyme formulations containing mixtures of pectinases (polygalacturonase, pectin lyase), xylanases and cellulases are being developed for debarking [50, 51]. Endoglucanase and pectinases are being used as pre-treatments agents in order to reduce beating time. Cellulases have also been applied for the treatment of chemical pulps after beating in order to facilitate the dewatering process, thereby increasing the maximum speed of the paper machine.

Lipases, lipoxygenases and laccase-mediator systems are being developed for the control of pitch, decreasing polyunsaturated fatty acids, conjugated resin acids, trilinolein and lipophilic extractives from different pulping processes [52, 53]. A complex mixture of enzymes including hemicellulases (xylanases, mannanases) is used for bleaching and removal of shives. Similar to the removal of residual hemicellulose in dissolving pulps, xylanases have been used to facilitate pulp bleaching since the 1980's. Xylanase bleaching is now a well-established technology in the pulp and paper industry. Although xylanases do not degrade or modify lignin directly they decrease the kappa number which reflects decrease in lignin content of pulp suggesting that xylanases contribute indirectly to the delignification of the pulps. Although bleaching with xylanases in most cases is performed with a crude mixture of different enzymes including mannanases, experiments with pure enzymes have shown that most of the effect is dependent of xylanase. The use of ligninolytic enzymes (laccases, lignin peroxidases and manganese peroxidases) to facilitate pulp bleaching has been extensively investigated especially during the 1990's [34]. Bleaching with ligninolytic enzymes from white rot fungi appears most promising. Next to xylanases and among the lignolytic enzymes, laccases are emerging as contenders for eco-friendly biobleaching of hard and soft wood pulps [54]. However, several critical challenges have to be faced before this enzyme can be implemented for pulp bleaching including finding the right-cost effective redox mediator, optimal pH and temperature conditions which prevent unwanted grafting reactions resulting from laccase oxidized ligin phenoxy radicals, semiquinones or quinones. Thus, although the use of laccase together with redox mediators and peroxidases shows great promise, they are more complicated than xylanase. Therefore, xylanases have remained the preferred bleaching agent with current efforts directed at producing alkalophilic and thermostable xylanases [55, 56]. Oxalate oxidase and oxalate decarboxylase based processes are being developed to address the serious problem of calcium oxalate deposition on Pulp and Paper processing machinery [57] while proteases are being investigated for the removal of microbial biofilms [58].

The obtained high quality dissolving pulps either from sulfite processing or upgraded kraft pulps processes are then used to make rayon (or: viscose) and acetate. Beside wood also cotton linter may be used as a raw material. Basically, the high quality dissolving pulps are dissolved in alkali to make a clear solution which is then used for spinning by pumping it through spinnerets in neutralizing spinning bath resulting in long strands of fibre. Lyocell fibre is made from a solution of pulp in the solvent NMMO. Lyocell made from hard wood (oak and birch) dissolving pulp is used to make high value products such as extremely strong fibres used in automotive filters, ropes, abrasive materials, bandages and protective suiting material etc.. Dissolving pulps from eucalyptus trees are used to make a specialty type of lyocell branded tencel produced by Lenzing (Austria), while bamboo dissolving pulps are used to produce rayon. Another high quality fibre produced from these pulps by a modified process with better wet resistance than viscose (rayon) is modal. All these cellulosic fibres including cotton exhibit structural differences related to crystallinity and polymerization degree which are responsible for differences in their chemical and textile properties [59].

2.5 Enzymes in cotton processing

Cotton, unlike bast fibres and wood cellulose fibres which have to be extracted, comes from seed hairs and therefore requires ginning, a mechanical process only, for its recovery. Cotton is approximately 88–96.5 % cellulose with pectin, flavonoids, waxes, sugars, proteins, organic molecules etc. comprising the remainder [60]. Cotton accounts for almost 40 % of the total textile fibre market [61]. After ginning, in the spinning process fibres are mechanically cleaned from impurities (leafs, stalks, seed coat fragments, etc.) and short fibres, aligned in parallel and twisted to form threads or yarns which may be then further transformed (woven or knitted) into fabrics. In the manufacture of plain woven fabrics the warp yarns are treated with sizing agents to prevent breaking during weaving. Most common used sizing agents are natural starch, modified starches, polyvinyl alcohol or carboxy-methyl cellulose together with lubrificating agents. During the sizing process, warp yarns are rolled on beams, driven through a warm sizing bath. Sizing agents glue the fibres together, decrease friction and strengthen the warp yarn. Historically the main sizing agent used for cotton fabrics is starch because of its excellent film forming capacity, availability and relatively low cost [26]. The sizing process is followed by the weaving process. After the weaving process, the sizing agents are removed (process called desizing), followed by scouring (to remove pectin, waxes, sugars, proteins, organic acids, ashes and coloring matter), bleaching to obtain white fabric for subsequent dyeing, printing and finishing processes.

Figure 2.3 shows the different cotton wet processing steps and major enzymes developed for each process.

2.5.1 Desizing

After weaving the sizing agents have to be removed. When starch is used as sizing agent, amylases have been used as traditional desizing agents. The removal of starch by soaking starch-sized fabric in water containing barley (1857) and using amylases in the same desizing process has been practiced since 1912, thus making it a standard technology in the textile industry. Amylases (α-amylases and β-amylases) are hydrolytic enzymes which catalyze the breakdown of starch into short chain sugars,

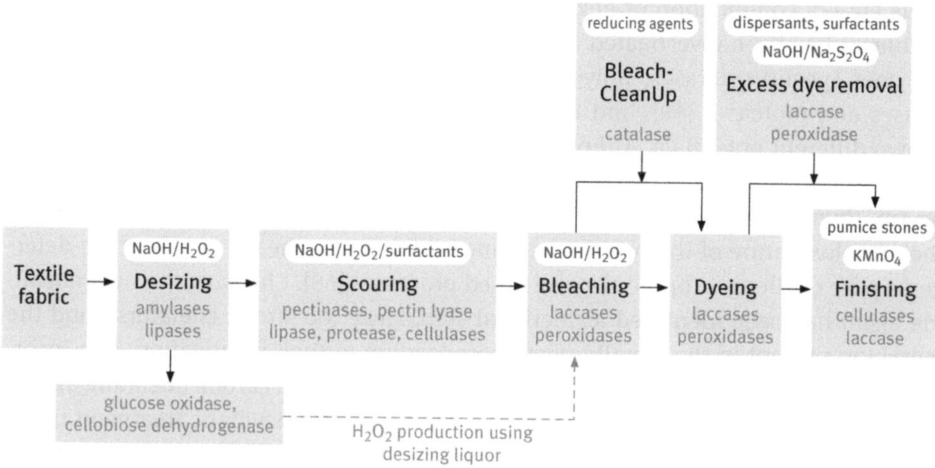

Fig. 2.3: Cotton processing steps with chemical (grey) and principal respective enzymes (green) developed for each process.

dextrin and maltose. The α-amylases are preferred for desizing since they are able to randomly hydrolyze the α(1–4) bonds in the starch backbone. The β-Amylases are not preferred for degradation of starch, because they only release maltose units from the chain ends of the starch polymer. The success of application of amylases is directly linked to their ability to selectively remove starch without damaging the fabric. The process unlike chemical based methods which use none specific toxic caustic soda (NaOH) together with hydrogen peroxide (H_2O_2) at high temperatures, amylases are used at low-temperatures between (30–60 °C) and low pH between 5.5–6.5 and can be operated as a batch or continuous process [62]. Although amylases technology in desizing is well established, studies aimed at improving the speed, economics and consistency of the process are ongoing. Among these a α-amylase-ultrasonic desizing system, screening for halotrophic, thermostable and highly efficient genetically engineered amylases are included [62–66].

2.5.2 Scouring

Scouring is used to remove non-cellulosic materials (pectin, waxes, sugars, proteins, organic acids etc.) which are mainly encountered in the outer layers of the cotton fibre (cuticle and primary cell wall) in order to prepare the fabrics for bleaching, dyeing, printing and finishing stages. Traditionally the scouring process uses strong sodium hydroxide solutions at high temperatures and eventually under pressure which may damage the fabric. Since the target of the scouring process is the removal of many structurally different chemicals in cotton (pectin, waxes, sugars, proteins, organic acids etc.), a combination of enzymes including pectinases, proteases, lipases and cellulases has been investigated for this process. Over the years several pectinolytic enzymes (pectin esterases, polygalacturonases and pectin lyases) [25, 67–70], cellulases and proteases [69], and lipases/cutinases, alone or combined [71, 72] have shown different potentials. Although, the use of both acidic and alkaline pectinases has proven to be effective during scouring, the former are able to selectively degrade pectin, thus making better candidates than the alkaline pectinases. However, due to the complex nature of the process, enzyme inhibitors e.g. heavy metals, ionic detergents pose challenges for the enzyme based processes [68]. Choosing the right chelating agent has been identified as a critical step, since too strong chelators bond the metal ion present in the metallo-pectinases leading to their inactivation. The incorporation of weak chelators such as phosphates, silicates and carbon chelating agents has been recommended. A combination of chemical and enzymatic treatment using *n*-hexane pre-treatment for the removal of wax improved pectinase activity [73]. Similarly, genetic engineering using directed evolution resulted in a novel pectate lyase with improved thermostability which had a 16 °C higher melting temperature than the wild-type and improved bioscouring activity as compared to the wild type [74]. This gives hope for the development of a robust and effective scouring enzyme formulation through genetic engineering. Current studies are also focusing on developing enzyme scouring formulations containing pectinases, lipases and cellulases in combination with surfactants (wetting agents and emulsifiers) and chelating agents. If successful enzymes are developed, scouring has the potential to reduce rinsing water consumption by 20–50 %, Biological Oxygen Demand (BOD) and Chemical Oxygen Demand (COD) during scouring by 20–45 %, Total Dissolved Solids (TDS) by 20–50 % and enable the production of high quality fabrics [75]. Enzymatic scouring has also the advantage that it can be carried out with existing scouring machinery (jets, overflows, winches, pad batchers, pad steamers, and pad rollers) [75].

2.5.3 Bleaching

Scouring is followed by a bleaching process aimed at removing coloring natural pigments (flavonoids) present in cotton fibres. Bleaching is therefore referred to as the last step in the pre-treatment process aimed at achieving the best possible whiteness of the fabric as well as increasing its hydrophilicity. Bleaching is basically an oxidation process. The traditional bleaching system contains H_2O_2 as the oxidizing agent, sodium hydroxide (NaOH) as the H_2O_2 activator and bleaching auxiliaries such as sequestering agents and H_2O_2 stabilizers. The chemical bleaching conditions involve incubating the fabric in an alkaline bath pH 10–12 at temperatures up to 120 °C for 10–20 minutes [75]. After this process the fabric is washed and then acidified to neutralize NaOH. Residual NaOH left on the fabric causes yellowing after drying. To achieve reproducible bleaching results, 10–15 % residual H_2O_2 of the initial quantity needs to remain in yarns or fabrics after bleaching but must then be completely removed before the textile is dyed [75]. Traditionally H_2O_2 removal techniques (bleach clean-up) used sulphur-containing reducing agents [75] accompanied with several rinsing steps. Enzymatic H_2O_2 removal using catalase adaptable for batch, semi-continuous and continuous processes and with both new and existing textile industrial installations is widely used since the late 1990s [75, 76]. Catalases catalyze the reduction of H_2O_2 to oxygen and water. Catalase mediated removal of H_2O_2 saves energy, water and does not affect the downstream dyeing process. Since catalases are tetrameric iron containing enzymes they are affected by chelating agents which can remove the iron from the active site. Bleach formulations are usually supplemented with silicates, phosphates or oxalates to prevent catalase inactivation. Many research activities have also directed efforts at developing enzyme based bleaching systems. For example, glucose oxidase alone or in combination with other highly reactive oxidants based processes has been investigated [77]. Recently Flitsch et al [78] developed a novel bleaching system based on cellobiose dehydrogenase (CDH). CDH is an extracellular flavocytochrome produced by wood degrading fungi [79] able to use a wide range of cello-oligosaccharides including cellulose oligomers as electron donors resulting in the production of H_2O_2 [78]. *In situ* production of H_2O_2 by CDH using desizing agents as electron donors was shown to enhance the bleaching effect of cotton fabrics [78]. Tzanov et al. [80] and Pereira et al [81] reported the enhancement of the bleaching effect on cotton fabrics treated with laccases. More recently combined laccase- ultrasound or laccase-H_2O_2 systems have been shown promising for bleaching [82–84] and a breakthrough looks highly likely in the near future. For example, in a pilot scale ultrasonic assisted enzyme bleaching reactor system a low frequency, high power (22 kHz, 2100 W) and high frequency, low power ultrasounds (850 kHz, 400 W) were determined as the required conditions to achieve satisfactory bleaching effects on cotton fabrics [85]. Nevertheless, as discussed before under bleaching of pulps, several critical challenges need to be addressed before these laccase based processes can be implemented for bleaching such as finding cheap and effective redox mediators, high pH and thermostable laccases and engineering reaction conditions which prevent unwanted grafting reactions resulting from laccase oxidized polyphenols. A breakthrough is likely in the near future given the recent progress in genetic engineering of laccase using site-directed mutagenesis, semi rational engineering, directed evolution and computational modelling tools aimed at active-center-redesigning, widening the pH activity profile and increasing thermostability [86–89].

2.5.4 Combined desizing, scouring and bleaching processes

Over the years there has been an increased effort in combining the desizing, scouring and bleaching processes in order to save time, energy, chemicals and water. Due to the non-specificity of the hydrolyzing and oxidizing reagents used in the combined chemical pretreatment of cotton fibres, only two to four chemicals such sodium hydroxide and hydrogen peroxide (H_2O_2) plus surfactants and stabilizing agents are necessary. In contrast the combination of the enzymatic pretreatment steps is more complex because of enzyme specifity and differences in optimum pH and temperature of the biocatalysts. Different enzyme combinations including amylases and glucose oxidases, enzyme mixtures of amylases, pectinases and glucose oxidases and tetra acetyl ethylene diamine (TAED), α-amylase and hemicellulase/pectinase in the desizing liquor, amyloglucosidase/pullanase enzyme [90–95] have been investigated. In CDH and GOX based systems, the sizing agents desized by amylase provide substrates for the enzyme which is then used to produce H_2O_2. The hydrogen peroxide is either used directly for bleaching or it is reacted in turn with tetra acetyl ethylene diamine (TAED) to produce peracetic acid as the bleaching agent. In addition to producing fabrics with acceptable qualities compared to current treatment methods the one-bath process requires less auxiliaries, saves energy and water. Ali et al [90] noted that the use of catalase to remove residual H_2O_2 before dyeing saved water consumption and thermal energy by at least 400 % and 50 % respectively when compared to conventional chemical process. Capitalizing on these recent developments, Novozyme successfully developed an enzyme formulation which allows combining scouring, bleaching and biopolishing with the benefit of saving time by 1–2 hours, saving water by 10–30 m^3 per ton treated fabric, reduce effluent water by 10–30 m^3 per ton treated fabric, save energy by up to 50 % with the additional benefit of increasing batch to batch reproducibility [96].

2.6 Enzymes in protein fibre processing

Protein fibres used in textiles include animal hair fibres mainly wool from sheep and other specialty hair like cashmere, alpaca, camel, angora and silk fibres secreted by a worm *Bombyx mori*.

2.6.1 Wool processing

Wool is composed of 82 % keratinous protein (characterized by high levels of cysteine), 17 % non-keratinous protein and a 1 % by mass of non proteinaceous material consisting mainly of waxes, lipids and polysaccharides [97]. The wool direct from the animal is often contaminated with vegetable matter and animal's excretions which need to be removed. Although wool fabrics possess excellent unique properties like resiliency, good isolation against heat and cold (cool in summer, warm in winter), good aesthetic look and comfort, the hydrophobic nature (due to the epicutical surface membranes containing fatty acids and hydrophobic impurities like wax and grease), felting-shrinkage and pilling tendency of wool after repeated laundry are its inherent problems [98]. To meet requirements for textile applications, the wool undergoes processing including carbonization, scouring, carding, gilling, combing, drafting, spinning and twisting. Current wool treatment technologies are exploring the possibility of using enzymes alone or in combination with chemicals. For example, researchers are investigating the possibility of replacing the traditional carbonization process with enzyme based methods. The carbonization process is employed to remove vegetable material, waxes and greasy material from wool using strong acids under baking conditions. During this process, wool is treated with 5–10 % sulphuric acid and then baked at 120 °C in order to form char. The char is then crushed and removed as dust. These severe conditions can reduce wool strength and also lower mean fibre length. Mixtures of hydrolases (pectinases, lyases,...) and oxidoreductases (lignolytic enzymes) are being investigated as potential candidates for the removal of this vegetable matter without damaging the wool [97, 99]. The chlorine-Hercosett process developed more than 30 years ago and used to modify the scales of the wool fibre in order to confer resistance to felt shrinking [100] is set to be replaced by proteases based treatments. Protease alone in combination with chemical or physical methods have been demonstrated to improve anti-shrinkage properties, remove impurities and increased subsequent dye affinity [101–103]. Jovancic et al. [104] noted changes on the wool surface and in the surface content of cysteic acid by combining H_2O_2 with proteases. El Sayed et al. [105] also developed an enzyme process involving the use of lipases followed by treating with glutathione reductase in the reduction step and then papain. The lipase removes lipids while the glutathione reductase reduced the disulfide bonds in the wool keratin and the papain smoothed and softened the wool. However, the main challenge of proteases is their small size, which leads to their penetration into the fibre cortex thereby destroying the inner parts of the wool structure [97]. This has been addressed by immobilizing the enzyme on surfaces of high molecular polymers [106–110]. Increasing the size of the enzyme restricted it at the surface as shown in Figure 2.4.

The pretreatment of wool fibres with H_2O_2 at alkaline pH in the presence of high concentrations of salts has also been shown to restrict the activity of the proteases

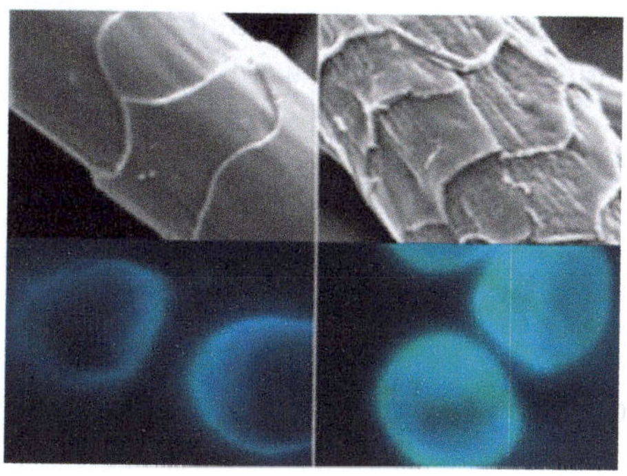

Fig. 2.4: Treatment of wool fibres with size-increased (left) and natural proteases. Electron microscopy (top) indicates considerable fibre damage while fluroescence labeling of proteases (bottom) and inspection of fibre cross-section indicates that larger proteases do not penetrate fibres.

to the outer surface of the wool [111]. Using genetic engineering, Araujo et al [112, 113] reported the construction of a novel high molecular weight subtilisin, based on the fusion of the DNA sequences coding for *Bacillus subtilis* prosubtilisin E and an elastin-like polymer. Likewise, Liu et al [114], using recombiant genetic engineering, expressed an extreme alkaline, oxidation-resistant keratinase from *Bacillus licheniformis* in recombinant *Bacillus subtilis* WB600 expression system stable at 10–50 °C and pH 7–11.5. Cortez et al [115] used a transglutaminase followed by protease treatment to increase wool strength. Transglutaminase (TGases, Fig. 2.5) is emerging as a versatile enzyme in wool processing, finding applications in repairing of damaged wool, upgrading shrink resistance (antifelt), mediating incorporation of amines or proteins into wool fibres, making the wool softer, controlling wettability, color fastness and tensile strength as well as imparting antimicrobial properties [116]. TGases catalyze the acyl transfer reaction between the γ-carboxyamide groups in glycine residues of peptide or protein and ε-amino groups in lysine residues, resulting in the formation of ε-(γ-glutamyl) lysine linkages and the release of ammonia [117].

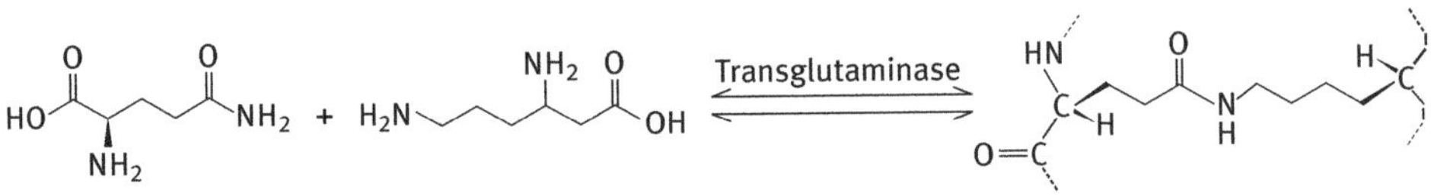

Fig. 2.5: Reaction mechanism of transglutaminases.

TGases are found in mammals, plants, crustaceans, fish and a wide range of invertebrates and microorganisms [116]. Unlike mammalian TGases which require calcium for activity, microbial transglutaminases (MTGases) do not require calcium and have a broader substrate specificity range making them attractive for industrial applications [116]. TGases have been shown to repair damaged wool during chemical and enzymatic treatment. Enzymatic grafting of amines or proteins to the wool fibres enhances the shrink resistance of the wool and improves the tensile strength of the wool fibres [116]. Treatment of leather with MTGases, together with keratin or casein, had beneficial effects on the subsequent dyeing and colour properties of leather [118]. In summary, MTGases improve shrink resistance, tensile strength, handle, softness, wettability and dye uptake, reduce felting tendency and protect wool from damage caused by common detergents. Therefore application of MTGases for leather and wool treatment seems to be a promising strategy although it is still at research level.

2.6.2 Silk processing

Silk, the queen of textiles, is a splendid gift of nature known for its elegance, refinement, beauty and luxury [119]. The most common silk is obtained from the mulberry silk worm although recently scientists are trying to produce silk from spiders. The fibroins of cocoon silk are glued together with wax, protein, sericin, pectin and pigments. Several steps are employed to recover the silk fibres including degumming. Degumming is a process used to remove sericin, which gives a dull appearance, harsh and stiff feeling to the fibre, and hides the typical shiny aspect, soft handle, and whiteness of silk [120]. Traditionally silk degumming involves treatment with soap, alkali and oxidizing conditions under high temperature for extended time [121]. The ability of enzymes to effectively degum silk is a current active subject of research. Several alkaline, acidic and neutral proteases [119, 122, 123] are being studied as degumming agents to selectively remove sericin. Alkaline proteases are showing great promise in improving silk surface properties like handle, shine and smoothness [123, 124]. Genetic engineering studies are also investigating the possibility of heterologous expression and using cocoonase [125] (serine protease produced by silk moths and used for softening the cocoons so that they can escape) as a silk degumming and bleaching agent. Mixtures containing proteases and lipases have been shown to efficiently degum silk without causing any damage with the additional benefit of imparting softness and increasing the efficiency of the dyeing process [120]. Treatment with MTGase followed by H_2O_2, protease and ultrasound improved the crease resistance of silk fabric and also enhanced its tensile strength [120].

2.7 Biotechnical processing of fibres made from synthetic polymers

Since the introduction of synthetic fibres in the middle of the last century their production and consumption has grown constantly and their market share has passed those of natural fibres, which is a general market trend observed in the last years [5]. Although synthetic fibres are originally derived from fossil fuels in some aspects they are more sustainable and greener than natural fibres, and in the future this scenario may further change in their favour if they or related fibres can be produced from renewable sources or in a bio-based process [126].

However, in textile applications there are some properties of synthetic materials which are not desirable for consumer goods, among them poor breathability and low moisture regain [127] (see also Tab. 2.1). The use of water based processes to modify fairly hydrophobic fibres or fabrics is a challenge by itself. In addition, most of the synthetic materials lack reactive groups on their backbone; which could be used for modifying their properties.

Over the last years, several enzyme based approaches have been described to process various synthetic materials. Since polyethylene terephthalate (PET), polyamide (PA) and polyacrilonitrile (PAN) account for around 90 % of all the synthetic fibres produced [5], it is completely reasonable that they are the most studied materials regarding enzymatic functionalization.

Traditionally methods used for the functionalization of synthetic materials include the use of harsh chemicals or plasma approaches. Chemical hydrolysis of the polymer backbone by using concentrated alkali or acids at high temperatures is a difficult to control process. These chemical processes typically lead to severe material loss of up to 15 %, which tremendously decrease the material properties, like tensile strength [128–130]. Plasma processes require a high amount of energy and are relatively more complex.

The development of small, yet insoluble, model substrates, which resemble the polymer backbone, can be regarded as a great progress to study and faster develop new variants with desired selectivity. The first of such model substrates developed by Heumann, Eberl [131], include bis(benzoyloxyethyl) terephthalate (3PET) and adipic acid bishexyl-amide (3PA) (Fig. 2.6). They were developed with the goal of studying the enzyme functionalization of polyethylene terephthalate (PET) and polyamide (PA).

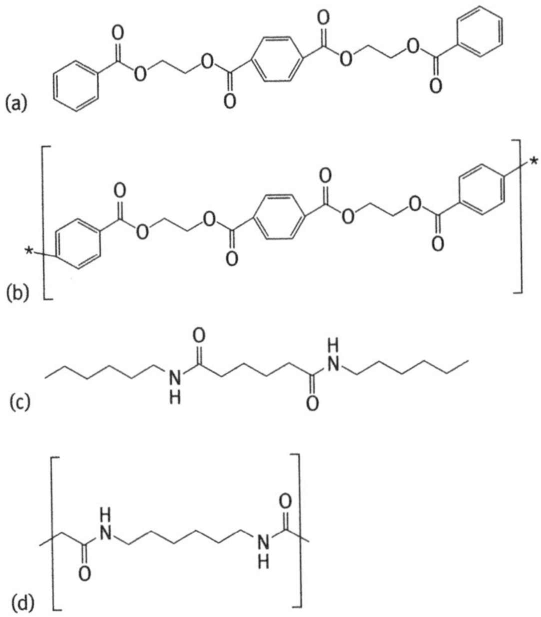

Fig. 2.6: Model substrates and their correspondent polymers typically used for screening purposes; (a) 3PET, (b) PET, (c) 3PA, (d) PA.

The presence of two benzoic acid moieties as final groups of the model substrate provides an easy way to distinguish more exowise type of enzymes from endowise ones. Enzymes, which preferentially hydrolyze the model substrate from the molecule end side, releasing relatively more benzoic acid (BA) or hydroxyethylbenzoate (HEB) than the rest of hydrolysis products can be regarded as more exowise than those in which the released products TA, MHET & BHET are in comparable ratios to BA & HEB. 3PET has been regularly used by the Guebitz group to study the hydrolysis pattern of their latest described cutinases [132–134] and mutants. Also worth mentioning as fast screening procedure is the approach described by Wei et al. in which a simple spectrophotometric method was developed to characterize the enzymes hydrolyzing PET [135].

2.7.1 Biofunctionalization of polyester fibres

In the case of polyesters of synthetic origin, PET has been undoubtedly the most studied one. High tenacity, chemical and environmental resistance together with high glass transition temperature make PET one of the most widespread polymers with applications ranging from textile to the most demanding high tenacity cords found as reinforcement materials in tires (see also Tab. 2.1). Mueller and coworkers were pioneers in the systematic study of enzymatic hydrolysis of different types of aliphatic aromatic polyesters, focusing on complete degradation for environmental and recycling purposes [136]. Since then, numerous cutinases, lipases and different esterases have been described as capable to catalyze the hydrolysis of the ester bond in PET. When used in a controlled manner, partial hydrolysis of the ester bonds present in the PET chains on the accessible fibre surface create defined carboxylic and hydroxyl groups on the polymer surface.

Cutinases are undoubtedly the most active and studied enzyme class. In nature, cutinases catalyze the hydrolysis of cutin, a complex polyester that forms part of the outer part of most plants. The chemical (ester bonds) and physical (hydrophobicity) similarities of cutin, with PET are obvious, and are one of the reasons for the high activity of cutinases towards this type of synthetic polymer. Several cutinases from fungal and bacterial origin have been described, like from *Fusarium solani* [137, 138], *Fusarium oxysporum* [137] *Humicola insolens, Fusarium solani* and *Pseudomonas mendocina* [138] the most commonly found in the literature.

Among the bacterial cutinases different *Thermobifida* species have been described [132, 134], but undoubtedly since its isolation and characterization in 2005 [139] *Thermobifida fusca* cutinases are by far the most extensively described in literature [133, 137, 140, 141]. Over the last years cutinase hydrolysis of PET has been thoroughly described in the reviews from Silva [5] and Zimmermann [142]. Worth mentioning are the PET degradation values reported for *Humicola insolens* cutinase achieving 96 % degradation of low crystallinity PET films [138], the 27 % degradation in 72 h. achieved by a polyesterase from *Saccharomonospora viridis* [143] or the 13 % weight loss after 48 h incubation of PET films with *Thermobifida fusca* [144].

In order to reach more efficient processes, increasing the enzymatic hydrolysis can lead to substantial increases of activity. Besides directed evolution several rational genetic engineering approaches were successfully applied to increase the hydrolytic rate. In their pioneering work Araujo and co-workers successfully increased the hydrolytic activity by enlarging the active site were the polymer chain is accommodated [113]. Lately, different biomimetic approaches showed potential to increase the enzymatic hydrolysis of PET. The fusion of two different types of binding modules to an already active cutinase from *Thermobifida cellulosilytica* not only increased the binding of enzyme to the polymer as shown by Quarz Crystal Microbalance (QCM-D) but also raised the amount of released products by 3.8 times. The increase in adsorption and hydrolysis was higher for the most hydrophobic binding module [145]. Zhang et al. showed that the interaction between binding modules and the polymer surface occurs via hydrogen bonds and hydrophobic interactions [146]. The fact that also the fusion of hydrophobins to already active cutinase can increase the hydrolytic rate [147] makes this approach rather powerful, which could be reapplied to other polymers for which enzyme adsorption could be the rate limiting factor. Also following this approach the study of different mutants derived from two native cutinases from *Thermobifida Cellolosylitica* in which only aminoacids located on the enzyme surface were interchanged, and an activity increase of up to three fold compared with the native enzyme was achieved when the aminoacid mutations lead to neutral area in the enzyme polymer interaction surface [140].

The use of lipases from different organisms like *Aspergillus sp.* [148] *Beauveria brongniartii* [149], *Burkholderia cepacia* [131], *Candida antarctica* [150, 151], *Pseudomonas mendocina* [152], *Thermomyces lanuginosus* and *Triticum aestivum* [153] has been reported to functionalize PET. However most of these studies describe indirect measurements like hydrophilicity increase, moisture regain or higher dye uptake without providing hydrolytic proof like hydrolysis products or increase of carboxylic and hydroxyl surface groups possible due to protein adsorption. On the other hand lipases from *T. lanuginosus* were reported to produce TA and MHET as main hydrolysis products from semi-crystalline PET, whereas hydrolysis of amorphous PET also produced BHET. In addition use of *T. lanuginosus* lead to an increase in the carboxyl groups on the surface of 4.9 % measured via XPS [153].

The commercial benefits of lipases in PET functionalisation that have been described range from the decrease of from 6 % to 0.5 % glue in the production of tarpaulins by using lipase, to create polar groups on the polymer surface which increase binding of PVC [149].

A very interesting approach undertaken by Eberl et al. was taking advantage of the interfacial activation of lipases by a non-ionic surfactant (triton X-100) to increase the enzymatic hydrolysis of PET by *Thermomyces lanuginosus* [154] which resulted in increased release of products. In the same study the use of the plasticizer N,N-diethyl-2-phenylacetamide (DEPA) was found to increase the amount of release products with *T. lanuginosus* lipase and *T. fusca* cutinase. In this case the increase on the mobility chains exposed to the plasticizer is supposed to expose in a higher extent the polymer chain to the enzyme.

Other enzymes that do not belong to any of the two groups and also hydrolyze PET include those from *Bacilus subtilis* [155], *Penicillium citrinum* [156] or *Cladosporium cladosporioides* [157].

Beside the functionalization through the introduction of additional reactive groups Feuerhack, Alisch-Mark [158] have also shown that treatments with *T. fusca* enzymes created surface modifications on the nanoscale which could be visualized by Atomic Force Microscopy and which may be responsible for improved wicking behaviour. Lipases and cutinases were also reported to be able to remove hydrolytically oligoesters, particularly the cyclic trimer, an oligomer typically found in PET fibres which migrates in hydrothermal treatments such as dyeing to the fibre surface and cause problems through deposit on the fibre or on the machinery [159, 160]. Furthermore, Yoon, Kellis [161] have shown in their pioneering work that enzymes can even remove in long term treatments pills from polyester fabrics.

2.7.2 Biofunctionalization of polyamides

Polyamides are a broad polymer family characterized for having a repetitive amide bond on their backbone. The synthesis of polyamides can be performed via two routes. The first one by polymerizing the corresponding dicarboxylic acid with the selected diamine, or by polymerisation of the lactams. Depending on the monomers used very different polyamides can be obtained, however, the most common are polyamides 6.6, with the starting monomers hexamethylenediamine and adipic acid, and polyamide 6, synthetized using caprolactam as precursor.

Polyamides exhibit very interesting properties like high strength, elasticity and resistance, which make them suitable for a wide variety of applications from textile, to automotive parts or ropes (Tab. 2.1). Polyamide enzymatic functionalization has been described to be possible via two mechanisms, either hydrolysis or oxidation. Hydrolysis of the amide bond between the monomers that form the polymer backbone leads to a primary amine and hydroxyl end group. The oxidative approach on the other hand due to the fact that it undergoes a radical mechanism results in a broader final group of products, which restrict subsequent chemical modifications.

There are several enzyme classes already described to hydrolyse PA, due to obvious resemblance to their natural substrate, being proteases the most described enzyme class. Among them enzymes from *Beauveria brongniartii* [162] as well as commercial preparations from *Bacilus subtilis* and *Aspergillus oryzae* [163], recombinantely expressed, have successfully been used to increase fibre hydrophilicity. Also for different types of esterases like the cutinase from *F. solani pisi* [113] or a lipase mutant from *Thermomyces lanuginosus* [164] expressed in *Aspergillus*, polyamide hydrolysis was shown by indirect methods like increase of hydrophilicity and direct chemical modifications via FTIR.

A wild type amidase from *Nocardia farcinica* was initially described [165] to increase hydrophilicity of PA fibres, and lately it was recombinantly expressed in *E.coli*. The amidase successfully hydrolyzed PA fabrics, and the new generated amines were used as grafting points to enzymatically couple ferulic acid, which was monitored in this case via FTIR [166]. Another hydrolase, likewise able to increase wettability of PA fabrics was an aminoacylase (EC 3.5.1.14) derived from *Aspergillus melleus* [167].

The enzymatic oxidation of PA has been described to be possible via two different approaches. Direct oxidation using a manganese peroxidase from *Bjerkandera adusta* [168] leads to cracks and breakage of PA fibres. With a laccase from *Trametes versicolor* in combination with the mediator 1-hydroxybenzotriazole (HBT) Fujisawa and co-workers reported oxidation of PA 6.6 and PE (polyethylene) membranes resulting in a decrease of the molecular weight of the polymer as well as disruption of the membranes [169]. As previously mentioned, the oxidation of the amide bond proceeds via a radical mechanism and the final products obtained are the same as in a thermal oxidation process in which a methylene group close to a Nitrogen atom is oxidized, and the chain reaction proceeds as shown by NMR [170, 171].

2.7.3 Biofunctionalization of acrylonitriles

Although acrylonitriles (PAN) have their share in the synthetic fibre market, there are by far much less enzymatic approaches described to functionalise this family of polymers. The term acrylonitrile involves a rather broad spectrum of polymers obtained by polymerization of acrylonitrile, which due to its high reactivity can polymerize with different monomers with an unsaturated ethylene group [5]. Despite of a low moisture regain (lower than polyamide fibres, but higher than polyester fibres), compared with PET and nylon, acrylonitriles permit much better water transport but have a lower tenacity and breaking elongation (Tab. 2.1).

When acrylic fibres are used in mixed fabrics they bring a distinct natural feel-like as opposed to the plastic perception of PET and polyamides [172]. Regarding the polymer structure of polyacrylonitriles, the main difference to polyesters and polyamides is that the PAN backbone has a pendant nitrile group, which is the target of most enzymatic modifications. The use of a nitrilase (E.C. 3.5.5) isolated from *Micrococcus luteus* [173] transformed the nitrile groups into carboxylic acid, increasing the adsorption of Methylene Blue, a cationic dye that selectively binds to anionic groups such as carboxylic acids. Similarly commercial nitrilases were studied in combination with different solvents and additives, with the purpose of increasing the carboxylic content on the surface [174]. Nitrile hydratases (EC 4.2.1.) on the other hand have been described to successfully convert the nitrile group into an amide. Tauber et al. [175] isolated and characterized a nitrile hydratase from *Arthrobacter sp.* Which was proved by XPS analyses of the fibre to be able to generate amides from the nitrile groups. Similarly the nitrile hydratase activity found in the lysate from *Brevibacterium imperiale* and *Corynebacterium nitrilophilus* showed also amide formation via XPS [176]. Nitrile hydratases can be used as a part of a tandem reaction together with amidase (EC 3.5.1) an enzyme that transforms the amide initially generated by the nitrile hydratase into a carboxylic group. This enzyme system has been described to increase hydrophilicity of PAN [173, 175]. Crude enzyme preparations from *Rhodococcus rhodochrous* [175] and *Agrobacterium tumefaciens* [177] showing both nitrile hydratase and amidase activity succesfully create amides or carboxylic groups on the PAN surface depending on the substrate type and composition and well as reaction conditions (time, temperature).

Tab. 2.1: Physical characteristics of important synthetic fibres.

Fibre	Moisture Regain (%)	T_g (°C)	Tenacity ($N\ tex^{-1}$) dry	Breaking Elongation (%)[a]	Rel. comparison Water transport (HS comfort)	Reference
PET	0.04[b]			12–55	3	178
	0.4[a]	70–80	0.3–0.8	15–55		179
	0.4					180
PBT		45				180
PTT		45				180
PA 6.6	4.0–4.5[b]		0.049–0.077	16–75	2	178
	4–5[a]	60–70				179
	3.5–4.5		0.2–0.6			181
PA 6	2.8–2.5[b]		0.042–0.077	30–90	2	178
	3.5–4.5					181
Acrylic	1.0–1.5[b]			35–55	5	178
	1.5–2.5		0.09–0.33	25–45		182
	1.5[a]	95–100	0.2–0.3	20–50		179
HDPE	0	−85	0.3–0.6	10–45		179
PP	<0.1					183
	0.04[a]	−5	0.3–0.8	15–35		179

[a] 20 °C and 65 % relative humidity, [b] 70 F; 65 % relative humidity

In summary, enzymes have shown an impressive potential to be used for surface polymer functionalization, without compromising the polymer bulk properties and representing a environmentally friendlier process that currently approaches.

2.8 Enzymes in textile dyeing and functionalization or finishing processes

Coloration is an integral part textiles processing which combines many types of chemicals and physico-chemical processes [184]. Enzymes have found application during the process of dyeing textile products and fibres as well as after dyeing for removing dyes from fibres and from textile dyeing effluents. Enzymes processes have also been investigated for finishing and functionalization of textiles.

2.8.1 Dyeing related enzyme processes

Enzymes are increasingly being used in modifying properties of textile materials by mediating the introduction of functional groups on the surface of textile fibres. Laccase mediated dyeing and grafting of functional molecules (antimicrobial agents, hydrophobic enhancing molecules etc.) is actively being pursued [185–187]. Some of the studies have even investigated combined fabric/fibre pre-treatment, dyeing and finishing of cotton [188, 189], wool [190], silk [188] with natural colorants using enzymes. Cavaco-Paulo and coworkers [191], used laccase for dyeing cotton with natural polyphenols. The flavonoids were oxidized by laccases and the resulting radicals reacted with the fibre producing different shades.

Amyloglycosidase, α-amylase and trypsin enzymes were successfully used to prepare fabrics for dyeing with saffron and curcumin [192–194]. Treatments of wool with protease and transglutaminase followed by dyeing with sappan produced colored wool [195]. Kumbasar et al [196] also noted that alkali proteases pretreatment of wool improved dyeing with tannin based natural dyes. Laccase treated madder dyed wool fabrics resulting in fabrics without any change in fastness. Tzanov and coworkers [197–201] worked extensively on the development of laccase-assisted wool and cellulose fibre dyeing process. Varying the concentration of the modifiers and the time of laccase treatment resulted in different hues and depths of shades on the wool fabric. Increasing laccase concentration during wool fabric dyeing process produced wool with varying colour due to the progressively higher oxidation of wool [190]. Herrero Acero et al [202] successfully used laccase in modifying the properties of flax and coconut fibres. As summarised in Nyanhongo et al [203, 204] many more strategies have been developed or are under development for producing natural fibres with antimicrobial, enhanced hydrophobic or hydrophilic properties as well as improving aesthetic look of natural fibres e.g. hair or wool dyeing. Tyrosinase is another enzyme actively being used to mediate the grafting of amino-compounds onto –tyrosine residues in silk fibroin leading to fabrics with increased dyeability, crinkling resistance, and antibacterial activity [185]. Another enzyme based method for wool treatment is the laccase mediated grafting of hydrophobic phenolic compounds such as lauryl gallate on wool fabric [187]. Enzymes are also being intensely investigated to replace H_2O_2 and phenylenediamines as oxidising agents in hair dyeing processes since they are less irritant [205].

An interesting approach is the use of oxidoreductases for the re-oxidation of leuco vat or sulfur dyes in textile dyeing. These dyes need to be solubilized for dyeing by chemical reduction, adsorb to the cellulose fibres and are then oxidized by hydrogen peroxide or oxygen. Different oxidases were found to accelerate the re-oxidation of the dyes [206].

Worth mentioning in this context is the possibility to produce dyes and pigments with microorganism. With the exception of indigo natural dyes find today little or almost no use in industrial dyeing. Indigo, although a natural dye obtained from the plant *indigofera tinctoria* is today almost exclusively of synthetic origin [207]. A biosynthetic route for its production from glucose or other simple precursos using a recombinant bacteria was developed. With a cluster of five genes comprising the tryptophan biosynthetic operon and an altered trpB gene the bacteria *Escherichia coli* is able to transform glucose in indole, which is further oxidized to indigo by a four genes encoded naphthalene dioxygenase enzyme system. Commercially significant rates and yields may be produced [208]. This process was further improved by removing side products suach as isatin and indirubin, once by introducing into the host organism a DNA fragment coding isatin-removing enzyme activity and second by adding a isatin hydrolase or an adsorbent like charcole to the fermentation broth [209]. More recently other strains were identified that can produce indigo from precursors such as indole or naphtalene [210–212]. A mutation of *Bacterium Vibrio gazogenes* a marine bacterium strain was reported to produce effectively prodiginine type pigments, that could dye wool, silk and synthetic fibres such as polyester and polyacrylic and also showed antibacterial properties against *Escherichia coli* and *Staphylococcus aureus* bacteria on the dyed products [213]. Pigments with polypyrrolic prodigiosin-like structures produced by *Streptomyces* strains were able to dye polyamide and polyacrilic fibres to deep shades and polyester and triacetate fibres noticeable [214]. An water soluble phenoxazine dye was also produced biocatalytically with a laccase from *Cerrena unicolor* starting with 3-amino-4-hydroxybenzensulfonic acid as precursor [215]. Altogether, in comparison to reports about microbial sources for dyes and pigments for food [216–219], little formation can be found about the microbial production of textile dyes or pigments [220, 221], and this seems to be a research field of high potential, especially concerning the sustainable and eco-friendly aspect of enzymes and renewable resources.

2.8.2 Cellulose fabric finishing or polishing

The formation of pilling on textile materials is a perennial problem encountered in the use of textiles and depends on fibre type and quality. This problem occurs in synthetic fibres such as polyesters, and natural or regenerated fibres such as cotton, wool and viscose, as well as yarn blends [222–226]. Pilling formation begins with the presence of mainly shorter fibres (microfibrils) that emerge from the yarn surface. External factors such as abrasion and friction during the process of washing and usage favour the formation of pilling [227, 228].

Biopolishing of cellulosic fibres aims at the removal of surface fibrils or protruding hairs of yarns and fabrics to create a smoother surface and thereby making the fabric less prone to forming pills. Besides, an increase in brightness, lustre and a softer feel is achieved. The treatment provides also an increase in hydrophilicity, an effect particularly desired in garments and towels. Biopolishing involves controlled hydrolysis of the protruding hairs or fuzz using cellulases and promotes a mass loss of 3 to 5 % and a strength loss of 2 to 7 % [229–233]. Pilling tendency and improvements in the pilling value after enzyme treatment are easily visualized after causing controlled abrasion (rubbing or shaking) in a standard pilling tester (Fig. 2.7).

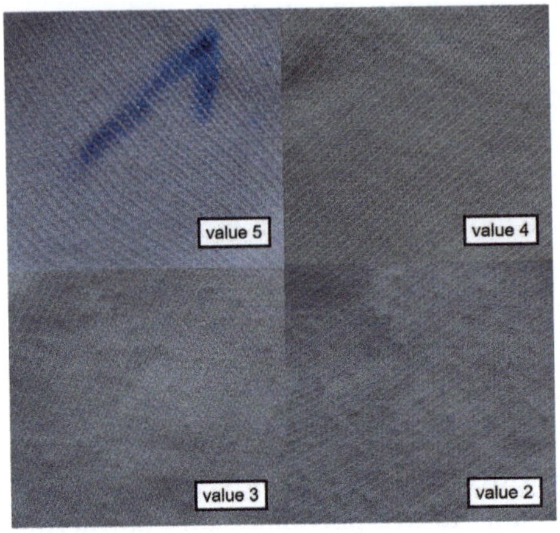

Fig. 2.7: Improvement in pilling tendency (from to 2 to 5) observed after enzymatic treatment (120 min, 48.78C) of a bleached knitted cotton fabric (worst pilling value 2) with cellulase preparations from different fungi (*T. reesei* – value 5; *P. echinulatum* – value 4; *A. nainiana* – value 3) as reported by Rau et al. [234]. Pilling was evaluated with a Martindale Abrasion tester according to ASTM D 4970–98.

Cellulases are a multicomponent enzyme system consisting mainly of three types of enzymes that act synergistically to degrade cellulose to glucose. These include endoglucanases (hydrolyzing cellulose polymers randomly along the chains, preferentially attacking the amorphous region thereby generating new ends), the exoglucanases or cellobiohydrolases (hydrolyze the non-reducing ends of the polymer chains to release cellobiose) and the β-glucosidases which then hydrolyze the released cellobiose into glucose. Biopolishing involves the action of cellulases and high mechanical agitation. The attack of cellulases occurs preferentially in the amorphous regions of the fibres, weakening thus the fibres and making them sensitive to the applied shear forces provided by mechanical action which lead to their break-off from the yarn surface. Endoglucanase enzymes are primarily responsible for these effects [62]. For bioposishing, endoglucanase rich formulations are important. Among the cellulases, "neutral cellulases working at pH between 5.5 and 8 pH and at 50–60 °C are most preferred as compared to acidic cellulases working within pH range of 4.5–5.5 and at a temperature of 45–55 °C [59]. A good control of the pH is necessary for optimum cellulase effect and enzyme deactivation after the process is essential to avoid fabric damage. The presence of high salt concentrations (provided by addition of electrolytes in dyeing with direct and reactive dyes) causes the decrease of cellulase activity. The presence of surfactants, especially ionic surfactants, also causes the decrease of cellulase activity due to enzyme denaturation and inactivation. However, small amounts of nonionic surfactants improve the action of cellulases, influencing the desorption process [235]. Large molecules of reactive and direct dyes, especially in larger quantities present in textiles, also promote the reduction of enzyme activity and consequently the desired effect on fabrics. In industrial processes, the mechanical agitation is an important factor because it accelerates adsorption and desorption effects of cellulases in cotton yarn, as well as providing the energy for fragmenting the fibrils partially hydrolysed, causing the loss of mass of textile. This effect is more pronounced with the use of acidic cellulases than with neutral cellulases [236–239]. Biopolishing is usually applied after dyeing, however, combined dyeing and finishing is of economic interest and was reported [240, 241]. A similar finishing process has also been shown to be necessary for treating lyocell and viscose fabric (produced from dissolving pulps) to prevent fibrillation and enhance their silky appearance. Particularly lyocell fibres have a strong pilling tendency when wet and cellulase treatment is used to remove the first non-uniform pilling of lyocell fabrics. Pilling as the result of subsequent wet processes such as dyeing produces the desired peach skin on lyocell garments [242, 243]. Treatment with cellulases rich in endoglucanases under mechanical stirring increases the dimensional stability of cellulosic fabrics with a low degree of shrinkage and strength loss. The exoglucanases produce no effect in biopolishing, even at high concentrations [244–248].

The influence of different process paramenters such as the increase in hydrolysis by the use of surfactants, mechanical agitation and the use of ultrasound has been thoroughly reviewed before [249], although interesting new insights were found for ultrasound [250] and mechanical agitation [251].

2.8.3 Enzymatic ageing of Denim fabrics

Ageing of (coloured) garments involves a partial and non-homogenous removal of colour and additional changes in the appearance and feel of textiles that makes them look worn-out and used without significant loss of mechanical resistance and functional properties. This effect, mostly desired on denim ware (jeans), a fabric composed of undyed white weft and indigo dyed warp yarns, can be achieved by abrasion of the outer layer of the ring-dyed warp yarn, so that part of the undyed yarn core becomes visible, or by decolorization or fading of the indigo dye. On denim fabrics, the dye removal causes the desired white-blue contrast of dyed and undyed yarns. Similar effects can also be obtained on superficially dyed or pigment dyed garments [59]. A classical success of enzyme application in textile finishing processes is the enzyme based ageing of Denim garments or biostoning which has replaced the traditional stonewashing process using pumice stones. Traditional stonewashing introduced for treating denim garments in the late 1970s is used to impart an artificial fashionable aged look on denim garments [99]. The stonewashing processes uses 1–2 kg stones per kg of jeans in 1 h hence the name stone-washing [252]. Blue jeans and other denim garments are some of the most popular wear and it was estimated that over 800 million pairs of blue jeans are produced worldwide every year [253]. The joint action of cellulases and mechanical agitation promotes defibrillation of surface fibres, formation of cotton powder and causing dye release from the dyed yarn. This gives the desired appearance of faded and aged garment [63]. Cellulase activity is not directly related to the abrasion or biostoning effect [254]. Instead, on endoglucaneses with good biostoning performance specific indigo-binding sites were identified, which seem to be responsible for indigo transport from the fibres [255]. Enzymes are slowly replacing, but not completely, the use of pumice stones. The concomitant use of pumice stones or other materials during biostoning may be explained by the need to increase the abrasion level. Early studies by Heikinhemo et al. [245] demonstrated that *Trichoderma reesei* endoglucanase II could effectively remove colour from denim, producing a good stonewashing effect with the lowest hydrolysis level. In order to produce enough cellulases for industrial application [245, 256], genetically engineered *T. reesei* strains able to produce elevated amounts of endoglucanase activity were developed.

The main advantages of enzyme ageing without stones are lower abrasion of the textile material and as a consequence less damage and and strength loss, preventing damage to the laundry machines, less solid waste and waste water problems and increased productivity, as more garment can be loaded in the machines [257].

One problem that occurs with the use of cellulases in the fading process is the redeposition of the indigo dye on the cotton fibres. This phenomenon, known as backstaining, reduces the desired white/blue contrast and can be attributed to the high affinity between indigo, cellulase proteins and cellulose. It was found to be caused by the interaction of cellulase and indigo, and it is influenced by the reversibility of cellulase adsorption to cellulose, which depends on the origin of the enzyme and the presence (high backstaining) or absence (low backstaining) and the size of a cellulose binding domain (CBD). Acid cellulases showed higher affinity for indigo than neutral cellulases [258–260]. As backstaining is more pronounced in the acid pH range, cellulases for biostoning are preferably neutral and rich in endoglucanases for removing indigo from denim and softening of cotton fabrics without significant damage to the fibres [232, 253, 261]. Treatment with neutral cellulases, such as those produced by *Humicola insolens*, resulted in less loss of fabric strength and less redeposition of the indigo dye on the fabric [258, 259, 262]. Cellulases of *P. echinulatum* with activity in the acid and neutral pH exhibited less backstaining in the neutral pH. The same enzymes displayed also a better biostoning effect with less backstaining when grown on cellulose instead of sugarcane bagasse [234]. Minimizing or preventing back staining is however current focus of research. It was also suggested that backstaining is caused by non-cellulase proteins present in the conventional whole acid cellulases. Many anti-backstaining agents based on chemicals or enzymes, like proteases have succeeded in minimizing back staining problems. Proteases were found to reduce backstaining and improve the contrast when applied at the end of the cellulase washing step or during rinsing or post-washing [263, 264]. Furthermore, the use of an additional washing with detergents or individual detergents components such as surfactants and dispersants in conjunction with strong mechanical stirring tends to improve the effect of fading [264]. The dosage of the enzyme depends on the type, density, porosity and hydrophilicity of the textile material as well as the treatment time (15 to 60 minutes), pH (4.5 to 7.0), temperature (45 to 65 °C), bath ratio (1:3 to 1:20), chemical auxiliaries (the presence of non-ionic surfactants and dispersing agents is recommended) and the type of equipment used [235, 257, 265].

Miettinen-Oinonen et al [266] reported the purification and characterization of three novel cellulases from *Melanocarpus albomyces* able to work under neutral conditions. Indeed combining endoglucanase with cellobiohydrolase demonstrated good biostoning performance and decreased the level of back staining. These studies were then followed by genetic engineering and expression of the respective cellulase genes in *T. reesei* for industrial production [246, 267, 268]. Thanks to the introduction of cellulases in the bio-stoning process, the jeans making is set to completely replace the stones leading to less damage of the garment and totally avoid the generation of pumice dust waste. Enzyme stonewashing is now a standard process used by most of the denim garment processors. Enzyme stonewashing has enabled the development of garments with many combinations of new looks, lower costs, shorter treatment times and reduced solid waste. Laccase based stoning system is also a fast emerging contender for denim [99, 269] with the additional advantage that unlike the cellulases which attack cellulose fibres, laccase only degrades indigo dye without affecting the fabric which results in even lesser damage to fabric and a unique wash-down of denim. It should be noted that laccases alone are not effective in decolorizing indigo on denim and therefore require a mediator, a molecule which when oxidized by laccase forms a highly reactive species able to react with/attack indigo.

2.9 Removal of unfixed dyes and dye effluent treatment

Color is one of the main features of fashionable garments and ecologic dyeing is one of the big challenges in the manufacture of textiles today. Despite of significant developments in dye structure, fibre chemistry and dyeing machinery, still much less than 100 % of the applied dye is fixed to the fibres. Textile dyes pose two main problems to the textile processing industry. Once, unfixed or excess dye that remains adsorbed to the fibres needs to be removed to achieve the desired quality standards. Second, dye residues from the dyeing liquor are discharged as effluent which has to be adequately treated to attend environmental legislation. Microbial (aerobic and anaerobic) systems using algae, yeast, filamentous fungi, bacteria, genetically modified microorganisms, microbial consortia and microbiological systems combined with Advanced Oxidation Processes [270] and/or enzymes based systems for the treatment of textile effluents containing dyes are intensively investigated. Due to many structural similarities between dyes and lignin (complex aromatic natural polymer found in plants), scientists have over the years focused onto exploring the possibility of developing a white rot fungi based dye decolourization systems of colored effluents including textile effluents. White rot fungi produce an impressive lignin degrading enzyme system containing among others; laccases, lignin peroxidases, manganese peroxidases and versatile peroxidases. These enzymes are able to degrade many recalcitrant substances such as chlorophenols, lignin-related structures, organophosphorous compounds, nonphenolic lignin model compounds, phenols and other aromatic compounds. Indeed, many researchers have demonstrated the ability of a number of fungal species from genera of *Pleurotus, Aspergillus, Bjerkandera, Pycnoporus, Sclerotium, Neurospora, Trametes, Chrysosporium, Dichomitus, Daedalea, Irpex and Polyporus* or their enzyme preparations to efficiently decolourize azo, triarylmethane, anthraquinonic, and indigoid textile dyes [271–283]. A number of studies have employed different reactor designs including stirred tanks, packed beds, airlifts, bubble columns, rotating disks resulting invarying decolourization effeciences [284–286]. Among these enzymes, laccases alone or in combination with redox mediators have been shown to be highly promising in extensively degrading many structurally different dyes. These enzymes decolourise textile dyes either through polymerisation or oxidative degradation. However, the low redox potential of the lignolytic enzymes has been shown to limit the degradation of some of the very stable dyes. Currently genetic engineering and screening activities are directing efforts at producing or finding thermostable high redox potential laccases or laccases that work under salt-alkaline conditions [279, 287, 288].

Coloring textiles requires uniform distribution of the dyes throughout the fibres, high exhaustion of the dyeing bath and good fastness of the colors to rubbing (crocking), laundring and home washing. Dye bath exhaustion is always below 100 % because of dye impurities (e.g. secondary products and unreacted reagents from dye production), unwanted side reactions during dyeing (e.g. hydrolysis of reactive dyes) and unfavorable distribution equilibria between dye liquor and fibre. Washing and rubbing fastness (crocking) depend on the fixation of colorants to the fibres through chemical or strong physical-chemical binding. Part of the dye molecules applied in dyeing, hydrolysed but also unreacted dye molecules, are not fixed to the fibres, and remain loosely attached or adsorbed to the fibre surface through weak hydrophobic interactions or hydrogen bonding. In order to achieve the desired quality standards and increase the fastness properties of the dyed goods, subsequent washing is necessary to remove this unfixed or excess dye. Efficiency of traditional postwashing procedures depends on the strength of dye/fibre interaction, the used washing chemicals and washing conditions. In the case of reactive dyed cellulosic fibres a cold and warm rinsing step with water to remove electrolytes, a soaping step with detergents and/or dispersing agents and a final warm rinsing step are used [289]. The limiting factor is however the distribution equilibrium of the dye between the washing solution (soaping bath) and the fibre, so usually more than one washing cycle is necessary. Dyeing of polyester fibres is almost invariably carried out with disperse dyes, hydrophobic organic molecules with very poor solubility in water. Predominantly azo and anthraquinone dyes are used For optimum fastness typical dyeing cycles include a reduction washing step with sodium dithionite and caustic soda at 80 °C to clear away surface deposits of dye which has not penetrated into the polymer matrix. Reduction clearing is not equally efficient on all dye structures and requires additional neutralizing waters [290]. Although The biodegradation of textile dyes has been widely investigated aiming an alternative treatment for textile effluents, little is reported about the use of oxidoreductases in afterwashes for the removal of unfixed or excess dye [291–293]. However, peroxidases are already used on industrial scale in the afterwash of cellulosics dyeing for some dyes e.g. Reactive Black 5. The main problem is that not all dyes are equally degraded by the same enzyme. The afterwash with different enzymes has also been reported after printing with reactive dyes to facilitate the removal of thickeners and excess dyes [294].

In summary, as shown in Tab. 2.2 many different enzymes are used for different purposes during the processing of natural fibres.

Tab. 2.2: Some of the major enzymes used in different processing steps of natural fibres.

Fibre	Cellulase	Xylanse	Laccase	Pectinases	Lipases	Proteases	Amylases	Catalase	Cellobiose dehydrogenase	Glucose oxidases	Transgluta-minase	Tyrosinases	Peroxidases
Bast	X	X		X									
Pulp	X	X	X	X	X	X	X	X					X
Cotton	X	X	X	X			X	X	X	X		X	X
Wool	X	X		X	X	X					X	X	
Silk	X	X		X	X	X					X	X	

2.10 Enzymes in textile after-care

A German chemist Otto Röhm was the first to patent proteases from the pancreas (trypsin and chymotrypsin) for use in detergent formulations during 1913. Later a bacterial protease from *Bacillus licheniformis was* introduced in the 1960s [295]. Since the early 1960s, enzymes have been gradually replacing or minimizing the use of phosphate based detergents. Proteases have now become standard ingredients in all kinds of detergents including those used for household laundry, cleaning contact lenses and heavy duty laundry. Proteases are classified by their catalytic mechanism into aspartic, glutamic, serine, cysteine or metalloproteases, by their ability to cleave terminal amino acids as exo- or endopeptidases, or by pH conditions for optimal activity (acid, neutral or alkaline proteases) [296]. Currently, serine proteases "so called subtilisins" from the genus *Bacillus* (*B. licheniformis, B. lentus, B. alcalophilus* or *B. amyloliquefaciens)* are predominantly used in detergents. Subtilisins are the most suitable proteases because they are stable at alkaline pH and show good washing performance at low temperatures, in the presence of sequestering agents, bleach and surfactants [297]. By 1985, >50 % of heavy-duty laundry detergents formulations in Europe contained enzymes. High-alkaline serine proteases have been successfully applied as protein degrading components of detergent formulations and are subject to extensive protein engineering efforts to improve their stability and performance [298, 299]. To meet the increasing demand for enzymes, heterologous expression systems are playing a critical role while genetic engineering of the enzymes through directed evolution and rational design techniques are being used to tailor the properties of the enzymes towards the needs of the detergent industry. An overview of the advances in genetic engineering studies are summarized by Vojcic et al. [300] The application of proteases in detergent used for wool or silk is problematic because of their ability to penetrate the fabric and hydrolyze wool or silk fibres thereby causing damage to the garments [115]. One approach to prevent this damage is through immobilizing the enzyme onto large molecules. The immobilized enzyme will not be able to penetrate and degrade the wool. Due to the large size the modified enzyme did not damage the wool and was even more effective in removing human blood stains [115]. Cortez et al. [115, 301] also showed that by incorporating TGases in the detergent formulation containing proteases wool, garments were protected from damage during laundry. Cellulases represent another group of enzymes that has been gaining increasing importance in the detergent industry since the 1970s. They now form part of many household washing powders, where they are used to remove fuzzy fibrils on fabric surfaces, improving fabric appearance and colour brightness. Due to the high pH of household detergents used for cellulosic fibres usually alkaline cellulases, especially from *Bacillus sp.*, are used in these formulations [59]. Alpha- amylases have been used in powder laundry detergents since 1975 and make 90 % of all liquid detergents [302, 303]. The α-amylase of *B. licheniformis* are used in dish-washing to remove starch-based dirties e.g. custards, gravies, chocolate etc. while lipases are incorporated in detergent formulations to remove troublesome oily and fatty stains [299]. Mannanases have also been introduced to remove stains containing guar gum, a commonly used stabilizing agent in food products. Especially the removal of chocolate stains, which are hard to clean, is boosted [304]. Thus nowadays many laundry detergent products contain cocktails of enzymes which include proteases, amylases, cellulases, mannanases and lipases [298, 305]. The enzymes used in detergents represent one of the largest and most successful applications of modern industrial biotechnology with a long history.

2.11 Conclusions

Advances in biotechnology and especially in enzyme technology (techniques in identifying and isolating ideal enzymes from nature, increasing understanding of enzyme reaction mechanisms, perfecting the art of industrial production of enzymes, ability to tailor or modify enzymes using genetic engineering techniques to meet desired applications) has witnessed a phenomenal increase in the application of enzymes in fibre processing and textile industry. The application of α-amylases is now a standard process in desizing and detergent industry. Cellulases are fast emerging as some of the most versatile enzymes which have found application in pretreatment of bast fibres, scouring, bio-polishing, denim finishing (stone-washing), wool carbonization process, wool scouring and textile after-care processes. Xylanases are now a well-established technology in bleaching of pulps while catalases are widely used to remove H_2O_2 in bleaching effluents. Thus, amylases, cellulases, catalases, proteases, xylanases, lipases and laccases are now household names in this industry. Their applications range from cellulose fibre and protein fibre extraction, improving purity of dissolving pulps, production and upgrading kraft pulps into dissolving pulps, wet processing of cotton (desizing, scouring, bleaching, polishing) and preparation of wool, finishing e.g. denim stone washing and garment after care (in detergent formulations, removing fuzz fibrils and repairing wool or silk garments). It is also envisaged that the highly promising multiple functions of microbial transglutaminase (MTGases) in wool and silk processing (repairing damaged wool during chemical or enzymatic processing, enhancing the shrink resistance of protein fibres (silk and wool) and improving their tensile strength will soon succeed. The enzyme based processes compared to conventional processes are less energy demanding, reduce water consumption and in addition, can be modified to act selectively thereby reducing fibre damage. The enzymes as compared to chemical catalysts have an additional advantage that they are readily biodegradable. The development of enzyme based textile effluent treatment systems containing oxidases (laccases, lignin peroxidases or manganese peroxidases etc.) will ultimately enable recycling of the process waters while current progress in developing enzyme formulations containing pectinases and xylanases, cellulases will enable the development of an effective bast fibre retting process. Advances in biotechnology will inevitably assist speed up the development of more versatile and robust enzymes and lead to the progressive decrease in the cost of enzymes thereby making them more competitive in fibre/textile industry.

Acknowledgements
The authors are grateful for the financial assistance provided by the Austrian Research Promotion Agency (FFG) through the K-project "Future Lignin and Pulp Processing Research" (FLIPPR) and Enzymes in Pulp and Paper processing (ENZPAP) projects. This work was supported by the Brazilian National Council of Scientific and Tech-

nological Development (CNPq), the National Council for the Improvement of Higher Education (CAPES) and the Regional University of Blumenau (FURB). J. Andreaus and M.D.D.J. Matuchaki are especially grateful for the grants received by CNPq and CAPES.

2.12 References

[1] Choi JM, Han SS, Kim HS. Industrial applications of enzyme biocatalysis: Current status and future aspects. Biotechnol Adv 2015.

[2] Es I, Vieira JDG, Amaral AC. Principles, techniques, and applications of biocatalyst immobilization for industrial application. Appl Microbiol Biot 2015;99:2065–82.

[3] Araujo R, Casal M, Cavaco-Paulo A. Application of enzymes for textile fibres processing. Biocatal Biotransfor 2008;26:332–49.

[4] Kozlowski R, Batog J, Konczewicz W, et al. Enzymes in bast fibrous plant processing. Biotechnol Lett 2006;28:761–5.

[5] Silva C, Cavaco-Paulo A, Fu JJ. Enzymatic biofinishes for synthetic textiles In: Paul R, ed. Functional Finishes for Textiles: Improving Comfort, Performance and Protection: Elsevier Ltd.; 2015:153–91.

[6] Brian M, Alejandro P. Natural fibres. Washington DC 20006 USA: International cotton advisory committee; 2009.

[7] Adrio JL, Demain AL. Microbial enzymes: tools for biotechnological processes. Biomolecules 2014;4:117–39.

[8] Mojsov K. Microbial cellulases and their applications in textile processing. International Journal of Marketing and Technology 2012;2:12–29.

[9] Kuhad RC, Gupta R, Singh A. Microbial Cellulases and Their Industrial Applications. Enzyme Research 2011;2011:1–10.

[10] Yan S, Wu G. Secretory pathway of cellulase: a mini-review. Biotechnol Biofuels 2013;6:177.

[11] Guimaraes LHS, Peixoto-Nogueira SC, Michelin M, et al. Screening of filamentous fungi for production of enzymes of biotechnological interest. Brazilian Journal of Microbiology 2006;37:474–80.

[12] Mohanty A, Misra M, Drzal L. Natural Fibers, Biopolymers, and Biocomposites. Boca Raton: CRC Press; 2005.

[13] Sfiligoj-Smole M, Hribernik S, Stana Kleinschek K, Kreže T. Plant Fibres for Textile and Technical Applications. In: Research AiA, ed. Advances in Agrophysical Research. InTech: InTech; 2013:369–98.

[14] Karimi S, Tahir PM, Karimi A, Dufresne A, Abdulkhani A. Kenaf bast cellulosic fibers hierarchy: a comprehensive approach from micro to nano. Carbohydr Polym 2014;101:878–85.

[15] Zhang J, Henriksson H, Szabo IJ, Henriksson G, Johansson G. The active component in the flax-retting system of the zygomycete Rhizopus oryzae sb is a family 28 polygalacturonase. J Ind Microbiol Biotechnol 2005;32:431–8.

[16] Ming L, Dinesh F, Geoffrey D, et al. Effect of harvest time and field retting duration on the chemical composition, morphology and mechanical properties of hemp fibers. Ind Crop Prod 2015;69:29–39.

[17] Bruhlmann F, Leupin M, Erismann KH, Fiechter A. Enzymatic degumming of ramie bast fibers. J Biotechnol 2000;76:43–50.

[18] Ling H, Ge J, Wei W, Ping W. Analysis of Bacterial Community in Water Retting of Flax. Chinese Journal of Applied and Environmental Biology 2009;15:703–7.

[19] Ahmed Z, Nizam SA. Jute – Microbiological and Biochemical Research. Plant Tissue Culture & Biotechnology 2008;18:197–220.

[20] Munshi TK, Chattoo BB. Bacterial population structure of the jute-retting environment. Microb Ecol 2008;56:270–82.
[21] Fu J, Mueller H, de Castro JV, Jr., et al. Changes in the bacterial community structure and diversity during bamboo retting. Biotechnol J 2011;6:1262–71.
[22] Sharma N, Rathore M, Sharma M. Microbial pectinase: sources, characterization and applications. Reviews in Environmental Science and Bio-Technology 2013;12:45–60.
[23] Zhang J, Pakarinen A, Viikari L. Synergy between cellulases and pectinases in the hydrolysis of hemp. Bioresour Technol 2013;129:302–7.
[24] Altschul SF, Madden TL, Schaffer AA, et al. Gapped BLAST and PSI-BLAST: a new generation of protein database search programs. Nucleic Acids Res 1997;25:3389–402.
[25] Jayani RS, Saxena S, Gupta R. Microbial pectinolytic enzymes: A review. Process Biochem 2005;40:2931–44.
[26] Hoondal GS, Tiwari RP, Tewari R, Dahiya N, Beg QK. Microbial alkaline pectinases and their industrial applications: a review. Appl Microbiol Biot 2002;59:409–18.
[27] Akin DE, Condon B, Sohn M, Foulk JA, Dodd RB, Rigsby LL. Optimization for enzyme-retting of flax with pectate lyase. Ind Crop Prod 2007;25:136–46.
[28] Jonn A.Foulk DEAaRBD. Processing techniques for improving enzyme-retting of flax. Ind Crop Prod 2001;13:239–48.
[29] Alix S, Lebrun L, Marais S, et al. Pectinase treatments on technical fibres of flax: Effects on water sorption and mechanical properties. Carbohyd Polym 2012;87:177–85.
[30] Pakarinen A, Zhang J, Brock T, Maijala P, Viikari L. Enzymatic accessibility of fiber hemp is enhanced by enzymatic or chemical removal of pectin. Bioresour Technol 2012;107:275–81.
[31] Arnoul-Jarriault B, Lachenal D, Chirat C, Heux L. Upgrading softwood bleached kraft pulp to dissolving pulp by cold caustic treatment and acid-hot caustic treatment. Ind Crop Prod 2015;65:565–71.
[32] Sixta H, Iakovlev M, Testova L, et al. Novel concepts of dissolving pulp production. Cellulose 2013;20:1547–61.
[33] Hiett LA. Dissolving cellulose: its present position and prospects for future development. Tappi Journal 1985;68:42–8.
[34] Bajpai P. Biotechnology for Pulp and Paper Processing. New York: Springer New York; 2012.
[35] Torres CE, Lenon G, Craperi D, Wilting R, Blanco A. Enzymatic treatment for preventing biofilm formation in the paper industry. Appl Microbiol Biot 2011;92:95–103.
[36] Torres CE, Negro C, Fuente E, Blanco A. Enzymatic approaches in paper industry for pulp refining and biofilm control. Appl Microbiol Biot 2012;96:327–44.
[37] Gubitz GM, Lischnig T, Stebbing D, Saddler JN. Enzymatic removal of hemicellulose from dissolving pulps. Biotechnol Lett 1997;19:491–5.
[38] Buchert J, Tenkanen M, Kantelinen A, Viikari L. APPLICATION OF XYLANASES IN THE PULP AND PAPER-INDUSTRY. Bioresource Technol 1994;50:65–72.
[39] Bernier R, Jr., Driguez H, Desrochers M. Molecular cloning of a Bacillus subtilis xylanase gene in Escherichia coli. Gene 1983;26:59–65.
[40] Goswami GK, Krishnamohan M, Nain V, Aggarwal C, Ramesh B. Cloning and heterologous expression of cellulose free thermostable xylanase from Bacillus brevis. SpringerPlus 2014;3:20–.
[41] Wong KK, Martin LA, Gama FM, Saddler JN, de Jong E. Bleach boosting and direct brightening by multiple xylanase treatments during peroxide bleaching of kraft pulps. Biotechnol Bioeng 1997;54:312–8.
[42] Bajpai P. Biological bleaching of chemical pulps. Crit Rev Biotechnol 2004;24:1–58.
[43] Christov LP, Prior BA. Repeated treatments with Aureobasidium pullulans hemicellulases and alkali enhance biobleaching of sulphite pulps. Enzyme Microb Tech 1996;18:244–50.

[44] Ibarra D, Kopcke V, Larsson PT, Jaaskelainen AS, Ek M. Combination of alkaline and enzymatic treatments as a process for upgrading sisal paper-grade pulp to dissolving-grade pulp. Bioresour Technol. England: 2010 Elsevier Ltd; 2010:7416–23.

[45] Gehmayr V, Schild G, Sixta H. A precise study on the feasibility of enzyme treatments of a kraft pulp for viscose application. Cellulose 2011;18:479–91.

[46] Cao Y, Tan H. Improvement of alkali solubility of cellulose with enzymatic treatment. Appl Microbiol Biotechnol 2006;70:176–82.

[47] Engstrom AC, Ek M, Henriksson G. Improved accessibility and reactivity of dissolving pulp for the viscose process: pretreatment with monocomponent endoglucanase. Biomacromolecules 2006;7:2027–31.

[48] Henriksson G, Christiernin M, Agnemo R. Monocomponent endoglucanase treatment increases the reactivity of softwood sulphite dissolving pulp. J Ind Microbiol Biotechnol 2005;32:211–4.

[49] Demuner BJ, Pereira Junior N, Antunes AMS. Technology Prospecting on Enzymes for the Pulp and Paper Industry. Journal of technology management & innovation 2011;6:148–58.

[50] Skals PB, Krabek A, Nielsen PH, Wenzel H. Environmental assessment of enzyme assisted processing in pulp and paper industry. International Journal of Life Cycle Assessment 2008;13:124–32.

[51] Evans JD, Akin DE, Foulk JA. Flax-retting by polygalacturonase-containing enzyme mixtures and effects on fiber properties. J Biotechnol 2002;97:223–31.

[52] Gutierrez A, del Rio JC, Martinez AT. Microbial and enzymatic control of pitch in the pulp and paper industry. Appl Microbiol Biotechnol 2009;82:1005–18.

[53] Zhang X, Renaud S, Paice M. The potential of laccase to remove extractives present in pulp and white water from TMP newsprint mills. Journal of Pulp and Paper Science 2005;31:175–80.

[54] Singh G, Kaur K, Puri S, Sharma P. Critical factors affecting laccase-mediated biobleaching of pulp in paper industry. Appl Microbiol Biotechnol 2015;99:155–64.

[55] Verma D, Satyanarayana T. Molecular approaches for ameliorating microbial xylanases. Bioresour Technol 2012;117:360–7.

[56] Juturu V, Wu JC. Microbial xylanases: engineering, production and industrial applications. Biotechnol Adv 2012;30:1219–27.

[57] Cassland P, Sjode A, Winestrand S, Jonsson LJ, Nilvebrant NO. Evaluation of oxalate decarboxylase and oxalate oxidase for industrial applications. Appl Biochem Biotechnol 2010;161:255–63.

[58] Marcato-Romain CE, Pechaud Y, Paul E, Girbal-Neuhauser E, Dossat-Letisse V. Removal of microbial multi-species biofilms from the paper industry by enzymatic treatments. Biofouling 2012;28:305–14.

[59] Andreaus J, Olekszyszen DN, Silveira MHL. Processing of cellulosic textile materials with cellulases. In: Fontana JD, Tiboni M, Grzybowski A, eds. Cellulose and Other Naturally Occurring Polymers. Tivandrum, Kerala, India: Rsearch Signpost; 2014:11–9.

[60] Karmakar SR. Chemical Technology in the Pre-Treatment Processes of Textiles: Elsevier; 1999.

[61] van Dam JEG. Environmental benefits of natural fibre production and use. Proceedings of the Symposium on Natural Fibres; 2008; Rome, Italy: Common Fund for Commodities; 2009. p. 3–17.

[62] Cavaco-Paulo A, Gübitz G. Catalysis and processing. In: Arthur Cavaco-Paulo, Gübitz G, eds. Textile processing with enzymes. Cambridge: Woodhead Publishing Ltd; 2003:86–199.

[63] Hao L, Wang R, Fang K, Liu J. Ultrasonic effect on the desizing efficiency of alpha-amylase on starch-sized cotton fabrics. Carbohydr Polym 2013;96:474–80.

[64] Kalpana BJ, Sindhulakshmi M, Pandian SK. Amylase enzyme from Bacillus subtilis S8–18: a potential desizing agent from the marine environment. Biotechnol Appl Biochem 2014;61:134–44.

[65] Battan B, Dhiman SS, Ahlawat S, Mahajan R, Sharma J. Application of Thermostable Xylanase of Bacillus pumilus in Textile Processing. Indian J Microbiol 2012;52:222–9.
[66] Fukuda T, Kato-Murai M, Kuroda K, Ueda M, Suye S. Improvement in enzymatic desizing of starched cotton cloth using yeast codisplaying glucoamylase and cellulose-binding domain. Appl Microbiol Biotechnol 2008;77:1225–32.
[67] Buchert J, Pere J, Puolakka A, Nousiainen P. Scouring of cotton with pectinases, proteases, and lipases. Text Chem Color Am D 2000;32:48–52.
[68] Klug-Santner BG, Schnitzhofer W, Vrsanska M, et al. Purification and characterization of a new bioscouring pectate lyase from Bacillus pumilus BK2. J Biotechnol 2006;121:390–401.
[69] Karapinar E, Sariisik MO. Scouring of cotton with cellulases, pectinases and proteases. Fibres Text East Eur 2004;12:79–82.
[70] Vigneswaran C, Ananthasubramanian M, Anbumani N, Kandhavadivu P. Ecofriendly Approach to Improve Pectinolytic Reaction and Process Optimization of Bioscouring of Organic Cotton Textiles. Journal of Engineered Fibers and Fabrics 2013;8:121–33.
[71] Degani O, Gepstein S, Dosoretz CG. Potential use of cutinase in enzymatic scouring of cotton fiber cuticle. Appl Biochem Biotech 2002;102:277–89.
[72] Sangwatanaroj U, Choonukulpong K, Ueda M. Cotton scouring with pectinase and lipase/ protease/cellulase. Aatcc Rev 2003;3:17–20.
[73] Agrawal PB, Nierstrasz VA, Klug-Santner BG, Gubitz GM, Lenting HB, Warmoeskerken MM. Wax removal for accelerated cotton scouring with alkaline pectinase. Biotechnol J 2007;2:306–15.
[74] Solbak AI, Richardson TH, McCann RT, et al. Discovery of pectin-degrading enzymes and directed evolution of a novel pectate lyase for processing cotton fabric. Journal of Biological Chemistry 2005;280:9431–8.
[75] Schönberger H, Schäfer T. Best Available Techniques in Textile Industry. In: (Umweltbundesamt) FEA, ed. Research Report 200 94 329 UBA-FB 000325/e. Berlin: Federal Environmental Agency (Umweltbundesamt): Berlin: Federal Environmental Agency (Umweltbundesamt); 2003:362.
[76] Loncar N, Fraaije MW. Catalases as biocatalysts in technical applications: current state and perspectives. Appl Microbiol Biotechnol 2015:Doi10.1007/s00253-015-6512–6.
[77] Tzanov T, Calafell M, Guebitz GM, Cavaco-Paulo A. Bio-preparation of cotton fabrics. Enzyme Microb Tech 2001;29:357–62.
[78] Flitsch A, Prasetyo EN, Sygmund C, Ludwig R, Nyanhongo GS, Guebitz GM. Cellulose oxidation and bleaching processes based on recombinant Myriococcum thermophilum cellobiose dehydrogenase. Enzyme Microb Tech 2013;52:60–7.
[79] Zamocky M, Schumann C, Sygmund C, et al. Cloning, sequence analysis and heterologous expression in Pichia pastoris of a gene encoding a thermostable cellobiose dehydrogenase from Myriococcum thermophilum. Protein Expr Purif 2008;59:258–65.
[80] Tzanov T, Costa SA, Gubitz GM, Cavaco-Paulo A. Hydrogen peroxide generation with immobilized glucose oxidase for textile bleaching. J Biotechnol 2002;93:87–94.
[81] Pereira L, Bastos C, Tzanov T, Cavaco-Paulo A, Guebitz GM. Environmentally friendly bleaching of cotton using laccases. Environmental Chemistry Letters 2005;3:66–9.
[82] Abou-Okeil A, El-Shafie A, El Zawahry MM. Ecofriendly laccase-hydrogen peroxide/ultrasound-assisted bleaching of linen fabrics and its influence on dyeing efficiency. Ultrason Sonochem. Netherlands2010:383–90.
[83] Davulcu A, Eren HA, Avinc O, Erismis B. Ultrasound assisted biobleaching of cotton. Cellulose 2014;21:2973–81.
[84] Basto C, Tzanov T, Cavaco-Paulo A. Combined ultrasound-laccase assisted bleaching of cotton. Ultrason Sonochem 2007;14:350–4.

[85] Goncalves I, Herrero-Yniesta V, Perales Arce I, Escrigas Castaneda M, Cavaco-Paulo A, Silva C. Ultrasonic pilot-scale reactor for enzymatic bleaching of cotton fabrics. Ultrason Sonochem 2014;21:1535–43.

[86] Toscano MD, De Maria L, Lobedanz S, Ostergaard LH. Optimization of a small laccase by active-site redesign. Chembiochem 2013;14:1209–11.

[87] Torres-Salas P, Mate DM, Ghazi I, Plou FJ, Ballesteros AO, Alcalde M. Widening the pH activity profile of a fungal laccase by directed evolution. Chembiochem 2013;14:934–7.

[88] Christensen NJ, Kepp KP. Stability mechanisms of a thermophilic laccase probed by molecular dynamics. PLoS One. United States2013:e61985.

[89] Pardo I, Camarero S. Laccase engineering by rational and evolutionary design. Cell Mol Life Sci 2015;72:897–910.

[90] Ali S, Khatri Z, Khatri A, Tanwari A. Integrated desizing-bleaching-reactive dyeing process for cotton towel using glucose oxidase enzyme. J Clean Prod 2014;66:562–7.

[91] Ibrahim NA, El-Hossamy M, Morsy MS, Eid BM. Development of new eco-friendly options for cotton wet processing. J Appl Polym Sci 2004;93:1825–36.

[92] Oner E, Sahinbaskan BY. A new process of combined pretreatment and dyeing: REST. J Clean Prod 2011;19:1668–75.

[93] Eren HA, Anis P, Davulcu A. Enzymatic One-bath Desizing – Bleaching – Dyeing Process for Cotton Fabrics. Text Res J 2009;79:1091–8.

[94] Maryan AS, Montazer M. A cleaner production of denim garment using one step treatment with amylase/cellulase/laccase. J Clean Prod 2013;57:320–6.

[95] Nina P, Forte TP. Complete enzymatic pre-treatment of cotton fabric with incorporated bleach activator. 2013:566–73.

[96] Sambanthan S. One-Step Biopolishing and bleach clean-up in the dye bath. www.novozymes.com; 2014.

[97] Ammayappan L. Eco-friendly Surface Modifications of Wool Fiber for its Improved Functionality: An Overview. Asian Journal of Textile 2013;3:15–28.

[98] Kaur A, Chakraborty JN. Optimization of Bromelain Treatment pH with Wool for Antifelting and Reduced Pilling Behaviour: Objective Assessment Approach. Journal of Textiles 2015;2015:7.

[99] Rehman MMA, Imran MA. Revolution of Biotechnology in Finishing Sector of Textile. Chemistry and Material Research 2014;6:92–103.

[100] Heiz H. Chlorine-Hercosett treatment of wool. Textil Veredlung 1981;16:43–53.

[101] Levene R, Cohen Y, Barkai D. Applying proteases to confer improved shrink-resistance to wool. J Soc Dyers Colour 1996;112:6–10.

[102] Parvinzadeh M. Effect of proteolytic enzyme on dyeing of wool with madder. Enzyme Microb Tech 2007;40:1719–22.

[103] Das T, Ramaswamy GN. Enzyme treatment of wool and specialty hair fibers. Text Res J 2006;76:126–33.

[104] Jovancic P, Jocic D, Molina R, Julia MR, Erra P. Shrinkage Properties of Peroxide-Enzyme-Biopolymer Treated Wool. Text Res J 2001;71:948–53.

[105] El-Sayed H, Hamed RR, Kantouch A, Heine E, Hocker H. Enzyme-based feltproofing of wool. Aatcc Rev 2002;2:25–8.

[106] Smith E, Shen JS. Surface modification of wool with protease extracted polypeptides. J Biotechnol 2011;156:134–40.

[107] Shen JS, Rushforth M, Cavaco-Paulo A, Guebitz G, Lenting H. Development and industrialisation of enzymatic shrink-resist process based on modified proteases for wool machine washability. Enzyme Microb Tech 2007;40:1656–61.

[108] Schroeder M, Lenting HBM, Kandelbauer A, Silva C, Cavaco-Paulo A, Gubitz GM. Restricting detergent protease action to surface of protein fibres by chemical modification. Appl Microbiol Biot 2006;72:738–44.
[109] Silva CM, Carneiro F, O'Neill A, et al. Cutinase – A new tool for biomodification of synthetic fibers. Journal of Polymer Science Part a-Polymer Chemistry 2005;43:2448–50.
[110] Silva C, Prabaharan M, Gubitz G, Cavaco-Paulo A. Treatment of wool fibres with subtilisin and subtilisin-PEG. Enzyme Microb Tech 2005;36:917–22.
[111] Lenting HB, Schroeder M, Guebitz GM, Cavaco-Paulo A, Shen J. New enzyme-based process direction to prevent wool shrinking without substantial tensile strength loss. Biotechnol Lett 2006;28:711–6.
[112] Silva C, Da S, Silva N, et al. Engineered Thermobifida fusca cutinase with increased activity on polyester substrates. Biotechnol J 2011;6:1230–9.
[113] Araujo R, Silva C, O'Neill A, et al. Tailoring cutinase activity towards polyethylene terephthalate and polyamide 6,6 fibers. J Biotechnol 2007;128:849–57.
[114] Liu B, Zhang J, Li B, Liao X, Du G, Chen J. Expression and characterization of extreme alkaline, oxidation-resistant keratinase from Bacillus licheniformis in recombinant Bacillus subtilis WB600 expression system and its application in wool fiber processing. World J Microbiol Biotechnol 2013;29:825–32.
[115] Cortez J, Bonner PL, Griffin M. Transglutaminase treatment of wool fabrics leads to resistance to detergent damage. J Biotechnol 2005;116:379–86.
[116] Tesfaw Asmamaw AF. Applications of Transglutaminase in Textile, Wool, and Leather Processing. International Journal of Textile Science 2014;3:64–9.
[117] Griffin M, Casadio R, Bergamini CM. Transglutaminases: nature's biological glues. Biochem J. England2002:377–96.
[118] Rachel NM, Pelletier JN. Biotechnological applications of transglutaminases. Biomolecules 2013;3:870–88.
[119] Rinsey Johnny VA, Karpagam Chinnammal S. Degumming of silk using protease enzyme from Bacillusspecies. International Journal of Science and Nature 2012;3:51–9.
[120] Li GH, Liu H, Li TD, Wang JY. Surface modification and functionalization of silk fibroin fibers/fabric toward high performance applications. Materials Science & Engineering C-Materials for Biological Applications 2012;32:627–36.
[121] Chellamani KP, Veerasubramanian D, Vignesh BRS. Textile Implants: Silk Suture Manufacturing Technology. Journal of Academia and Industrial Research (JAIR) 2014;3:127–31.
[122] Freddi G, Kundu SC. 6 – Silk fibroin microfiber and nanofiber scaffolds for tissue engineering and regeneration. Silk Biomaterials for Tissue Engineering and Regenerative Medicine: Woodhead Publishing; 2014:157–90.
[123] Arami M, Rahimi S, Mivehie L, Mazaheri F, Mahmoodi NM. Degumming of Persian silk with mixed proteolytic enzymes. J Appl Polym Sci 2007;106:267–75.
[124] Freddi G, Anghileri A, Sampaio S, Buchert J, Monti P, Taddei P. Tyrosinase-catalyzed modification of Bombyx mori silk fibroin: Grafting of chitosan under heterogeneous reaction conditions. J Biotechnol 2006;125:281–94.
[125] Unajak S, Aroonluke S, Promboon A. An active recombinant cocoonase from the silkworm Bombyx mori: bleaching, degumming and sericin degrading activities. J Sci Food Agric 2014.
[126] Lapkin A, Constable DJC. Green Chemistry Metrics: Measuring and Monitoring Sustainable Processes: Blackwell Publishing Ltd.; 2009.
[127] Guebitz GM, Cavaco-Paulo A. Enzymes go big: surface hydrolysis and functionalisation of synthetic polymers. Trends in Biotechnology 2008;26:32–8.
[128] Guebitz GM, Cavaco-Paulo A. New substrates for reliable enzymes: enzymatic modification of polymers. Curr Opin Biotech 2003;14:577–82.

[129] Kisner A, Rainert KT, Ferrari F, et al. Chemical functionalization of polyamide 6.6 fabrics. Reactive & Functional Polymers 2013;73:1349–56.
[130] Zeronian SH, Collins MJ. Surface modification of polyester by alkaline treatments. Textile Progress 1989;20:1–34.
[131] Heumann S, Eberl A, Pobeheim H, et al. New model substrates for enzymes hydrolysing polyethyleneterephthalate and polyamide fibres. Journal of Biochemical and Biophysical Methods 2006;69:89–99.
[132] Ribitsch D, Acero EH, Greimel K, et al. Characterization of a new cutinase from Thermobifida alba for PET-surface hydrolysis. Biocatal Biotransfor 2012;30:2–9.
[133] Herrero Acero E, Ribitsch D, Steinkellner G, et al. Enzymatic Surface Hydrolysis of PET: Effect of Structural Diversity on Kinetic Properties of Cutinases from Thermobifida. Macromolecules 2011;44:4632–40.
[134] Ribitsch D, Acero EH, Greimel K, et al. A New Esterase from *Thermobifida halotolerans* Hydrolyses Polyethylene Terephthalate (PET) and Polylactic Acid (PLA). Polymers-Basel 2012;4:617–29.
[135] Wei R, Oeser T, Billig S, Zimmermann W. A high-throughput assay for enzymatic polyester hydrolysis activity by fluorimetric detection. Biotechnol J 2012;7:1517–21.
[136] Mueller R-J. Biological degradation of synthetic polyesters—Enzymes as potential catalysts for polyester recycling. Process Biochem 2006;41:2124–8.
[137] Alisch-Mark M, Herrmann A, Zimmermann W. Increase of the hydrophilicity of polyethylene terephthalate fibres by hydrolases from Thermomonospora fusca and Fusarium solani f. sp. pisi. Biotechnol Lett 2006;28:681–5.
[138] Ronkvist AM, Xie WC, Lu WH, Gross RA. Cutinase-Catalyzed Hydrolysis of Poly(ethylene terephthalate). Macromolecules 2009;42:5128–38.
[139] Kleeberg I, Welzel K, Vandenheuvel J, Muller RJ, Deckwer WD. Characterization of a new extracellular hydrolase from Thermobifida fusca degrading aliphatic-aromatic copolyesters. Biomacromolecules 2005;6:262–70.
[140] Herrero Acero E, Ribitsch D, Dellacher A, et al. Surface engineering of a cutinase from Thermobifida cellulosilytica for improved polyester hydrolysis. Biotechnol Bioeng 2013;110:2581–90.
[141] Roth C, Wei R, Oeser T, et al. Structural and functional studies on a thermostable polyethylene terephthalate degrading hydrolase from Thermobifida fusca. Appl Microbiol Biot 2014;98:7815–23.
[142] Zimmermann W, Billig S. Enzymes for the Biofunctionalization of Poly(Ethylene Terephthalate). In: Nyanhongo GS, Steiner W, Gübitz G, eds. Biofunctionalization of Polymers and their Applications: Springer Berlin Heidelberg; 2011:97–120.
[143] Kawai F, Oda M, Tamashiro T, et al. A novel Ca2+-activated, thermostabilized polyesterase capable of hydrolyzing polyethylene terephthalate from Saccharomonospora viridis AHK190. Appl Microbiol Biot 2014;98:10053–64.
[144] Then J, Wei R, Oeser T, et al. Ca2+ and Mg2+ binding site engineering increases the degradation of polyethylene terephthalate films by polyester hydrolases from Thermobifida fusca. Biotechnol J 2015;10:592–8.
[145] Ribitsch D, Yebra AO, Zitzenbacher S, et al. Fusion of binding domains to Thermobifida cellulosilytica cutinase to tune sorption characteristics and enhancing PET hydrolysis. Biomacromolecules 2013;14:1769–76.
[146] Zhang Y, Wang L, Chen J, Wu J. Enhanced activity toward PET by site-directed mutagenesis of Thermobifida fusca cutinase–CBM fusion protein. Carbohyd Polym 2013;97:124–9.

[147] Ribitsch D, Herrero Acero E, Przylucka A, et al. Enhanced cutinase-catalyzed hydrolysis of polyethylene terephthalate by covalent fusion to hydrophobins. Appl Environ Microbiol 2015;81:3586–92.
[148] Fischer-Colbrie G, Heumann S, Liebminger S, Almansa E, Cavaco-Paulo A, Guebitz GM. New enzymes with potential for PET surface modification. Biocatal Biotransfor 2004;22:341–6.
[149] Almansa E, Heumann S, Eberl A, et al. Enzymatic surface hydrolysis of PET enhances bonding in PVC coating. Biocatal Biotransfor 2008;26:365–70.
[150] Kim H, Song W. Lipase treatment of polyester fabrics. Fiber Polym 2006;7:339–43.
[151] Khoddami A, Morshed M, Hossein T. Effects of enzymatic hydrolysis on drawn polyester filament yarns. Iran Polym J 2001;10:363–70.
[152] Hsieh Y-L, Cram LA. Enzymatic Hydrolysis to Improve Wetting and Absorbency of Polyester Fabrics. Text Res J 1998;68:311–9.
[153] Brueckner T, Eberl A, Heumann S, Rabe M, Guebitz GM. Enzymatic and chemical hydrolysis of poly(ethylene terephthalate) fabrics. Journal of Polymer Science Part a-Polymer Chemistry 2008;46:6435–43.
[154] Eberl A, Heumann S, Brückner T, et al. Enzymatic surface hydrolysis of poly(ethylene terephthalate) and bis(benzoyloxyethyl) terephthalate by lipase and cutinase in the presence of surface active molecules. J Biotechnol 2009;143:207–12.
[155] Ribitsch D, Heumann S, Trotscha E, et al. Hydrolysis of polyethyleneterephthalate by p-nitrobenzylesterase from Bacillus subtilis. Biotechnol Prog 2011;27:951–60.
[156] Liebminger S, Eberl A, Sousa F, et al. Hydrolysis of PET and bis-(benzoyloxyethyl) terephthalate with a new polyesterase from Penicillium citrinum. Biocatal Biotransfor 2007;25:171–7.
[157] Sato M. Deterioration of filaments and films of polyethyleneterphthalate with enzyme of *Cladosporium cladosporioides* FERM J-8. Sen-ito Kogyo 1983;39:T209–T19.
[158] Feuerhack A, Alisch-Mark M, Kisner A, Pezzin SH, Zimmermann W, Andreaus J. Biocatalytic surface modification of knitted fabrics made of poly (ethylene terephthalate) with hydrolytic enzymes from Thermobifida fusca KW3b. Biocatalysis and Biotransformation 2008;26:357–64.
[159] Nechwatal A, Blokesch A, Nicolai M, et al. A contribution to the investigation of enzyme-catalysed hydrolysis of poly(ethylene terephthalate) oligomers. Macromol Mater Eng 2006;291:1486–94.
[160] Hooker J, Hinks D, Montero G, Icherenska M. Enzyme-catalyzed hydrolysis of poly(ethylene terephthalate) cyclic trimer. J Appl Polym Sci 2003;89:2545–52.
[161] Yoon MY, Kellis J, Poulose AJ. Enzymatic modification of polyester. Aatcc Rev 2002;2:33–6.
[162] Almansa E, Heumann S, Eberl A, Kaufmann F, Cavaco-Paulo A, Gubitz GM. Surface hydrolysis of polyamide with a new polyamidase from Beauveria brongniartii. Biocatal Biotransfor 2008;26:371–7.
[163] Parvinzadeh M, Assefipour R, Kiumarsi A. Biohydrolysis of nylon 6,6 fiber with different proteolytic enzymes. Polym Degrad Stabil 2009;94:1197–205.
[164] Kiumarsi A, Parvinzadeh M. Enzymatic hydrolysis of nylon 6 fiber using lipolytic enzyme. J Appl Polym Sci 2010;116:3140–7.
[165] Heumann S, Eberl A, Fischer-Colbrie G, et al. A novel aryl acylamidase from Nocardia farcinica hydrolyses polyamide. Biotechnol Bioeng 2009;102:1003–11.
[166] Herrero Acero E, Ribitsch D, Rodriguez RD, et al. Two-step enzymatic functionalisation of polyamide with phenolics. Journal of Molecular Catalysis B: Enzymatic 2012;79:54–60.
[167] Kim HR, Seo HY. Enzymatic hydrolysis of polyamide fabric by using acylase. Text Res J 2013;83:1181–9.

[168] Friedrich J, Zalar P, Mohorčič M, Klun U, Kržan A. Ability of fungi to degrade synthetic polymer nylon-6. Chemosphere 2007;67:2089–95.
[169] Fujisawa M, Hirai H, Nishida T. Degradation of Polyethylene and Nylon-66 by the Laccase-Mediator System. Journal of Polymers and the Environment 2001;9:103–8.
[170] Deguchi T, Kakezawa M, Nishida T. Nylon biodegradation by lignin-degrading fungi. Appl Environ Microb 1997;63:329–31.
[171] Deguchi T, Kitaoka Y, Kakezawa M, Nishida T. Purification and Characterization of a Nylon-Degrading Enzyme. Appl Environ Microb 1998;64:1366–71.
[172] Masson JC. Acrylic fiber technology and applications. New York1995.
[173] Fischer-Colbrie G, Matama T, Heumann S, Martinkova L, Cavaco Paulo A, Guebitz G. Surface hydrolysis of polyacrylonitrile with nitrile hydrolysing enzymes from Micrococcus luteus BST20. J Biotechnol 2007;129:62–8.
[174] Matamá T, Carneiro F, Caparrós C, Gübitz GM, Cavaco-Paulo A. Using a nitrilase for the surface modification of acrylic fibres. Biotechnol J 2007;2:353–60.
[175] Tauber M, Cavaco-Paulo A, Robra K-H, Gübitz GM. Nitrile Hydratase and Amidase from *Rhodococcus rhodochrous* Hydrolyze Acrylic Fibers and Granular Polyacrylonitriles. Appl Environ Microb 2000;66:1634–8.
[176] Battistel E, Morra M, Marinetti M. Enzymatic surface modification of acrylonitrile fibers. Applied Surface Science 2001;177:32–41.
[177] Fischer-Colbrie G, Herrmann M, Heumann S, et al. Surface modification of polyacrylonitrile with nitrile hydratase and amidase from Agrobacterium tumefaciens. Biocatal Biotransfor 2006;24:419–25.
[178] Frushour BG, Knorr RS. Acrylic Fibers. In: Lewin M, ed. Handbook of fiber chemistry. 3rd ed: CRC Press, Taylor & Francis Group; 2007:811–973.
[179] Mather RR. Polyolefin fibres. In: McIntyre JE, ed. Synthetic fibres: nylon, polyester, acrylic, polyolefin. 1 ed. Cambridge CB1 6AH, England: Woodhead Publishing Ltd; 2005:235–92.
[180] East AJ. Polyester fibres. In: McIntyre JE, ed. Synthetic fibres: nylon, polyester, acrylic, polyolefin. 1 ed. Cambridge CB1 6AH, England: Woodhead Publishing Ltd; 2005:95–166.
[181] Richards AF. Nylon fibres. In: McIntyre JE, ed. Synthetic fibres: nylon, polyester, acrylic, polyolefin. 1 ed. Cambridge CB1 6AH, England: Woodhead Publishing Ltd; 2005:20–94.
[182] Cox R. Acrylic fibres. In: McIntyre JE, ed. Synthetic fibres: nylon, polyester, acrylic, polyolefin. 1 ed. Cambridge CB1 6AH, England: Woodhead Publishing Ltd; 2005:167–234.
[183] Zhu M-F, Yang HH. Polypropylene Fibers. In: Lewin M, ed. Handbook of fiber chemistry. 3rd ed: CRC Press, Taylor & Francis Group; 2007:139–260.
[184] Shahid M, Shahid ul I, Mohammad F. Recent advancements in natural dye applications: a review. J Clean Prod 2013;53:310–31.
[185] Wang P, Yu ML, Cui L, Yuan JG, Wang Q, Fan XR. Modification of Bombyx mori silk fabrics by tyrosinase- catalyzed grafting of chitosan. Eng Life Sci 2014;14:211–7.
[186] Wang Q, Fan X, Hu Y, Yuan J, Cui L, Wang P. Antibacterial functionalization of wool fabric via immobilizing lysozymes. Bioprocess and Biosystems Engineering 2009;32:633–9.
[187] Hossain K, Gonzalez MD, Lozano GR, Tzanov T. Multifunctional modification of wool using an enzymatic process in aqueous-organic media. J Biotechnol 2009;141:58–63.
[188] Vankar PS, Shanker R, Verma A. Enzymatic natural dyeing of cotton and silk fabrics without metal mordants. J Clean Prod 2007;15:1441–50.
[189] Vankar PS, Shanker R. Ecofriendly ultrasonic natural dyeing of cotton fabric with enzyme pretreatments. Desalination 2008;230:62–9.
[190] Montazer M, Dadashian F, Hemmatinejad N, Farhoudi K. Treatment of Wool with Laccase and Dyeing with Madder. Appl Biochem Biotech 2009;158:685–93.

[191] Kim S, Moldes D, Cavaco-Paulo A. Laccases for enzymatic colouration of unbleached cotton. EnzMicrobTechnol 2007;40:1788–93.
[192] Tsatsaroni E, Liakopouloukyriakides M. Effect of Enzymatic Treatment on the Dyeing of Cotton and Wool Fibers with Natural Dyes. Dyes Pigments 1995;29:203–9.
[193] Tsatsaroni E, Liakopoulou-Kyriakides M, Eleftheriadis I. Comparative study of dyeing properties of two yellow natural pigments – Effect of enzymes and proteins. Dyes Pigments 1998;37:307–15.
[194] Liakopoulou-Kyriakides M, Tsatsaroni E, Laderos P, Georgiadou K. Dyeing of cotton and wool fibres with pigments from Crocus sativus – Effect of enzymatic treatment. Dyes Pigments 1998;36:215–21.
[195] Zhang RP, Cai ZS. Study on the Natural Dyeing of Wool Modified with Enzyme. Fiber Polym 2011;12:478–83.
[196] Kumbasar EPA, Atav R, Bahtiyari MI. Effects of Alkali Proteases on Dyeing Properties of Various Proteinous Materials with Natural Dyes. Text Res J 2009;79:517–25.
[197] Tzanov T, Costa SA, Guebitz GM, Cavaco-Paulo A. Effect of the temperature and bath composition on the dyeing with catalase treated bleaching effluents. Color Technology 2001;(in press).
[198] Tzanov T, Silva CJ, Zille A, Oliveira J, Cavaco-Paulo A. Effect of some process parameters in enzymatic dyeing of wool. Appl Biochem Biotech 2003;111:1–13.
[199] Zille A, Munteanu FD, Guebitz GM, Cavaco-Paulo A. Laccase kinetics of degradation and coupling reactions. Journal of Molecular Catalysis B: Enzymatic 2005;33:23–8.
[200] Hadzhiyska H, Calafell M, Gibert JM, Daga JM, Tzanov T. Laccase-assisted dyeing of cotton. Biotechnol Lett 2006;28:755–9.
[201] Tzanov T, Costa SA, Guebitz GM, Cavaco-Paulo A. Dyeing With Catalase Treated Bleaching Baths. Color Technology 2001;117.
[202] Acero EH, Kudanga T, Ortner A, et al. Laccase Functionalization of Flax and Coconut Fibers. Polymers-Basel 2014;6:1676–84.
[203] Nyanhongo GS, Prasetyo EN, Herrero Acero E, Guebitz GM. Reaction engineering strategies for successful development of functional polymers using oxidative enzymes. In: Liebner F, Rosenau T, eds. Functional Materials from Renewable Sources: ACS Books; 2012:329–43.
[204] Nyanhongo G.S. SWGGM. Biofunctionalization of polymers and their applications: Springer Berlin Heidelberg; 2011.
[205] Jeon JR, Kim EJ, Murugesan K, et al. Laccase-catalysed polymeric dye synthesis from plant-derived phenols for potential application in hair dyeing: Enzymatic colourations driven by homo- or hetero-polymer synthesis. Microb Biotechnol 2010;3:324–35.
[206] Xu F, Salmon S. Potential applications of oxidoreductases for the re-oxidation of leuco vat or sulfur dyes in textile dyeing. Eng Life Sci 2008;8:331–7.
[207] Dawson TL. Biosynthesis and synthesis of natural colours. Color Technol 2009;125:61–73.
[208] Ensley BD. Biosynthesis of the Textile Dye Indigo by a Recombinant Bacterium. Chimia 1994;48:491–2.
[209] Dodge TC, Lauff JJ, Wendt DJ, Weyler W. Improved microbial production of indigo. Google Patents; 1997.
[210] Qu Y, Zhang X, Ma Q, et al. Indigo biosynthesis by Comamonas sp. MQ. Biotechnol Lett 2012;34:353–7.
[211] Bhushan B, Samanta SK, Jain RK. Indigo production by naphthalene-degrading bacteria. Lett Appl Microbiol 2000;31:5–9.
[212] Mercadal JPR, Isaac P, Sineriz F, Ferrero MA. Indigo production by Pseudomonas sp J26, a marine naphthalene-degrading strain. J Basic Microb 2010;50:290–3.

[213] Alihosseini F, Lango J, Ju KS, Hammock BD, Sun G. Mutation of Bacterium Vibrio gazogenes for Selective Preparation of Colorants. Biotechnol Progr 2010;26:352–60.

[214] Kramar A, Ilic-Tomic T, Petkovic M, et al. Crude bacterial extracts of two new Streptomyces sp isolates as bio-colorants for textile dyeing. World J Microb Biot 2014;30:2231–40.

[215] Forte S, Polak J, Valensin D, et al. Synthesis and structural characterization of a novel phenoxazinone dye by use of a fungal laccase. J Mol Catal B-Enzym 2010;63:116–20.

[216] Stolarzewicz I, Kapturowska A, Bialecka-Florjanczyk E. Microbiological sources of colorants in food technology. Postep Mikrobiol 2012;51:167–76.

[217] Chandi GK, Gill BS. Production and Characterization of Microbial Carotenoids as an Alternative to Synthetic Colors: a Review. Int J Food Prop 2011;14:503–13.

[218] Dufosse L. Microbial production of food grade pigments. Food Technol Biotech 2006;44:313–21.

[219] Mapari SAS, Nielsen KF, Larsen TO, Frisvad JC, Meyer AS, Thrane U. Exploring fungal biodiversity for the production of water-soluble pigments as potential natural food colorants. Curr Opin Biotech 2005;16:231–8.

[220] Venil CK, Zakaria ZA, Ahmad WA. Bacterial pigments and their applications. Process Biochem 2013;48:1065–79.

[221] Vigneshwaran N, Saxena S, Kathe AA, Gayal SG, Balasubramanya RH. Bacterial pigments for eco-friendly textile dyeing. Quality Textiles for Quality Life, Vols 1–4 2004:765–8.

[222] Busiliene G, Lekeckas K, Urbelis V. Pilling Resistance of Knitted Fabrics. Materials Science-Medziagotyra 2011;17:297–301.

[223] Canoglu S, Yukseloglu S. Hairiness Values of the Polyester/Viscose Ring-Spun Yarn Blends. Fibres & Textiles in Eastern Europe 2008;16:34–8.

[224] Okubayashi S, Bechtold T. A pilling mechanism of man-made cellulosic fabrics – Effects of fibrillation. Textile Research Journal 2005;75:288–92.

[225] Wang L, Wang X, Yao M. Recent research on pilling of wool knitwear. Proceedings of 2006 China International Wool Textile Conference & IWTO Wool Forum 2006:252–9.

[226] Tusief M, Mahmood N, Saleem M. Effect of Different Anti Pilling Agents to reduce Pilling on Polyester/Cotton Fabric. Journal of the Chemical Society of Pakistan 2012;34:53–7.

[227] Hearle JWS, Wilkins AH. Mechanistic modelling of pilling. Part I: Detailing of mechanisms. Journal of the Textile Institute 2006;97:359–68.

[228] Bui HM, Ehrhardt A, Bechtold T. Pilling in cellulosic fabrics, Part 2: A study on kinetics of pilling in alkali-treated lyocell fabrics. Journal of Applied Polymer Science 2008;109:3696–703.

[229] Galante YA, Formantici C. Enzyme applications in detergency and in manufacturing industries. Current Organic Chemistry 2003;7:1399–422.

[230] Bajaj P. Finishing of textile materials. Journal of Applied Polymer Science 2002;83:631–59.

[231] Balci O, Gencer U. Cellulase enzyme application for the cotton based woven fabrics Part I. Determination of effect of enzyme on the performance. Industria Textila 2013;64:20–6.

[232] Bai G, Fu K, Jin N, et al. Bio-polishing of Cotton Fabric with Cellulase. Automation Equipment and Systems, Pts 1–4 2012;468–471:46–9.

[233] Csiszar E, Szakacs G, Rusznak I. Bioscouring of cotton with cellulase enzyme. Abstracts of Papers of the American Chemical Society 1997;213:27-CELL.

[234] Rau M, Heidemann C, Pascoalin AM, et al. Application of cellulases from Acrophialophora nainiana and Penicillium echinulatum in textile processing of cellulosic fibres. Biocatalysis and Biotransformation 2008;26:383–90.

[235] Zilz L, Rau M, Budag N, Scharf M, Cavaco-Paulo A, Andreaus J. Nonionic surfactants and dispersants for biopolishing and stonewashing with Hypocrea jecorina cellulases. Color Technol 2013;129:49–54.

[236] Andreaus J, Cavaco-Paulo A. Enzimas no processamento de fibras têxteis. In: Bon EPS, Ferrara MA, Corvo ML, eds. Enzimas em Biotecnologia: produção, aplicações e mercado. Rio de Janeiro: Interciência; 2008:179–204.
[237] Cavaco-Paulo A. Mechanism of cellulase action in textile processes. Carbohydrate Polymers 1998;37:273–7.
[238] Montazer M, Maryan A. Influences of Different Enzymatic Treatment on Denim Garment. Applied Biochemistry and Biotechnology 2010;160:2114–28.
[239] Hao L, Men Y, Qi Z, et al. Influences of Acid Cellulase on Indigo Backstaining on Cotton. Advances in Applied Science and Industrial Technology, Pts 1 and 2 2013;798–799:35–8.
[240] Dong YC, Xu K, Tang XQ. Combined dyeing and bio-finishing of cotton fabrics with reactive dyes and acid cellulase. Cell Chem Technol 2004;38:445–55.
[241] Kokol V, Golob V. The effect of single bath treatment on reactive dyeing and bio-finishing of cotton fabric. Quality Textiles for Quality Life, Vols 1–4 2004:757–61.
[242] Morgado J, Cavaco-Paulo A, Rousselle MA. Enzymatic treatment of lyocell - Clarification of depilling mechanisms. Text Res J 2000;70:696–9.
[243] Carrillo F, Colom X, Lopez-Mesas M, Lis MJ, Gonzalez F, Valldeperas J. Cellulase processing of lyocell and viscose type fibres: kinetics parameters. Process Biochem 2003;39:257–61.
[244] Duran N, Duran M. Enzyme applications in the textile industry. Review of Progress in Coloration and Related Topics 2008;30:41–4.
[245] Heikinheimo L, Buchert J, Miettinen-Oinonen A, Suominen P. Treating denim fabrics with Trichoderma reesei cellulases. Text Res J 2000;70:969–73.
[246] Pazarlioglu NK, Sariisik M, Telefoncu A. Treating denim fabrics with immobilized commercial cellulases. Process Biochem 2005;40:767–71.
[247] Cortez JM, Ellis J, Bishop DP. Using Cellulases to Improve the Dimensional Stability of Cellulosic Fabrics. Textile Research Journal 2002;72:673–80.
[248] Ibrahim N, El-Badry K, Eid B, Hassan T. A new approach for biofinishing of cellulose-containing fabrics using acid cellulases. Carbohydrate Polymers 2011;83:116–21.
[249] Andreaus J, Filho EXF, Bon EPS. Biotechnology of Holocellulose-Degrading Enzymes. In: Hou CT, Shaw J-F, eds. Biocatalysis and Bioenergy. Hoboken, New Jersey: John Wiley & Sons, Inc.; 2008:197–230.
[250] Subhedar PB, Gogate PR. Enhancing the activity of cellulase enzyme using ultrasonic irradiations. J Mol Catal B-Enzym 2014;101:108–14.
[251] Silveira MHL, Rau M, Andreaus J. Influence of mechanical agitation on the pH profile of total, soluble and insoluble filter paper activity of Hypocrea jecorina cellulase preparations. Biocatal Biotransfor 2012;30:63–70.
[252] Nielsen P, Kuilderd H, Zhou W, Lu X. Enzyme biotechnology for sustainable textiles. Boca Raton: CRC Pres; 2009.
[253] Bhat MK. Cellulases and related enzymes in biotechnology. Biotechnol Adv 2000;18:355–83.
[254] Gusakov AV, Berlin AG, Popova NN, Okunev ON, Sinitsyna OA, Sinitsyn AP. A comparative study of different cellulase preparations in the enzymatic treatment of cotton fabrics. Appl Biochem Biotech 2000;88:119–26.
[255] Gusakov AV, Sinitsyn AP, Markov AV, Sinitsyna OA, Ankudimova NV, Berlin AG. Study of protein adsorption on indigo particles confirms the existence of enzyme-indigo interaction sites in cellulase molecules. J Biotechnol 2001;87:83–90.
[256] Miettinen-Oinonen A, Suominen P. Enhanced production of Trichoderma reesei endoglucanases and use of the new cellulase preparations in producing the stonewashed effect on denim fabric. Appl Environ Microb 2002;68:3956–64.

[257] Nierstrasz VA, Warmoeskerken MMCG. Process engineering and industrial enzyme applications. In: Cavaco-Paulo A, Gübitz G, eds. Textile processing with enzymes. Cambridge: Woodhead Publishing Ltd; 2003:120–57.
[258] Andreaus J, Campos R, Gubitz G, Cavaco-Paulo A. Influence of cellulases on indigo backstaining. Textile Research Journal 2000;70:628–32.
[259] Campos R, Cavaco-Paulo A, Andreaus J, Gubitz G. Indigo-cellulase interactions. Textile Research Journal 2000;70:532–6.
[260] Cavaco-Paulo A, Morgado J, Almeida L, Kilburn D. Indigo backstaining during cellulase washing. Text Res J 1998;68:398–401.
[261] Sinitsyn A, Gusakov A, Grishutin S, Ankudimova N. Application of microassays for investigation of cellulase abrasive activity and backstaining. Journal of Biotechnology 2001;89:233–8.
[262] Rousselle M-A, Bertoniere NR, Howley PS, Pere J, Buchert J. Effect of Purified Trichoderma reesei Cellulases on the Supramolecular Structure of Cotton Cellulose. Textile Research Journal 2003;73:921–8.
[263] Yoon MY, McDonald H, Chu K, Garratt C. Protease, a new tool for denim washing. Text Chem Color Am D 2000;32:25–9.
[264] Andreaus J, Campos R, Cavaco-Paulo A. Reduction of indigo backstaining by post-washing. Melliand International 2001;7:318–20.
[265] Esfandiari A, Firouzi-Pouyaei E, Aghaei-Meibodi P. Effect of enzymatic and mechanical treatment on combined desizing and bio-polishing of cotton fabrics. Journal of the Textile Institute 2014;105:1193–202.
[266] Miettinen-Oinonen A, Londesborough J, Joutsjoki V, Lantto R, Vehmaanpera J. Three cellulases from Melanocarpus albomyces for textile treatment at neutral pH (vol 34, pg 332, 2004). Enzyme Microb Tech 2004;34:624-.
[267] Haakana H, Miettinen-Oinonen A, Joutsjoki V, Mantyla A, Suominen P, Vehmaanpera J. Cloning of cellulase genes from Melanocarpus albomyces and their efficient expression in Trichoderma reesei. Enzyme Microb Tech 2004;34:159–67.
[268] Anish R, Rahman MS, Rao M. Application of cellulases from an alkalothermophilic Thermomonospora sp. in biopolishing of denims. Biotechnol Bioeng 2007;96:48–56.
[269] Campos R, Kandelbauer A, Robra KH, Cavaco-Paulo A, Guebitz GM. Indigo degradation with purified laccases from Trametes hirsuta and Sclerotium rolfsii. J Biotechnol 2001;89:131–9.
[270] Solis M, Solis A, Ines Perez H, Manjarrez N, Flores M. Microbial decolouration of azo dyes: A review. Process Biochem 2012;47:1723–48.
[271] Abadulla E, Robra KH, Gubitz G, Silva L, Cavaco-Paulo A. Enzymatic Decolorization of Textile Dyeing Effluents. TextileResJ 2000;70(5):409–14.
[272] Abadulla E, Tzanov T, Costa S, Robra KH, Cavaco-Paulo A, Gubitz GM. Decolorization and detoxification of textile dyes with a laccase from Trametes hirsuta. Appl Environ Microbiol 2000;66:3357–62.
[273] Hou HM, Zhou JT, Wang J, Du CH, Yan B. Enhancement of laccase production by Pleurotus ostreatus and its use for the decolorization of anthraquinone dye. Process Biochem 2004;39:1415–9.
[274] Rodriguez Couto S, Sanroman M, Gubitz GM. Influence of redox mediators and metal ions on synthetic acid dye decolourization by crude laccase from Trametes hirsuta. Chemosphere 2005;58:417–22.
[275] Couto SR, Rosales E, Sanroman MA. Decolourization of synthetic dyes by Trametes hirsuta in expanded-bed reactors. Chemosphere 2006;62:1558–63.

[276] Nyanhongo GS, Gomes J, Guebitz GM, Zvauya R, Read J, Steiner W. Decolorization of textile dyes by laccases from a newly isolated strain of Trametes modesta. Water research 2002;36:1449–56.
[277] Rodriguez Couto S. Dye removal by immobilised fungi. Biotechnology Advances 2009;27:227–35.
[278] Khlifi R, Belbahri L, Woodward S, et al. Decolourization and detoxification of textile industry wastewater by the laccase-mediator system. J Hazard Mater 2010;175:802–8.
[279] Lu L, Wang TN, Xu TF, Wang JY, Wang CL, Zhao M. Cloning and expression of thermo-alkali-stable laccase of Bacillus licheniformis in Pichia pastoris and its characterization. Bioresour Technol 2013;134:81–6.
[280] Khouni I, Marrot B, Moulin P, Ben Amar R. Decolourization of the reconstituted textile effluent by different process treatments: Enzymatic catalysis, coagulation/flocculation and nanofiltration processes. Desalination 2011;268:27–37.
[281] Chander M, Arora DS. Evaluation of some white-rot fungi for their potential to decolourise industrial dyes. Dyes Pigments 2007;72:192–8.
[282] Rodriguez-Couto S. Enzymatic biotransformation of synthetic dyes. Curr Drug Metab 2009;10:1048–54.
[283] Ryan S, Schnitzhofer W, Tzanov T, Cavaco-Paulo A, Gubitz GM. An acid-stable laccase from Sclerotium rolfsii with potential for wool dye decolourization. Enzyme Microb Tech 2003;33:766–74.
[284] Wesenberg D, Kyriakides I, Agathos SN. White-rot fungi and their enzymes for the treatment of industrial dye effluents. Biotechnology Advances 2003;22:161–87.
[285] Diwaniyan S, Kharb D, Raghukumar C, Kuhad RC. Decolorization of Synthetic Dyes and Textile Effluents by Basidiomycetous Fungi. Water Air and Soil Pollution 2010;210:409–19.
[286] Kanmani P, Kumar RS, Yuvaraj N, Paari KA, Pattukumar V, Arul V. Microbial Decolorization of Synthetic Dyes and Reactive Dyes of Industrial Effluents by Using a Novel Fungus Aspergillus proliferans. Water Environment Research 2011;83:2099–106.
[287] Moreira-Neto SL, Mussatto SI, Machado KMG, Milagres AMF. Decolorization of salt-alkaline effluent with industrial reactive dyes by laccase-producing basidiomycetes strains. Lett Appl Microbiol 2013;56:283–90.
[288] Asgher M, Kamal S, Iqbal HM. Improvement of Catalytic Efficiency, Thermo-stability and Dye Decolorization Capability of Pleurotus ostreatus IBL-02 laccase by Hydrophobic Sol Gel Entrapment. Chem Cent J 2012;6:110.
[289] Broadbent AD. Basic principles of textile coloration. England: SDC; 2001.
[290] Hawkyard C. Synthetic Fibre Dyeing: Society of Dyers and Colorist; 2004.
[291] Lund H, Wu G. Removing excess disperse dye from dyed textiles by peroxidase or laccase. 5th International Conference on Textile Biotechnology; 2007; Wuxi, China.
[292] Pedersen G, Schmidt M. Removal of excess dye from new textiles. Google Patents; 1994.
[293] Schmitt S, de Souza R, Bettin F, Dillon AJP, Valle JAB, Andreaus J. Decolorization of aqueous solutions of disperse textile dyes by oxidoreductases. Biocatal Biotransfor 2012;30:48–56.
[294] Salem AA, Nassar SH, Rekaby MM. Improving the Removal of Printing Paste Thickener and Excess Dye by Enzymatic Washing of Printed Cotton Fabrics. J Nat Fibers 2008;5:1–18.
[295] Maurer KH. Detergent proteases. Curr Opin Biotechnol 2004;15:330–4.
[296] Gupta R, Beg QK, Khan S, Chauhan B. An overview on fermentation, downstream processing and properties of microbial alkaline proteases. ApplMicrobBiotech 2002;60:381–95.
[297] Saeki K, Ozaki K, Kobayashi T, Ito S. Detergent alkaline proteases: Enzymatic properties, genes, and crystal structures. Journal of Bioscience and Bioengineering 2007;103:501–8.
[298] Niyonzima FN, More S. Detergent-Compatible Proteases: Microbial Production, Properties, and Stain Removal Analysis. Preparative Biochemistry & Biotechnology 2015;45:233–58.

[299] Zhang J, Zhang Y, Li W, Li XL, Lian X. Optimizing Detergent Formulation with Enzymes. Journal of Surfactants and Detergents 2014;17:1059–67.
[300] Vojcic L, Pitzler C, Korfer G, et al. Advances in protease engineering for laundry detergents. N Biotechnol 2015:doi:10.1016/j.nbt.2014.12.010.
[301] Cortez J, Anghieri A, Bonner PLR, Griffin M, Freddi G. Transglutaminase mediated grafting of silk proteins onto wool fabrics leading to improved physical and mechanical properties. Enzyme Microb Tech 2007;40:1698–704.
[302] Hmidet N, Ali NEH, Haddar A, Kanoun S, Alya SK, Nasri M. Alkaline proteases and thermostable alpha-amylase co-produced by Bacillus licheniformis NH1: Characterization and potential application as detergent additive. Biochem Eng J 2009;47:71–9.
[303] Mitidieri S, Martinelli AHS, Schrank A, Vainstein MH. Enzymatic detergent formulation containing amylase from Aspergillus niger: A comparative study with commercial detergent formulations. Bioresource Technol 2006;97:1217–24.
[304] Novozymes. Novozymes' Mannaway® – New power for your detergent brand. In: A/S N, ed. Bagsvaerd, Denmark: Novozymes A/S; 2006.
[305] Niyonzima FN, More SS. Detergent-Compatible Bacterial Amylases. Appl Biochem Biotech 2014;174:1215–32.

Parisa Rahimi Tamandegani, Sedigheh Karimi Dorcheh, and Khabat Vahabi

3 Microbial peptides and peptibols

3.1 Introduction

Madly growing human population points health and food safety as critical aspect of our civilization. Therefore, having high efficiency of crop harvest is a necessity in our world. Crop loss by pathogens is an important threatening factor in human health and imposes yearly billions of dollars on human society (Klosterman et al. 2009). Today, a serious trouble we routinely face is resistance of aggressive pathogenic microbes to traditional antibiotics. In addition, application of chemicals to reduce pathogens harmcauses environmental and health problems. Accordingly, an unavoidable step in near future is finding the safe and ecofriendly alternatives for harmful methods. In this respect, biological based methods showed promising results in recent years to deal with reducing consequence of pathogen attacks on human and agricultural crops (Dawson and Hilton, 2011; Powles and Yu, 2010). Their potential in control of pathogens as alternative for chemical methods is almost endless. One of new rising branches of bio-based methods is application of ribosomal and non-ribosomal antimicrobial peptides (AMPs).

Pathogens employ diverse life strategies to overcome host defense systems. Also these systems dynamically have been adapted to their alteration via co-evolution (Hoeksema 2010). Plant pathogens use different ways to bypass first defense line such as entering through stomata, hydathodes (water pores at the end of leaves), wounds and penetrating by hyphal structures such as haustoria. All of these trespassing strategies for plant protection layers are recognized by plant (Dangl and Jones 2001; Ausubel 2005; Chisholm et al., 2005). This occurs via perception of different pathogen effector molecules (virulence factors) on the plant cell such as pathogen-associated molecular patterns (PAMPs), which are part of essential molecules for microbes (Janeway 1989). Depend on the sort of pathogen and effector molecules emanating from them, plant responds to treat by its own innate immune system in two distinct ways (Jones and Dangl 2006) including PTI (PAMP-triggered immunity) and ETI (effector-triggered immunity). PTI happening via mitogen-activated protein kinase (MAPK) cascade and reactive oxygen species (ROS) generation followed by enhancing ethylene (ET) and jasmonic acid (JA) level (Felix et al., 1999; Thomma et al., 1998) and activation of defense and pathogen responsive genes (Schwessinger and Zipfel 2008). ETI represents a stronger response like PCD (programmed cell death) (Jones and Dangl 2006) and results in SAR (systemic acquired resistance) through salicylic acid (SA) induction and signaling (Metraux et al., 1990; Delaney et al., 1994) which leads to hypersensitive response of host plant and extensive resistance against biotic stresses (Ward et al., 1991). However, plant immunity system is mosaic-like system that relies on the individual response of each cell to the systemic infection (Dangl and Jones, 2001; Ausubel 2005; Chisholm et al., 2005). Expression of AMPs in confrontation with biotic stresses is the final outcome of plant response to the pathogens (Stotz et al., 2009; Wang et al., 2009).

Multicellular organisms like fungi, plants and animals protect themself from invasion of different pathogens including bacteria, fungi and viruses via production of AMPs. Perturbing of pathogen cell membrane integrity and consequently cell lysis and death is the main role of AMPs during defense response of host organism (Buţu & Buţu, 2011).

3.2 AMPs

AMPs are cationic short peptides (12<AMPs<100, aa) with high content of arginine, lysine and histidine, that exist in all organisms and make positive network charge (network charge +2 to +9). Moreover, their amphipathic structure (including hydrophobic and hydrophilic domains) as an advantage enables them to move in biphasic environment like lipid rich membrane and cellular spaces (Ebenhen 2014 and references in it). AMPs synthesis is normally through ribosomal and non-ribosomal (peptaibols) approaches. Unlike the ribosomes, non-ribosomal peptide synthetases are mRNA independent. Each non-ribosomal peptide synthetase can synthesize only one type of peptide (Brogden et al., 2005). Also AMP secondary structure includes α-helical, cysteine-containing β-sheet, and flexible specific amino acids rich structure. Due to the relative ease of synthesis and characterization, 30 to 50 % of identified AMPs are α-helical peptides. Their length is between 12 to 40 amino acid residues. β-Sheet AMPs contain disulfide bonds mostly belong to the conserved defensin family which forms by disulfide bonds between antiparallel sheets (2–3 β-sheets). Till now more than 2706 AMPs have been characterized among different organisms (Ebenhen 2014). Based on the amino acid composition and structure, Brogden et al. (2005) (http://aps.unmc.edu/AP/main.php) classified AMPs in three sub- groups include, First Anionic antimicrobial peptides which are produced against Gram-positive and Gram-negative bacteria including cofactor dependent small peptides (721.6–823.8 Da). Second medium size cationic peptides (~290 Da, <40 amino acid) and finally small size cationic peptides (~44 Da).

3.3 AMPs antimicrobial activity

Wide range of targets and mechanisms of action across the AMPs is one of their well described characteristic (Kalfa et al., 2001). Interference with cell membrane, enzymes, ribosome and protein function and structure, DNA binding, inhibition of cell wall, DNA, RNA and protein synthesis, protein folding prohibition via chaperone blockage, mitochondria damaging and ROS induction cellular restricting are some probable mechanisms of AMPs action against pathogens (Peters et al., 2010)

AMPs mediated cell death happens as a process consisting different steps such as attraction and attachment to the microbial surface (Brogden, 2005, Lee et al.; 2004) and peptide insertion (Brogden, 2005). Confronting with these steps is critical for microbial pathogens to acquire resistance against AMPs-based defenses for their survival (Yeaman and Younth 2003; Brogden, 2005). Reduction of the negative surface charges and peptide attachment rate to the outer membrane by *Staphylococcus aureus* reduces their susceptibility to AMPs (Peschel et al., 1999). Also, *Klebsiella pneumonia* uses a polysaccharide capsule to restrict AMPs interaction via membrane (Brogden 2005). *S. typhimurium* (Groisman et al., 1992) and *Pseudomonas aeruginosa* (Mcfee et al., 2003) use hydrophobic interaction to prevent pore formation and peptide insertion. It has been suggested that proteolysis of AMPs by *S. aureus* lytic enzyme, aureolysin, might be a candidate for its resistance against antimicrobial fragments (Guina et al., 2000; Oren et al., 1999).

Role of AMPs is not limited to biotic responses (Hegedüs and Marx, 2013). They involve in response to abiotic stress, nutrition limitations (Meyer et al., 2002; Meyer and Stahl, 2002), asexual spore production, H_2O_2 sensing and signaling (Eigentler et al., 2012).

3.4 AMPs source

AMPs have been found in different organisms including plants, microbes, insects, amphibians and mammals (Ka-Yee Choi, 2012). They play important role in non-specific host defense system (Pelegrini and Franco 2005; Hegedüs and Marx 2013) by their antimicrobial immunomodulatory effects (Ka-Yee Choi, 2012; Nawrot et al., 2014). AMPs role in prokaryotes and lower eukaryotes is limited to increase the potential of organism in competing for nutrients. In contrast, they are part of defense machinery against pathogens in higher eukaryotes (Hegedüs & Marx, 2013). Root and all aerial parts of plant produce AMPs in different plant species with anti-microbial effects against plant and human pathogens, parasites and neoplastic cells (Montesinos, 2007). Defensins, thionins, lipid transfer proteins, cyclotides, snakins, and hevein-like proteins are just some examples of plants large AMPs pool (Nawrot et al., 2014; Pelegrini and Franco, 2005; Segura et al., 2007; Giudici et al., 2006; Buţu and Buţu, 2011; Kader, 1996). Thionins and Viscotoxins have lytic activity against fungi (Pelegrini and Franco, 2005; Giudici et al., 2006). Snakins (Segura et al., 2007) and defensins play antifungal and antibacterial activity (Vriens et al., 2014; Mendez et al., 1990).

Insects AMPs are typically cationic (Bulet and Stöcklin 2005) and act in/or outside of body during complete and incomplete metamorphosis to control microbial infections. Also, different AMPs such as sarcotoxins, apidaecins, hyphancin, enbocin, spodopsin, ponericins, melittin, stomoxyn, spinigerin and ceratotoxins are the main part of insect defense system that show antimicrobial characteristic against plant-associated bacteria as well as human pathogens (Casteels et al., 1989; Bulet and Stöcklin 2005).

Mammalian AMPs mainly include defensins, cathelicidins, bactenecins and indolicidins which are front line of defense system (Lehrer et al., 1991). They confront with microbes via non-oxidative actions like destruction of the pathogen membrane (Boman, 1995).

Different viruses produce variety of AMPs to develop their own propagation mechanism and oppose to the defense machinery of host cells. HIV1 (human immunodeficiency virus type 1) produces an arginine rich cytolytic transmembrane protein, LLPs (Lentivirus lytic peptides) as a potential AMP (Tencza, 1997) that its antimicrobial mechanisms against *S. marcescens* have been well charactrized (Phadke et al., 2002).

3.5 Microbial AMPs

3.5.1 Fungal AMPs

Fungi have been considered as one of the best sources for secondary metabolites specifically peptide antibiotics which are very diverse in structure (Degenkolb and Brückner 2008). Different ascomycete fungi produce cysteine-rich defensin-like AMPs (Wnendt et al., 1994; Binder et al.; 2011; Marx et al., 1995; Seibold et al., 2011; Gun Lee et al., 1999; Geisen, 2000; Skouri-Gargouri and Gargouri, 2008; Kovacs et al., 2011). All these proteins exclusively exhibit antifungal activity against filamentous ascomycetes including opportunistic plant and animal pathogens such as *Fusarium* sp., *Botrytis* sp. and *Aspergillus fumigatus* (Marx, 2004; Meyer 2008). These proteins are encoded by orthologous genes and exhibit both similarities and differences concerning their species-specificity, primary structure, protein activity and target sites (Marx 2004).

Also, activity against microbes (Hajji et al., 2010) and yeasts (Gun Lee et al., 1999; Seibold et al., 2011) have been reported. Main AMPs producer ascomycetes are *Aspergillus* spp. (Wnendt et al., 1994; Binder et al., 2011; Gun Lee et al., 1999; Skouri-Gargouri and Gargouri, 2008) and *Penicillium* spp. (Marx et al., 1995; Seibold et al., 2011; Geisen 2000) which produce variety of ribosomal synthesis based AMPs with antifungal and antibacterial activities.

Aspergillus species produce antimicrobial peptides such as AFP from *A. giganteus* (Meyer, 2007), AFPNN5353 produced by *A. giganteus* (Binder et al., 2011), AcAFP and AcAMP from *A. clavatus* (Skouri-Gargouri and Gargouri, 2008), ANAFP from *A. niger* (Gun Lee et al., 1999) and finally AcAMP from *A. clavatus* (Hajji et al., 2010) which are well-known for their general antibiotic activity against fungi, Gram-positive and

Gram-negative bacteria (Hajji et al., 2010). *Penicillium* spp. are the next main producers of AMPs among fungi. They synthesize peptides like PAF (Penicillium antifungal protein) originated from *Penicillium chrysogenum* (Marx et al., 1995), PgAFP produced by *P. chrysogenum* (Batta et al., 2009), NAF from *Penicillium nalgiovense* (Geisen 2000) and BP from *Penicillium brevicompactum* (Seibold et al., 2011). Also, variety of different ribosomal synthesis based AMPs are produced by other fungi including NFAP from *Neosartorya fisheri* (Kovacs et al., 2011), Trtesin from endophytic *Fusarium tricinctum* (Tejesvi et al., 2013), Copsin from *Coprinopsis cinerea* (Essig et al., 2014) and plectasin from *Pseudoplectania nigrella* (Mygind et al., 2005) with antifungal and antibacterial propreties.

3.5.1.1 Fungal ribosomal AMPs mechanism of action and application

Interaction with cell wall and plasma membrane molecules involving in signaling cascades have been suggested as antibiotic startup for many fungal AMPs (Hegedüs and Marx 2013). Nowadays fungal cell wall targeting antimycotics are highly preferable (V. Meyer 2008). Mushroom uses Copsin to interfere with bacteria cell wall synthesis by binding to the peptidoglycan precursor lipid II (Essig et al., 2014). Calcium influx resulted from hyperpolarization of the *A. nidulans* plasma interaction with PAF have been reported (Leiter et al., 2005) which mainly results in apoptosis originated from G-protein coupled signaling (Batta et al., 2009). Some fungal AMPs show dose-dependent activity like BP (Seibold et al., 2011) and Anafp (Gun Lee et al., 1999) that affect filamentous fungi and yeast (Seibold et al., 2011). AFP antifungal activity, is based on Ca^{2+} signaling (Binder et al., 2011). AFP possesses very important characteristics like sustainability, high efficacy, organism limited toxicity and economic significance (Meyer, 2007).

Demand for efficient antifungals is increasing as a consequence of human and plant fungal pathogen prevalence. Application of fungal AMPs covers a wide range of human problems from agriculture to medical aspects. AcAMP have been suggested as potent candidate in biological control of plant and food pathogens (Hajji et al., 2010). PAF (Penicillium antifungal protein) can be a good choice for treating lethal Aspergillus infections of human because of its specific antibiotic effects against fungi without harm for our cells (Batta et al., 2009). Comparable results of plectasin with vancomycin and penicillin in curing pneumonia suggests it as a promising alternative for traditional antibiotics with almost no toxic effect in mice cells (Mygind et al., 2005)

3.6 Peptaibiotics

Peptaibiotics are peptides with antibiotic properties, unique bioactivities and conformations (Degenkolb and Brückner 2008). Fungi are normal producer of peptaibiotics. They are normally linear, amphipathic polypeptides (Szekeres et al., 2005), contain-

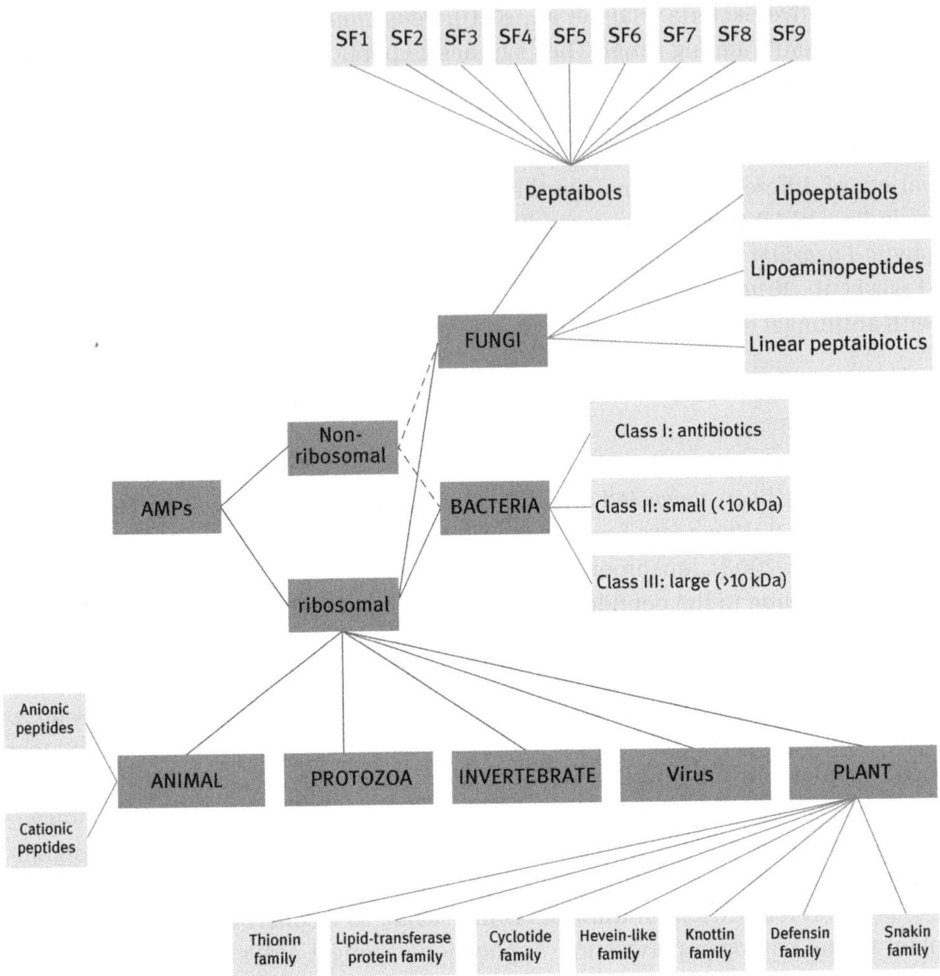

Fig. 3.1: Schematic hierarchy of different AMPs across organisms.

ing 4–21 residues with molecular mass between 500 to 2200 u, restricted between an acetyled N-terminus and C-terminus amino alcohol. Also they contain non-proteinogenic amino acids (α-aminoisobutyric acid, Aib) and/or lipoamino acids, isovaleric acid (Iva) and the amino acid hydroxyproline (Hyp) (Degenkolb et al., 2007; Wiest et al., 2002). Depend on C-terminus and N-terminus, five subgroups of peptaibiotics can be defined (Fig. 3.1).

1. **Peptaibols** with amine, amide, free amino acid, 2, 5-dioxopiperazine, or sugar alcohol located in C-terminus.
2. **Lipopeptaibols**, octanoic, decanoic, or cis-dec-4-enoic acid attach via acetylation reaction to the N-terminus (Toniolo et al., 2001).

3. **Lipoaminopeptides** (aminolipopeptides), substitution of the N-terminus by unbranched, a- or g-methyl-branched, saturated or unsaturated C4–C15 fatty acids (Degenkolb et al., 2002).
4. **Linear peptaibiotics** which are not members of three first group (Degenkolb et al., 2007).
5. A recently introduced and natural origin **cyclic, Aib- or Iva** containing peptides (Degenkolb and Brückner, 2008; Degenkolb et al., 2007)

3.7 Peptaibols

Peptaibols have been reported originally from fungi (Kenner and Sheppard 1958). Their production have been exclusively reported in filaments fungi from different genera including Trichoderma, Acremonium (Schiell et al., 2001), Tylopilus, Boletus (Lee et al., 1999), Bionectria (Figueroa et al., 2013), Paecilomyces, Emericellopsis, Cephalosporium, Stilbella and Gliocladium (Shenkarev et al., 2013). First peptaibol, alameticine, isolated form *Trichoderma viride* (Meyer and Reusser 1967) and structurally recognized (Payne et al., 1970) which induces action potentials in target cell membranes (Mueller and Rudin 1968). Peptaibols have short lengths (5 to 20 residues) and have been classified in 9 subfamilies (Fig. 3.1) (SF1 to SF9) according to their sequence alignment (Chugh and Wallace 2001). Peptaibol Database (http://www.cryst.bbk.ac.uk/peptaibol) is different from normal peptides and describes sequence, structure, biological source and bibliographical data for the naturally occurring unusual class of peptides. Peptaibols have not generally been introduced in databases of protein sequence like Swiss-Prot and TrEMBL because of their unusual amino acids, short lengths and absence of genetic coding sequences. (Whitmore and Wallace, 2004; Neumann et al., 2015).

3.8 Peptaibol biosynthesis

Due to their unusual amino acid content, non-ribosomal biosynthesis of peptaibiotics and peptaibols is possible by peptide synthetases, large multifunctional enzymes (Stein et al., 1996). Peptide synthetases genes are the responsible genes for peptaibiotics synthesis in fungi. They synthesizes peptibiotics molecules from different precursors (nonproteinogenic amino acids and hydroxy or carboxyl acids) via N-methylation, acylation and reduction reactions (Szekeres et al., 2005, Zocher and Keller, 1997). Each multifunctional modular peptaibiotics synthesis enzyme is responsible for assembly of different size peptaibiotics. (Szekeres et al., 2005, Zocher and Keller, 1997).

3.8.1 Peptiabols activity

Peptaibols and peptaibiotics have demonstrate broad-spectrum antibiotic action against Gram-positive bacteria (Jen et al., 1987; Wiest et al., 2002), fungi, viruses (Szekeres et al., 2005), plants and their producer hosts (Degenkolb, et al., 2003, Schiell et al., 2001). It has been suggested that biological control ability (BCAs) of Trichoderma genus against pathogenic fungi is a synergistic effect of several mechanisms like lytic enzyme production, colonization of target microbe (Wiest et al., 2002) and secondary metabolite production along with production of peptaibols (Brito et al., 2014).

Peptaibols mechanism of action is combination of their physico-chemical and biological properties (Szekeres et al., 2005) including the membrane spanning due to their amphipathic nature (Mueller and Rudin, 1968; Szekeres et al., 2005; Chugh & Wallace, 2001), forming voltage-dependent ion channels in plasma membranes (Mueller and Rudin, 1968), membrane modification and disturbance (Sansom, 1993; Rebuffat et al., 1992), increasing membrane permeability (Castro and Micaêlo 2014) and consequently cytoplasmic leakage mediated cell death such as alameticin mechanism of action (Chugh and Wallace, 2001). In addition, inhibition of mitochondrial ATPase, oxidative phosphorylation, immunosuppression (Csermely et al., 1994), platelet aggregation (Chikanishi et al., 1997) and neuroleptic effects (Ritzau et al., 1997) have been described as mechanism of different peptiboles action against target cells. The exogenous application of peptaibols enhances plant resistance against pathogens via triggering plant defense response (Engelberth et al., 2000). However, interaction between cell wall degrading enzymes (endochitinase, exochitinase and glucan 1,3-β- glucosidase) and peptaibiotics could be the possible biocontrol processes and plant self-defense mechanisms (Leiter et al., 2005). Trichogin (Iftemi et al., 2014) and alamethicin (Chugh and Wallace, 2001) span the membrane and enhance ion diffusion leading to cell death. Antagonistic effects of Sepedonium (anamorphic ascomycetes) originated from chalciporin A, chalciporin B and chrysospermins against basidiomycetes have been described (Neuhof et al., 2007). Also LP237-F8, F5, and F7, produced by *Tolypocladium geodes* in liquid cultivation condition, antimicrobial activity have been discribed , (Tsantrizos et al., 1996). *T. viride* and *T. asperellum* synthetize asperelines A and E and five trichotoxins including T5D2, T5E, T5F, T5G and 1717A showed promising result in restricting different bacteria pathogens. (Brito et al., 2014).

Furthermore, toxic peptaibols like acrebol (produced by *Acremonium exuviarum*) with high risk for targeting of mitochondria and consequence paralysis of energy producing cells in mammalian have been reported (Andersson et al., 2009). Accumulation of solvable peptaibols in filter-feeder mollusks (*Mytilus edulis*) could end up with health hazards for shellfish and their subsequent consumers (Poirier et al., 2007).

3.9 Bacterial AMPs

Bacteriocin peptides including cationic and neutral AMPs (small size between 20–50 aa) produced by Gram-positive and -negative bacteria (Joerger 2003) are targeting other bacteria via cell membranes permeabilizing mechanism (Baba and Schneewind, 1998; Drider et al., 2006). Their genes are located on mobile genetic materials such as conjugative plasmids and transposons (Jack et al., 1995). Some bacterial AMPs like microcin C7 and mersacidin inhibit bacterial molecular functions such as protein synthesis and peptidoglycan biosynthesis, respectively. Production of acteriocins increases bacterial potential to compete in complex microclimates by restricting other bacterial growth or killing them (Hassan et al., 2012; Joerger 2003). Bacterial AMPs are mainly restricted to specific targets with no broad spectrum (Hassan et al., 2012). Some small ribosomal originated bacteriocins (low molecular weight) are mainly active against close relative species like microcins form Enterobacteriaceae. Responsible genes for post-translationally modified microcins are located on plasmids (Baquero et al., 1978; Severinov et al., 2007).

The lantibiotics (class I, like Nisin, subtilin, lacticin 3147, mersacidin and thuricin CD) and the non-lantibiotics (class II) are main Gram-positive bacteria originated bacteriocins. Post-translational modifications are present in small lantibiotics (19–38 amino acids) like lanthionine or b-methyllanthionine (Bonelli et al., 2006). Some lantibiotics form pores in the cytoplasmic membrane and other mechanisms of action are involving in inhibition of cell wall biosynthesis (Brotz et al., 1995). Mersacidin produced by *Bacillus* sp. (Brotz et al., 1995) has a chromosomal gene (Altena et al., 2000). Its target is not the cell membrane and mainly disturbs cell wall via peptidoglycan metabolism inhibition (Brotz et al., 1995). Nisin, a heat resistant 34 aa bacteriocin produced by *Lactococcus lactis*, is resulted from ribosomal synthesis followed by post translational modification. It targets a broad spectrum of Gram-positive bacteria and their spores. Nisin activity against food-borne pathogens and heat stability makes it the best antimicrobial peptide for food preservation (Eigentler et al., 2011). Non-lantibiotics are a large group of bacteriocins with four subgroups, they are small cationic (25–60 amino acids), heat-stable and non-modified antimicrobial peptides that mainly produce by lactic acid bacteria (LAB) (Cotter et al., 2005b). Class IIa bacteriocins contains pediocin-like bacteriocins with strong antilisterial activity (e.g. pediocin PA-1, enterocin A and P, mesentericin Y105 and leucocin A). Class IIb or two-peptide bacteriocins whose complementary action of their two different peptides is required for their full activity (e.g. ABP-118, lactococcin G and plantaricin EF and JK).

Cyclic bacteriocins resulted from head-to-end fusion, make Class IIc bacteriocins such as enterocin AS-48, lactocyclicin Q and garvicin ML. Finally Class IId including linear single-peptide and non-pediocin-like bacteriocins (e.g. lactococcins A and B and enterocin B) (Cotter et al., 2005). Bacteriocin genes show high linkage to specific genes like genes encoding immunity proteins, two-component regulators, and specific transporters involved in their production (Holtsmark et al., 2008).

3.9.1 Mechanisms of bacteriocins action

There are several mechanisms of bacterial AMPs action in target cells. Pore formation in plasma membrane of target cell is a general mechanism of LAB bacteriocins activity (Hassan et al., 2012). Docking of target cell by cell wall precursor molecule lipid II in some of lantibiotics (class I bacteriocins) is one of the probable mechanisms. Nisin inhibits the cell wall biosynthesis (Wiedemann et al., 2001). Also, blocking of cell wall synthesis by lantibiotics through removing lipid II from the cell division site has been reported (Hasper et al., 2006). *Lactococcus lactis* bacteriocin targets mannose phosphotransferase system (man-PTS) of target cells and kills it by increasing permeability of its membrane (Diep et al., 2007). It has been suggested that antimicrobial activity and specificity of bacteriocins needs a receptor on target cell (Hassan et al., 2012). In contrast, it has been documented that LnqQ (Lacticin Q) increases membrane permeability independently from Lipid II (receptor of nisin) (Yoneyama et al., 2009).

3.10 AMP application in medicine

Antibiotics with their critical role in new medicine have been considered as most important therapeutics (Steckbeck et al., 2014). Treatment of infectious diseases is new emerging worldwide problem because of pathogens resistance against antibiotic. there is no promising perspective for developing new and effective antibiotics in recent years (Hassan et al., 2012). AMPs are naturally produced and have potential to be the alternatives for traditional antibiotics against difficult-to-treat bacteria (Steckbeck et al., 2014). They have variation in structure and chemical properties and show antimicrobial activity against broad spectrum of pathogens with higher specific and wide range of activities. (Marshall, 2003). Easiness of amplification and transfer of AMPs small coding genes make them good candidate for producing transgenic lines and large scale production (Thomma et al., 2002). Moreover, some bacteriocin have very narrow spectrum of activity against different microbes which can be used as a specific antibiotic against a pathogen among community of other beneficial microbes. For instance, thuricin CD particularly targets Cl. difficile in distal colon models without effect on beneficial microbial community (Rea et al., 2010).

3.11 AMPs in plant protection

Potential of bacteriocins in control of plant pathogens attracts attentions toward AMPs as new and effective option in plant disease management (Holtsmark et al., 2008). They have been broadly used in production of transgenic resistant plant against biotic stresses (Thomma et al., 2002). Using AMPs in agricultural transgenic crops provides higher resistance against plant pathogens (Thomma et al., 2002). The size of AMPs, ease of handling, high antibiotic effect in low concentration enhances their efficiency as good alternative of classical methods in plant protection. Despite their proven abilities in plant protection, low stability and toxicity, expensive production, bioethics and social concerns restricts marketing of commercial cultivars (Montesinos, 2007). Bacteriocins derived from *Pseudomonas* (Syringacin M, syringacin L1 and pyocin L1) and *Pectobacterium* (pectocins containing ferredoxin) cause resistance against *P. syringae* pathogens in *Arabidopsis thaliana* and *Nicotiana benthamiana* (Grinter 2014). Therefore AMPs are a good biologic based potential in protecting plants against plant pathogens for external application or transformation in different crops.

3.12 AMP application in pharmacy and biotechnology

Efficient way to deliver the drugs medications in human body to their target cell and tissue is a challenge in designing and formulation of therapeutic biomolecules (Nasrollahi et al., 2012). Transposing skin as a first layer of human protection system by dermatological therapeutic molecules have been possible by different methods like liposomes, vesicular systems, nanoparticulate delivery system such as solid lipid nanoparticles, nanostructured lipid carriers, polymer-based nanoparticles and magnetic nanoparticles for dermatological applications (Desai et al., 2010). The CPPs (Cell-penetrating peptides) emerge as the novel transdermal delivery systems to traffic different bimolecular cargos like small interfering RNA nucleotides, drug-loaded nanoparticles, proteins, and peptides. Receptors and membrane damage free delivery increase research and industrial application of CPPs particularly in cosmetic and pharmacy. CPPs have been known as a best option in drug delivery and vaccination because of their ability in transferring of high molecular weight molecules (such as RNA, DNA, nanoparticles and medical macromolecules) across the cells and skin. The low toxicity for biotic systems results in increasing CPPs application in pharmaceutical formulations specifically in transdermal drugs (Nasrollahi et al., 2012). In addition, nisin has been used as enhancer of bio-ethanol production by its specific antibiotic activity to restrict unwanted microbial growth during fermentation process (Jozala et al., 2008). So it is not irrational to expect marvels and unexpected achivments of human in therapeutic fields resulted from combination of biotechnology and pharmaceutical application of AMPs.

3.13 AMP in food industry

Resistance of food pathogens due to massive use of preservation methods emerges the urgent need for efficient and safe alternatives. Enteric bacteria resistance to salty and low pH conditions as common approaches in food preservation restricts efficiency and safety of food industry. Nowadays it is necessary to have natural, safe, efficient with wide spectrum activity and simple alternative like AMPs to fulfill demands for preserve free, naturally based processed foods and ecofriendly methods. Potential of AMPs has been documented already as food preservatives, e.g. nisin is now an FDA-approved peptide (Rydlo et al., 2006).

3.14 References

[1] Altena, K., A. Guder, et al. (2000). "Biosynthesis of the Lantibiotic Mersacidin: Organization of a Type B Lantibiotic Gene Cluster." Applied and Environmental Microbiology 66(6): 2565–2571.

[2] Andersson, M. A., R. Mikkola, et al. (2009). "Acrebol, a novel toxic peptaibol produced by an Acremonium exuviarum indoor isolate." Journal of Applied Microbiology 106(3): 909–923.

[3] Ausubel, F. M. (2005). "Are innate immune signaling pathways in plants and animals conserved?" Nat Immunol 6(10): 973–979.

[4] Baba, T. and O. Schneewind (1998). "Instruments of microbial warfare: Bacteriocin synthesis, toxicity and immunity." Trends in Microbiology 6(2): 66–71.

[5] Batta, G., T. Barna, et al. (2009). "Functional aspects of the solution structure and dynamics of PAF – a highly-stable antifungal protein from Penicillium chrysogenum." FEBS Journal 276(10): 2875–2890.

[6] Binder, U., M. Bencina, et al. (2011). "The Aspergillus giganteus antifungal protein AFPNN5353 activates the cell wall integrity pathway and perturbs calcium homeostasis." BMC Microbiol 11(1): 209.

[7] Binder, U., M. Bencina, et al. (2011). "The Aspergillus giganteus antifungal protein AFPNN5353 activates the cell wall integrity pathway and perturbs calcium homeostasis." BMC Microbiol 11(1): 209.

[8] Boman, H. (1995). "Peptide Antibiotics and Their Role in Innate Immunity." Annual Review of Immunology 13(1): 61–92.

[9] Bonelli, R. R., T. Schneider, et al. (2006). "Insights into In Vivo Activities of Lantibiotics from Gallidermin and Epidermin Mode-of-Action Studies." Antimicrobial Agents and Chemotherapy 50(4): 1449–1457.

[10] Brito, J. P. C., M. H. S. Ramada, et al. (2014). "Peptaibols from Trichoderma asperellum TR356 strain isolated from Brazilian soil." SpringerPlus 3(1): 600.

[11] Brogden, K. A. (2005). "Antimicrobial peptides: pore formers or metabolic inhibitors in bacteria?"
Nat Rev Micro 3(3): 238–250.

[12] Brotz, H., G. Bierbaum, et al. (1995). "Mode of action of the lantibiotic mersacidin: inhibition of peptidoglycan biosynthesis via a novel mechanism?" Antimicrobial Agents and Chemotherapy 39(3): 714–719.

[13] Bulet, P. and R. Stocklin (2005). "Insect Antimicrobial Peptides: Structures, Properties and Gene Regulation." Protein & Peptide Letters 12(1): 3–11.

[14] Casteels, P., C. Ampe, et al. (1990). "Isolation and characterization of abaecin, a major antibacterial response peptide in the honeybee (Apis mellifera)." Eur J Biochem 187(2): 381–386.

[15] Castro, T. G. and N. M. Micaêlo (2014). "Modeling of Peptaibol Analogues Incorporating Nonpolar α,α-Dialkyl Glycines Shows Improved α-Helical Preorganization and Spontaneous Membrane Permeation." J. Phys. Chem. B 118(3): 649–658.

[16] Chikanishi, T., K. Hasumi, et al. (1997). "Clonostachin, a Novel Peptaibol That Inhibits Platelet Aggregation." J. Antibiot. 50(2): 105–110.
[17] Chisholm, S. T., G. Coaker, et al. (2006). "Host-Microbe Interactions: Shaping the Evolution of the Plant Immune Response." Cell 124(4): 803–814.
[18] Choi, K.-Y. and N. Mookherjee (2012). "Multiple Immune-Modulatory Functions Of Cathelicidin Host Defense Peptides." Frontiers in Immunology 3.
[19] Chugh, J. K. and B. A. Wallace (2001). "Peptaibols: models for ion channels." Biochemical Society Transactions 29(4): 565.
[20] Chugh, J. K. and B. A. Wallace (2001). "Peptaibols: models for ion channels." Biochemical Society Transactions 29(4): 565.
[21] Cotter, P., C. Hill, et al. (2005). "Bacterial Lantibiotics: Strategies to Improve Therapeutic Potential." CPPS 6(1): 61–75.
[22] Cotter, P. D., C. Hill, et al. (2005). "Food Microbiology: Bacteriocins: developing innate immunity for food." Nat Rev Micro 3(10): 777–788.
[23] Csermely, P., L. Radics, et al. (1994). "The nonapeptide leucinostatin A acts as a weak ionophore and as an immunosuppressant on T lymphocytes." Biochimica et Biophysica Acta (BBA) – Molecular Cell Research 1221(2): 125–132.
[24] Dangl, J. L. and J. D. G. Jones (2001). "Plant pathogens and integrated defence responses to infection." Nature 411(6839): 826–833.
[25] Dawson, C. J. and J. Hilton (2011). "Fertiliser availability in a resource-limited world: Production and recycling of nitrogen and phosphorus." Food Policy 36: S14–S22.
[26] Degenkolb, T., A. Berg, et al. (2003). "The occurrence of peptaibols and structurally related peptaibiotics in fungi and their mass spectrometric identification via diagnostic fragment ions." J. Peptide Sci. 9(11-12): 666–678.
[27] Degenkolb, T. and H. Brückner (2008). "Peptaibiomics: Towards a Myriad of Bioactive Peptides Containing Cα-Dialkylamino Acids?" Chemistry & Biodiversity 5(9): 1817–1843.
[28] Degenkolb, T., S. Heinze, et al. (2002). "Formation of New Lipoaminopeptides, Acremostatins A, B, and C, by Co-cultivation of Acremonium sp. Tbp-5 and Mycogone rosea DSM 12973." Bioscience, Biotechnology and Biochemistry 66(4): 883–886.
[29] Degenkolb, T., J. Kirschbaum, et al. (2007). "New Sequences, Constituents, and Producers of Peptaibiotics: An Updated Review." Chemistry & Biodiversity 4(6): 1052–1067.
[30] Delaney, T. P., S. Uknes, et al. (1994). "A Central Role of Salicylic Acid in Plant Disease Resistance." Science 266(5188): 1247–1250.
[31] Desai, P., R. R. Patlolla, et al. (2010). "Interaction of nanoparticles and cell-penetrating peptides with skin for transdermal drug delivery." Molecular Membrane Biology 27(7): 247–259.
[32] Diep, D. B., M. Skaugen, et al. (2007). "Common mechanisms of target cell recognition and immunity for class II bacteriocins." Proceedings of the National Academy of Sciences 104(7): 2384–2389.
[33] Drider, D., G. Fimland, et al. (2006). "The Continuing Story of Class IIa Bacteriocins." Microbiology and Molecular Biology Reviews 70(2): 564–582.
[34] Ebenhan, T., O. Gheysens, et al. (2014). "Antimicrobial Peptides: Their Role as Infection-Selective Tracers for Molecular Imaging." BioMed Research International 2014: 1–15.
[35] Ehrenstein, G. and H. Lecar (1977). "Electrically gated ionic channels in lipid bilayers." Quart. Rev. Biophys. 10(01): 1.
[36] Eigentler, A., I. Pócsi, et al. (2011). "The anisin1 gene encodes a defensin-like protein and supports the fitness of Aspergillus nidulans." Archives of Microbiology 194(6): 427–437.
[37] Engelberth, J., T. Koch, et al. (2000). "Channel-Forming Peptaibols Are Potent Elicitors of Plant Secondary Metabolism and Tendril Coiling." Angew. Chem. Int. Ed. 39(10): 1860–1862.

[38] Essig, A., D. Hofmann, et al. (2014). "Copsin, a Novel Peptide-based Fungal Antibiotic Interfering with the Peptidoglycan Synthesis." Journal of Biological Chemistry 289(50): 34953–34964.

[39] Felix, G., J. D. Duran, et al. (1999). "Plants have a sensitive perception system for the most conserved domain of bacterial flagellin." The Plant Journal 18(3): 265–276.

[40] Figueroa, M., H. Raja, et al. (2013). "Peptaibols, Tetramic Acid Derivatives, Isocoumarins, and Sesquiterpenes from a Bionectria sp. (MSX 47401)." J. Nat. Prod. 76(6): 1007–1015.

[41] Geisen, R. (2000). "P. nalgiovense carries a gene which is homologous to the paf gene of P. chrysogenum which codes for an antifungal peptide." International Journal of Food Microbiology 62(1-2): 95–101.

[42] Giudici, M., J. Antonio Poveda, et al. (2006). "Antifungal effects and mechanism of action of viscotoxin A3." FEBS Journal 273(1): 72–83.

[43] Grinter, R. and D. Walker (2011). Lipid II-Degrading M-Class Bacteriocins. Encyclopedia of Inorganic and Bioinorganic Chemistry, Wiley-Blackwell: 1–10.

[44] Groisman, E. A., C. Parra-Lopez, et al. (1992). "Resistance to host antimicrobial peptides is necessary for Salmonella virulence." Proceedings of the National Academy of Sciences 89(24): 11939–11943.

[45] Guina, T., E. C. Yi, et al. (2000). "A PhoP-Regulated Outer Membrane Protease of Salmonella enterica Serovar Typhimurium Promotes Resistance to Alpha-Helical Antimicrobial Peptides." Journal of Bacteriology 182(14): 4077–4086.

[46] Gun Lee, D., S. Yub Shin, et al. (1999). "Isolation and Characterization of a Novel Antifungal Peptide from Aspergillus niger." Biochemical and Biophysical Research Communications 263(3): 646–651.

[47] Hajji, M., K. Jellouli, et al. (2010). "A highly thermostable antimicrobial peptide from Aspergillus clavatus ES1: biochemical and molecular characterization." Journal of Industrial Microbiology & Biotechnology 37(8): 805–813.

[48] Hasper, H. E., N. E. Kramer, et al. (2006). "An Alternative Bactericidal Mechanism of Action for Lantibiotic Peptides That Target Lipid II." Science 313(5793): 1636–1637.

[49] Hassan, M., M. Kjos, et al. (2012). "Natural antimicrobial peptides from bacteria: characteristics and potential applications to fight against antibiotic resistance." J Appl Microbiol 113(4): 723–736.

[50] Hegedüs, N. and F. Marx (2013). "Antifungal proteins: More than antimicrobials?" Fungal Biology Reviews 26(4): 132–145.

[51] Heng, N. C. K., P. A. Wescombe, et al. (2007). The Diversity of Bacteriocins in Gram-Positive Bacteria. Bacteriocins, Springer Science + Business Media: 45–92.

[52] Hoeksema, J. D. (2010). "Ongoing coevolution in mycorrhizal interactions." New Phytologist 187(2): 286–300.

[53] Holtsmark, I., V. G. H. Eijsink, et al. (2008). "Bacteriocins from plant pathogenic bacteria." FEMS Microbiology Letters 280(1): 1–7.

[54] Janeway, C. A. (1989). "Approaching the Asymptote? Evolution and Revolution in Immunology." Cold Spring Harbor Symposia on Quantitative Biology 54(0): 1–13.

[55] Jen, W. C., G. A. Jones, et al. (1987). "The antibacterial activity of alamethicins and zervamicins." Journal of Applied Bacteriology 63(4): 293–298.

[56] Joerger, R. (2003). "Alternatives to antibiotics: bacteriocins, antimicrobial peptides and bacteriophages." Poultry Science 82(4): 640–647.

[57] Jones, J. D. G. and J. L. Dangl (2006). "The plant immune system." Nature 444(7117): 323–329.

[58] Jozala, A. F., A. M. Lopes, et al. (2008). "Liquid–liquid extraction of commercial and biosynthesized nisin by aqueous two-phase micellar systems." Enzyme and Microbial Technology 42(2): 107–112.

[59] Kader, J.-C. (1996). "LIPID-TRANSFER PROTEINS IN PLANTS." Annu. Rev. Plant. Physiol. Plant. Mol. Biol. 47(1): 627–654.

[60] Kalfa, V. C., D. Palmquist, et al. (2001). "Suppression of Mannheimia (Pasteurella) haemolytica serovar 1 infection in lambs by intrapulmonary administration of ovine antimicrobial anionic peptide." International Journal of Antimicrobial Agents 17(6): 505–510.

[61] Kenner, G. W. and R. C. Sheppard (1958). "α-Aminoisobutyric Acid, β-Hydroxyleucine, and γ-Methylproline from the Hydrolysis of a Natural Product." Nature 181(4601): 48–48.

[62] Klosterman, S. J., Z. K. Atallah, et al. (2009). "Diversity, Pathogenicity, and Management of Verticillium Species." Annual Review of Phytopathology 47(1): 39–62.

[63] Kovács, L., M. Virágh, et al. (2011). "Isolation and characterization of Neosartorya fischeri antifungal protein (NFAP)." Peptides 32(8): 1724–1731.

[64] Lee, M.-T., F.-Y. Chen, et al. (2004). "Energetics of Pore Formation Induced by Membrane Active Peptides †." Biochemistry 43(12): 3590–3599.

[65] Lee, S.-J., B.-S. Yun, et al. (1999). "Tylopeptins A and B, New Antibiotic Peptides from Tylopilus neofelleus." J. Antibiot. 52(11): 998–1006.

[66] Lehrer, R. I., T. Ganz, et al. (1991). "Defensins: Endogenous antibiotic peptides of animal cells." Cell 64(2): 229–230.

[67] Leiter, E., H. Szappanos, et al. (2005). "Antifungal Protein PAF Severely Affects the Integrity of the Plasma Membrane of Aspergillus nidulans and Induces an Apoptosis-Like Phenotype." Antimicrobial Agents and Chemotherapy 49(6): 2445–2453.

[68] Marahiel, M. A., T. Stachelhaus, et al. (1997). "Modular Peptide Synthetases Involved in Nonribosomal Peptide Synthesis." Chem. Rev. 97(7): 2651–2674.

[69] Marahiel, M. A., T. Stachelhaus, et al. (1997). "Modular Peptide Synthetases Involved in Nonribosomal Peptide Synthesis." Chem. Rev. 97(7): 2651–2674.

[70] Marshall, S. H. and G. Arenas (2003). "Antimicrobial peptides: A natural alternative to chemical antibiotics and a potential for applied biotechnology." Electronic Journal of Biotechnology 6(3).

[71] Marx, F. (2004). "Small, basic antifungal proteins secreted from filamentous ascomycetes: a comparative study regarding expression, structure, function and potential application." Applied Microbiology and Biotechnology 65(2).

[72] Marx, F., U. Binder, et al. (2007). "The Penicillium chrysogenum antifungal protein PAF, a promising tool for the development of new antifungal therapies and fungal cell biology studies." Cellular and Molecular Life Sciences 65(3): 445–454.

[73] McPhee, J. B., S. Lewenza, et al. (2003). "Cationic antimicrobial peptides activate a two-component regulatory system, PmrA-PmrB, that regulates resistance to polymyxin B and cationic antimicrobial peptides in Pseudomonas aeruginosa." Molecular Microbiology 50(1): 205–217.

[74] Mendez, E., A. Moreno, et al. (1990). "Primary structure and inhibition of protein synthesis in eukaryotic cell-free system of a novel thionin, gamma-hordothionin, from barley endosperm." Eur J Biochem 194(2): 533–539.

[75] Metraux, J. P., H. Signer, et al. (1990). "Increase in Salicylic Acid at the Onset of Systemic Acquired Resistance in Cucumber." Science 250(4983): 1004–1006.

[76] Meyer, C. E. and F. Reusser (1967). "A polypeptide antibacterial agent isolated fromTrichoderma viride." Experientia 23(2): 85–86.

[77] Meyer, V. (2007). "A small protein that fights fungi: AFP as a new promising antifungal agent of biotechnological value." Applied Microbiology and Biotechnology 78(1): 17–28.

[78] Meyer, V. and U. Stahl (2002). "New insights in the regulation of the afp gene encoding the antifungal protein of Aspergillus giganteus." Current Genetics 42(1): 36–42.

[79] Montesinos, E. (2007). "Antimicrobial peptides and plant disease control." FEMS Microbiology Letters 270(1): 1–11.

[80] Mueller, P. and D. O. Rudin (1968). "Action Potentials induced in Biomolecular Lipid Membranes." Nature 217(5130): 713–719.
[81] Mygind, P. H., R. L. Fischer, et al. (2005). "Plectasin is a peptide antibiotic with therapeutic potential from a saprophytic fungus." Nature 437(7061): 975–980.
[82] Nasrollahi, S. A., C. Taghibiglou, et al. (2012). "Cell-penetrating Peptides as a Novel Transdermal Drug Delivery System." Chemical Biology & Drug Design 80(5): 639–646.
[83] Nawrot, R., J. Barylski, et al. (2013). "Plant antimicrobial peptides." Folia Microbiol 59(3): 181–196.
[84] Neuhof, T., A. Berg, et al. (2007). "Peptaibol Production bySepedonium Strains Parasitizing Boletales." Chemistry & Biodiversity 4(6): 1103–1115.
[85] Neumann, N. K. N., N. Stoppacher, et al. (2015). "The Peptaibiotics Database – A Comprehensive Online Resource." Chemistry & Biodiversity 12(5): 743–751.
[86] Oren, Z., J. C. Lerman, et al. (1999). "Structure and organization of the human antimicrobial peptide LL-37 in phospholipid membranes: relevance to the molecular basis for its non-cell-selective activity." Biochem. J. 341(3): 501.
[87] Pelegrini, P. B. and O. L. Franco (2005). "Plant γ-thionins: Novel insights on the mechanism of action of a multi-functional class of defense proteins." The International Journal of Biochemistry & Cell Biology 37(11): 2239–2253.
[88] Peschel, A., M. Otto, et al. (1999). "Inactivation of the dlt Operon inStaphylococcus aureus Confers Sensitivity to Defensins, Protegrins, and Other Antimicrobial Peptides." Journal of Biological Chemistry 274(13): 8405–8410.
[89] Peters, B. M., M. E. Shirtliff, et al. (2010). "Antimicrobial Peptides: Primeval Molecules or Future Drugs?" PLoS Pathogens 6(10): e1001067.
[90] Phadke, S. M., V. Lazarevic, et al. (2002). "Lentivirus Lytic Peptide 1 Perturbs both Outer and Inner Membranes of Serratia marcescens." Antimicrobial Agents and Chemotherapy 46(6): 2041–2045.
[91] Poirier, L., J.-C. Amiard, et al. (2007). "Determination of peptaibol trace amounts in marine sediments by liquid chromatography/electrospray ionization-ion trap-mass spectrometry." Journal of Chromatography A 1160(1-2): 106–113.
[92] Powles, S. B. and Q. Yu (2010). "Evolution in Action: Plants Resistant to Herbicides." Annu. Rev. Plant Biol. 61(1): 317–347.
[93] Rea, M. C., C. S. Sit, et al. (2010). "Thuricin CD, a posttranslationally modified bacteriocin with a narrow spectrum of activity against Clostridium difficile." Proceedings of the National Academy of Sciences 107(20): 9352–9357.
[94] Rebuffat, S., H. Duclohier, et al. (1992). "Membrane-modifying properties of the pore-forming peptaibols saturnisporin SA IV and harzianin HA V." FEMS Microbiology Letters 105(1–3): 51–60.
[95] Ritzau, M., S. Heinze, et al. (1997). "Ampullosporin, a New Peptaibol-type Antibiotic from Sepedonium ampullosporum HKI-0053 with Neuroleptic Activity in Mice." J. Antibiot. 50(9): 722–728.
[96] Rydlo, T., J. Miltz, et al. (2006). "Eukaryotic Antimicrobial Peptides: Promises and Premises in Food Safety." Journal of Food Science 71(9): R125–R135.
[97] Sansom, M. S. P. (1993). "Alamethicin and related peptaibols ? model ion channels." Eur Biophys J 22(2).
[98] Schiell, M., J. Hofmann, et al. (2001). "Cephaibol's, New Peptaibol Antibiotics with Anthelmintic Properties from Acremonium tubakii DSM 12774." J. Antibiot. 54(3): 220–233.
[99] Schwessinger, B. and C. Zipfel (2008). "News from the frontline: recent insights into PAMP-triggered immunity in plants." Current Opinion in Plant Biology 11(4): 389–395.

[100] Segura, A., M. Moreno, et al. (1999). "Snakin-1, a Peptide from Potato That Is Active Against Plant Pathogens." Molecular Plant-Microbe Interactions 12(1): 16–23.
[101] Seibold, M., P. Wolschann, et al. (2011). "Properties of the bubble protein, a defensin and an abundant component of a fungal exudate." Peptides 32(10): 1989–1995.
[102] Severinov, K., E. Semenova, et al. (2007). "Low-molecular-weight post-translationally modified microcins." Molecular Microbiology 65(6): 1380–1394.
[103] Shenkarev, Z. O., A. S. Paramonov, et al. (2013). "Peptaibol Antiamoebin I: Spatial Structure, Backbone Dynamics, Interaction with Bicelles and Lipid-Protein Nanodiscs, and Pore Formation in Context of Barrel-Stave Model." Chemistry & Biodiversity 10(5): 838–863.
[104] Skouri-Gargouri, H. and A. Gargouri (2008). "First isolation of a novel thermostable antifungal peptide secreted by Aspergillus clavatus." Peptides 29(11): 1871–1877.
[105] Steckbeck, J. D., B. Deslouches, et al. (2014). "Antimicrobial peptides: new drugs for bad bugs?" Expert Opin. Biol. Ther. 14(1): 11–14.
[106] Stotz, H. U., B. Spence, et al. (2009). "A defensin from tomato with dual function in defense and development." Plant Mol Biol 71(1-2): 131–143.
[107] Szekeres, A., B. Leitgeb, et al. (2005). "Peptaibols and Related Peptaibiotics of Trichoderma." Acta Microbiologica et Immunologica Hungarica 52(2): 137–168.
[108] Tejesvi, M. V., D. R. Segura, et al. (2013). "An antimicrobial peptide from endophytic Fusarium tricinctum of Rhododendron tomentosum Harmaja." Fungal Diversity 60(1): 153–159.
[109] Tencza, S. B. (1999). "Lentivirus-derived antimicrobial peptides: increased potency by sequence engineering and dimerization." Journal of Antimicrobial Chemotherapy 44(1): 33–41.
[110] Thomma, B., B. Cammue, et al. (2002). "Plant defensins." Planta 216(2): 19–202.
[111] Thomma, B. P. H. J., K. Eggermont, et al. (1998). "Separate jasmonate-dependent and salicylate-dependent defense-response pathways in Arabidopsis are essential for resistance to distinct microbial pathogens." Proceedings of the National Academy of Sciences 95(25): 15107–15111.
[112] Toniolo, C., M. Crisma, et al. (2001). "Lipopeptaibols, a novel family of membrane active, antimicrobial peptides." Cellular and Molecular Life Sciences 58(9): 1179–1188.
[113] Tsantrizos, Y. S., S. Pischos, et al. (1996). "Peptaibol metabolites of Tolypocladium geodes." Canadian Journal of Chemistry 74(2): 165–172.
[114] V, M., W. M, et al. (2002). "Transcriptional regulation of the Antifungal Protein in Aspergillus giganteus." Molecular Genetics and Genomics 266(5): 747–757.
[115] Vater, J. (1996). "The Multiple Carrier Model of Nonribosomal Peptide Biosynthesis at Modular Multienzymatic Templates." Journal of Biological Chemistry 271(26): 15428–15435.
[116] Viterbo, A. D. A., A. Wiest, et al. (2007). "The 18mer peptaibols from Trichoderma virens elicit plant defence responses." Mol Plant Pathol 8(6): 737–746.
[117] Vriens, K., B. Cammue, et al. (2014). "Antifungal Plant Defensins: Mechanisms of Action and Production." Molecules 19(8): 12280–12303.
[118] Wang, G., X. Li, et al. (2009). "APD2: the updated antimicrobial peptide database and its application in peptide design." Nucleic Acids Research 37(Database): D933–D937.
[119] Ward, E. R., S. J. Uknes, et al. (1991). "Coordinate Gene Activity in Response to Agents That Induce Systemic Acquired Resistance." The Plant Cell 3(10): 1085.
[120] Whitmore, L. (2004). "The Peptaibol Database: a database for sequences and structures of naturally occurring peptaibols." Nucleic Acids Research 32(90001): 593D–594.
[121] Wiedemann, I., E. Breukink, et al. (2000). "Specific Binding of Nisin to the Peptidoglycan Precursor Lipid II Combines Pore Formation and Inhibition of Cell Wall Biosynthesis for Potent Antibiotic Activity." Journal of Biological Chemistry 276(3): 1772–1779.

[122] Wiest, A., D. Grzegorski, et al. (2002). "Identification of Peptaibols from Trichoderma virens and Cloning of a Peptaibol Synthetase." Journal of Biological Chemistry 277(23): 20862–20868.
[123] Wnendt, S., N. Ulbrich, et al. (1994). "Molecular cloning, sequence analysis and expression of the gene encoding an antifungal-protein from Aspergillus giganteus." Current Genetics 25(6): 519–523.
[124] Yeaman, M. R. (2003). "Mechanisms of Antimicrobial Peptide Action and Resistance." Pharmacological Reviews 55(1): 27–55.
[125] Yoneyama, F., Y. Imura, et al. (2008). "Lacticin Q, a Lactococcal Bacteriocin, Causes High-Level Membrane Permeability in the Absence of Specific Receptors." Applied and Environmental Microbiology 75(2): 538–541.
[126] Zocher, R. and U. Keller (1996). Thiol Template Peptide Synthesis Systems in Bacteria and Fungi. Advances in Microbial Physiology, Elsevier BV: 85–131.

Swapnil Ganesh Sanmukh, Krishna Suresh Khairnar, Sérgio Luis Felisbino, and Waman Narayan Paunikar

4 Introspecting Bacteriophage Specificity and Decoding Phage Enzyme Non-Specificity for Antimicrobial Applications

4.1 Introduction

The Bacteriophages are still debated for their inclusion as a living or non-living nature by various biologists. The bacterial host specificity of phages is a matter of further research in the scientific community. The phage enzymes which have bacterial lytic properties have been studied for many decades by various researchers and produced a number of interesting findings that question the specificity of both bacteriophage host and enzyme specificity (Paunikar *et al.*, 2011; Sanmukh *et al.*, 2012; Khairnar *et al.*, 2013; Khairnar *et al.*, 2014).

Bacteriophages are reported to produce five major enzymes which include: peptidoglycan hydrolases, endosialidases, endorhamnosidases, alginate lyases and hyaluronate lyases, all have broad application potential (Drulis-Kawa *et al.*, 2015). The proteins encoded by lysis cassette genes viz. holins, endolysins, spanins are responsible for progeny release during the phage lytic cycle. The holin expressed in the late stages of phage infection forms pores in the host cell, allowing lysins to gain access to the peptidoglycan in the cell wall for the release of progeny phages (Loessner, 2005; Fischetti, 2004; Vincent, 2005). Phage lytic enzymes have been demonstrated to destroy bacterial cell wall, which has resulted in a new term “enzybiotics” for these proteins (Nelson *et al.*, 2001; Schuch *et al.*, 2002). The phage and phage-derived proteins have been explored for their various applications in therapy, diagnostics and biotechnology (Drulis-Kawa *et al.*, 2015).

The majority of endolysins have narrow range of lytic activity against bacterial cells when applied exogenously (Loessner, *et al.*, 1997) and its enzyme activity is restricted to the host bacterial species or genus specific from which it is derived (Zimmer *et al.*, 2002; Loeffler, *et al.*, 2001; Loessner, *et al.*, 1997, 2002). The broad spectrum lysins such as amidase PlyV12 (broad anti-bacterial range; Navarre, *et al.*, 1999) of *Enterococcus faecalis* phage Φ1 exhibit anti-bacterial activity against *Streptococcus pyrogenes*, group B and C *Streptococci* and *S. aureus* (Yoong, 2004). Similarly, Mur-LH (muramidase) of *Lactobacillus helviticus* phage Φ 303 shows lytic activity against 10 different bacterial species (Deutsch, *et al.*, 2004). The other example includes *Streptococcus agalactiae* B30 phage lysin (Pritchard *et al.*, 2004) and *Clostridium perfringens* phage Φ3626 Ply3626 amidase (Zimmer *et al.*, 2002).

The phage enzymes are highly evolved molecules which can quickly destroy the bacterial cell wall (Vincent, 2005). As described by Sulakvelidze *et al.* (2001), the lytic activity of phage enzymes is observed in growth media, phosphate buffer and even human blood. As the lytic activity is expressed in growth media, the effect of phage enzymes in combination can be more devastating for the pathogenic bacterial strains and can be used as a potential alternative or complimentary method for phage therapy.

Despite of the tremendous potential and wide range of applications of phage therapy, there is a need for further introspection about their nature and detailed mechanisms to enhance their uses. In the National Environmental Engineering Research Institute (CSIR-NEERI), research has been carried out on *bacteriophage therapy*. Phage therapy is widely used for the treatment of aquacultural pathogens, veterinary treatment and some are even approved for human applications (Sanmukh *et al.*, 2012; Khairnar *et al.*, 2013; Sanmukh *et al.*, 2014a, b; Khairnar *et al.*, 2014). Although the use of whole phage has the capability of generating the required enzymes continuously upon optimum nutrient and environmental conditions, the phages may undergo lysogenic conversions with their hosts and create a new problem of enzymatic resistance or provide phage immunity. Our major objective is to target the crucial component of phages that is responsible to carry out phage lytic activity, to avoid the involvement of the whole phage for the antimicrobial activity. We employed different lytic enzymes of the phages in consortia for the treatment of pathogenic bacteria. A new method has been developed to avoid the direct involvement of phages for the phage therapy by separate isolation and quantification of lytic enzymes against their specific hosts. In this chapter, we are going to discuss our new approach, which will benefit not only as a novel method but also it has proven to be less laborious, time efficient and effective.

4.2 Case studies

We selected six Gram-negative bacterial strains namely *Escherichia coli* (ATCC 15597), *E.coli* (ATCC 13706), *E.coli* (MTCC 1585), *E.coli* (MTCC 1588), *E.coli* (MTCC 521) & *E.coli* (MTCC 1650) for the study of phage lytic activity for antimicrobial application. The enrichment of different *E.coli* hosts was carried out in EC media (Eaton *et al.*, 1998).

The enriched host *E. coli* ATCC 13706 was inoculated with the ATCC 13706-B1 phage at 37 °C for 24 hours. In addition, enriched hosts MTCC 1585 and MTCC 1650 were separately inoculated with the MTCC 1742 phage at 37 °C for 24 hours (Standard methods, 1998; Paunikar *et al.*, 2012). The 1 ml of host (10,000 cfu/ml) was inoculated with 0.1 ml (1000 pfu/ml) of phage in 10 ml centrifuge tube, EC media (Eaton *et al.*, 1998) was used for each inoculation to make up the final volume 10 ml. The prepared samples were incubated at 37 °C till complete pelletization of host debris was observed in the each tube representing bacterial lysis (which is mostly 36–48 hours). The debris is an indication that the phage enzyme has effectively lysed the host cells resulting into the release of new phage progenies in the broth. The broth now includes host debris, phage particles and phage enzymes. The debris can be separated by normal centrifugation at 5000 g for 5 minutes whereas, the phage particles can be removed by coal bed method in which phages are adsorbed to the coal bed and eluted afterwards with buffer. These enriched phages separated by our indigenous coal bed method (Dafale *et al.*, 2008; Khairnar *et al.*, 2014) are representing approximately the same concentration. The remaining filtrate will be just phage enzyme mixture. The filtrate can be stored in vials at 4 °C for further studies.

4.2.1 Procedure for studying antimicrobial action of mixture of phage enzymes

To study the effect of antimicrobial action of mixture of phage enzymes on the hosts the experiment was designed to facilitate the effective activity of lytic enzymes. The concentrations of the above hosts were so adjusted during pour plating so that the clear zone formation due to surrounding cell lysis can be seen. The EC media was added with 0.1 % 2,3,5-Triphenyltetrazolium chlôride (TPTZ) solution for colour differentiation and lysis visualization for plaque assay (Eaton *et al.*, 1998) as shown in Figure 4.1. The concentration of bacterial consortia was 1000 cfu/ml. The crude enzyme mixture where added separately on two different plates to study the effect of its lytic activities (Fig. 4.2). The phage enzyme mixture can be added either by paper disc method or by bore well method for proper percolation of crude enzymes (Paunikar *et al.*, 2011; Khairnar *et al.*, 2014). We also kept control (plaques observed by using plaque assay method) to have a comparative study of both the methods.

We first tested the susceptibility of all six hosts for phage susceptibility by plaque assay method (Standard methods 1998). *E.coli* (ATCC 15597) and *E.coli* (ATCC 13706) were susceptible to the ATCC 13706-B1 phage, whereas *E.coli* (MTCC 1585) and *E.coli* (MTCC 1650) were susceptible to the MTCC 1742 phage. Out of the six hosts under consideration, *E.coli* (MTCC 1588) and *E.coli* (MTCC 521) showed no susceptibility for either phage. To test the non-specificity of phage enzymes, the consortia of six hosts were inoculated with phages separately to observe plaque formation as represented in Figure 4.1.

In our current studies, based on natural bacteriophage and prostate cancer cell interactions, we found that the bacteriophages can interact with mammalian cell surface proteins and can induce internal cellular signalling and gene expression alterations. After exposing 80% confluent (4×10^3) epithelial prostatic tumoral cell line PC3 to M13 and T4 bacteriophages for 24 hours with concentration of about 10×10^7 pfu (plaque forming unit), we observed that integrins, which includes integrin alpha V, alpha 5, alpha 3 and Beta 5; mRNAs expression were amplified significantly as analysed though quantitative PCR experiments. Moreover, bacteriophages exposure also induced an increase in the phosphorylated forms of AKT and PI3K by this cell, verified by western blotting.

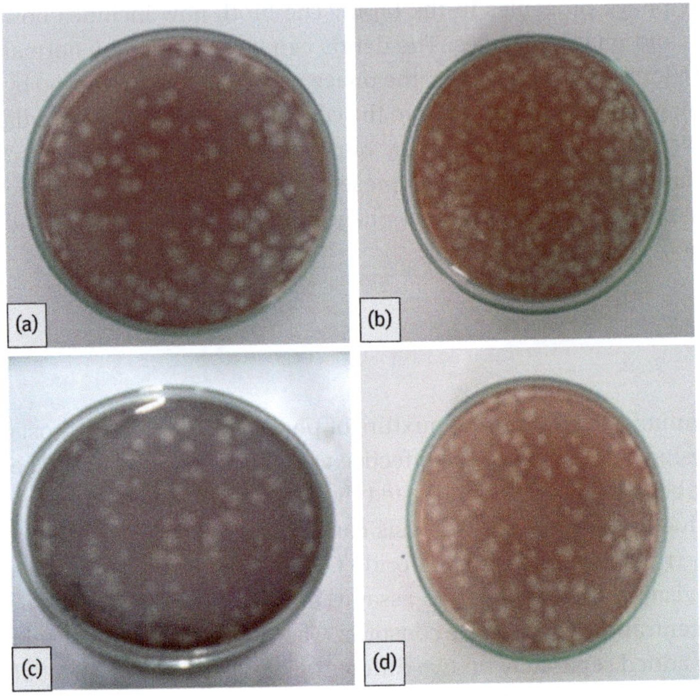

Fig. 4.1: The comparative study of plaque assay to show the non-specificity of phage enzymes as observed from plaque formation: **(a)** Mixed culture of *E.coli* (ATCC 15597) and *E.coli* (ATCC 13706) were susceptible to for ATCC 13706-B1 phage; **(b)** Consortia of *E.coli* (six strains), showing non-specificity of phage enzymes when inoculated against both the ATCC 13706-B1 phage and MTCC 1742 phage; **(c)** Mixed culture of *E.coli* (MTCC 1585) and *E.coli* (MTCC 1650) were susceptible to for MTCC 1742;
(d) Consortia of *E.coli* (six strains), showing non-specificity of phage enzymes when inoculated against single ATCC 13706-B1 phage.

These effects on the specific gene expression reveals that phages have a broad range binding properties through their Proteinous external coat, which may be representing as a natural peptide display for inducing this effect. Moreover, previous works from Andrzej Gorski's group (Groski *et al.*, 2003; Dabrowska *et al.*, 2004(a); Dabrowska *et al.*, 2004(b); Dabrowska *et al.*, 2014) have verified similar properties in both in vitro and in vivo studies for melanoma cells. Our sole aim for conducting these study was to demonstrate that phages, in absence or presence of its host, have significant influence on their surrounding environment and, being a natural nanoparticles, are capable of demonstrating various novel properties, which we might had unable to identify.

Fig. 4.2: The comparative study for understanding the lytic activity and efficacy of crude phage enzyme was demonstrated as observed from clear zone formation. Consortia of *E.coli* (ATCC 15597), *E.coli* (ATCC 13706), *E.coli* (MTCC 1585), *E.coli* (MTCC 1588), *E.coli* (MTCC 521) & *E.coli* (MTCC 1650) showing clear zone of lysis as observed from paper disc test due to lytic enzymes from ATCC 13706-B1 phage; MTCC 1742 phage.

Our preliminary results suggest that natural bacteriophages can interacts with cell surface receptors integrins and mediate gene expression changes. Future in vitro as well as in vivo studies would be interesting to understand various properties associated with natural bacteriophages exposure (unpublished data).

Our work clearly aims for introspecting bacteriophage specificity and understanding phage enzyme non-specificity for antimicrobial applications for treating antibiotic resistant bacterial cells.

4.2.2 Results

The comparative assay for the study of lytic activities of phages and their enzymes demonstrated and verified the non-specificity of phage encoded enzymes as well as the anti-microbial activity exhibited by them. We found that by the use of paper disc method, the application of phage enzyme (crude enzymes) was quite easy, simple and effective as compared with phage plaque assay approach. The effectiveness can be observed from the clear zone formed around the paper discs represented in Figure 4.2. The result implies that the phage enzyme or combination of different lytic enzymes can be effective as they have activity against cell wall components. This clearly indicates that the crude or combined application of phage enzymes has more successful results as compared with the traditional phage therapy approach, considering its various advantages over phage therapy as mentioned in Table 4.1.

The quality and effectiveness of phage enzymes for their anti-microbial action can be optimized or increased by its selective purification and quantification. The development of enzyme application method compared with phage therapy will boost the targeted antimicrobial action against any pathogenic or antibiotic resistant microorganisms. Establishment of this method will be helpful in assisting various other assays (such as comet assay, plaque assay and phage therapy) and for their modification as well as increasing their efficacy (Khairnar *et al.*, 2013; Khairnar *et al.*, 2014; Sanmukh *et al.*, 2012; Sanmukh *et al.*, 2015).

4.3 Discussion

In this case study, we have demonstrated comparatively the plaque assay method and crude phage enzyme approach for their potential lytic activity useful for antimicrobial applications against pathogens. The uses of phage enzymes directly have more advantages over the phage therapy (see Table 4.1) due to its minimum requirement and non-specific activity towards microbial cell wall. The susceptible host in the media was lysed due to the lytic activity of the phage enzymes; they also target the adjacent host cells which are not susceptible towards these phages. The adjacent host which have non-specificity towards the phages under considerations are targeted and get lysed due to the substrate specificity of the enzymes in action. The phages may be showing specific or broad specific activity towards their hosts, buts the enzymes released followed by such lyses event are acting on the cell wall components of encountered hosts and are irrespective of their phage specificity.

Tab. 4.1: Comparison of traditional phage therapy and combined phage therapy approach for understanding its potential advantages.

Traditional Bacteriophage Therapy	Combined application of phage enzymes
Attachment with receptors for infection	Not applicable as enzyme is already released
Lysogenic formation within host	Not applicable
Phage resistance	Not applicable as phages are not involved
Environmental factors like temperature, pH, etc are determinative for proper infection and lyses event to occur.	Optimum temperature of enzymes, pH, etc is required for proper lyses event to occur.
A good nutrient condition is required for phage infection and lyses by phages.	Nutrient condition is not required as phages are not involved.
Complicated, laborious and time consuming	Simple, easy and efficient

The experiments conducted on consortia of different bacterial hosts and their phages showed the same results with delayed incubation time. This experiment thus demonstrates the collective activity of phage enzymes against microbial pathogens. As endolysins are classified depending upon enzymatic specificity towards cell wall components (Loessner *et al.*, 1997; Young *et al.*, 2000, Loessner *et al.*, 2002; Loessner *et al.*, 2005, Fischetti, 2004), their combination will prove very helpful in combined application of phage enzymes against pathogens showing anti-microbial actions. It has been already reported that repeated exposure of such lytic enzyme(s) to bacteria did not lead to the resistant strains (Loeffler *et al.*, 2001). The lytic enzymes target the cell wall components and resistance has not been reported to these enzymes (Loessner *et al.*, 2005; Vincent, 2005). Hence, phage enzymes can prove to be very useful even for multi-drug resistant as well as antibiotic resistant pathogens.

Our experiment thus shows that the activities of phage enzymes are totally cell wall components specific and is independent of host type (Paunikar *et al.*, 2011) and reveal the combined application of phage enzymes is effective for controlling pathogenic organisms. It would be interesting to find out whether different phage enzymes can be used in combinations to inhibit pathogenic bacteria in natural environment and living organisms. The different phage enzymes varying in enzyme activity against various hosts can be applied in combination to lyse any pathogenic bacterial cell and thereby prevent outbreak of diseases.

Acknowledgements

We are thankful to Council for Scientific and Industrial Research-National Environmental Engineering Research Institute (CSIR-NEERI), Nagpur (India) for providing facilities to carry out this research activity. We would also like to acknowledge and thanks all the research, project and trainee staffs for their help and assistance to make this work possible.

4.4 References

[1] Dafale, N., Lakhe, S., Yadav, K., Purohit, H. & Chakrabarti, T. (2008). Concentration and recovery of coliphages from water with bituminous coal. Water Environment Research , 80(3), 282–288.

[2] Deutsch, S.M., Guezenec, S., Piot, M., Foster, S., Lortal, S. & Mur, L.H. (2004). The broad spectrum endolysin of *Lactobacillus helveticus* temperate bacteriophage φ- 0303. Applied Environmental Microbiology, 70. 96–103.

[3] Dabrowska K, Opolski A, Wietrzyk J, Switala-Jelen K, Boratynski J, Nasulewicz A, Lipinska L, Chybicka A, Kujawa M, Zabel M, Dolinska-Krajewska B, Piasecki E, Weber-Dabrowska B, Rybka J, Salwa J, Wojdat E, Nowaczyk M, Gorski A. (2004a) Antitumour activity of bacteriophages in murine experimental cancer models caused possibly by inhibition of ß 3 integrin signaling pathway. Acta Virol; 48:241–248.

[4] Dabrowska K, Opolski A, Wietrzyk J, Switala-Jelen K, Godlewska J, Boratynski J, Syper D, Weber-Dabrowska B, Gorski A. (2004b). Anticancer activity of bacteriophage T4 and its mutant HAP1 in mouse experimental tumour models. Anticancer Res; 24:3991–3995.

[5] Dabrowska Krystyna, Miernikiewicz Paulina, Piotrowicz Agnieszka, Hodyra Katarzyna, Owczarek Barbara, Lecion Dorota, Kazmierczak Zuzanna, Letarov Andrey, Górski Andrzej. (2014) Immunogenicity studies of protein forming the T4 phage head surface. J Virol; 88: 12551–12557.

[6] Drulis-Kawa Z., Majkowska-Skrobek G. & Maciejewska B (2015). Bacteriophages and phage-derived proteins – application approaches. Feb 9. [Epub ahead of print]. To be published in Current Medicinal Chemistry Available at: http://europepmc.org/abstract/MED/25666799.

[7] Eaton, A.D, Clesceri, L.S. & Greenberg, A.E. (1998). Standard method for examination of water and waste water. 20th edn. APHA, AWWA and WEF Washington, USA.

[8] Fischetti, V.A. (2004). The use of phage lytic enzymes to control bacterial infections. In Kutter E. & Sulakvelidze A. (Eds) Bacteriophages: Biology and Applications. (pp. 321–334). Boca Raton: CRC Press.

[9] Fuhrman, J.A. (1999). Marine viruses and their biogeochemical and ecological effects. Nature, 399, 541–548.

[10] Gorski A, Dabrowska K, Switala-Jelen' K, Nowaczyk M, Weber-Dabrowska B, Boratynski J, Wietrzyk J, Opolski A. (2003). New insights into the possible role of bacteriophages in host defense and disease. Med Immunol; 14: 2

[11] Khairnar, K., Raut, M.P., Chandekar, R.H., Sanmukh, S.G. & Paunikar, W.N. (2013). Novel bacteriophage therapy for controlling metallo-betalactamase producing *Pseudomonas aeruginosa* infection in catfish. BMC Veterinary Research;9, 264. Doi:10.1186/1746-6148-9-264.

[12] Khairnar K., Sanmukh, S.G., Chandekar, R.H. & Paunikar, W.N. (2014). A simple and novel modification of comet assay for determination of bacteriophage mediated bacterial cell lysis. Journal of Virological Methods, 203, 33-8. doi:10.1016/j.jviromet.2014.02.028.

[13] Loeffler, J.M., Nelson, D. & Fischetti, V.A. (2001). Rapid killing of *Streptococcus pneumoniae* with a bacteriophage cell wall hydrolase. Science, 94, 2170–2172.

[14] Loessner, M.J., (2005). Bacteriophage endolysins-current state of research and applications. Current Opinion in Microbiology, 8, 480– 487.

[15] Loessner, M.J., Kramer, K., Ebel, F. & Scherer, S. (2002). C-terminal domains of *Listeria monocytogenes* bacteriophage murein hydrolases determine specific recognition and high-affinity binding to bacterial cell wall carbohydrates. Molecular Microbiology, 44, 335–349.

[16] Loessner, M.J., Maier, S.K., DaubekPuza, H., Wendlinger, G. & Scherer, S. (1997). Three *Bacillus cereus* bacteriophage endolysins are unrelated but reveal high homology to cell wall hydrolases from different bacilli. Journal of Bacteriology, 179, 2845–2851.

[17] Lopez, R., Garcia, E. & Garcia, P. (1997). The pneumococcal cell wall degrading enzymes: a modular design to create new lysins? Microbial Drug Resistance, 3, 199–211.

[18] Navarre, W.W., Ton-That, H., Faull, K.F. & Schneewind, O. (1999). Multiple enzymatic activities of the murein hydrolase from staphylococcal phage φ11. Identification of a D-alanyl-glycine endopeptidase activity. Jounal of Biological Chemistry, 274, 15847–15856.

[19] Nelson, D., Loomis, I. & Fischetti, V.A. (2001). Prevention and elimination of upper respiratory colonization of mice by group A streptococci by using a bacteriophage lytic enzyme. Proceedings of the National Academy of Sciences, United States of America, 98, 4107–12.

[20] Paunikar W.N., Sanmukh S.G. & Ghosh T.K. (2011). Non-specificity of phage enzymes. International Journal of Pharma and Biosciences, 2, B546–B552.

[21] Pritchard, D.G., Dong, S., Baker, J.R. & Engler, J.A. (2004). The bifunctional peptidoglycan lysin of *Streptococcus agalactiae* bacteriophage B30. Microbiology, 150, 2079–2087.

[22] Sanmukh, S.G., Khairnar, K.S., Swaminathan, S. & Paunikar, W.N. (2015). Novel applications of bacterial and algal viruses in advancement of molecular biology and for enhancement of bio-fuel production, In A.K. Roy (Ed.), Emerging Technologies of the 21st Century (pp 199–212). New Delhi, India: New India Publishing Agency.

[23] Sanmukh, S., Bruno, B., Ramakrishnan, U., Khairnar, K., Swaminathan, S. & Paunikar, W.N. (2014a). Bioactive compounds derived from microalgae showing antimicrobial activities. Jounal of Aquatic Research Development, 5, 224. doi:10.4172/2155-9546.1000224.
[24] Sanmukh, S.G., Khairnar, K. & Paunikar, W.N. (2014b). Novel application of algal viruses in increasing the extraction efficiency of algal lipid for potential biodiesel production. African Journal of Biotechnology, 13, 1666–1670.
[25] Sanmukh, S.G, Meshram, D.B, Paunikar, W.N. & Swaminathan, S. (2012). Interaction of fishes with pathogenic bacteria and application of phages for their control-a review. Reviews in Fish Biology & Fisheries, 22, 567–574.
[26] Schuch, R., Nelson, D. & Fischetti, V.A. (2002). A bacteriolytic agent that detects and kills *Bacillus anthracis*. Nature, 418, 884–9.
[27] Sulakvelidze, A., Alavidze, Z. & Morris, Jr, JG. (2001). Bacteriophage therapy. Antimicrobial Agents & Chemotherapy, 45, 649–659.
[28] Vincent, A.F. (2005). Using phage lytic enzymes to control pathogenic bacteria. BMC Oral Health; 6 (Suppl 1):S16. Doi:10.1186/1472-6831-S1-S16.
[29] Yoong, P., Schuch, R., Nelson, D. & Fischetti, V.A. (2004). Identification of a broadly active phage lytic enzyme with lethal activity against antibiotic resistant *Enterococcus faecalis* and *Enterococcus faecium*. Journal of Bacteriology, 186, 4808–4812.
[30] Young, R.Y., Wang, I.N. & Roof, W. D. (2000). Phages will out: strategies of host cell lysis. Trends of Microbiology, 8, 120–128.
[31] Zimmer, M., Vukov, N., Scherer, S. & Loessner, M.J. (2002). The murein hydrolase of the bacteriophage φ3626 dual lysis system is active against all tested *Clostridium perfringens* strains. Applied Environmental Microbiology; 68, 5311– 5317.

Avnish Kumar, Monika Asthana, Krishn Gopal Jain, and Vinod Singh

5 Microbial production of enzymes: An Overview

5.1 Abstract

Microbes are either harmful or beneficial to us naturally. However, we can get benefitted from every type of microbes using their metabolic activities for the production of industrial enzymes. Enzymes are one of the world's most widely used catalyst applied in industrial and laboratory tests for screening and production of important products like detergent, food, medicines etc. This chapter provides an overview on commercial production and application of some enzymes important for industrial, analytical and clinical purposes.

5.2 Introduction

The special characteristics of enzymes which include: enormous catalytic activity, tolerance to a varied range of pH, stability of enzyme activity over a range of temperature and pH, and other harsh reaction conditions are exploited for their commercial interest and industrial applications. Various Biotechnology industries like dairy industry, leather industry, textile industry etc. have exploited these properties of enzymes for their applications in processing of raw materials and production of goods and in bioconversions and bio-remediation. Enzymes are biocatalysts which lower the activation energy required for a reaction, without being used up. They are proteins (except ribozyme) in nature that is synthesized as intra and extra cellular compounds. They are important for specific metabolic activities in all types of living cells. Side reactions leading to the wasteful formation of byproducts are rare in enzyme catalyzed reactions in contrast with uncatalyzed ones. Enzyme accelerate rate of reactions by factors of as much as a million or more. Indeed, most reactions in biological system do not take place at perceptible rates in the absence of enzymes. Carbonic anhydrase, which catalyzes the hydration of CO_2 is one of the fastest enzymes known (Fisher *et al.*, 2007). In vertebrates blood 70 % of the CO_2 is transported in the form of bicarbonate ions. The CO_2 reacts with H_2O to produce H_2CO_3 which dissociate into H^+ and HCO_3^-. This reaction occurs at very slow rate in the plasma, but inside RBCs this reaction occurs 10^7 times faster than that in the plasma, because, inside RBC's one carbonic anhydrase molecule can hydrate 10^6 molecules of CO_2 per second (Berg *et al.*, 2002).

Excessive study and knowledge of certain enzymes creates a special interest to utilize them as catalysts in numerous processes on an industrial scale. Today, out of 4000 known enzymes, about 200 are in commercial use. Microbial enzymes are known to be superior enzymes obtained from different microorganisms; bacteria, fungi and yeasts, particularly for applications in industries on commercial scales.

Specific microorganisms have been isolated from natural sources under extreme culture conditions, with the objective that such isolated microbes would possess the capability to bio-synthesize special enzymes (Nigam, 2013). The first industrially produced enzyme was the fungal amylase takadiastase, which was employed as a pharmaceutical agent in the United States. Since then several enzymes have been produced from microbes for industrial purposes. By 1969, 80 % of all laundry detergents contained enzymes, chiefly proteases. The development of amylases and amyloglucosidases for the production of glucose from starch has led to a new industrial application of enzymes. Also, the use of glucose isomerase for the production of fructose has become wide spread since 1967. Microbial rennin is also have high significance instead of calf's rennin in cheese production since 1965. Today about 95 % of laundry detergent in Europe contains proteases, although such detergents are much less widely used in other countries. Along with these, additional enzymes such as lipases, amylases, pectinases, and oxidoreductases were used experimentally in the detergent industry (Wrotnowski, 1997).

The industrially important enzymes are amylase, protease, catalase, isomerase, pencillin acylase, etc., for analytical purposes glucose oxidase, galactose oxidase, alcohol dehydrogenase, muramidase, cholesterol oxidase, for clinical purposes, asparaginase, protease, lipase and streptokinase. The majority of the industrial enzymes are of microbial origin.

5.3 Production of enzymes

Enzymes production from microorganism at commercial level is carried out in large vessels called fermenter and the process is called fermentation. This reaction, which takes place in the absence of oxygen, is called fermentation. Fermentation has been used for thousands of years to produce food and beverages, such as cheese, yoghurt, beer and wine. Yeast is a fungus whose enzymes aid the breakdown of glucose into ethanol and carbon dioxide anaerobically.

Glucose + yeast → Ethanol + Carbon dioxide

Fermentation is influenced by numerous factors, including temperature, pH, nature and composition of the medium (carbon source; glucose, maltose, lactose, sucrose and starch, nitrogen source; $NaNO_3$, $(NH_4)_2SO_4$, NH_4Cl, NH_4NO_3 and KNO_3), dissolved O_2, dissolved CO_2, operational system (e.g. batch, fed-batch, continuous), feeding with precursors, mixing (cycling through varying environments), and shear rates in the fermenter. Variation in these factors may affect: the rate of fermentation; the product spectrum and yield; the organoleptic properties of the product (appearances, taste, smell, and texture), the generation of toxins; nutritional quality; and other physio-chemical properties (Chisti, 1999).

5.3.1 Submerged fermentation

Fed batch is the most common type of submerged fermentation process used in production of industrial enzymes. This involves growing carefully selected strains of microorganisms in closed large vessels with volumes of up to 1,000 cubic meters containing a rich broth of nutrients (the fermentation medium) and a high concentration of oxygen. As the microorganisms break down the nutrients, they release the desired enzymes into solution. In this process, sterilized fresh media added intermittently to the bioreactor and the product harvested when either reactor is full or after completion of fermentation. Parameters like temperature, pH, oxygen consumption and carbon dioxide formation are measured and controlled to optimize the fermentation process. Enzymes harvesting steps include removal of insoluble products, e.g. microbial cells by centrifugation. As most industrial enzymes are extracellular (secreted by cells into the external environment), they remain in the fermented broth. The enzymes in the remaining broth are then concentrated by evaporation, membrane filtration or crystallization depending on their intended application. If pure enzyme preparations are required, they are usually isolated by gel or ion exchange chromatography (Chisti, 1999).

5.3.2 Solid state fermentation

Solid-state fermentation (SSF) involves the cultivation of microorganisms on a solid substrate, such as grains, rice and wheat bran, sugar beet pulp and wheat and corn flour. The selection of substrate depends on many factors, which is mainly related to the cost and the availability of the substrate. Other factors include particle size and the level of moisture. Smaller substrate particles have a larger surface area for the proliferation of the microorganisms, but if too small the efficiency of respiration will be impeded and poor growth and hence poor production of enzymes will result. Larger particles provide more efficient aeration and respiration, but there is a reduction in the surface area. A compromise must be reached, regarding the particle size of the substrate for a particular process. SSF requires moisture to be present on the substrate, for the microorganisms to produce enzymes. As a consequence the water content of the substrate must also be optimized, as a higher or lower presence of water may adversely affect the microbial activity. Water also has implications for the physicochemical properties of the solid substrate. Industrially important enzymes, viz. proteases, pectinases, glucoamylases and cellulases have been produced by SSF (Chisti, 1999).

5.4 Enzymes of industrial Importance

Production of new microbial enzymes starts with screening of microorganism for desirable activity using appropriate selection procedure. The level of enzyme activity produced by an organism from a natural environment is often low and needs to be elevated for industrial production. Increase in enzyme levels is achieved either by mutation of the organism or by using recombinant organism. The microbial enzyme could be characterized by its pI, molecular weight and action pattern.

5.4.1 Amylases

The history of amylases began in 1812 when the first starch degrading enzyme was discovered by Gottlieb Sigismund Kirchhoff (Gurung *et al.*, 2013). Amylases are significant enzymes for their specific use in the industrial starch conversion process in food, fermentation, textile to paper industries. One of the main uses of amylases is the production of sweeteners for the food industry. The hydrolysis of starch with amylase results first in the production of short chain polymers called dextrin, then the disaccharide maltose and finally glucose (Pandey *et al.*, 2000). Amylolytic enzymes act on starch and related oligo- and polysaccharides. The amylolytic and related enzymes have been classified as glycoside hydrolases. The most important enzymes in the starch-saccharification process are α-amylases (α-1,4-glucan-4-glucanohydrolase), β-amylases (α-1,4-glucano maltohydrolase), glucoamylases, glucose isomerases, pullulanases and isoamylases. The enzymes have been produced by a wide range of microorganisms and substrates and categorized as exo-, endo-, de-branching and cyclodextrin producing enzyme. Alpha amylase can hydrolyse amylose into smaller dextrins and a little maltose. On further proceeding of reaction some more amount of dextrin converted to maltose. Then beta amylase precedes the complete conversion. Beta amylase splits maltose molecules from the ends of the branches of amylopectin until its action blocked by 1,6-glycoside linkage or branch points, leaving branched molecules of residual dextrin. Among amylases α-Amylase is in maximum demand due to its wide range of applications in the industrial front. With consumers growing increasingly aware of environmental issues, industries find enzymes as a good alternative over other chemical catalysts. α-Amylase can be produced by plant or microbial sources. Due to the advantages that microbial production offers, α-Amylase from microorganisms has been focused upon and preferred to other sources for production.

5.4.1.1 Microbial production of α-Amylases

α-Amylases (1,4-α-glucan-glucanohydrolases) are extracellular enzymes which hydrolyze α-1,4-glycosidic bonds. These enzymes are endoenzymes as they split the substrate in the interior of the molecule. α-Amylases are produce by my bacteria and fungi (Tab. 5.1). α-Amylases are classified according to their starch-liquefying and/or saccharogenic effect, pH, temperature range, and stability.

Tab. 5.1: Bacteria and fungi which produce α-Amylases.

Bacteria	*Bacillus subtilis, B. cereus, B. amyloliquefaciens, B. coagulans, B. polymyxa, B. stearothermophilus, B. caldolyticus, B. acidocaldarius, B. subtilis* var. *amylosaccharaticus, B. licheniformis, Lactobacillus, Micrococcus, Pseudomonas, Arthrobacter, Escherichia, Proteus, Thermomonospora*, and *serratia*.
Fungi	*Aspergillus, Penicillium, Cephalosporium, Mucor, Candida, Neurospora* and *Rhizopus*

Saccharogenic amylases produce free sugars, whereas starch-liquefying amylases break down the starch polymer but do not produce free sugars. The substrate concentrations determine the extent to which the enzymes act in a liquefying or saccharogenic manner. Three strains of *Bacillus, B. subtilis* Marburg, *B. subtilis* var. *amylosaccharaticus* and *B. natto* produce saccharogenic amylases whereas *B. amyloliquefaciens* produces liquefying amylases.

The most important α-amylases are produced by *B. amyloliquefaciens, B. licheniformis* and *Aspergillus oryzae*. *Bacillus* amylases are used much more extensively than those of *Aspergillus*.

5.4.1.1.1 Bacterial production

In industry α-amylases are produce either in batch or fed-batch fermentation. A medium for the production of α-amylases in a 100 m^3 fermenter with *B. subtilis* consists of 5 % starch, 0.56 % NH_4NO_3, 0.28 % sodium citrate, 0.13 % KH_2PO_4, 0.05 % $MgSO_4 \cdot 7H_2O$, 0.01 % $CaCl_2 \cdot 2H_2O$, 0.5 % peptone, 0.2 % yeast extract ; pH 6.8. At 45 °C, the maximal specific enzyme formation rate is reached with the test strain after 18 hours; however the maximal amount of enzyme produced (up to 3000 units/ml) is obtained at considerably lower temperatures (27 to 30 °C). The enzyme formation rate is very low during exponential growth in many strains, but just before the growth rate decreases and spore formation begins, amylase production increases. Amylase is excreted as the nascent protein from the ribosome through the membrane. The nascent protein contains an additional peptide, the signal peptide, which participates in the transfer of nascent amylase from ribosome to the membrane. The signal peptide is removed during folding.

5.4.1.1.2 Fungal Production of alpha amylase

The nutrient solution used for amylase production using *Aspergillus oryzae* consist of 8 % starch, 1.2 % $NaNO_3$, 0.1 % K_2HPO_4, 0.1 % $MgSO_4$, 0.05 % KCl, 0.003 % $FeSO_4$,

0.08 % $Mg(NO_3)_2$, 0.05 % $Mg(H_2PO_4)_2$, 2 % malt extract. The activity of enzyme produced varied and depends on the medium composition. The optimal temperature lies in a narrow range between 28–30 °C, and the duration of the fermentation process is 3–4 days.

Beta amylases are usually plant origin, but some microbial producers are also known: *Bacillus polymyxa*, *Bacillus cerus*, *Bacillus megaterium*, *Streptomyces* sp., *Pseudomonas* sp. and *Rhizopus japanicus*. Although yield in a wild strain are usually low, mutants have been discovered which produce 200 times more enzyme then the wild type. Bacterial beta amylases have greater heat resistance (>70 °C) than plant beta amylases and the pH optimum is also higher (about pH7). In contrast to alpha amylase calcium is not necessary for stabilization and activation of bacterial beta amylases. In the future it is likely that beta amylases will be used for the production of maltose syrup.

Glucoamylase (alpha 1,4 glucan gulcanohydrolases) act on starch by splitting glucose unit from the non-reducing ends. Maltose is broken down only slowly, while 1,4 bonds in the branch polysaccharides are hardly attacked. Thus glucose, maltose and limit dextrins are the end product of glucoamylase action. Microorganism uses to produce glucoamylase are *Aspergillus niger*, *Aspergillus oryzae*, *Aspergillus awarmori*, *Rhizopus niveus*, *Rhizopus delemar*, *Rhizopus formosaensis* and *Rhizopus Javenicus*. Production strains in Western Europe and in the United States are mutants of *Aspergillus niger* and *Aspergillus niveus*. Today industrial fermentation of glucoamylase is carried out almost exclusive in submerged fermenter up to 150 m^3 in volume. Due to the increasing demand for glucoamylase in fructose syrup production, smaller system which involve growth on the surface of liquid or solid are no longer economical. However, in genetic study, strains are frequently isolated which produce higher yield in surface culture than in submerged culture. Experiments with batch fed fermentation have been successfully carried out over period as long as 320 hours.

1,6 Glycoside-Splitting enzymes can split the side chain of amylopectin via hydrolyses alpha 1,6 bond are either act directly on amylopectin, such as pullulanases and isoamylase or work (indirectly) after enzymatic modification of the substrate. Both batch fermentation and continuous process are in use for production of these 1,6 Glycoside-Splitting enzymes. Pullulanases are isolated majorly from the fungus *Pullularia pullulans*, *Agrobacter aerogene* ATCC 15050. Several other pullulanase producers *Bacillus* polymyxa, *Pseudomona saccharophila*, *Sterptococcus* sp., and *Steptomyces* sp. have been reported by researchers. Isoamylase produced by *Pseudomonas amyloderamosa*, *Agrobacterium*, *Erwinia*, *Staphylococcus*, *Serratia*, *Nocardia*, *Bacillus*, *Pediococcus*, *Lactobacillus*, *Leuconostoc*.

5.4.2 Glusose isomerase (D glucose ketoisomerase)

Glusose isomerase causes the isomerization of the glucose to fructose and it is a reversible reaction. Due to higher consumption of sugar in the world, price of sugar are higher than starch which leads to a high commercial importance of this enzyme. It is estimated that sugar consumption rate is 50 kg/person/year. The application of these enzymes has been established in starch liquefaction, paper, food, sugar and pharmaceutical industries.

Glucose has 70–75 % sweetening strength of beet sugar (sucrose) but beta D fructo pyranose (fructose), the sweetest monosaccharide has twice the sweetening strength of sucrose. Thus, process for the manufacture of fructose is of considerable value. Chemical production of fructose is not acceptable to the food industry. Hence, glucose isomerases are used which has additional activity with D glucose and D ribose. The industrially useful glucose isomerase must have low pH optimum to avoid side reactions, high specific activity, and high temperature optimum.

5.4.2.1 Microbial production of Glusose isomerase

The most important Glusose isomerase producers are Bacillus coagulans, Streptomyces rubiginosus, Actinoplanes missouriensis, Flavobacterium arborescens, Streptomyces phaeochromogens, Streptomyces olivochromogens, Arthrobacter sp., Microbispora rosea, Micromonospora coerula, Microellobospora flavea, Nocardia asteroides, Nocardia dassonvillei, Bravibacterium imperial.

This enzyme is produced by batch and continuous fermentation. In batch processes, *B coagulans* does not form enzyme during the log phase. As soon as the glucose content in the nutrient solution approaches zero, growth ceases. In a typical diauxy, additional carbon source present in the medium are then metabolize and enzyme production begins. Maximal enzyme activity obtained after 24 hrs incubation. The enzyme produced can be purified from solution by filtration, carbon treatment, and ion exchange chromatography to prevent loss of activity during isomerization. In a continuous fermentation process the growth limiting substrate must be glucose for optimal enzyme yields, oxygen limitation is also necessary since microaerophillic condition insides the cells established the condition.

5.4.3 Proteases

Proteases are useful in the detergent industry, dairy industry, pharmaceutical industry, leather industry, manufactures of protein hydrolyses, food industry, film industry, and waste processing. Proteases are produced commercially both from bacteria and fungi. An important distinguishing feature is the pH optimum of the proteases. Subtilisin, trypsin, thrombin and DNA polymerase are few examples of protease enzymes. The proteases on the market include alkaline, neutral, and acid proteases.

5.4.3.1 Microbial production of Alkaline proteases

Alkaline proteases are used as detergents enzymes due to stability at high temperature, at alkaline range 9–11, Stability in association with chelating agents and perborates. Its bacterial producers are *Bacillus licheniformis, Bacillus amyloliquifacience, Bacillus firmis, Bacillus megaterium, Bacillus pumilis, Streptomyces fradiae, Streptomyces graiseus*, and *Streptomyces rectus* and the fungal producers are *Aspergillus niger, Aspergillus sojae, Aspergillus oryzae* and Aspergillus *flavus*. Enzymes used in detergents are mainly proteases from *Bacillus* strains (bacillopeptidases). The best known protease is Subtilisin Carlsberg from *Bacillus licheniformis*, Subtilisin BPN and Subtilisin novo from *Bacillus amyloliquifaciens*.

The fed batch process is used to maintain a lower concentration of NH_4 ions and amino acids (nitrogenous materials) that can suppress protease production. Continuous processes are not used commercially. High oxygen partial pressure is generally necessary for optimum protease titers. Aeration rates are 1 ppm and the time spend of the fermentation is 48 to 72 hours, depending on the organism. Proteases must be converted into the particulate from before they are added to detergent, since if dry enzymes powder is inhaled by production workers or users, allergic reaction may result. Enzyme concentrates are marketed in microencapsulated form. To make a suitable encapsulated product, a wet paste of enzyme is melted at 52–70 °C with the hydrophobic substances such as polyethylene glycol and then converted into tiny particles. These solidified spherical particles are not hazardous when added directly to the detergent.

5.4.3.2 Microbial production of Neutral Proteases

Neutral proteases are also produced by bacteria and fungi. Bacterial producer are *Bacillus subtilis, Bacillus cereus, Bacillus megaterium, Pseudomonas aeruginosa, Streptomyces griseus, Aspergillus oryzae, Aspergillus sojae* and, *Pericularea oryzae*. These enzymes are unstable at higher temperature at very narrow pH range. Their relative stability is enhanced by adding calcium, sodium, and chloride. Neutral proteases are quickly inactivated in presence of alkaline proteases, thus have limited application in leather and food industry.

5.4.3.3 Microbial production of Acid proteases

Acid proteases are isolated from fungi and used in cheese and curd production. These enzymes worked well at optimum pH 2–4 however, stable at pH 4–5.5. The enzyme produced industrially by submerged culture process in a stirred bioreactor. Several genera of fungi produces acid proteases are *Aspergillus, Candida, Coriolus, Endothia, Enthomophthora, Irpex, Mucor, Penicillin, Rhizopus, Sclerotium*, and *Torulopsis* and bacterial species are *Streoptococcus lactis, streptococcus cremoris, Streptococcus thermophiles, Lactobacillus helveticus, Lactobacillus bulgaricus*. Fermentation media contains 3 % soya meal, 1 % glucose, 1 % skimmed milk, 0.3 % $NaNO_3$, 0.05 % K_2HPO_4, 0.025 % $MgSO_4 \cdot 7H_2O$; pH 6.8. The fermentation takes 48 hrs at 28 °C. The purification steps included removal of myelin debris and evaporate to enhance concentration. Acid proteases are used in medicine, in the digestion of soy protein, for soy sauce production, and to break down wheat gluten in the baking industry. The organism selection is depends upon the desired curd structure and flavor, it is reported that acid proteases enhance bitterness of curd and cheese.

5.4.4 Lactase

Lactase, are known as beta galactosidase that convert lactose into glucose and galactose. These enzymes are produced by bacteria and yeast. This enzyme is intracellular in origin, but, in case of fungi these are extracellular in origin and has an optimum pH of 4 to 5.4 (Tanaka *et al.*, 1975).

5.4.4.1 Microbial production of Lactase

For industrial production of enzymes it is especially important to improve conditions so as to obtain higher yield of lactase. Only 5 % enzymes are produced by solid state fermentation. Submerged fermentation appears to be the method of choice where sterilization and process control is easier to achieve (Frost and Moss, 1987). Commercially lactases are produced from yeast and fungi. The latases are extracted from fungi *Aspergillus oryzae* and *Aspergillus niger*, and the yeast *Kluyveromyces lactis, K. fragilis* and *Torula cremoris* and bacteria *Bacillus* sp. The yeast is cultured in a submerged process but the *Aspergillus* must be cultivated on the surface of wheat bran at 30 °C. Lactose is generally added to induce enzyme formation. Media for lactase production from fungi included Soya bean meal 30 g, Milk powder 10 g, KH_2PO_4 1.8 g, Yeast extract 1.8 g, $MgSO_4$ 0.1 g, $FeSO_4$ 5 mg, $MnSO_4$ 5 mg/Liter). Media for Lactase from bacteria utilizes (10 % lactose, 10 % peptone, 10 % yeast extract, 0.5 % $(NH_4)_2SO_4$, 0.1 % NaH_2PO_4, 0.05 % $MgSO_4 7H_2O$, pH7). Lactase is used as digestive enzyme in the case of lactose intolerance (usually due to lactase deficiency) and as the feed additive to increase the nutritive values of some animal feeds. In the dairy industry, lactase is used to breakdown lactose in milk. For example, in Europe, lactose low milk is produced by pasteurization and then keeping with yeast lactase (4 Hrs at 35 °C, which result in 70–80 % hydrolysis of the lactose). The resulting product is then sterilized by the ultra-high temperature process and marketed. Lactases are also used commercially to process whey.

5.4.5 Pectinases

Pectinase is a well known term for commercial enzyme preparation that break down pectin; a polysaccharide substrate, found in the cell wall of plants. Pectinase (E.C.3.2.1.15) constitutes a complex enzymatic system responsible for the degradation of pectic substances (Kittur, *et al.*, 2003). Pectic substances are glycosidic macromolecules with high molecular weight. They form the major components of the middle lamella and primary plant cell wall. The pectinase preparation contains at least six enzymes, splitting pectin at different sites for the molecules. The basic structure of pectin is alpha 1,4 linked galacturonic acid, with up to 95 % of its carboxylic groups esterified with methanol. Pectinase is used primarily to clarify fruit juices and grapes must, for the maceration of vegetables and fruits, and for the extraction of olive oil. Pectin like substance is present at concentration ranging from 0.2–1 % in grapes to 30 % in sugar beet pulp. By treatment with pectinase (1–2 hrs at 54 °C or 6–8 hrs at 18 °C) the yield of fruit juice during processing is considerably increased.

5.4.5.1 Microbial production of Pectinase

A number s of commercial firms produce fungal pectinase using *Aspergillus niger* or *Aspergillus wentii* and one fermentation plant uses *Rhizopus*. Both surface and submerged processes are used. Pectinases have been discovered in some other fungi and also in bacteria, protozoa, insects and higher plants. The fermentation with *Aspergillus niger* runs for 60–80 hours in fed batch culture at pH 3–4 and 37 °C, using 2 % sucrose and 2 % pectin. The purification of the enzyme is simple: the biomass is removed by filtration or centrifugation, stabilizing agents are added, the enzyme is precipitate with organic solvent and the crude protein dried.

5.4.6 Lipases

Lipases (glycerolester hydrolases) spit fat (glycerol ester) into di or monoglycerides and fatty acids. They are usually extracellular enzymes. Isoenzyme with varying pH optima, temperature optima or substrate specificity is frequently produced. The commercial use of lipases has been limited. They are primarily marked for the therapeutic purposes as the digestive enzymes to supplement pancreatic lipases. The enzyme also finds some use in the dairy industry.

Lipases are one of the highly commercialized enzymes; have an important role found and ranked after proteases and amylases. Lipases are widely used in industrial application such as, detergent industry, pharmaceutical industry, pulp and paper industry, production of biodiesel, dairy and bakery foods, fats and oils. Lipases are fond in animal, plant and microorganisms. Besides the stability, selectivity and broad substrate specificity microbial lipases are more promising in terms of availabilities and productivities.

5.4.6.1 Microbial production of Lipase

Among the various sources of lipases, fungi is recognized as the best enzyme producer and also used for industrial application (Sánchez-Pérez *et al.*, 2014). Lipases have been widely studied for their properties and utilization in many industries. Fungi producing lipase include *Aspergillus, Mucor, Rhizopus, Penicillium, Geotrichum,* and the yeast *Torulopsis* and *Candida*. Bacteria producing lipase include *Pseudomona, Achromobacter* and *Staphylococcus*. More than 40 microbial commercial lipases are available (White and White, 1997). Among them *Candida antarctica, Candida rugosa, Penicillium sp., Mucor miehei, Rhizopus oryzae, Aspergillus niger, Thermomyces lanuginosa, Staphylococcus xylosus, Bacillus sp.,* and *Acinetobacter* are the main lipase producing microorganisms.

In most cases, enzyme production must be induced by adding oils and fats. However, there are some cases in which fats have no effects on lipase production and with *Penicillium roqueforti*, they actually repress enzyme production. Glycerol, a product of lipase action represses lipase formation. This is also true for glucose, so that lipase production of *Geotricum candidum* is first induced at the end of log phase (around 50 hrs). Lipases are generally bound to the cells and hence inhibit an overproduction, but addition of a cation, such as magnesium ions liberates the lipase and leads to a higher enzyme titer in this production process. Since free fatty acids determine the odor and taste of the cheese, and the cheese ripening process is effected by lipase, microbial affects during the aging process can be due to lipase action. For instance, in the production of Roquefort cheese (blue cheese) spores of *Penicillium roqueforte* are added and sort of lipase preparation. In the soap industry, the lipase from *Candida cylinderaceae*, is used to hydrolyze oil. A potentially important future use of lipases is for the synthesis of esters from acid and alcohol in non aqueous media. Another potential use is improvement of fats quality by the exchange of one fatty acid by another.

5.4.7 Penicillin acylase

Many derivative of penicillin used in medicine are produced semi-synthetically by chemical modification of the basic penicillin ring structure that is, 6 aminopenicillanic acid (6 APA). 6APA is commercially produced by the splitting off of the acyl side chain of microbially produced penicilllin G with the enzyme peniciilin acylase (also called penicillin amidase or benzyl penicillin amydohydrolase). *Penicillin acylase* thus have commercial use in the production of semisynthetic penicillins.

5.4.7.1 Microbial production of Penicillin acylase

Penicillin acylases are produced by yeast, fungi and bacteria. They can be subdivided into two types. Type I acylase also called the fungal type, phenoxymethyl penicillin (penicillin V) but attack benzyl penicillin (Penicillin G) much less efficiently. Known producers of these acylases are *Penicillium sp.*, *Penicilliun chrysogenum*, *Aspergillus ochraceus*, *Trichophyton mentagrophytes*, *Apidermophyto floccosum*, *Cephalosporium* sp and *Flusarium semitactum*. Aside from fungi the bacterium *Strptomyces lavendulae* also produce this type of acylase. The Pencillin V acylase produced by yeast *Rhodotorula aurantiaca* (NCIM 3425) and *R glutinis (Kumar et al., 2000)*. In addition of *Penicillin* V *Penicillin* K (heptyl Penicillin), dihydro penicillin F (pentyl Penicillin), and a number of synthetic penicillins are split. These acylase, which are usually extracellular, have a pH optimum of 10 and a temperature optimum of 50 °C.

Type II acylase also designated as bacterial type are produced by Aerobacter, Alcaligenes, borditella, Cellulomonas, Corynebacteruium, Erwinia, Escherichia, Flavobacterium, micrococcus, Nocardia, Proteus, Pseudomonas, Salmonella, Sarcinia, Xanthomonas. The temperature optimum 40 °C and pH optimum is 8, which is lower than of the type I acylase. Phenyl acetic acid act as a competitive inhibitor and 6 APA as a non-competitive inhibitor.

The extracellular and intracellular acylase of various microorganisms differ in substrate specificities. The substrate specificity is determined by the nature of the acyl moiety and not by the 6-APA.

Penicillin acylase from *E. coli* are used almost exclusively for 6-APA production. The production strains are mutants of *E. coli* ATCC 11105 and *E. coli* ATCC 9637. Intracellular acylase production is induced in wild type strains by addition of phenylacetic acid. Glucose represses enzyme production and must therefore be present in the culture medium only in low concentrations. The production rate is also strongly affected by the O_2 partial pressure. Although *E. coli* an grow under anaerobic condition, enzyme yields are low; however enzyme production is also suppressed by high aeration rates.

A suitable nutrient medium consist of 2 % corn steep water, at pH 7.0. The inoculum consists of 0.25 % of an 18 hrs pre-culture. To induce enzyme formation, after 8 hrs of fermentation at 24 °C, 0.1 % of sterile ammonium phenyl acetate solution is added at hourly intervals for the next 13 hrs. The cells are concentrated 20 folds on a separator and the enzyme is then released by use of a pressure disintegrator (500 Kg/cm^2 pressure). The crude enzyme preparation so obtained is subsequently purified by conventional biochemical procedure.

The enzyme yield in commercial production has been substantially improved by optimizing the culture medium and by use of classical genetic techniques. Recombinant DNA techniques promise to increase the yields even more. For instant by cloning the penicillin acylase gene on a multicopy plasmid, strains have been obtain having a 28 folds increase in enzyme formation rate is non-induced fermentation and a further 6 fold increase in induced fermentation.

5.4.8 Catalases

Catalases are metalloenzyme which catalyzes the degradation of hydrogen peroxide (H_2O_2) into dioxygen and water (Shin *et al.*, 2008). This enzyme was isolated from broad range of prokaryotic and eukaryotic micro-organisms (Daniela *et al.*, 2003).

The main uses of catalase include decomposition of H_2O_2 added to process streams to act as an antimicrobial agent. It is very important to remove the impurities of cotton fabric before dyeing. Refining can remove wax, pection, and various pigments. Bleaching can remove residual impurities and cottonseed hulls. After bleaching, the residual hydrogen peroxide should be effectively removed. Otherwise, it will lead to color difference between batches and uneven dyeing. However, the traditional method requires a lot of water.

The dyeing and printing industry have many advantages of using catalase. Maintain the consistency of color between each batch; Uniform dyeing effect; Reduce water consumption and energy consumption; Environment friendly; Shorten the process time and increase production.

5.4.8.1 Microbial production of catalases

The production of catalase by bacteria has been well known since the discovery of this enzyme, and the early recognition of its almost universal occurrence in living cells and tissues. Solid state fermentation (SSF) is one of the methods for producing catalase enzyme in submerged fermentation, because of its simple low cost investment and also for maximum yield. (Zubeyde *et al.*, 2003). Solid-state (substrate) fermentation (SSF) has been defined as the fermentation process occurring in the absence or near-absence of free water.

Catalase is produced by *Staphylococci* and *Micrococci* and identified as catalase-positive. Other catalase-positive organisms include *Listeria*, *Corynebacterium diphtheriae*, *Burkholderia cepacia*, *Nocardia*, the family *Enterobacteriaceae* (*Citrobacter*, *E. coli*, *Enterobacter*, *Klebsiella*, *Shigella*, *Yersinia*, *Proteus*, *Salmonella*, *Serratia*, *Pseudomonas*), *Mycobacterium tuberculosis*, *Aspergillus*, and *Cryptococcus*. The mixture produces bubbles or froth, the organism is said to be 'catalase-positive', otherwise the organism is 'catalase-negative'. *Streptococcus* and *Enterococcus spp.* are catalase-negative. Catalase-positive bacteria include strict aerobes as well as facultative anaerobes. They all have the ability to respire using oxygen as a terminal electron acceptor. Catalase-negative bacteria may be anaerobes, or they may be facultative anaerobes that only ferment and do not respire using oxygen as a terminal electron acceptor. Catalase removes hydrogen peroxide in the following reaction

$$(2\,H_2O_{2^-} \rightarrow 2\,H_2O + O_2).$$

Commercial sources of catalases are mammalian liver and *Aspergillus niger*, used for sterilization or bleaching. 10 gm of agro wastes were suspended in 7.5 ml of minimal media in a 250 ml flask and autoclaved at 15 lbs pressure, 121 °C for 20 minutes. It was cooled before using. The culture media consists of malt extract agar (Oxoid) at 4±6 °C on a basal medium (BM) containing: glucose, 8.0 %; peptone, 0.3 %; $NaNO_3$, 0.05 %; KH_2PO_4, 0.1 %; $CaCO_3$, 3.5 % that must be sub-cultured every month. The aerated (at a rate of 2 l air min^{-1}) and stirred (300 rev min^{-1}) fermentor cultures were run at 30 °C for 72 h; occasionally, antifoam A emulsion (Sigma) was added to break the foam. The pH was kept at the level of 6.0±6.5 by the initial addition of 35.0 g l^{-1} $CaCO_3$.

5.4.9 Alcohol Dehydrogenase

Alcohol dehydrogenases belong to a class of oxidoreductases and catalyze the NAD(P) H-dependent reduction of a variety of endogenous and xenobiotic carbonyl compounds (Reid *et al.*, 1994). There is considerable interest in the use of alcohol dehydrogenases for the production of chiral alcohols in the pharmaceutical and fine chemicals industries (Panke *et al.*, 2004; Schmid, *et al.*, 2001; Schoemaker *et al.*, 2003). For the inherent advantages over chemocatalysts in terms of their high chemo-, enantio-, and region-selectivity, stereospecific alcohol dehydrogenases are very interesting from both scientific and industrial perspectives. Alcohol dehydrogenases for asymmetric reduction are ubiquitous in nature and have been characterized from diverse sources, including bacteria (Itoh *et al.*, 1999), yeasts (Katz *et al.*, 2003), plants (Melis *et al.*, 1999), and tissues from several mammalian species (Wsól *et al.*, 1999). They display different physical and enzymatic properties, show a wide variety of substrate specificities, and are mainly classified into three superfamilies, zinc-dependent alcohol dehydrogenases, short-chain dehydrogenase/reductases (SDRs), and aldo-keto reductases, based on their catalytic properties and sequence information. Among the various enzyme sources, *Candida species* are attractive as highly stereospecific oxidoreductase donors (Engelking *et al.*, 2004; Kataoka *et al.*, 2006; Kataoka *et al.*, 2004). In many cases, the stereospecific oxidoreductases from strains of the genus Candida are dissimilar in structure and classified in different superfamilies (Kataoka *et al.*, 2006; Kataoka *et al.*, 2004).

Alcohol dehydrogenase is useful in biotransformation, and for the synthesis of enantiomerically pure stereoisomers of chiral alcohols. In fuel cells, alcohol dehydrogenases can be used to catalyze the breakdown of fuel for an ethanol fuel cell. Scientists at Saint Louis University have used carbon-supported alcohol dehydrogenase with poly(methylene green) as an anode, with a nafion membrane, to achieve about 50 $\mu A/cm^2$ (Moore *et al.*, 2005). Enzymes that have activity using butanol and ethanol as substrates. sulphide-free leather processing, Eco-friendly waste management strategies for greener environment towards sustainable development in leather industry.

5.4.9.1 Microbial production of Alcohol dehydrogenase

For the synthesis of chiral alcohols, only lab-scale syntheses with commercially available alcohol dehydrogenases have been described yet. Bacterial species useful for enzyme Alcohol dehydrogenase production are *Zymomonas mobilis, Clostridium acetobutylicum. Clostridium kluyveri, Citrobacter freundii* and *Klebsiella pneumoniae Bacillus methanolicus, E. coli, Thermoanaerobium brockii, Lactobacillus* and *Salmonella typhimurium*, while fungal species are *Saccharomyces cerevisiae, Pleurotus ostreatus, Tricholoma matsutake* and *Agaricus blazei, Flammulina velutipes,. Saccharomyces cerevisiae* is the most popular industrial microorganism because it utilises cheap substrates for growth and production using batch growth (Submerged fermentation) at 30 °C for 24 h with medium $FeSO_4 \cdot 7H_2O$ 0.012 g dm^{-3}, $(NH_4)_{2-}$ SO_4 7.5 g dm^{-3}, $CaCl_2$ 0.03 g dm^{-3}, $CuSO_4 \cdot 5H_2O$ 0.001 g dm^{-3}, KH_2PO_4 1.5 g dm^{-3}, $ZnSO_4 \cdot 7H_2O$ 0.006 g dm^{-3} and $MgSO_4 \cdot 7H_2O$ 0.691 g dm^{-3} in distilled water. As a carbon source, different initial concentrations of glucose (c_{GO} = 5, 10, 30 and 50 g dm^{-3}) were used (Thatipamala *et al.*, 1992).

5.4.10 Glucose oxidase

Glucose oxidase belongs to oxidoreductase and is also called as glucose aerodehydrogenase. The fungal glucose oxidase is generally glycosylated flavoprotein with molecular weight about 155.000, consisting two identical subunits of 80 kDa linked by disulfide bonds. Glucose oxidase (β-D-glucose: oxygen-1-oxidoreductase) oxidize β-D-glucose ($C_6H_{12}O_6$) to gluconolactone ($C_6H_{10}O_6$) and reduces FAD to $FADH_2$, releasing H_2O_2 in the presence of O_2· (Witteveen *et al.*, 1999). The gluconolactone then hydrolyzes spontaneously to gluconic acid ($C_6H_{12}O_7$).

Cell growth:

$C_6H_{12}O_6 + \text{biomass} + O_2 \rightarrow \text{biomass}$

Glucose oxidation:

$C_6H_{12}O_6 + O_2 \rightarrow C_6H_{10}O_6 + H_2O_2$ (enzyme glucose oxidase)

Gluconolactone hydrolysis:

$C_6H_{10}O_6 + H_2O \rightarrow C_6H_{12}O_7$ (GA)

H_2O_2 decomposition:

$H_2O_2 \text{ catalysis} \rightarrow 2H_2O + O_2$

Optimum pH for activity of the enzyme is 5.5 with a broad range from 4 to 7 (Bentley, 1959) with a special application in biotechnology and industries. The glucose oxidase enzyme has considerable industrial importance and used in different food products (dried egg, bread, beverages and vegetables) for processing and preservation.

5.4.10.1 Microbial production of Glucose oxidase

Glucose oxidase was first isolated from mycelia of *A. niger* and *Penicillium glaucum* by Muller (1928) and later they were detected in different sources (insects, honey, algae and micro fungi) and recently obtained from *Penicillum amagasakiense*. The most common sources of glucose oxidase are *A. niger*, *P. notatum*, *P. glaucum*, *P. amagasakiense*, *P. purpurogenum*, *P. variabile* and *Alternaria alternate* (Caridis *et al.*, 1991). Glucose and proteose peptone are utilized as carbon and nitrogen source for maximum production of glucose oxidase by fungus. Glucose oxidase production generally carried out by liquid state fermentation using a medium containing glucose, urea, $CaCO_3$, KH_2PO_4 and CSL for *A. Niger*, for *P. notatum*, the medium contained KH_2PO_4, NH_2NO_3, $(NH_4)_2SO_4$, $MgSO_4 \cdot 7H_2O$, $Na_3C_6H_5O_7$, peptone, yeast extract and glucose (Sabir *et al.*, 2007; Rasul *et al.*, 2011). An inoculum size of 5 % was added aseptically to Erlenmeyer flasks and incubated at orbital shaker (120 rpm) for 72 h at 30 °C (Zia *et al.*, 2010). The pH 5.0–6.0 and the temperature between 30 and 35 °C were found to be optimum for glucose oxidase production. The crude enzyme isolated from fermented biomass using filtration followed by centrifugation at 7826 x g at 0 °C for 15 minutes. The activity of glucose oxidase could be checked by o-dianisidine (Blaedel & Uhl, 1975). Addition of di-ammonium hydrogen phosphate, potassium di-hydrogen phosphate and magnesium sulphate supports the glucose oxidase production. Glucose oxidase production could be increased with presence of calcium carbonate. Calcium lactate is effective for stabilizing glucose oxidase activity. In industry, it is used to produce gluconic acid and as a source of hydrogen peroxide in food preservation. In diagnostics, it is the basis of sensors for the estimation of glucose concentration in blood, serum or plasma.

5.4.11 Galactose oxidase

Galactose oxidase was discovered in the extracellular culture medium of the fungus *Polyporus circinatus* by Cooper *et al.*, (1959) however later discoveries indicated that it was actually *Dactylium dendroides* (Nobles *et al.*, 1963). Galactose oxidase (D-galactose:oxygen 6-oxidoreductase) is a member of the free radical copper oxidase family, and contains a novel metalloradical complex (Whittaker, 2005) consisting of a copper atom and a tyrosine residue covalently attached to the sulphur of a cystein. This enzyme is secreted by filamentous fungi such as *Dactylium dendroides* (Markus *et al.*, 1965), *Gibberella fujikuroi* (Aisaka and Terada, 1981) and *Fusarium spp.* (Barbosa-Tessmann *et al.*, 2001). GAO catalyses the oxidation of primary alcohols to the corresponding aldehydes, while concomitantly reducing oxygen to hydrogen peroxide in its catalytic reaction (Whittaker & Whittaker, 2000, 2001).

$$\text{D-galactose} + O_2 \rightleftharpoons \text{D-galacto-hexodialdose} + H_2O_2$$

5.4.11.1 Microbial production of galactose oxidase

The enzyme is secreted by a number of fungal species, of which Fusarium graminearum (formerly classified as Dactylium dendroides), Aspergillus oryzae and Fusarium venenatum, Fusarium sambucinum, Fusarium acuminatum.

Efficient amount of galactose oxidase could be produced by submerged fermentation when glucose (1.0 % w/v) used as the sole carbon source at 25 °C, with orbital shaking (100 rpm) and an initial medium of pH 7.0, for 96 h, using a 2 % (v/v) inoculum made from a homogenized four-day-old liquid culture, in the presence of copper, manganese, and magnesium. The enzyme purification is a one-step process of affinity chromatography, with a recovery of 42 % of the initial activity. Optimal pH and temperature for the enzyme activity were 8.0 and 30 degrees C, respectively. The enzyme was thermo inactivated at temperatures above 60 °C. The purified enzyme was active toward various substrates, including galactose, dihydroxyacetone, guar gum, lactose, melibiose, methyl-galactopyranoside, and raffinose.

5.4.12 Hexokinase

The first step in metabolism of glucose (Glc) is usually phosphorylation, catalyzed by hexokinase (Fig. 5.1).

It involves nucleophilic attack of the C6 hydroxyl oxygen of glucose on the phosphorous of the terminal phosphate of ATP. ATP binds to the enzyme as a complex with $\mathbf{Mg^{++}}$.

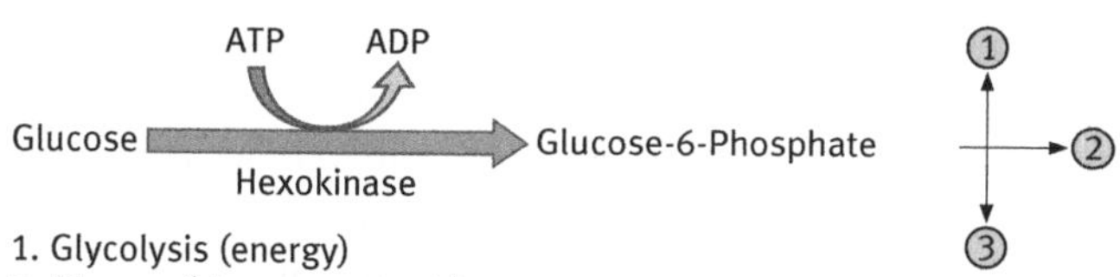

1. Glycolysis (energy)
2. Storage (glycogen, starch)
3. Pentose phosphate pathway (NADPH, anabolism)

Fig. 5.1: First step in metabolism of glucose: phosphorylation.

5.4.13 Muramidase

Bacterial source can provide four types of enzymes that able to hydrolyze peptidoglycan from cell wall i.e. muramidases (lysozyme-like enzymes), glucosaminidases, amidases and peptidases. Muramidase is an enzyme that attack and damage the cell wall by generating osmotic pressure within them to rise with resultant bursting of the cell. The muramidase enzymes were not detectable in cultures grown in the absence of bacterial cells, and were apparently repressed when glucose was present in the growth medium. The low pH enzyme was purified and shown to be a muramidase with a molecular weight of about 22 000. Muramidase has wide application like treatment of virus, in treatment of bacterial infection, and mucolytic activity. It is also present in tears so considered as synonym of lysozme. Cellosyl is a bacterial muramidase from *Streptomyces coelicolor*. Similar to other lysozymes, the enzyme cleaves the beta-1,4-glycosidic bond between N-acetylmuramic acid and N-acetylglucosamine units, but it also exhibits a beta-1,4-N,6-O-diacetylmuramidase activity. *Lactococcus lactis* muramidase AcmA can hydrolyze the cell wall of *Streptococcus thermophilus*, without affecting viability. This intercellular activity of the lactococcal muramidase results in chain disruption of streptococci in vivo. Lysozyme, muramidase, and aprotinin, a protease inhibitor, both exhibit antimicrobial activities against different microorganisms, were chosen as model proteins to develop more potent bactericidal agents with broader antimicrobial specificity. Muramidase activity is higher in axenic culture.

5.4.13.1 Microbial production of muramidase

Muramidase was first crystallized form *Micrococcus lysodeikticus* and characterised by Herbert and Pinsent (1948). Muramidase is produced by microbes *Gliomastix murorum* (fungi), *Aspergillus bisporus* (fungi), *Schizophyllum commune* (mushroom), *Chalaropsis sp.* (fungi), *Agaricus bisporus* (cultivated mushroom), *Aspergillus flavus* (fungi), and *Micrococcus lysodeikticus* (Gram positive bacteria). Muramidases (pH optimum of 3.4) was produced extracellularly by fungus *Gliomastix murorum* on media enriched with heat-killed cells of *Bacillus subtilis* at pH5, but not at pH7 (Rhodes & Grant, 1990). A second, much weaker, wall degrading enzyme activity with a broad pH optimum of 6–9 was produced under both growth conditions. *Aspergillus bisporus* muramidase activity showed at optimum pH 6.77, 6.07 and 4.01. Muramidase from *Schizophyllum commune* was inducible by bacterial cells or cell components and repressed by glucose (Grant *et al.*, 1990). *Chalaropsis* muramidase is a major extracellular enzyme having activity during growth of fungus at low pH.

The optimum temperature as well as pH for the muramidase activity were found to lie at 50 °C and pH 5 respectively (Hayashi *et al.*, 1964). Muramidase is used in production of many cheeses to prevent “late blowing” at high pH, brined cheese type.

5.4.14 Cholesterol oxidase

Cholesterol oxidase is a bacterial FAD-containing flavooxidase that catalyzes the first reaction in cholesterol catabolism. Cholesterol oxidase is an oxidoreductase enzyme that utilizes cholesterol and O_2, to produce cholest-4-en-3-one and H_2O_2. They act on the CH-OH group of donor with oxygen as acceptor. Other names of this enzyme are cholesterol-O_2 oxidoreductase, 3beta-hydroxy steroid oxidoreductase, and 3beta-hydroxysteroid:oxygen oxidoreductase. Cholesterol oxidases catalyse bile acid biosynthesis. The various substrates of this enzyme are Cholesterol, 7α- or 7β-OH Cholesterol, 7-Ketocholesterol, Ergosterol, β-Sitosterol, Pregnenolone, Cholestanol. Cholesterol oxidase has been used in biocatalysis for the production of a number of steroids, as an insecticidal protein against boll weevil larvae and, in particular, as a diagnostic enzyme for determining serum levels of cholesterol. These applications prompted various laboratories worldwide to isolate this flavooxidase from different sources and to improve its properties by protein engineering (Pollegioni *et al.*, 2009).

5.4.14.1 Microbial production of Cholesterol oxidase

Several microorganisms of diverse environments have ability to produce cholesterol oxidase. In 1944, Turfitt was the first to isolate the enzyme from the microbe *Rhodococcus erythropolis*. Other microorganisms are *Arthrobacter, Rhodococcus equi, Nocardia erythropolis, Mycobacterium* and *Nocardia rhodochrous, Arthrobacter rhodochrous, Corynebacterium cholesterolicum, Mycobacterium spp., Nocardia rhodochrous, Nocardia rhodochrous* produce an intracellular/membrane cholesterol oxidase, while *Schizophyllum commune, Pseudomonas spp., Rhodococcus equi, Arthrobacter simplex, Brevibacterium sterolicum, Bacillus spp., Pseudomonas, Schizophyllum commune, Brevibacterium sterolicum, Streptoverticillum cholesterolicum, Micrococcus spp* and *Streptomyces violascens* produce an extracellular enzyme. Some microbes *Streptomyces violascens, Streptomyces spp., Streptomyces griseocarneus, Actinomyces lavendulae,* and *Nocardia erythropolis* produces both extra and intra cellular enzyme. γ-*Proteobacterium* Y-134 is also able to produce this enzyme. Cholesterol oxidase enzyme requires the prosthetic group FAD for their activity (both as a covalently or noncovalently bound cofactor). An exception to this could be the cholesterol oxidase isolated from γ-*Proteobacterium* Y-134, which was reported to contain covalently bound FMN as a cofactor (Isobe *et al.*, 2003). This enzyme also differs in substrate specificity (it is not active on pregnenolone and ergosterol) compared with other known cholesterol oxidase.

It was found that cholesterol oxidase production was affected by the amount of oleic acid in the culture broth, and that it was the level of insoluble oleic acid, rather than the soluble component adsorbed onto the cell, that was directly influencing the formation of cholesterol oxidase. In addition, an increase in the rate of agitation correlated with the extent of cholesterol oxidase production. Bacteria produces enzyme at 35 °C with optimum pH of 7.0. The bacterial cholesterol oxidase had a molecular mass of 55 kDa (Lashkarian *et al.*, 2010). The extraction and purification procedures are generally required in order to obtain a sufficient yield of enzyme from cholesterol oxidase producing bacteria to improve yield and/or clarification. The techniques adopted for extraction are largely dependent on the nature of the secretion of the enzyme, i.e. whether intracellular or extracellular. The intracellular cholesterol oxidase is believed to be an essential membrane-bound enzyme located on the outside of the cell. The enzyme may be extracted from bacterial cells by treatment with Triton X-100 or trypsin, or to a lesser extent with phosphate buffer, but not by mechanical cell disruption or by changes in pH or ionic strength of the buffer (Cheetham, 1979).

The purification of cholesterol oxidase involved differential ammonium sulphate precipitation, ion-exchange and gel filtration chromatography. Extra-cellular cholesterol oxidase was purified to 21-fold, by affinity-like chromatography on the Kieselguhr-cholesterol column (Doukyu & Aono, 1998). Impurities in the applied supernatant were eluted from the column throughout the sample application by washing with the phosphate buffer. The enzyme was retained on the Kieselguhr-cholesterol and was subsequently eluted with buffer containing Triton X-100.

cholesterol oxidase from *Streptomyces sp.* produces insecticidal proteins; from *Rhodococcus equi* membrane damage in host, opportunistic infection in severally immune-suppressed humans (HIV infection) required cholesterol oxidase for their infection in host, *Nocardia* Serum cholesterol measurement, Bacterial Cells surface cholesterol can be estimated with the help of cholesterol oxidase, cholesterol oxidase-encoding gene (pimE), cholesterol oxidase act as a signaling protein for the biosynthesis of polyene macrolide pimaricin which produce antifungal antibiotic used in food industry; biosynthesis of polyene macrolid pimracin, *Streptomyces natalensis* Alters lipid deposition in smooth muscle cells, *Burkholeria* and *Chromobacterium* sp. Thermal, detergent and organic solvents tolerance, *Rhodochorus erythropolis* Biocatalysis, oxidations of cyclic allylic, bicyclic and trycyclic alcohol to synthesize several ergot alkaloids, *Streptomyces virginiae* Cloned cholesterol oxidase choL, fragment from *Streptomyces virginiae* is used in the oxidation of diosgein to 4-ene-3-keto steroids.

5.4.15 Asparaginase

L-asparaginase is catalysed the hydrolysis of L-asparagine into L-aspartatic acid and ammonia. As L-asparagine is an important factor of tumor cell development so Asparaginase is used as anti tumor/cancer agent in treatment of leukemia and lymphomas as the malignant cell can be destroyed by this enzyme.

5.4.15.1 Microbial production of Asparaginase

Microorganism producing L asparaginase are bacteria: Acinetobacter calcoaceticus, Bacillus sp., B. mesentericus, B. polymyxa, Citrobacter sp., Corynebacterium glutamicum, Escherichia coli, Enterobacter aerogenes, Enterobacter colaceae, Erwinia aroideae, Erwinia carotovora, Erwinia chrysanthemi, Helicobater pylori, Klebsiella pneumoniae, Mycobacterium pheli, Pseudomonas ovalis, Serattia marcescens, Staphyllococcus sp., S aureus, Streptococcus albus, Tetrahymena pyriformis, Thermus thermophiles, Thermus aquaticus, Vibrio succenogenes, Yeast: Candida utilis, C guilliermondii, Pichia polymorpha, Sachharomyces cerevisiae, Actinomycetes sp., Streptomyces karnatakensis, S venezuelae, S collinus, Theroactinomyces vulgaris, Fungi: Aspergillus nidulans, Aspergillus terreus, Cylidrocapron obtusisporum, Mucor sp., Algae: Chlamydomonas sps., Its commercial production is through mutants of Escherichia coli ATCC 9637, Serratia marcescens ATCC 60 and Erwinia carotovora are used. Its media contain 3 % corn steep liquor, 0.6 % sodium acetate and 0.2 % ammonium sulphate at pH 7.0–8.0, at 28–37 °C, used in reactors up to 80 m^3 in volume. Oxygen supply is critical during the 16–40 hrs fermentation. At the end of fermentation process the cells are broken open and the enzyme is crystallized.

5.4.16 Streptokinase

The extracellular enzyme streptokinase (EC 3.4.99.22) is produced by various strains of h-hemolytic streptococci. Fibrinolytic action indirectly by activating the circulatory plasminogen. The complete amino acid sequence of streptokinase was first established by Jackson and Tang (1982). Streptokinase has a molar mass of 47 kDa and is made up of 414 amino acid residues (Malke and Ferretti, 1984). The protein exhibits its maximum activity at a pH of approximately 7.5 and its isoelectric pH is 4.7 (De Renzo *et al.*, 1967; Taylor & Botts, 1968; Brockway & Castellino, 1974). The enzyme streptokinase producing streptococci were first identified in 1874 by Billroth in exudates of infected wounds (Billroth, 1874; Jones, 1978). Later, the blood of scarlet fever patients was shown to contain similar microorganisms. By 1919, *Streptococcus* sp. had been classified into alpha, beta and gamma variants based on the distinct types of hemolytic reactions the variants produced on blood agar plates (Schottmüller, 1903; Brown, 1919; Jones, 1978). The activation of plasminogen by streptokinase is species specific (McCoy et al., 1991). It is formed by several structural domains [α, β and γ] with different functional properties. Streptokinase is known to activate plasminogen both by fibrin-dependent and fibrin independent mechanisms (Reed *et al.*, 1999).

5.4.16.1 Microbial production of streptokinase

Most of the streptokinases are obtained from h-hemolytic streptococci of the Lancefield groups A, C and G. The group C strain *Streptococcus equisimilis* H46A (ATCC 12449) isolated from a human source in 1945 has been widely used for producing streptokinase.

Group A hemolytic streptococci commonly require complex and rich media supplemented with various nutritional factors for growth (Bernheimer *et al.*, 1942). A modification of the medium originally developed by Bernheimer *et al.*, (1942) was used by Christensen (1945) for producing streptokinase from *S. equisimilis* H46A. This medium contained peptone, phosphate salts, glucose, biotin, riboflavin, tryptophan, glutamine and nucleic acids (thiamine, adenine and uracil). The glutamine content was only 25 % of that in the original medium of Bernheimer *et al.*, (1942). A low glucose concentration at inoculation allowed growth to become well established without excessive generation of acid. Alkali (5 M sodium hydroxide) was added at regular intervals to maintain a neutral pH. In the absence of pH control, both the biomass growth and streptokinase production were adversely affected. Batch cultures of streptococci have been characterized as having a rather extended lag phase followed by a relatively short period of exponential growth.

Several schemes have been described for recovery and purification of streptokinase either from the commercially available crude preparations or the fermentation broths of various bacteria purified streptokinase from a relatively crude commercial preparation. Methods included column chromatography on DEAE-cellulose was followed by column electrophoresis in sucrose density gradients to obtain a five- to six fold increase in purity. Repeated chromatography was necessary to remove the last detectable traces of impurities. A combination of ion exchange (DEAE-Sephadex A-50) and gel permeation (Sephadex G-100) chromatography was used to purify streptokinase. Streptokinase was fractionated either by hydroxyapatite chromatography or ammonium sulfate fractionation. Precipitation with 40 –50 % ammonium sulphate resulted in a two- to threefold increase in specific activity of streprtokinase. The precipitate was recovered by centrifugation and dialyzed against 0.09 M sodium chloride. The dialyzed solution was further purified by gradient elution from a DEAE-cellulose chromatography column to concentrate 10-fold by ultrafiltration.

The earliest chromatographic like methods used insolubilized di-isopropyl fluorophosphates (DIP) plasmin as the affinity ligand (Castellino *et al.*, 1976). This ligand was produced by the conversion of plasminogen to plasmin with urokinase and inhibition of the proteolytic activity of plasmin by DIP (Castellino *et al.*, 1976). Affinity column method can decrease streptokinase activity by 30 %. The various safety considerations relevant to production of biopharmaceutical proteins have been recomonded (Chisti, 1998).

Various enzymes other than those described above (Jegannathan & Nielsen *et al.*, 2013) have a significant place in the list of microbial enzymes, which have established their applications in bio-industries. Economic importance of desired enzyme is regularly sought by industries to develop better and cheaper process/methods for enzyme production (Tab. 5.2).

Tab. 5.2: Showing enzymes and their application, comparative environmental assessments of enzymatic and conventional processes in various industries.

Industry	Application process	Enzyme	Function of enzyme	Savings obtained by enzyme use	Scale of production	Reference, publication type
Pulp and paper	Thermo-mechanical pulping	Cellulase	Softens wood chips	Energy	Full-scale	(1), peer reviewed journal paper
					Full-scale	(2), report
	Deinking	Cellulase	Acts on recycled fibers and facilitates ink loosening	Chemicals	Full-scale	(1), peer reviewed journal paper
	Bleaching	Laccase	Oxidizes lignin and enhances lignin removal	Bleaching chemicals, energy	Full-scale	(3), peer reviewed journal paper
		Xylanase	Hydrolyzes xylan and enhances lignin extraction	Bleaching chemicals	Lab-scale	(1), peer reviewed journal paper
	Pitch control	Lipase	Hydrolyzes pitch	Cleaning agent, talc, energy	Full-scale	(1), peer reviewed journal paper
	Stickies control	Esterase	Hydrolyzes glue and controls stickies	Talc, solvent, energy	Full-scale	(1), peer reviewed journal paper
Leather	Beam house	Protease, lipase	Facilitates hair and fat removal from hides	Chemicals, energy	Full-scale	(4), technical journal paper[c]
Textile	Scouring	Pectate lyase	Degrades pectin and assist in removal of wax etc. from raw cotton	Energy, water, chemicals, cotton	Full-scale	(5), chapter in a book[c]

Tab. 5.2: (continued)

Industry	Application process	Enzyme	Function of enzyme	Savings obtained by enzyme use	Scale of production	Reference, publication type
	Bleaching	Catalase	Hydrolyzes hydrogen peroxide to oxygen and water	Heat, electricity, water	Full-scale	(5), chapter in a book[c]
	Bleaching	Arylesterase	Perhydrolyzes to form peracetic acid (bleaching activator)	Chemical, cotton, energy	Full-scale	(6), conference poster[c]
Detergent	Laundry washing	Protease, lipase, amylase, cellulase	Removes stains from laundry	Chemicals, energy	Model detergent	(7), technical journal paper[c]
Food and beverage	Degumming of soybean oil	Phospolipase	Hydrolyzes phospholipids	Oil, chemicals	Full-scale	(8), technical journal paper[c]
	Hard stock production	Lipase	Interesterification of vegetable oil	Chemicals, energy	Lab/full-scale	(9), peer reviewed journal paper
	White bread production	Amylase	Degrades starch and delays hardening of bread	Bread	Full-scale	(10), technical journal paper[c]
	Steamed bread production	Amylase, lipase				(11), report[c]
	Mozzarella cheese production	Phospholipase	Hydrolyzes phospholipids in milk	Milk	Full-scale	(12), peer reviewed journal paper[c]
	100 % Barley beer production	Amylase, protease etc.	Converts starch to fermentable sugars	Barley, energy	Full-scale	(13), technical journal paper[c]
		Amylase, protease		Energy		(14), conf. presentation

Tab. 5.2: (continued)

Industry	Application process	Enzyme	Function of enzyme	Savings obtained by enzyme use	Scale of production	Reference, publication type
Animal feed	Pig feed production	Xylanase	Depolymerizes xylans and enables better digestion	Feed	Full-scale	(15), peer reviewed journal paper
	Pig feed production	Phytase	Hydrolyzes phytate and releases phosphorus bound in feed	Inorganic phosphorus	Full-scale	(16), peer reviewed journal paper
	Poultry feed production	Phytase			Full-scale	(17), technical journal paper
	Poultry feed production	Protease	Hydrolyzes protein in the feed	Feed protein	Full-scale	(18), peer reviewed journal paper[c]
Fine chemicals	Amino-butanoic acid production	Lipase	Aminolysis in (*S*)-3-aminobutanoic acid production	Chemicals, waste	Lab-scale	(19), peer reviewed journal paper
	α-Naphthol production	Toluene ortho-monooxygenase	Oxidizes naphthalene to alpha-naphthol	Chemicals, waste	Lab-scale	(20), conference paper
Pharmaceuticals	(*S*)-2-3-Dihydro-1H-indole-2-carboxlic acid production	Phenylalanine ammonia lyase	Formation of C–N bond	Chemicals, energy	Lab-scale	(21), peer reviewed journal paper
	γ-Amino-butyric acid production	Lipolase	Resolution of cyanodiester	Chemicals, energy, waste	Full-scale	(22), peer reviewed journal paper
	6-Aminopenicillanic acid production	Penicillin amidase	Deacylates penicillin molecule	Chemicals, energy	Lab scale	(23), peer reviewed journal paper

Tab. 5.2: (continued)

Industry	Application process	Enzyme	Function of enzyme	Savings obtained by enzyme use	Scale of production	Reference, publication type
	7-Amino-penicillanic acid production	D-Amino acid oxidase, glutaryl 7-ACA acylase	Oxidizes cephalosporin C salt and deacylates glutaryl 7-aminocephalosporic acid	Chemicals, energy	Full-scale	(24), peer reviewed journal paper
Cosmetics	Oleo chemical ester production	Lipase	Transesterification of vegetable oil	Chemicals, energy, raw material	Full-scale	(25), technical journal paper[c]
Biodiesel	Methyl ester production	Lipase	Catalyzes the reaction of triglyceride and menthol to form methyl ester	Energy, chemicals, raw material	Lab-scale	(26), peer reviewed journal paper
					Lab-scale	(27), peer reviewed journal paper

(1) Skals *et al.*, 2008, (2) Kallioinen *et al.*, 2003, (3) Fu *et al.*, 2005, (4) Nielsen, 2006, (5) Nielsen & Høier, 2009, (6) Dettore, 2011, (7) Nielsen & Skagerlind, 2007, (8) Cowan *et al.*, 2008, (9) Holm & Cowan, 2008, (10) Oxenbøll and Ernst, 2008, (11) Oxenbøll *et al.*, 2010, (12) Nielsen & Høier, 2009, (13) Kløverpris & Spillane, 2010, (14) Yon-Miaw, 2011, (15) Nielsen *et al.*, 2008, (16) Nielsen and Wenzel, 2006, (17) Nagaraju & Nielsen, 2011, (18) Oxenbøll *et al.*, 2011, (19) Weiß *et al.*, 2010, (20) Osborne-Lee *et al.*, 2008, (21) Poechlauer *et al.*, 2010, (22) Dunn, 2011, (23) Biwer & Heinzle, 2004, (24) Henderson *et al.*, 2008, (25) Thum and Oxenbøll, 2008, (26) Harding *et al.*, 2007, (27) Jegannathan *et al.*, 2011.

5.5 Future recommendation

The development in molecular biology, enzyme immobilization, downstream processing, microbial genetics and machinery has helped us tools to enhance the enzyme production. New strains of microbes can be developed by genetic engineering which can produce multiple folds of enzyme than the wild strains. For example, the yeast *Hansenula polymorpha* has been genetically modified so that 35 % of its total protein consists of enzyme alcohol oxidase. Development of high quality column for affinity chromatography, ion exchange chromatography, molecular sieve chromatography and modern purification process like release of enzyme by means of new cell breaking methods, counter current distribution and precipitation has increase the enzyme yield. Optimization of conditions like use of low cost nutrients, use of chemical which induce microbes for enzyme production and introduction of fed batch fermentations has improved the production of enzymes.

5.6 References

[1] Fisher, S.Z., Tu, C., Bhatt, D., Govindasamy, L., Agbandje-McKenna, M., McKenna, R., Silverman, D.N. (2007). Speeding up proton transfer in a fast enzyme: kinetic and crystallographic studies on the effect of hydrophobic amino acid substitutions in the active site of human carbonic anhydrase II. Biochemistry, 46(12), 3803–3813.

[2] Berg, J.M., Tymoczko, J.L., Stryer, L. (2002). Chapter 9 Catalytic Strategies. Section 9.2 Making a Fast Reaction Faster: Carbonic Anhydrases. In: Biochemistry. 5th edition. W H Freeman, New York.

[3] Nigam, P.S. (2013). Microbial Enzymes with Special Characteristics for Biotechnological Applications. Biomolecules 3, 597–611.

[4] Wrotnowski, C. (1997). Unexpected niche applications for industrial enzymes drives market growth, Genetic Engineering News, 1, 14–30.

[5] Chisti, Y. (1999). Fermentation (Industrial): Basic Consideration, In: Encyclopedia of food microbiology, Robinson R, Batt C, and Patel P Editors, Academic Press London, pp. 663–674.

[6] Pandey, A., Nigam, P., Soccol, C.R., Soccol, V.T., Singh, D., Mohan, R. (2000). Advances in microbial amylases. Biotechnology Application and Biochemistry, 31, 135–152.

[7] Tanaka, Y., Kagamiishi, A., Horiuchi, T. (1975). Purification and Properties of β-Galactosidase from *Aspergillus oryzae*. Journal of Biochem, 77(1), 241–247.

[8] Frost, G.M., Moss, D.A. (1987), Enzyme technology, (Ed. J.F. Kennedy), In Biotechnology 7a, VCH, pp. 62–213.

[9] Kittur, F.S., Kumar, A.B.V., Gowda, L.R., Tharanathan, R.N. (2003). Chitosanolysis by a Pectinase isozyme of Aspergillus niger – A non-specific activity. Cabohydrate polymers, 53, 191–196.

[10] Kumar, A., Prabhune, A., Suresh, C.G., Pundle, A.V. (2008). Characterization of smallest active monomeric penicillin V acylase from new source: A yeast, *Rhodotorula aurantiaca* (NCIM 3425). Process Biochemistry, 43, 961–967.

[11] Renge, V.C., Khedkar S.V., Nandurkar, N.R., (2012). Enzyme synthesis by fermentation method: A review, Scientific Reviews & Chemical Communications, 2(4), 585–590.

[12] Nigam, P.S., (2013). Microbial enzymes with special characteristics for biotechnological applications. Biomolecules, 3(3), 597–611. doi: 10.3390/biom3030597

[13] Mukherjee, A.K., Adhikari, H., Rai, S.K., (2008). Production of alkaline protease by a thermophilic *Bacillus subtilis* under solid-state fermentation (SSF) condition using *Imperata* cylindrical grass and potato peel as low-cost medium: Characterization and application of enzyme in detergent formulation. Biochemical Engineering Journal, 39, 353–361. doi: 10.1016/j.bej.2007.09.017.

[14] Rahman, R.N.Z.R.A., Basri, M., Salleh, A.B., (2003). Thermostable alkaline protease from *Bacillus stearothermophilus* F1; Nutritional factors affecting protease production. Annals of Microbiology. 53, 199–210.

[15] Chudasama, C.J., Jani, S.A., Jajda, H.M., Pate, H.N., (2010). Optimization and production of alkaline protease from *Bacillus thuringiensis* CC7. J. Cell & Tissue Research, 10, 2257–2262.

[16] Genckal, H., Tari, C. (2006). Alkaline protease production from alkalophilic *Bacillus* sp. isolated from natural habitats. Enzyme and Microbial Technology. 39, 703–710. doi: 10.1016/j.enzmictec.2005.12.004. [Cross Ref]

[17] Gupta, R., Beg, Q.K., Lorenz P. (2002). Bacterial alkaline proteases: Molecular approaches and Industrial Applications. Applied Microbiology and Biotechnology. 59, 15–32. doi: 10.1007/s00253-002-0975-y.

[18] Shin, D.H., Choi, Y.S., Cho, Y.H. (2008). Unusual properties of catalase a (Kata) of *Pseudomonas aeruginosa* Pa14 are associated with its biofilm peroxide resistance. Journal of Bacteriology, 190, 2663–2670.

[19] Daniela, M., Eva, B., Sergio, R. (2003). Separation and characterization of two catalase activities isolated from the yeast *Trigonopsis variabilis*. Enzyme and Microbial Technology, 32(5), 596–605.

[20] Zubeyde, B., Fikret, U., Cetin, A. (2003). Solid state fermentation for production of α-amylase by a thermotolerant Bacillus subtilis from hot-spring water. Process Biochemistry, 38(12), 1665–1668.

[21] Reid, M.F., Fewson, C.A. (1994). Molecular characterization of microbial alcohol dehydrogenases. Critical Reviews in Microbiology, 20, 13–56.

[22] Panke, S., Held, M., Wubbolts, M. (2004). Trends and innovations in industrial biocatalysis for the production of fine chemicals. Current Opinion in Biotechnology, 15, 272–279.

[23] Schmid, A., Dordick, J.S., Hauer, B., Kiener, A., Wubbolts, M., Witholt, B., (2001). Industrial biocatalysis today and tomorrow. Nature, 409, 258–268.

[24] Schoemaker, H. E., Mink, D., Wubbolts, M.G. (2003). Dispelling the myths—biocatalysis in industrial synthesis. Science, 299, 1694–1697.

[25] Itoh, N., Mizuguchi, N., Mabuchi, M., (1999). Production of chiral alcohols by enantioselective reduction with NADH-dependent phenylacetaldehyde reductase from *Corynebacterium* strain, ST-10. Journal of Molecular Catalysis B, 6, 41–50.

[26] Katz, M., Hahn-Hagerdal, B., Gorwa-Grauslund, M.F., (2003). Screening of two complementary collections of *Saccharomyces cerevisiae* to identify enzymes involved in stereo-selective reductions of specific carbonyl compounds: an alternative to protein purification. Enzyme and Microbial Technology, 33, 163–172.

[27] Melis, L.E.D., Whiteman, P.H., Stevenson, T.W., (1999). Isolation and characterisation of a cDNA clone encoding cinnamyl alcohol dehydrogenase in *Eucalyptus globulus* Labill. Plant Science, 143, 173–182.

[28] Wsól, V., Skalova, L., Szotakova, B., Trejtnar, F., Kvasnickova, E., (1999). Sex differences in stereospecificity of oracin reductases in rat in vitro and in vivo. Chirality, 11, 505–509.

[29] Engelking, H., Pfaller, R., Wich, G., Weuster-Botz, D., (2004). Stereoselective reduction of ethyl 4-chloro acetoacetate with recombinant *Pichia pastoris*. Tetrahedron Asymmetry, 15, 3591–3593.

[30] Kataoka, M., Hoshino-Hasegawa, A., Thiwthong, R., Higuchi, N., Ishige, T., Shimizu, S., (2006). Gene cloning of an NADPH-dependent menadione reductase from Candida macedoniensis, and its application to chiral alcohol production. Enzyme and Microbial Technology, 38, 944–951.

[31] Kataoka, M., Delacruz-Hidalgo, A.R.G., Akond, M.A., Sakuradani, E., Kita, K., Shimizu, S., (2004). Gene cloning and overexpression of two conjugated polyketone reductases, novel aldo-keto reductase family enzymes of *Candida parapsilosis*. Applied Microbiology and Biotechnology, 64, 359–366.

[32] Moore, C.M., Minteer, S.D., Martin, R.S., (2005). Microchip-based ethanol/oxygen biofuel cell. Lab on a Chip, 5(2), 218–225.

[33] Thatipamala, R., Rohani, S., Hill, G.A., (1992). Effects of high product and substrate inhibitions on the kinetics and biomass and product yields during ethanol batch fermentation. Biotechnology Bioengineering, 40, 289–297.
[34] Witteveen, C.F.B., van de Vondervoort, V., Swart, K., Visser, J., (1990). Glucose oxidase over producing and negative mutants of *Aspergillus nidulans*. Applied and Environmental Microbiology, 33: 683–86.
[35] Bentley, R., (1959). Glucose aerodehydrognase (glucose oxidase). Methods in Enzymology, 1, 340–345.
[36] Caridis, K.-A., Christakopoulos, P., Macris, B.J., (1991). Simultaneous production of glucose oxidase and catalase by *Alternaria alternate*. Applied Microbiology and Biotechnology, 34(6), 794–797.
[37] Sabir, S., Bhatti, H.N., Zia, M.A., Sheikh, M.A., (2007). Enhanced production of glucose oxidase using *Penicillium notatum* and rice polish. Food Technology and Biotechnology, 45(4), 443–446.
[38] Rasul, S., Zia, M.A., Sheikh, M.A., Iftikhar, T., (2011). Enhanced production and characterization of a novel β-D-glucose: oxygen 1 oxidoreductase by using *Aspergillus niger* UV-180-C mutant strain. African Journal of Biotechnology, 10(65), 14522–14533.
[39] Zia, M.A., Rehman, K., Sheikh, M.A., Khan, I.A., (2010). Chemically treated strain improvement of *Aspergillus niger* for enhanced production of glucose oxidase. International Journal of Agriculture and Biology, 12(6), 964–966.
[40] Blaedel, W.J., and Uhl, J.M., (1975). Nature of Materials in Serum that Interfere in the Glucose Oxidase-Peroxidase-o-Dianisidine Method for Glucose, and Their Mode of Action, Clinical Chemistry, 21(1), 119–124.
[41] Gornall, A.G., Bradwill, C.J., David, M.M., (1949). Determination of serum proteins by means of biuret reagent. Journal of Biological Chemistry, 177, 751–776.
[42] Cooper, J.A.D., Smith, W., Bacila, M., Medina, H., (1959). Galactose Oxidase from Polyporus circinatus, Journal of Biological Chemistry, 234(3), 445–448.
[43] Nobles, M., Madhosingh, C., (1963). *Dactylium dendroides* (Bull.) Fr. Misnamed as *Polyporus circinatus* Fr, Biochemical and Biophysical Research Communications, 12(2), 146–147.
[44] Whittaker, J.W., (2005). The radical chemistry of galactose oxidase: minireview. Archives of Biochemistry and Biophysics, 433,227–239.
[45] Markus, Z., Miller, G., Avigad, G., (1965). Effect of culture condtions on the production of D-galactose oxidase by *Dactylium dendroides*. Applied Microbiology, 13, 686–693.
[46] Aisaka, K., Terada, O., (1981). Production of galactose oxidase by *Gibberella fujikuroi*. Agricultural and biological chemistry,10, 2311–2316.
[47] Barbosa-Tessmann, I.P., da Silva, D.A., Peralta, R.M., Kemmelmeier, C., (2001). A new species of *Fusarium* producer of galactose oxidase. Journal of Basic Microbiology, 3–4, 143–148.
[48] Whittaker, M.M., Whittaker, J.W., (2000). Expression of recombinant galactose oxidase by *Pichia pastoris*. Protein Expression and Purification, 20, 105–111.
[49] Whittaker, M.M., Whittaker, J.W., (2001). Catalytic reaction profile for alcohol oxidation by galactose oxidase. Biochemistry, 40, 7140–7148.
[50] Rhodes, L.L., Grant, W.D., (1990). A muramidase from the fungus *Gliomastix murorum*. New Zealand Natural Sciences, 17, 55–60.
[51] Grant, W.D., Prosser, B.A., Asher R.A., (1990). A bacteriolytic muramidase from the basidiomycete *Schizophyllum commune*. Journal of General Microbiology, 136, 2267–2273.
[52] Hayashi, K., Yamasaki, N., Funatsu, M., (1964). Muramidase Catalyzed Hydrolysis of Glycol Chitin. Agricultural and Biological Chemistry, 28(8), 517–523.
[53] Pollegioni, L., Piubelli, L., Molla, G., (2009). Cholesterol oxidase: biotechnological applications. FEBS Journal, 276(23), 6857–6870.

[54] Isobe, K., Shoji, K., Nakanishi, Y., Yokoe, M., Wakao, N., (2003). Purification and some properties of cholesterol oxidase stable in detergents from γ-*Proteobacterium* Y-134. Journal of Bioscience and Bioengineering, 95, 257–263.
[55] Lashkarian, H.E., Raheb, J., Shahzamani, K., Shahbani, H., Shamsara M., Hajipour, M., (2010). Isolation and Identification of a Native *Rhodococcus* Strain producing cholesterol oxidase from Soil. International Journal of Biotechnology and Biochemistry, 6, 205–217.
[56] Cheetham, P.S.J., (1979). The identification, properties, and immobilization of the 3P-hydroxysteroid oxidase and 3-oxosteroid A"-A'-isomerase components of cholesterol oxidase from *Nocardia rhodochrous*. Journal of Applied Biochemistry, 1, 51–59.
[57] Doukyu, N., Aono, R., (1998). Purification of extracellular cholesterol oxidase with high activity in the presence of organic solvents from *Pseudomonas* sp. Strain ST-200. Applied and Environmental Microbiology, 64(5), 1929–1932.
[58] de Renzo, E.C., Siiteri P.K., Hutchings, B.L., Bell, P.H., (1967). Preparation and certain properties of highly purified streptokinase. Journal of Biological Chemistry, 242, 533–542.
[59] Taylor, F.B., Botts, J., (1968). Purification and characterization of streptokinase with studies of streptokinase activation of plasminogen. Biochemistry, 7, 232–242.
[60] Brockway, W.J., Castellino, F.J., (1974). A characterization of native streptokinase and altered streptokinase isolated from a human plasminogen activator complex. Biochemistry 13, 2063–2070.
[61] Reed, G.L., Houng, A.K., Liu, L., Parhami-Seren, B., Matsueda, L.H., Wang, S., Hedstrom, L., (1999). A catalytic switch and the conversion of streptokinase to a fibrin-targeted plasminogen activator. Proceedings of the National Academy of Sciences of the United States of America, 96(16), 8879–8883.
[62] Bernheimer, A.W., Gillman, W., Hottle, G.A., Pappenheimer, A.M., (1942). An improved medium for the cultivation of hemolytic *Streptococcus*. Journal of Bacteriology, 43, 495–498.
[63] Billroth, A.W., (1874) Untersuchungen über die Vegetationsformen von *Coccobacteria septica*. Georg Reimer, Berlin, Germany.
[64] Jones, D., (1978) Composition and differentiation of the genus *Streptococcus*. In: Streptococci (Society for Applied Bacteriology Symposium Series No. 7) (eds Skinner, F.A. and Quesnel, L.B.). Academic Press, London, UK, pp. 1–49.
[65] Schottmüller, H., (1903). Die Artunterscheidung der für den Menschen pathogenen Streptokokken durch Blutagar. Münchener Medizinische Wochenschrift, 50, 849–853.
[66] Brown, J.H., (1919). The Use of Blood Agar for the Study of Streptococci. (Monograph No. 9). The Rockefeller Institute for Medical Research, New York.
[67] Christensen, L.R., (1945). Streptococcal fibrinolysis: a proteolytic reaction due to serum enzyme activated by streptococcal fibrinolysin. Journal of General Physiology, 28, 363–383.
[68] Chisti, Y., (1998). Biosafety. In: Subramanian G, editor. Bioseparation and Bioprocessing: A Handbook vol. 2. New York: Wiley-VCH,pp. 379–415.
[69] Castellino, F.J., Sodetz, J.M., Brockway, W.J., Siefring, G.E., (1976). Streptokinase. Methods in Enzymology, 45, 244–257.
[70] Jegannathan, K.R., Nielsen, P.H. (2013). Environmental assessment of enzyme use in industrial production e a literature review. *Journal of Cleaner Production*, 42, 228–240.
[71] Skals, P.B., Krabek, A., Nielsen, P.H., Wenzel, H., (2008). Environmental assessment of enzyme assisted processing in pulp and paper industry. International Journal of Life Cycle Assessment. 13, 124–132.
[72] Kallioinen, A., Pere, J., Siika-aho, M., Lehtilä, A., Mälkki, H., Syri S., Thun, R., (2003). Biotechnical Methods for Improvement of Energy Economy in Mechanical Pulping. Espoo. VTT Tiedotteita e Research Notes 2183. VTT Technical Research Centre, Finland.

[73] Fu, G.Z., Chan, A.W., Minns, D.E., (2005). Preliminary assessment of the environmental benefits of enzyme bleaching for pulp and paper making. International Journal of Life Cycle Assessment, 10, 136–142.
[74] Nielsen, P.H., (2006). Environmental assessment of enzyme application in tanning industry. Leather International (August/September), 18–22.
[75] Nielsen, P.H., Høier, E., (2009). Environmental assessment of yield improvements obtained by the use of the enzyme phospholipase in mozzarella cheese production. International Journal of Life Cycle Assessment, 14, 137–143.
[76] Dettore, C., (2011). Comparative life-cycle assessment (LCA) of textile bleaching systems: gentle power bleach vs. conventional bleaching System. In: Scientific Poster. AATCC International Conference, 22–24 March.
[77] Nielsen, P.H., Skagerlind, P., (2007). Cost-neutral replacement of surfactants with enzymes. Household and Personal Care Today 4, 3–7.
[78] Cowan, D., Oxenbøll, K.M., Holm, H.C., (2008). Enzymatic bioprocessing of oil and fats. Oil Mill Gazetteer, 113, 10–13. http://www.iomsa.org/leads/omgmar08.pdf accessed on 06/09/2015.
[79] Holm, H.C., Cowan, W.D., (2008). The evolution of enzymatic interesterification in the oils and fats industry. European Journal of Lipid Science and Technology, 110(8), 679–691.
[80] Oxenbøll, K.M., Ernst, S., (2008). Environment as a new perspective on the use of enzymes in the food industry. Food Science and Technology, 22, 45–47.
[81] Oxenbøll, K.M., Li, Y., Axelsen, A.G., (2010). Life Cycle Assessment of Using Novamyl Steam for Industrial Production of Steamed Bread in China. Novozymes A/S, Krogshoejvej 36, 2880 Bagsvaerd, Denmark.
[82] Kløverpris, J.H., Spillane, D., (2010). Comparative LCA of malt-based beer and 100 % barley beer. Brewing Beverage Industry International, 2, 34–36.
[83] Yon-Miaw, S., (2011). Barley brewing e an opportunity for choice and sustainability. In: Brewing Conference, 15 June, Bangkok.
[84] Nielsen, P.H., Dalgaard, R., Korsbak, A., Pettersson, D., (2008). Environmental assessment of digestibility improvement factors applied in animal production: use of Ronozyme WX CT xylanase in Danish pig production. International Journal of Life Cycle Assessment. 13, 49–56.
[85] Nielsen, P.H., Wenzel, H., (2006). Environmental assessment of Ronozyme P5000 CT phytase as an alternative to inorganic phosphate supplementation to pig feed used in intensive pig production. International Journal of Life Cycle Assessment. 12, 514–520.
[86] Nagaraju, R.K., Nielsen, P.H., (2011). Environmental advantages of phytase over inorganic phosphate in poultry feed. Poultry Punch 27, 57–58/74.
[87] Oxenbøll, K.M., Pontoppidan, K., Fru-Nji, F., (2011). Use of a protease in poultry feed offers promising environmental benefits. International Journal of Poultry Science, 10, 842–848.
[88] Weiß, M., Brinkmann, T., Groger, H., 2010. Towards a greener synthesis of (S)-3-aminobutanoic acid: process development and environmental assessment. Green Chemistry, 12, 1580–1588.
[89] Osborne-Lee, I., Ajiboye, G., Deason, J., (2008). Economic and environmental impact assessment of alpha-naphthol production by traditional and enzymatic routes. In: Proceedings of AIChE Annual Meeting, 16–21 Nov, Philadelphia.
[90] Poechlauer, P., Braune, S., De Vries, A., May, O., (2010). Sustainable route design for pharmaceuticals. Chemistry Today, 28, 14–17.
[91] Dunn, P.J., (2011). The importance of green chemistry in process research and development. Chemical Society Reviews, 41, 1452–1461.
[92] Biwer, A., Heinzle, E., (2004). Environmental assessment in early process development. Journal of Chemical Technology and Biotechnology, 79, 597–609.
[93] Henderson, R.K., Jiménez-González, C., Preston, C., Constable, D.J.C., Woodley, J.M., (2008). EHS & LCA assessment for 7-ACA synthesis. Indian journal of Biotechnology. 4, 180–192.

[94] Thum, O., Oxenbøll, K.M., (2008). Biocatalysis e a sustainable method for the production of emollient esters. International Journal of Applied Sciences, 134, 44–47.
[95] Harding, K.G., Dennis, J.S., Von Blottnitz, H., Harrison, S.T.L., (2007). A life-cycle comparison between inorganic and biological catalysis for the production of biodiesel. Journal of Cleaner Production, 16, 1368–1378.
[96] Jegannathan, K.R., Wang Ting, V.F., Pogaku, R., (2011). LCA of biodiesel production using alkali, soluble and immobilized enzyme catalyst processes. Biomass & Bioenergy, 35, 4221–4299.
[97] Gurung, N., Ray, S., Bose, S., & Rai, V., (2013). A Broader View: Microbial Enzymes and Their Relevance in Industries, Medicine, and Beyond. BioMed Research International, 2013, 329121. http://doi.org/10.1155/2013/329121.
[98] Sánchez-Pérez de C.L., Barranco-Florido J.E., Rodríguez-Navarro S., Cervantes-Mayagoitia J.F., Ramos-López, M.Á., (2014). Enzymes of entomopathogenic fungi, Advances and Insights Advances in enzyme research, 2(2), Article ID:46652,12 pages DOI:10.4236/aer.2014.22007.
[99] White, J.S. and White, D.C. (1997). Source book of enzymes, CRC Press, New York. p. 1008. ISBN-13 978-0-849-39470–6.

Marcela Claudia Pagano, Luiz Henrique Rosa,
and Natalia V. Fernández

6 Microbial Pigments

6.1 Introduction

The food and beverage industrial communities search for all natural colorants and, especially for uncommon colors such as blue, in natural plant, animal, macroscopic fungi and microbial sources (Newsome *et al.*, 2014). This is due to recent increasing concern regarding the consumption of natural ingredients. As well, the need to search for novel antibiotics to face the advent of strains presenting antimicrobial resistance as well as new emerging pathogens have lead to found promising compounds showing bioactivities in pigments (Tuli *et al.*, 2014). Additionally, several artificial colorants result in hyperallergenicity or carcinogenicity and thus have being disqualified, which have leads to increase research on natural compounds (Reviewed by Tuli *et al.*, 2014).

With regard to plant pigments, a long reported history is available (Tab. 6.1). For example, Tanaka *et al.* (2008) reviewed the major types of pigments synthetized by plants, except for chlorophylls. Some pigments such as Curcumin, a notorious vegetal yellow pigment used for longtime as medicinal, and also in food (Mayell, 2001), have been recently tested as larvicidal against *Culex quinquefasciatus* vector of filariasis (Madhu *et al.*, 2010) and more recently to kill *Aedes aegypti* mosquito larvae (Anstrom *et al.*, 2012).

Among the natural sources, pigment producing microorganisms have potential to meet actual new defies. Natural colors develop the market tendencies of the product including associated features such as antioxidant, anticancer properties, to name just a few (Reviewed by Tuli *et al.*, 2014). Several microorganisms are promising for natural colors prospection due its increased management flexibility for cultivation (Tuli *et al.*, 2014).

Most significant natural pigments produced by bacteria, algae and fungi were compiled by Danshan and Manonmani (2015). Nowadays, researchers continue to bio-explore for novel compounds from microbes and fungi obtaining several drugs which valuated the fungal bio-prospecting market. Moreover they have stressed the urgent need to search for sites and substrates with high fungal richness towards protection of fungal resources for unique compound detection (Tang *et al.*, 2007). In this review, a general summary and most notorious reports on this topic are mentioned. Moreover, various sources of microbial pigments and the needs to explore their biological and medicinal properties like antimicrobial, antioxidant, anticancer and anti-inflammatory are presented. The study also emphasizes upon key parameters to improve the bioactivity and production of microbial pigments for their commercial use.

6.2 Microbial sources of natural color

Two primary sources of natural pigments are plants and microorganisms. However, vegetal pigments have disadvantages (low water solubility, instability to light, heat or adverse pH) being only available in some seasons (Reviewed by Darshan and Manonmani, 2015). In contrast, microbial pigments present more pigment stability in addition to their easily cultivation (Parekh *et al.*, 2000). Moreover, fast grow microorganisms can be cultivated in cheap culture medium, and can produce different shades of colours (Reviewed by Darshan and Manonmani, 2015). Thus, the industrial production of natural food colorants by microbial has a pivotal role in the world *presently.*

Microorganisms produce a diversity of antibiotics, enzymes and pigments being a hopeful source for natural colors. Pigments are present in plants, bacteria, fungi and most organisms (Tab. 6.1) providing attractive colors which gratify human beings. A huge collection of colorants obtained from natural sources such as animals (and especially insects), plants and microorganisms have been examined in order to obtain different compounds (Shahid *et al.*, 2013). For a review of background and history of pigments and production technologies see Venil *et al.* (2013).

Most microorganisms can be isolated and purified from different environmental sources such as water bodies, soil, plants, insects and animals. Some food colorants pigments are particularly studied (Delgado-Vargas *et al.*, 2000). Pagano and Dhar (2014) reviewed recent reports on fungal pigments focusing on their potential use and significance for human benefit. They highlighted the importance of research on fungal pigments both for humans and natural systems. Additionally, there is enormous demand for coloring agents in industries like textile, plastic, paint, paper and printing (Tuli *et al.*, 2014). Much interest was dedicated to specific pigments, such as natural blue or red colorant for food industries; however, novel chemical classes from pigment produced by marine fungi are also investigated (Dufossé *et al.*, 2014). Moreover, there is increasing interest to improve the bioactivity and production of microbial pigments for their commercial use in pharmacological and medical fields (Reviewed by Tuli *et al.*, 2014).

With regard to Bacteria, red-pigmented compounds, named Prodigiosins, synthezised by both Gram-negative and Gram-positive bacteria have potential applications as therapeutic drugs against cancer with little or no toxicity to organism (Darshan and Manonmani, 2015). However, the biological role of these pigments in the microbe itself remains unclear.

A recent review by Babitha (2009) (Tab. 6.2) focused on fungal microorganisms of the genus *Monascus*, which produce red pigments, used for foods and named "*Monascus* pigments". Tuli *et al.* (2014) revised the microbial pigments as natural color sources highlighting the potential of pigment producing microorganisms. They compiled sources of microbial pigments and their biological and clinical properties (antimicrobial, antioxidant, anticancer and anti inflammatory). Additionally, they showed a list of pigment producing microorganisms together with anticipated bioactivity.

The technology for microbial pigment production at industrial scale is costly. However, low cost processes are increasingly investigated (Babitha, 2009). Efforts are being made to reduce the production cost of fermentation based microbial pigments. In order to assess the production cost, economic comparison has been drawn between natural and synthetic pigments (Tuli *et al.*, 2014). Cheaper substrates are necessary for the growth of selected microorganisms. Additionally, researchers demand for studies on chemical structure, mechanism of action and activity of microbial pigments in order to rethink strategies of *terminal diseases* (Tuli *et al.*, 2014).

In this regard, filamentous fungi have high potential for the production of food pigments due to their chemical and color diversity in their pigments and simple large scale cultivation.

Tab. 6.1: Journal articles dealing with pigments and different organisms.

Key words	Number of journal articles
Pigments	147,279
Plant pigments	16,175
Bacterial pigments	6,196
Microbial pigments	2,459
Fungal pigments	1,687
Microbial natural pigments	297

Database survey conducted on May 2016 (SCOPUS).

Moreover, a variety of strains are well known. This will depend on market and consumer's expectations, industrial prices and availability (Dufossé *et al.*, 2014). Another expectative is to use cheaply available organic residues from industries as substrate (See Babitha, 2009).

6.3 Microbial pigments in natural sites

The high number of ecosystems and plant diversity described worldwide suggest that, most habitats are underexplored for microbial occurrence. Interest in natural sites to be prospected is increasing in parallel with the recognition that novel compounds from fungi to obtain several drugs will valuate the fungal bio-prospecting market. Moreover, researchers have stressed the urgent need to search for sites and substrates with high fungal richness towards protection of fungal resources for unique compound detection

(Tang *et al.*, 2007). Preserving ecosystem biodiversity will help to maintain its multifunctionality, which can be decisive to obtaining new compounds against microorganisms. For example, Dark septate endophytes (DSE) are frequently associated to roots of numerous endemic and endangered plants, still more in tropical regions where many fungi remain new to science (Hawksworth, 2001).

The kingdom Fungi (yeasts and filamentous fungi) are a promising source able to produce different both primary and secondary metabolites, which include peptides, enzymes, acids, antibiotics and pigments. Fungi called "black fungi" includes different species are able to produce a mixture of several different pigments to survive adverse conditions (carotenoids, lycopene and xanthophylls can predominate). For example, carotenoids protect the fungus from damaging sunlight and UV light; melanin and sporopollenin found in the spores and hyphal walls protect from radiation and drought (Reviewed by Pagano and Dhar, 2015).

Additionally, fungal pigments have been reported as bioactive compounds with antioxidative, free radical scavenging, anticancer, immunomodulatory, antiviral and antibacterial activities, many of these pigments are quinones or similar conjugated structures (Velísek and Cejpek, 2011).

One of the most important pigments produced by fungi are the melanins (Fig. 6.1), which are high molecular weight pigments and displayed different biological functions (Hill, 1992). Several fungal species are able to produce melanin, which seems to confer advantage to survive in the environment by protecting against environmental predators, heavy metals toxicity, and physical insults such as UV and solar radiation (Nosanchuk and Casadevall, 2003). Melanized fungi were found on the walls of damaged reactor in Chernobyl, which were consequently exposed to high levels of radiation (Mironenko *et al.*, 2000). After that, further studies were made to investigate the radioprotective activity of microbial melanins (Dadachova *et al.*, 2007).

Researchers have bio-explored for stressed and extreme habitats worldwide. Vaz *et al.* (2011) investigated yeasts diversity from different sites in Antarctica. They studied rhizospheric *Deschampsia antartica* soil, ornithogenic (penguin guano) soil, water

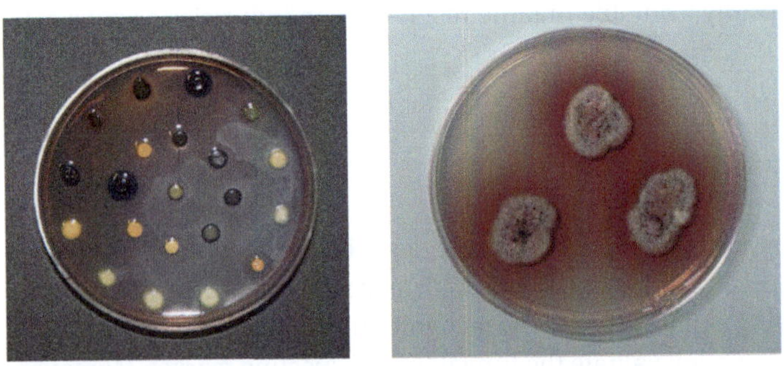

Fig. 6.1: Fungal colonies isolated from environmental samples showing different pigments (Photographs by L H Rosa).

from lakes, seawater and sediments. Besides possible novel species identification, they found that ten yeast isolates produced mycosporines and that more than 40 % produced pigments and could be considered adapted to survive in the continent. The adaptation of yeasts to cold habitats reveals structural and biochemical characteristics, such as the production of photoprotective compounds (carotenoids and mycosporines), as possible strategy to survive in extreme environments (Libkind *et al.*, 2009; Thomas-Hall *et al.*, 2010). Mycosporines (watersoluble, UV-absorbing compounds) in yeasts from high-altitude lakes have been correlated with the survival of some species adapted to these environments (Libkind *et al.*, 2005; Libkind *et al.*, 2009).

In addition, Fernández *et al.* (2012) analyzed the yeast and yeast-like fungi present on *Nothofagus nervosa* (raulí) dry fruits. In this work, a total of 171 isolates corresponding to 17 species were recovered, most of which belonged to the phylum Ascomycota (78 % of the isolates and 61 % of the species). All the ascomycetous species were dark pigmented melanin containing yeast-like fungi, while only one basidiomycetous species was pigmented (*Rhodotorula colostri*, a pink carotenogenic yeast). Besides, most of the species present on *N. nervosa* fruits (71 %) were capable of synthesizing mycosporines. It is important to mention that the three most abundant species (which accounted for 73 % of the isolates) corresponded to melanin and mycosporine producing yeast-like fungi (Ascomycetous yeast sp. 1, *Aureobasidium pullulans* and *Dothichiza* sp. 1), being two of them novel species (Tab. 6.2). It has been described that mycosporines, as well as pigments such as melanin and carotenes, play an antioxidant role in addition to UV protection (Pagano and Dhar, 2015; Moliné *et al.* 2011, Fernández *et al.* 2012). Altogether, these results highlight how important these components (carotenes, melanin, mycosporines) are for surviving in harsh environments, usually characterized by reduced access to nutrients, high fluctuations in temperature and exposure to desiccation and/or high UV radiation. Due to the different adaptation strategies (carotenes, melanin, mycosporines) of the microorganisms present in this type of environments and to the high proportion of novel species usually present in them, they should be considered one of the main research targets for looking for species capable of synthetizing molecules of biotechnological importance, appropriate for the bio-prospecting market.

Tab. 6.2: Fungal species isolated from *Nothofagus nervosa* dry fruits surface. % = percentage out of the total number of isolates, Pig = pigments, M = melanin, C = carotenes, Myc = mycosporines, +/− = presence/absence.

Taxa	Total (%)	Pig	Myc
Ascomycota			
Ascomycetous yeast sp. 1	46 (27 %)	$+^{M}$	+
Ascomycetous yeast sp. 2	1 (<1 %)	$+^{M}$	−

Tab. 6.2: (continued)

Taxa	Total (%)	Pig	Myc
Aureobasidium pullulans	31 (18 %)	$+^{M}$	+
Coniozyma leucospermi	1 (<1 %)	$+^{M}$	+
Dothichiza sp. 1	47 (27 %)	$+^{M}$	+
Dothichiza sp. 2	1 (<1 %)	$+^{M}$	+
Lecythophora mutabilis	3 (2 %)	$+^{M}$	–
Phaeomoniella sp.	1 (<1 %)	$+^{M}$	+
Phaeomoniella zymoides	1 (<1 %)	$+^{M}$	+
Taphrina wiesneri	1 (<1 %)	$+^{M}$	+
Basidiomycota			
Cryptococcus adeliensis	2 (1 %)	–	–
Cryptococcus diffluens	1 (<1 %)	–	+
Cryptococcus heveanensis	20 (12 %)	–	+
Cryptococcus wieringae	1 (<1 %)	–	+
Rhodotorula colostri	4 (2 %)	$+^{C}$	–
Rhodotorula fujisanensis	9 (5 %)	–	–
Trichosporon dulcitum	1 (<1 %)	–	+

6.4 Pigments and plant endophytes

It is known that fungal melanins can increase fungal survival in some environments (Rizner and Wheeler, 2003); however, some endophytic fungi cause asymptomatic infections in aerial tissues of woody plants. The increasing study in this field is desired due to possible increased fitness of the plant hosts synthesis and of abundant secondary metabolites, which can acquire potential economic significance (Redman *et al.*, 2002; Schulz *et al.*, 2002). Suryanarayanan *et al.* (2004) found that melanin in the hyphae of *Phyllosticta capitalensis* (teleomorph *Guignardia mangiferae*) can be related to the success of this woody tree leave endophyte with cosmopolitan (mangroves, dry deciduous forest, moist deciduous forest and semi-evergreen forest of temperate and tropical regions) geographic distribution.

Dark septate endophytes colonized plant roots being identified on the basis of commonly septate hyphae, usually dark pigmented, with facultatively developed sclerotia (Jumpponen, 2001). Similarly to Arbuscular mycorrhizal fungi (AMF), DSE may play an important role in improving plant performance (Strobel, 2003).

Interestingly, endophytic fungal pigment can surpass the effect of some commercial antibiotics against human pathogenic bacteria such as *Staphylococcus aureus, Klebsiella pneumoniae, Salmonella typhi* and *Vibrio cholera* (Visalakchi and Muthumary, 2010). However, most plant species have not been studied for their endophytic root colonization. Some studies, as part of mycorrhizal targeted investigations, have revealed the common presence of DSE in roots. In a more recent example of target study, Zubek *et al.* (2011) reported for the first time the root colonization of 16 endemic plant species from Asia, belonging to never studied *Asyneuma, Clementsia*, and *Eremostachys* plant genera. They found that the mycelium of DSE complemented the AMF colonization in most studied roots, as it was suggested that DSE could enhance plant performance in the case of plants which are rarely or not colonized by AMF (Zubek *et al.*, 2011).

Looking for belowground interactions from a Valdivian Temperate Rainforest in Patagonia, Argentina, Fernández *et al.* (2012) reported the dominant presence of DSE with conspicuous melanized structures in the roots of 21 species of Pteridophytes. It was observed in these plants that DSE were growing on the root surface as well as inside it. Within the root, DSE formed different types of structures, which sometimes occupied the entire volume of the cell (Fig. 6.2). Similar structures have been also reported in high Andean flora (Bruzone, 2008) and in different *Nothofagus* species (Salgado-Salomón *et al.*, 2013; Fernández *et al.*, 2015). Melanin is a recalcitrant polymer that provides structural strength and protection from environmental stress (Henson *et al.*, 1999), so the occurrence of highly melanised fungal structures associated with different plant taxa in several environments worldwide emphases the importance of this pigment for DSE. It has been also described that these fungi may provide important benefits to the host plant, such as protection against pathogens or abiotic stresses (water availability, extreme temperatures, frosts), being this capability probably related with the high amount of melanin in their mycelia.

Tab. 6.3: Selected book, book chapters, reviews or papers dealing with fungal and microbial pigments * Review.

Ecosystem type/Focus	References
Fungal biomolecules	Gupta *et al.* (2015)
Bacterial prodigiosins	Darshan and Manonmani (2015)*
Microbial pigments as natural color sources	Tuli *et al.* (2014)*
Natural Blue Colorants	Newsome *et al.* (2014)*
Natural colorants	Shahid *et al.* (2013)*
Bacterial pigments	Venil *et al.* (2013)
Pigments of higher fungi	Velišek and Cejpek (2011)
Microbial Pigments	Babitha (2009)

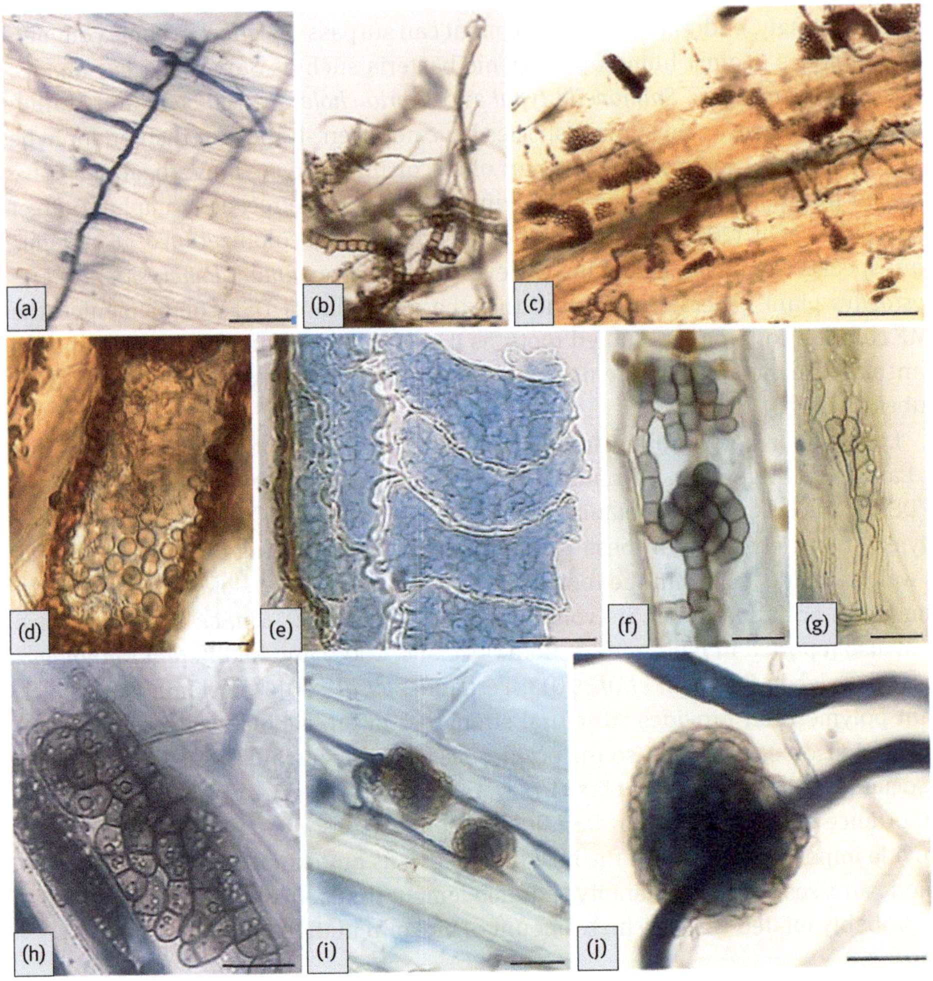

Fig. 6.2: Colonization of Dark Septate Endophytes in pteridophytes of a Valdivian temperate forest. **1.** Runner hyphae on root surface of *Lycopodium paniculatum*; **2.** Extraradical dark-pigmented hyphae in *Grammitis magellanica*; **3.** Highly melanised fungal structures within *Hymenophyllum seselifolium* roots; **4.**, **5.** DSE occupying most of the cell volume in *Equisetum bogotense*; **6.**, **7.** Moniliform and irregularly lobed melanised hyphae in *H. tortuosum* and *Blechnum penna-marina* respectively; **8.** Intracellular thick-walled and darkly pigmented microsclerotia in *E. bogotense*; **9.**, **10.** Cerebriform microsclerotia in *Megalastrum spectabile* and *B. chilense*, respectively (Scale bars: 1 = 200 µm; 2,3,5 = 100 µm; 4,6–9 = 20 µm; 10 = 10 µm) (Photographs by N. V. Fernández).

6.5 Conclusion

This chapter explored the current information relevant to microbial pigments in relation to sources, cost and new challenges we have to deal with. Throughout the chapter, the importance of microbial pigment production especially the benefits from human health has been one of the emerging fields of research to display its potential for a variety of applications.

The appreciation is that the occurrence of novel compounds in unexplored regions is the most probable fact and the focus in those habitats must be increased. Recent studies showed that better knowledge of the wide variation in native yeast as well as plant endophytes associate with different endemic plants is important for understanding the diversity of pigments of different regions and for management purposes. This review would provide a basis for the systemic study of microbial pigments and for evaluating and screening efficient microbial species. Further studies are required to understand pigment mechanisms in the vast microorganisms associations.

6.6 References

[1] Anstrom DM, Zhou X, Kalk CN, Song B, Lan Q (2012) Mosquitocidal Properties of Natural Product Compounds Isolated From Chinese Herbs and Synthetic Analogs of Curcumin J Med Entomol. 49(2): 350–355.

[2] Babitha S, Microbial Pigments. In: Biotechnology for Agro-Industrial Residues Utilisation. P. Singh nee' Nigam, Pandey A (eds), 2009, pp. 147–162.

[3] Babitha S, Sandhya C, Pandey A (2004) Natural food colorants. Appl Bot Abstr 23(4):258–66.

[4] Babitha S, Soccol CR, Pandey A (2007) Solid-state fermentation for the production of Monascus pigments from jackfruit seed. Bioresour Technol 98(8):1554–60.

[5] Bruzone C. Micorrizas y hongos septados marrones de la Flora Altoandina. Tesis para optar al Título de Licenciada en Biología. Centro Regional Universitario Bariloche, Universidad Nacional del Comahue, 2008.

[6] Dadachova E, Bryan RA, Howell RC, Schweitzer AD, Aisen P, Nosanchuk JD, Casadevall A (2007) The radioprotective properties of fungal melanin are a function of its chemical composition, stable radical presence and spatial arrangement. Pigment Cell Melanoma Res. 21: 192–199.

[7] Darshan N, Manonmani H K (2015) Prodigiosin and its potential applications. J Food Sci Technol. DOI 10.1007/s13197-015-1740-4.

[8] Delgado-Vargas F, Jiménez AR, Paredes-López O. Natural Pigments: Carotenoids, Anthocyanins, and Betalains — Characteristics, Biosynthesis, Processing, and Stability Critical Reviews in Food Science and Nutrition 2000; 40(3):173–289.

[9] Dufossé L, Fouillaud M, Caro Y, Mapari SAS, Sutthiwong N (2014) Filamentous fungi are large-scale producers of pigments and colorants for the food industry. Current Opinion in Biotechnology 26:56–61.

[10] Fernández N, Fontenla S, Messuti MI. Co-ocurrence of Arbuscular Mycorrhizas and Dark Septate Endophytes in Pteridophytes from a Valdivian Temperate Rainforest in Patagonia, Argentina. In: Pagano, M.C., ed. Mycorrhiza: Occurrence in Natural and Restored Environments, Nova Science Publishers, New York, USA 2012 pp. 99–126.

[11] Fernández N, Marchelli P, Gherghel F, Kost G, Fontenla S (2015) Ectomycorrhizal fungal communities in *Nothofagus nervosa* (raulí): a comparison between domesticated and naturally established specimens in a native forest of Patagonia, Argentina. Fungal Ecology 18: 36–47.

[12] Fernández N, Mestre MC, Marchelli P, Fontenla S (2012) Yeast and yeast-like fungi associated with dry indehiscent fruits of *Nothofagus nervosa* in Patagonia, Argentina. FEMS Microbiology Ecology 80:176–192.

[13] Gupta VK, Mach RL, Sreenivasaprasad S. (2015) Fungal Biomolecules: Sources, Applications and Recent Developments. Wiley-Blackwell, Hoboken, 408 pages. ISBN: 978-1-118-95829-2.

[14] Hawksworth DL (2001) The magnitude of fungal diversity: the 1.5 million species estimate revisited. Mycological Research 105: 1422–1432.

[15] Henson JM, Butler MJ, Day AW (1999) The dark side of the mycelium: Melanins of Phytopathogenic Fungi. Annual Review of Phytopathology 37: 447–471.

[16] Hill HZ (1992) The function of melanin or six blind people examine an elephant. Bioessays 14, 49–56.

[17] Jumpponen A (2001) Dark septate endophytes —are they mycorrhizal? Mycorrhiza 11:207–211.

[18] Libkind D, Moliné M, Sampaio JP, Van Broock M (2009) Yeasts from high-altitude lakes: influence of UV radiation. FEMS Microbiol. Ecol. 69 (3), 353–362.

[19] Libkind D, Sommaruga R, Zagarese H, van Broock M (2005) Mycosporines in carotenogenic yeasts. Syst. Appl. Microbiol. 28 (8), 749–754.

[20] Madhu SK, Shaukath AK, Vijayan VA (2010) Efficacy of bioactive compounds from *Curcuma aromatica* against mosquito larvae. Acta Tropica 113, 1, 7–11.

[21] Mayell M. Maitake extracts and their therapeutic potential: a review (2001) Altern Med Rev, 6: 48–60.

[22] Mironenko NV, Alekhina IA, Zhdanova NN, Bulat SA (2000) Intraspecific variation in gamma-radiation resistance and genomic structure in the filamentous fungus Alternaria *alternata*: a case study of strains inhabiting Chernobyl reactor no. 4. Ecotoxicol. Environ. Saf. 45, 177–187.

[23] Moliné M, Arbeloa EM, Flores MR, Libkind D, Farías ME, Bertolotti SG, Churio MS, van Broock MR (2011) UVB photoprotective role of mycosporines in yeast: photostability and antioxidant activity of mycosporine-glutaminolglucoside. Radiat Res 175: 44–50.

[24] Newsome A G., Culver CA, van Breemen R B. Nature's Palette: The Search for Natural Blue Colorants. J. Agric. Food Chem., 2014, 62 (28), 6498–6511.

[25] Nosanchuk JD, Casadevall A (2003) The contribution of melanin to microbial pathogenesis. Cell Microbiol. 5, 203–223.

[26] Parekh S, Vinci VA, Strobel RJ (2000) Improvement of microbial strains and fermentation processes.ApplMicrobiol Biotechnol 54:287–301.

[27] Redman RS, Sheehan KB, Stout RG, Rodriguez RJ, Henson JM (2002) Thermotolerance generated by plant/fungal symbiosis. Science 298: 1581.

[28] Rizner TL, Wheeler MH (2003) Melanin biosynthesis in the fungus *Curvularia lunata* (teleomorph: *Cochliobolus lunatus*). Canadian Journal of Microbiology 49: 110–119.

[29] Salgado-Salomón ME, Barroetaveña C, Rajchenberg M (2013) Occurrence of dark septate endophytes in *Nothofagus* seedlings from Patagonia, Argentina. Southern Forests 75: 97–101.

[30] Schulz B, Boyle C, Draeger S, Römmert A-K, Krohn K (2002) Endophytic fungi: a source of novel biologically active secondary metabolites. Mycological Research 106: 996–1004.

[31] Shahid M, Shahid-ul-Islam, Mohammad F (2013) Recent advancements in natural dye applications: a review. Journal of Cleaner Production 53: 310–331.

[32] Strobel G, Daisy B (2003) Bioprospecting for microbial endophytes and their natural products. Microbiol. Mol.Biol. Rev. 67:491–502.

[33] Strobel G, Daisy B, Castillo U, Harper J (2004) Natural products from endophytic microorganisms, J. Nat. Prod.67:257–268.

[34] Strobel GA (2003) Endophytes as sources of bioactive products. Microb. Infect. 5:535–544.
[35] Suryanarayanan TS, Ravishankar JP, Venkatesan G, Murali TS (2004) Characterization of the melanin pigment of a cosmopolitan fungal endophyte Mycol. Res. 108 (8): 974–978.
[36] Tanaka Y, Sasaki N, Ohmiya A. Biosynthesis of plant pigments: anthocyanins, betalains and carotenoids. The Plant Journal – Harnessing plant biomass for biofuels and biomaterials (2008) 54, 733–749.
[37] Tang AMC, Shenoy BD, Hyde KD (2007) Fungal Diversity. Hodkinson TR, Parnell JAN, Eds. Reconstructing the Tree of Life Taxonomy and Systematics of Species Rich Taxa.
[38] Thomas-Hall SR, Turchetti B, Buzzini P, Branda E, Boekhout T, Theelen B, Watson K (2010) Cold-adapted yeasts from Antarctica and the Italian Alps—description of three novel species: *Mrakia robertii* sp. nov., *Mrakia blollopis* sp. nov. and *Mrakiella niccombsii* sp. nov. Extremophiles. 14 (1), 47–59.
[39] Tuli HS, Chaudhary P, Beniwal V, Sharma AK. Microbial pigments as natural color sources: current trends and future perspectives J Food Sci Technol. DOI 10.1007/s13197-014-1601-6.
[40] Vaz ABM, Rosa LH, Vieira MLA, Garcia V, Brandão LR, Teixeira LCRS, Moliné M, Libkind D, van Broock M, Rosa CA (2011) The diversity, extracellular enzymatic activities and photoprotective compounds of yeasts isolated in Antarctica. Brazilian Journal of Microbiology 42: 937–947.
[41] Velíšek J, Cejpek K (2011) Pigments of higher fungi: a review. Czech J. Food Sci. 29 (2):87–102.
[42] Venil CK, Zakaria ZA, Ahmad WA. Bacterial pigments and their applications. Process Biochemistry 48 (2013) 1065–1079.
[43] Zubek S, Nobis M, Błaszkowski J, Mleczko P, Nowak A. (2011) Fungal root endophyte associations of plants endemic to the Pamir Alay Mountains of Central Asia. Symbiosis 54:139–149.

Shadia M. Abdel-Aziz, Vijai K. Gupta, D Sukmawati, and M. Fadel

7 Role of nutrient in microbial developments and microbial metabolic diversity

7.1 Abstract

Any natural ecological entity such as water, air, leaves, and trees has provided a tremendous intensity of organisms. Soil and marine environments are the most preferred sources for organisms because they are a vast reservoir of diverse habitats that can produce bioactive microbial metabolites or "bio-mining". Nutritional resources are a focal point of microbial competition, and there is a relationship between limiting nutrients and microbial growth, development, and diversity. Microbial Diversity is the variety of different types of life found on earth. It is a measure of the variety of organisms present in different ecosystems. Biodiversity can refer to an ecosystem variation, genetic variation, or species variation within an area, biome, or planet. There is a continual change in species diversity through time, and metabolic diversity vary with environmental factors such as nutrient, temperature, ammonium and CO_2 concentrations. Improvement of the industrial strains can be put down by manipulation of the existing genetic material or introducing new genetic properties into the microorganisms by recombinant DNA technology or genetic engineering.

7.2 Introduction

Microorganisms are existing everywhere and can manipulate all the environments in which they live. They were present on Earth 3.5 to 4 billion years ago, and have been evolving and expanding into new environments ever since. Their presence has driven the development of new ecosystems - some of which allowed for the evolution of more complex organisms. There is a continual change in species diversity through time, and biodiversity vary with environmental factors such as nutrient, temperature, ammonium concentration, and CO_2 concentration (Okasha 2010). Without microbes, the recycling of essential nutrients on Earth would halt. Microbes communicate, i.e., some generate signals for the formation of metabolically diverse communities. Some use sophisticated signaling to establish complex relationships with higher organisms. Microorganisms have become ubiquitous due to the quick replicating and exchanging genetic material between each other. Many reports have dealt with bacteria, transform of chemicals, and how their activities can help human life and living (Parales *et al.* 2002). Nutritional resources are a focal point of microbial competition. Studies on microbial growth kinetics, verified a relationship between limiting nutrients and microbial growth, microbial development, and metabolic diversity, is linearly dependent on the initial concentration of a limiting nutrient and carbon source (Hibbing *et al.* 2010).

Microorganisms have to face extreme environmental conditions such as nutrient unavailability, low or high temperatures, and extreme pH values. Microorganisms have evolved special metabolites to survive and proliferate during the evolution and development. Many microorganisms require special growth parameters (physical and chemical) that are hard or even impossible to reproduce in the laboratory (Hibbing *et al.* 2010). Microbial development due to starvation for nutrient and abiotic-stress conditions has shown that microorganisms are capable of forming endospores, multicellular fruiting bodies, and surface-attached microbial communities, known as biofilms, all are adapted to survival in a variety of harsh environments. On other side, microorganisms are known to produce a large number of secondary metabolites. While these compounds help the producing organisms to survive under unfavorable conditions, they are not always essential for growth and reproduction. Many of these metabolites have been known to possess antimicrobial properties, which may provide an edge to the producing organism in the competition for food and space in its habitat. The immense metabolic diversity among the bacterial population from different ecological habitats is a well- recognized fact. Microbial metabolites confer the ability of microbes to cope with a range of adverse conditions, thereby facilitating the adaptation required to thrive under such conditions. Moreover, presence of such structural metabolites is a strong indication of their functionally diverse nature (Singla et al. 2014). This chapter deals with the role of nutrient in microbial development and the broad microbial diversity of microorganisms, and presents some of the bacterial metabolic diversity and their practical applications. The chapter highlights the gut microbiome and strategies for manipulation of the gut microbiota as a potential therapeutic option to treat chronic gastrointestinal diseases.

7.3 Microbial Diversity

Biological diversity means the variability among living organisms from all sources and the ecological complexes of which they are part and it includes diversity within species or between species and of eco-systems. Diversity generally means “variation”, “differentiation”, or “diversification”, in contrast to “uniformity”. It may also be understood as “variety” and “differences” (Kratochwil 1999). Diversity is a measure of the complexity of the community structure and is increased or decreased by physical, chemical and biological factors. Microbial diversity can refer to ecosystem variation, genetic variation, or species variation within an area, biome, or planet (Gaston 2000). High diversity generally indicates a balanced, stable responsive community, while low diversity occurs in an area where the community is dominated by a few species (Das et al. 2006). Beginning of the life is not certainly known. Before the ozone layer formed, the surface of Earth was exposed to a strong radiation. It's likely that early microbes adapted to the high temperatures associated with abundant volcanic activity and geological turmoil which resulted in accumulation of carbon dioxide in the atmosphere. Biodiversity is the result of 3.5 billion years of evolution (Sahney et al. 2010). Until approximately 600 million years ago, all life consisted of archaea, bacteria, protozoans and similar single-celled organisms. During the Cambrian explosion, almost every phylum of multicellular organisms was first appeared (Sahney et al. 2010).

7.3.1 Microbial Nutrition

Nutrients are chemical substances acquired from the environment and used for cellular activities. Essential microbial nutrients include: i) macronutrients, required in large quantities, such as oxygen, hydrogen, carbon, nitrogen, phosphorus, and sulphur, and play principal roles for cell growth, energy needs, cell structure, survival as well as to build organic molecules. About 95 per cent of cellular dry weight in microorganisms is accounted for these six elements; ii) micronutrients, required in lesser quantities, such as potassium, magnesium, calcium, sodium, iron, manganese, cobalt copper, molybdenum and zinc. Microorganisms differ greatly with respect to the chemical forms in which these elements are utilized as nutrients. ***Autotrophs*** are microorganisms that uses CO_2 as carbon source, and not nutritionally dependent on other living organisms (***Phototrophs*** gain energy through Photosynthesis, whereas ***Chemotrophs*** gain energy from chemical compounds). ***Heterotrophs*** must obtain carbon in an organic form such as proteins, carbohydrates, lipids and nucleic acids, made by other living organisms. Some organic compounds such as amino acids and vitamins, that cannot be synthesized by an organism because they lack the genetic and metabolic mechanisms to synthesize them, must be provided for microbial growth as essential organic nutrients. Organic nutrients-contain carbon and hydrogen atoms and are usually the products of living things (organisms, plant and animal residues). Inorganic nutrients are atom or molecule that contains a combination of atoms other than carbon and hydrogen.

Activity and production of microbial metabolites such as exopolysaccharides (EPSs) and bioflocculants are greatly affected by the nutrient source. Critical factors for EPSs and bioflocculant production include, but are not limited to, parameters such as bacterial growth phase, carbon source, nitrogen source, oxygenation rate, temperature, and pH. On other side, EPSs play a key role for the understanding of structure, function, properties and development of microbial aggregates (Biofilm). A huge variety of biopolymers, such as polysaccharides, polyesters, and polyamides, are naturally produced by microorganisms (Staudt 2009). The composition of EPS material produced varies widely from one bacterial species to another (Staudt 2009). Additionally, EPS biosynthesis and composition are a function of a variety of environmental factors. Specific culture conditions can dramatically impact the yield of EPSs and their chemical composition.

The bacterial strain *Bacillus alvei*, relocated to be *Paenibacillus alvei* according to the 16S rRNA gene, has proved to adapt with various stress conditions. This strain, isolated from soil, is capable of living under extensive environmental stress (Abdel-Aziz *et al* 2013, 2014). It has adopted special metabolic pathways to survive in extreme conditions and so has a better capacity to produce special bioactive compounds (Metabolites). Effect of carbon source and nutrient medium on the growth of *Paenibacillus alvei* NRC-14 are extensively studied. This strain produces a variety of enzymes and diversity of metabolites with different carbon sources used (Abdel-Aziz *et al* 2011), under normal conditions (Table 7.1, Fig. 7.3). When exposed to abiotic stress conditions, the metabolic pathway of this strain is obviously differed, i.e., it produces enzymes, exopolysaccharides, and biopolymer flocculants as a form of self-protection (Table 7.2, Fig. 7.3).

Tab. 7.1: Enzymes and metabolites produced by strain *P. alvei* NRC14, under normal conditions, during growth on different carbon sources.

Carbon source	pH	T/°C	Metabolites produced	Effect	Enzymes produced	Reference
1. sugarcane bagasse	6.0	30	–	–	Cellulase, xylanase, chitosanase, inulinase	Abdel-Aziz and Mouafi 2005.
2. Flaked chitosan	6.0	30	Biopolymer flocculant	A great flocculating activity	Chitosanase, chitinase, and β-1,4 glucanase.	Abdel-Aziz *et al.* 2011.
3. Flaked chitosan + sugarcane bagasse	5.5	30	Chitosan-oligosaccharides	Show inhibitory effects	Chitosanase, cellulose, β-1,4 glucanase, and xylanase.	Abdel-Aziz *et al.* 2012*a*.
4. Fungal mycelium (*Aspergillus niger*)	7.0	30	–	The culture super-natant showed inhibitory effect	Chitinase, chitosanase, β-1,3 glucanase, and protease	Abdel-Aziz *et al.* 2012*b*.
5. *N*-acetyl-glucosamine	7.0	30	–	–	Chitinase, chitosanase, and β-glucanases.	Abdel-Aziz *et al.* 2012*c*.

Tab. 7.2: Enzymes and metabolites produced by strain *P. alvei* NRC14, under abiotic-stress conditions, during growth on different carbon sources at various temperatues and pH values.

Carbon source	pH	T/°C	Metabolites produced	Effect	Enzymes produced	Reference
1. Flaked chitin	7.0	40	Antifungal compound	Show inhibitory effect	Culture broth was free of enzymes.	Abdel-Aziz *et al.* 2012*b*.
1. Flaked chitin + sugarcane molasses	6.5	40	Antmicrobial compound	Show inhibitory effects		Abdel-Aziz *et al.* 2012*c*.
2. Flaked chitosan	6.0	40	Antifungal compound	Show inhibitory effect	Chitosanase, chitinase, β-glucanase.	Abdel-Aziz *et al.* 2012*d*.
3. Flaked chitosan	3.0	30	Antibacterial compound	Show inhibitory effect	Culture broth contained chitosanase, chitinase protease	Abdel-Aziz *et al.* 2013.
4.Sucrose/Starch	6.0	10	Biopolymer flocculant	A great flocculating activity	Not detected	Abdel-Aziz *et al.* 2014.

7.3.2 Microbial Communities

7.3.2.1 Soil Ecosystem

The soil ecosystem can be defined as an interdependent life-support system composed of air, water, minerals, macro- and micro-organisms. These organisms can be distinguished into microflora (*e.g.*, bacteria, archaea, and fungi), earthworms, slugs and snails, protozoa, nematodes, and different arthropods. These soil biota, one of the most important soil habitats, play a major role in many essential natural processes, which determine nutrient recycling and nutrient and water availability for agricultural productivity (Bardgett 2005). The interactions between soil organisms and soil functions are known to be incredibly complex. Soil organisms break down organic matter, making nutrients available for plant uptake and other soil microbiota. Microbial metabolites act to maintain soil structure and viability. The decomposition of organic matter by soil organisms has an immense influence on soil fertility, plant growth, soil structure, and carbon storage. Terrestrial biodiversity tends to be highest near the equator, which seems to be the result of the warm climate and high primary productivity (Nannipieri et al. 2003). Biodiversity is not distributed evenly on Earth. It is the richest in the tropics. Marine biodiversity tends to be highest along coasts in the

Western Pacific, where sea surface temperature is highest and in the mid-latitudinal band in all oceans.

7.3.2.2 Marine Ecosystem

Marine systems differ from terrestrial systems in so many ways. Patterns of biodiversity in terrestrial systems may not be applicable to marine situations. The diversity of microorganisms associated with water, soil, and fauna is also incredibly rich. Marine ecosystem may be the largest one on earth and has been used for a variety of purposes by humans for millennia (Ellingsen 2002). Marine micoorganisms play a crucial role in almost all biogeochemical processes that sustain the biosphere, and provide a variety of products (goods) and functions (services) which are essential to sustain human well-being. Goods include production of food, natural substances, fine chemicals for biotechnology and pharmaceuticals. These goods are mainly metabolites produced by microorganisms. Aquatic organic matter comprises one of the largest reservoirs of reduced carbon on Earth, and it provides an important base for microbial life in the water column. Organic matter mineralization by heterotrophic microbes plays a crucial role in the cycling of carbon and other elements, thereby providing an important ecosystem service and affect microbial metabolism and diversity. It is a major challenge to establish links between the phylogenetic and metabolic diversity of microbes (Vincx et al. 2009). Marine organic matter comprises a heterogeneous matrix of dissolved molecules, colloids, gels and particles (Eva 2015). In sea-grass ecosystems, on other side, photosynthetic activity by the sea-grass themselves creates conditions in the rhizosphere that support varying metabolic pathways in the remineralization process, increasing the availability of nutrients required for sea-grass meadow development (Bourque *et al.* 2105).

7.3.2.3 Hot springs Ecosystem

Hot springs have been investigated since the XIX century, but isolation and examination of their thermophilic microbial inhabitants did not start until the 1950s (Olalla *et al.* 2013). Many thermophilic microorganisms have since been discovered and more inclusive understanding of microbial communities: their diversity, structure, the interactions existing between their components, and the factors shaping the nature of these communities (Ward *et al.* 1998, Kemp and Aller 2004). Thereafter, possibility of cloning and sequencing the total environmental DNA, defined as metagenome, have been reported (Olalla *et al.* 2013, Chan *et al.* 2015, Mangrola *et al.* 2015). Hot springs have been a source of thermophilic enzymes of industrial interest, encouraging further study of the poorly understood diversity of microbial life in these habitats (Olalla *et al.* 2013, Chan *et al.* 2015). Currently there is a great interest in hot springs, which are the natural habitat of thermophilic and hyperthermophilic microorganisms with optimal growth temperatures of >55°C and >80°C, respectively. Enzymes obtained from hot springs microbiota have been proved to be extremely valuable as biocatalysts for industrial and biotechnological purposes (Suman et al. 2015). Each hot spring differs from others in temperature, chemical composition and its gradients of temperature or light. Hot springs comprise several habitats, such as thermal fluids, microbial mats and sediments (Ward *et al.* 2002). Less diversity and richness in hot springs than in other aquatic environments have been reported (Kemp and Aller 2004, Olalla *et al.* 2013). Many studies, however, have been reported regarding cyanobacteria (Ward *et al.* 1998, Miller and Richard 2000, Klatt et al. 2011, Mandal and Rath 2015). Analysis of thermal communities from all over the world has found the recurrent presence of certain groups, although in some hot springs there are specific strains that contribute significantly to the composition of the microbial community, and whose presence is correlated to the geochemical properties of hot springs.

7.3.2.4 Gut Microbiota

The human gut microbiota has become the subject of extensive research in recent years. Human gut harbors an extremely complex and dynamic microbial community with symbiotic functional capabilities that directly affect human physiology, metabolism, nutrition as well as immune function and overall human health (Guinane and Cotter 2013). Most of the microorganisms residing in the gastrointestinal tract are bacteria, with a density of approximately 10^{13}–10^{14} cells/g fecal matter, in which 70 % of the total microbes colonize the colon (Wang et al. 2015). Gut microbial community plays an important role in protecting the host against pathogenic microbes [Fukuda et al. 2012], modulating immunity [Maynard et al. 2012], regulating metabolic processes [Tremaroli and Backhed 2012, Cani 2014], and is regarded as a neglected endocrine organ [Clarke et al. 2014]. Disruption to the gut microbiota has been linked with gastrointestinal conditions such as irritable bowel syndrome, ulcerative colitis, *Clostridium difficile*-associated diarrhea, inflammatory bowel disease, and obesity (David *et al.* 2015). The contribution of beneficial components of the gut microbiome to the host physiology, metabolism and immune function has become the focus of ever more attention, and will undoubtedly lead to new therapeutic approaches (Guinane and Cotter 2013, David *et al.* 2015). Dietary control of the intestinal microbiota is essential. The significant role of the gut microbiota in human physiology and disease is important issue to be understood. Probiotic and prebiotic products are marketed in response to consumer demand, and the science and technology around these products are progressing rapidly. Mechanistic understanding and systems biology approaches of studying functional interactions between diet composition, gut microbiota, host-microbial co-metabolism, the complex signaling pathways involved as well as shaping of innate immune development are recently reported (Duffy *et al.* 2015). Confirming the safety and mechanisms of action of probiotic/prebiotic formulations is critical. Beneficial modulation of the human microbiota by using these nutritional and biotherapeutic strategies holds considerable promise as next-generation drugs, vaccinomics, metabolic agents, and in novel food discovery (Duffy *et al.* 2015).

7.4 Microbial Development

The study of microbial development due to starvation for nutrient or other abiotic-stress factors has, however, shown that microorganisms are capable of forming complex differentiation and behaviors. Examples include: i) *Bacillus subtilis* spore formation, in which individual vegetative cells integrate multiple external and internal signals to successfully synthesize a new morphological structure. Endospores adapt to survival in a variety of harsh environments; ii) the behavior of *Myxococcus xanthus* complexity is an entire cell population, which forms a multicellular fruiting body that contains resistant spores; iii) formation of surface-attached microbial communities, known as biofilms, serves as another excellent model system for the study of microbial development. Development is marked by these features: *a*) Cells of the developing organism maintain the same DNA content as the original cell; *b*) As cells reproduce, they change into different kinds of cells with different complementary functions. This process of change is called differentiation; *c*) the differentiated cells form tissues and organs with different functions that cooperate with each other for the good of the overall organism (John 2015). Some lineages of cells are even programmed to die, to enable maturation of organs; this programmed cell death is called apoptosis (*i.e.*, the death of cells which occurs as a normal and controlled part of an organism's growth or development). In the evolutionary context, however, multicellular development forms a continuum with developmental processes in microbes–many of which have major significance for the environment and for human health (John 2015).

Microbial adaptation to the nutrient deprivation and starvation are extensively studied. The relationship between nutrition, community structure and function during ecosystem development, particularly for microorganisms, is well known. Shifts in nutrient availability may, in part, evolve microbial community development (Joseph *et al.* 2014). Bacteria have developed novel mechanisms for the regulation of both catabolic and anabolic pathways to encounter starvation and abiotic stress in the surrounding environments. Unlike plant and animal cells, most bacteria are exposed to a constantly changing physical and chemical environment. Within limits, bacteria can react to changes in their environment through changes in patterns of structural proteins, transport proteins, toxins, enzymes, etc., which adapt them to a particular ecological situation. For example, *E. coli* does not produce fimbriae for colonization purposes when living in a planktonic (free-floating or swimming) environment. *Vibrio cholerae* does not produce the cholera toxin that causes diarrhea unless it is in the human intestinal tract. *Bacillus subtilis* does not make the enzymes for tryptophan biosynthesis if it can find pre-existing tryptophan in its environment. If *E. coli* is fed glucose and lactose together, it will use the glucose first because it takes two less enzymes to use glucose than it does to use lactose. The bacterium *Neisseria gonorrhoeae* will develop a sophisticated iron gathering and transport system if it senses that iron is in short supply in its environment (Todar 2012). Nutrients alone act as an important control over microbial community development, greatly accelerating the rate of succession. Nutrient availability may be important in controlling successional changes in microbial development and community composition (Joseph *et al.* 2014). Examples for microbial development are discussed below.

7.4.1 Spore Forming

Microorganisms sense and adapt to changes in their environment. When favored nutrients are exhausted, some bacteria may become motile to seek out nutrients, forming endospores, or they may produce enzymes to exploit alternative resources. Spores are minute, typically one-celled, reproductive unit capable of giving rise to a new individual without sexual fusion, characteristic of lower plants, bacteria, fungi, and protozoans. Grain mold fungi also produce spores capable of aerial dispersal in the field as well as within a grain storage bin. One example of an extreme survival strategy employed by certain Gram-positive bacteria is the formation of endospores. This complex developmental process is often initiated in response to nutrient deprivation. It allows the bacterium to produce a dormant and highly resistant cell to preserve the cell's genetic material in times of extreme stress. Endospores can survive environmental assaults that would normally kill the bacterium. These stresses include high temperature, high UV irradiation, desiccation, chemical damage and enzymatic destruction. The extraordinary resistance properties of endospores make them of particular importance because they are not readily killed by many antimicrobial treatments. A variety of different microorganisms form "spores" or "cysts", but the endospores of low Gram-positive bacteria are by far the most resistant to harsh conditions. Endospore formation is usually triggered by a lack of nutrients, and usually occurs in Gram-positive bacteria. In endospore formation, the bacterium divides within its cell wall. One side then engulfs the other. Endospores enable bacteria to lie dormant for extended periods, even centuries. Revival of spores millions of years old has been claimed. When the environment becomes more favorable, the endospore can reactivate itself to the vegetative state.

7.4.2 Formation of Fruiting Body

The strain *Myxococcus xanthus* is a Gram-negative rod-shaped bacterium. Under starvation conditions, it undergoes a magnificent developmental process in which roughly 100,000 individual cells aggregate to form a structure called the fruiting body over the course of several hour (Kuner and Kaiser 1982). When the food supply runs low,

M. xanthus initiates a complex developmental program, and tens of thousands of cells aggregate to form dome-like structures called fruiting bodies (Fig. 7.1). When the building movements complete, a portion of cells within the fruiting body differentiates into heat-resistant spores, which can re-germinate under favorable environments. The behavior of *Myxococcus xanthus* complexity is an example for bacterial development due to nutrient unavailability. An important feature of *M. xanthus* behavior is its ability to move on a solid surface by a mechanism called "gliding". Gliding motility is a method of locomotion that enables bacterial movement on a solid surface without the aid of flagella (Kuner and Kaiser 1982). The study of gliding motility is important since it is widespread in nature and has been shown to be central for biofilm production and pathogenesis.

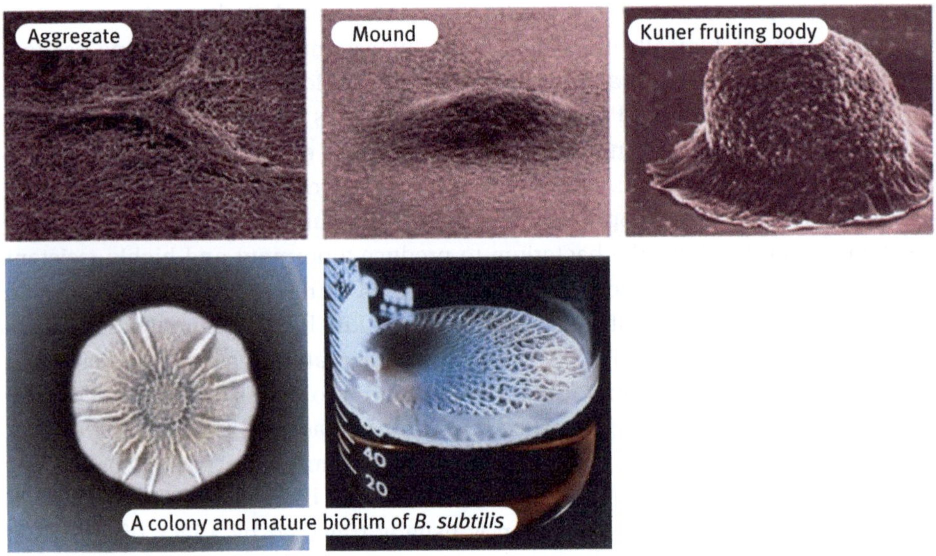

Fig. 7.1: Forms of bacterial development: Fruiting body formation by Myxcoccus xanthus (Above image) and highly structured biofilm formed by B. subtilis strain (Down images). (Kuner and Kaiser 1982, *Steven et al. 2001*).

7.4.3 Formation of Biofilm

Bacterial biofilm is a structured community of bacterial cells enclosed in a self-produced polymeric matrix and adherent to an inert or living surface, which constitutes a protected mode of growth that allows survival, where bacteria in hostile environment. Biofilm-forming microorganisms have been shown to elicit specific mechanisms for initial attachment to a surface, formation of microcolony leading to development of mature biofilm (Fig. 7.1). They differ from their free-living counterparts in their growth rate, composition and increased resistance to biocides, antibiotics and antibodies. This makes them highly difficult to eradicate with therapeutic doses of antimicrobial agents. Understanding of mechanism of their formation and survival under sessile environments may help in devising control strategies. Efficient biofilm disruption and removal requires better knowledge of the structural and rheological properties of the biofilm matrix and the transport mechanisms may help in devising control strategies by which antibiotics can penetrate the colony.

7.4.3.1 History

Antoni van Leeuwenhoek, using his simple microscope, first observed "animalcule" on tooth surfaces and can be credited with the discovery of microbial biofilms (Heukelekiann and A. Heller 1940). The first recorded observation concerning biofilms, however, comes from 1933 papers by ZoBell and Henrici about bacteria in submerged surfaces" (Henrici 1933, ZoBell and Allen 1933). The bacterial growth and activity were substantially enhanced by the incorporation of a surface to which these organisms could attach, and the number of bacteria on the surfaces was dramatically higher than in the surrounding medium. Characklis (1973), studied microbial slimes in industrial water systems and showed that they were not only tenacious, but also highly resistant to disinfectants such as chlorine. Costerton (1978) was instrumental in alerting the world about the importance of biofilms, coining the term in 1978. He put forth a theory of biofilms that explained the mechanisms whereby microorganisms adhere to living and nonliving materials and the benefits accrued by this ecologic niche. The two major thrusts in the last decade which dramatically impacted our understanding of biofilms are, the utilization of the confocal laser scanning microscope to characterize biofilm ultrastructure, and an investigation of the genes involved in cell adhesion and biofilm formation (Prakash et al. 2003, Yoko et al. 2015).

7.4.3.2 Biofilm Development

Phases of biofilm development include initial attachment to a surface, followed by formation of micro-colonies. Subsequently, biofilms grow and mature (Figure 7.2). The molecules that connect the cells in a biofilm are predominantly exopolysaccharide, teichoic acids, and some proteins such as the accumulation-associated protein. Finally, some cells can detach and remove outside the biofilm to re-establish and form new aggregation of biofilms (Figure 7.2). Detachment is a prerequisite for the dissemination and it is facilitated by expression of the surfactant-like peptides, which are important in producing the 3-dimensional structure of the biofilm. Biofilms can comprise a single microbial species or multiple microbial species and can form on a range of biotic and abiotic surfaces (O'Toole et al. 2000). Although mixed-species biofilms predominate in most environments, single-species biofilms exist in a variety of infections and on the surface of medical implants (Nwodo et al. 2012, Henning et al. 2015).

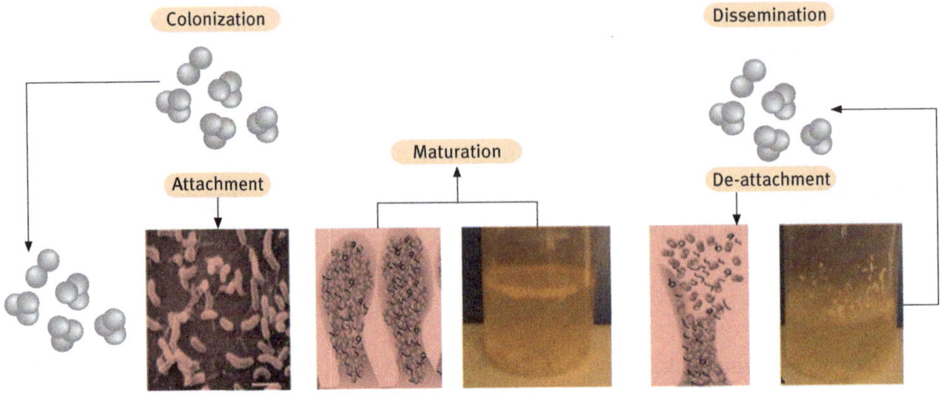

Fig. 7.2: Scheme for the phases of biofilm development by the soil bacterial strain Paenibacillus alvei NRC-14 (The design is adapted from Otto, 2008).

These single-species biofilms are the focus of most current research. *Pseudomonas aeruginosa* has emerged as the most studied single-species, biofilm-forming gram-negative bacterium. In addition, *Pseudomonas fluorescens*, *Escherichia coli*, and *Vibrio cholera* have also been studied in detail (Hobley et al. 2015). The gram-positive biofilm-forming bacteria that have been studied include *Staphylococcus epidermidis*, *Staphylococcus aureus*, and the enterococci (Henning et al. 2015).

Recent genetic and molecular approaches used to study bacterial biofilms have identified genes, regulatory for initial cell-surface interactions, biofilm maturation, and the return of biofilm microorganisms to a planktonic mode of growth. Biofilms of *Pseudomonas aeruginosa* in the lungs of cystic fibrosis patients can be life-threatening. Cystic fibrosis and the thick mucus, occurred due to inherited disorder, provides a suitable environment for an infection to become established and the infectious bacteria become less susceptible to antibiotics. For patients, the result is a prolonged infection that is very difficult to be treated. The structural nature of the biofilm and characteristics of the sessile cells produce resistance towards the antimicrobial agents, leading to a protected environment against adverse conditions (Patil 2005). Some mechanisms of bacterial survival inside the biofilm include: i) resistance to antibiotics due to persistence against much higher concentrations of antibiotics and disinfectants than free-living organisms because many antibiotics act only on actively growing cells, i.e., bacteria inside the biofilm is dormant. Such dormant cells can serve to reestablish a biofilm once the antimicrobial agent is no longer present; ii) the layered nature of a biofilm, i.e., the effectiveness of a disinfectant, such as bleach, is depleted as it acts on outer layers of the film; bacteria located in inner layers may survive; iii) the generation of proteins that provide antimicrobial resistance, such as enzymes that inactivate hydrogen peroxide. Some biofilms are able to manufacture larger quantities of such enzymes so they become more resistant than planktonic (free-floating) bacteria.

7.5 Microbial Metabolic Diversity

Immense metabolic diversity among the bacterial population from different ecological habitats is a well-recognized fact. During such phases of their life cycle, microbes produce a large number of secondary metabolites. The primary function of these metabolites is to mediate critical functions in the life cycle of the producing organism under the given conditions. However they have also been demonstrated to possess varied properties of potential therapeutic applications, such as, anti-diabetic adjuvant, anti-oxidant, anti-inflammatory, anti-cancer, enzyme inhibitors, immunosuppressents, immunostimulant, antitrypanosomal (Singla *et al.* 2014). Secondary metabolite production in microbes is strongly influenced by nutritional factors and growth conditions. Metabolism processes describe the sum of biochemical reactions required for energy generation and the use of energy to synthesize cell material from small molecules in the environment. While cell physiology describes the role of metabolic reactions in the life processes of organisms, the metabolic process describes anabolism and catabolism of a cell. Anabolism involves the biosynthesis and building of complex molecules from simple ones, and requires energy (ATP). Catabolism involves the degradation and breaking down of complex molecules into simple ones, and generates energy (ATP).

Microbial metabolism is classified into nutritional groups on the basis of three major criteria: a) Source of energy, used for growth; b) Source of carbon; and c) Sours of electron donors used for growth. This strain *Paenibacillus alvei* NRC-14, in fact, is particularly interesting because it: 1) produces large quantities and a variety of extracellular metabolites; 2) controlling its growth conditions using minimum nutrient substrates; 3) synthesizes biofilm as a form of self-protection when exposed to high or low temperatures; 4) produces a tremendous variety of carbohydrate-active enzymes and exhibits multiple-adaptive response when exposed to abiotic stress conditions; 5) synthesizes a biopolymer flocculant has chitosan-like structure, when grown at 30 °C, under growth shaking conditions; 6) produces antimicrobial agents when exposed to 40 °C under shaking culture condition; 7) synthesizes a biofilm when exposed to a high temperature (39±2 °C), under static conditions. Interestingly, when a 24-hrs old of this culture was kept at refrigerator condition (4 °C), a viscous biopolymer was synthesized, whose viscosity increased by time (Abdel-Aziz *et al.* 2014). This strain is an excellent model for synthesis of natural products and a diversity of metabolites under either normal or abiotic stress conditions (Figure 7.3).

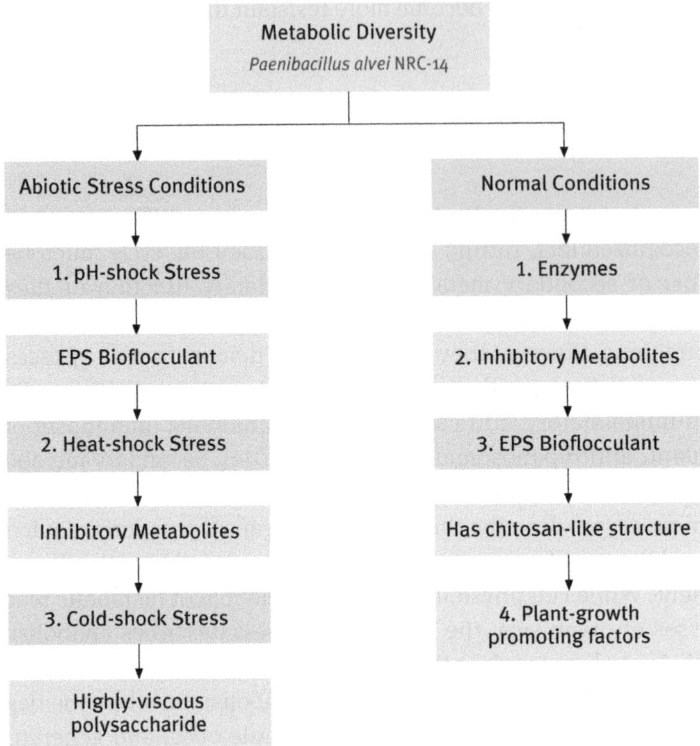

Fig. 7.3: The broad metabolic diversity produced by the soil bacterial strain Paenibacillus alvei NRC-14.

7.5.1 Microbial Metabolites

Metabolites are the intermediates and products of metabolism. The term metabolite is usually restricted to small molecules. Bacteria have extensively been used in biological research due to their abundant, easily growth, and are relatively simplified in cellular organization. Bacterial metabolism have extensively been studied to know how to inhibit or stop their growth and harmful metabolites. Metabolism products can be categorized into primary and secondary metabolites. Microorganisms can be used in industrial microbiology to obtain amino acids, develop vaccines, antibiotics, and fine chemicals for pharmaceutical fields (Boundless. “Primary and Secondary Metabolites.” *Boundless Microbiology*. Boundless, 03 Jul. 2014).

7.5.1.1 Primary Metabolites

Primary metabolites are directly involved in normal growth or energy-yielding catabolism, development, and reproduction. Primary metabolite is typically a key component in maintaining normal physiological processes; thus, it is often referred to as a central metabolite. Primary metabolites are typically formed during the growth phase as a result of energy metabolism, and are deemed essential for proper growth. Amino acids, nucleotides, proteins, carbohydrates, lipids, vitamins are primary metabolites that needed for cell growth, whereas ethanol, acetone-butanol, and organic acids are those required for deriving energy (Boundless.). Within the field of industrial microbiology, alcohol is one of the most common primary metabolites used for large-scale production. Specifically, alcohol is used for processes involving fermentation which produce products like beer and wine. Additionally, primary metabolites such as amino acids–including L-glutamate and L-lysine, which are commonly used as supplements–are isolated via the mass production of a specific bacterial species, *Corynebacteria glutamicum*. Another example of a primary metabolite commonly used in industrial microbiology includes citric acid. Citric acid, produced by *Aspergillus niger*, is one of the most widely used ingredients in food production (Boundless 2014). It is commonly used in pharmaceutical and cosmetic industries as well.

7.5.1.2 Secondary metabolites

Secondary metabolites don't play a role in microbial-cell growth, development, and are typically formed during the end or near the stationary phase of growth. Many of the identified secondary metabolites have a role in ecological function, including defense mechanism(s), by serving as antibiotics and by producing pigments. Toxins, gibberellins, alkaloids, antibiotics, and biopolymers are examples of secondary metabolites. When there is depletion of one or more nutrients in the culture medium, secondary metabolites are produced by microorganisms. Secondary metabolites usually have an important ecological function and serve a diverse survival functions in nature. Although not essential for microbial growth, secondary metabolites are very important for the health, nutrition, and economics of our societies (Berdy, 2005). Most of the secondary microbial metabolites exhibit a great numbers of diverse and versatile biological effects, first of all antimicrobial activities (Berdy 2005). Already hundreds of different pathogenic and other microbes such as Gram-positive, Gram-negative bacteria, fungi, yeasts, *etc.*, are described as test organisms in the direct activity-based screenings.

7.5.2 Natural Microbial Products

Nature is an attractive source of new therapeutic candidates as a tremendous chemical diversity is found in millions of species of plants, animals, marine organisms and microorganisms. Natural products, derived from microbes due to the metabolic diversity, remain an important source of new drugs and new chemical entities. Microorganisms are being employed, since several decades, for large scale production of a variety of bio-chemicals ranging from alcohol to antibiotics, and in processing of foods and feeds. Microorganisms have great potential as natural sources of drugs for the treatment and prevention of diseases like cancer, anaemia, diarrhoea, obesity, diabetes, atopic dermatitis, Crohn's disease, etc. They are also potential sources of natural antioxidants, colours, immuno-suppressants, enzyme inhibitors, hypocholesterolemic agents, vitamins, enzymes, and antibiotics (Gupta et al. 2014). The term "biopolymers" usually describes polymers produced in a natural metabolic way by living microorganisms. The molecular backbones of these biopolymers are composed of repeating units of polysaccharides, nucleic acids, or amino acids which are quite suitable for high-value medical applications such as tissue engineering and drug delivery (Gupta et al. 2014). Nowadays, green biotechnology, green products, and green chemicals have gain more and more importance in the industrial practice. The development of such green biotechnology has transformed microbes into useful factories for producing valuable polymers from renewable biomass. The most important features of the bioactive microbial metabolites are their microbial origin, their interaction with the environment and their unique chemical structures. Natural products including the microbial metabolites may be practically utilized in three different ways: i) Applying the natural/fermentation product directly in the medicine, agriculture, or in any other fields; ii) Using as a starting material for subsequent chemical or microbiological modification (Derivatization); or iii) They can be used as lead compounds for chemical synthesis of new analogs (Gupta et al. 2014). Recent progress at the interface of chemistry and biology has enabled the production of a variety of new biopolymers with properties that substantially differ from synthetic polymers. Some of the more biopolymers such as γ-polyglutamic acid, bacterial cellulose and xanthan gum are mainly produced by microorganisms.

7.5.3 Gut Microbiota and Human Health

The intestinal microbiota of healthy individuals is known to confer a number of health benefits relating to, for example, protection against pathogens, nutrition assimilation, host metabolism and immune modulation. Gut microbiota plays a central role in human gastrointestinal health and disease. It is, therefore, logical that modulating the gut microbiota should be considered as a therapeutic strategy to treat chronic disease. Microorganisms quickly adapt to a changing environment by sensing and responding to environmental change or stress, such as nutrients. Understanding how pathogens sense and utilize nutrients during pathogen-host interactions provides an important insight to control diseases caused by these microbes (Xue 2015). There is growing evidence that the metabolic diversity products of gut microbiota play a central role in human gastrointestinal health, and can reduce or prevent diseases. Modulation of gut microbiota include supplementation with probiotics to fortify the immune system and eliminate pathogens (Guinane and Cotter 2013, Zhang et al. 2015). Prebiotics and probiotics are becoming increasingly popular due to their conferred benefits upon host well-being and health. Probiotics can be defined as live microorganisms which, when administered in adequate amounts, confer a health benefit on the host. Prebiotics are selectively fermented ingredients that allow specific changes, both in the composition and in the activity of gastrointestinal microflora. Prebiotics typically are fibers that cannot be digested by the host, but are metabolized by the colonic microbiome resulting in expansion of certain bacterial species and the release of metabolites, such as short-chain fatty acids and thus have the potential to improve gastrointestinal health. The combination of probiotics and prebiotics is termed "synbiotic". The increasing use of probiotic and prebiotic foods has led to the studies on their actual functions in the human body. In the environment, microorganisms are extremely important in recycling of nutrients, balance of trophic chains, vital physiological activities in plants and animals as well as the conservation of natural habitats (Panizzon et al. 2015). In human food, these microorganisms contribute from flavoring products to the synthesis of antimicrobial substances and vitamins essential to living beings (Conlon and Bird 2015).

7.5.3.1 Function of Metabolic Diversity

Microbial colonization occurs throughout the length of the human gut from the oral cavity to the rectum. The stomach and proximal small intestine support relatively low numbers of microorganisms that can tolerate the pH conditions, oxygen exposure, and the relatively rapid transit rates that prevail in these regions. In contrast, conditions in the large intestine generally favor the establishment of an extremely dense microbial community that is dominated by obligate anaerobic bacteria. These organisms gain energy mainly through the fermentation of non-digested dietary components and of host secretions. When the host-bacteria system is in balance or homeostasis, it can contribute for many aspects of host physiology. However, when the tight balance between host and microbiota is broken it can result in the development of certain pathologies. Supplementation with probiotics and prebiotics provide great benefits for human health sustainability. Health promoting functions of gut microbiota and probiotics include: a) Inhibition of growth of exogeneous and/or harmful bacteria; b) aid in digestion and/or absorption of food ingredients/minerals; c) Stimulation of immune functions; d) prevention of pathogens colonization; e) restore the normal intestinal flora during antibiotic therapy (Gibson and Roberfroid 1995). Gut bacterial communities or "microbiota" improve the digestion of nutrients and provide new metabolic functions to their host such as the short chain fatty acids and greatly improve the human health. (Figure 7.4).

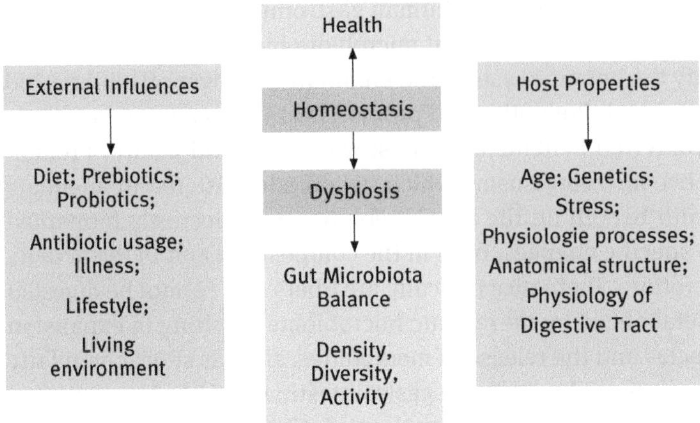

Fig. 7.4: Balance of gut microbiota by metabolic diversity of probiotics to sustain human health and well-being (Adapted from Zhang et al. 2015).

Recently, colonic health has been linked to the maintaining overall health status and reducing the risk of diseases by changes in lifestyle (Conlon and Bird 2015, Neis et al. 2015). Functional foods, such as probiotics, prebiotics, and dietary fibers target the colon and affect its environment and enhance short fatty acid production. Short chain fatty acids, also referred to as volatile fatty acids, are fatty acids with an aliphatic tail of less than six carbon atoms (Brody 1999, Mayu *et al.* 2015). Short chain fatty acids include formic, acetic, propionic, butyric, and valeric acids. These fatty acids are rapidly absorbed in the cecum and colon with only 5 % to 10 % being excreted in the feces (Hijova and chremalova 2007). The major short chain fatty acids: acetate, propionate, and butyrate, are absorbed at comparable rates in different regions of the colon. Once absorbed, short chain fatty acids are metabolized at three major sites in the body: i) colonic epithelium cells that use butyrate as a major substrate for the maintenance of energy; ii) liver cells that metabolize residual butyrate and propionate; 50 to 70 % of acetate is also taken up by the liver; iii) muscle cells that generate energy the oxidation of residual acetate (Hijova and chremalova 2007). Role and benefits of short chain fatty acids has expanded to include their role as nutrients for the colonic epithelium, as modulators of the colonic and intracellular pH, cell volume, and other functions associated with ion transport. Thus, overall of such diversity in microbial metabolites and modulation of gut microbiota can verify sustainability of human health.

7.5.3.2 Antimicrobials and Human Health

Use of oral probiotic, provided a great metabolic cultures to restore the gut microbiota has led to: i) promising results in the treatment of intestinal disorders; ii) manipulation of the gut microbiota in a way that will benefit host health; and/or iii) synthesis and production of antimicrobials such as bacteriocins to exclude the pathogens (Guinane and Cotter 2013). The ability of bacteria to produce bacteriocins is a common feature among gut microbes and is a good approach to overcome the negative effect of the classical antibiotics on beneficial gut microbiota. Bacteriocins are ribosomally synthesized small antimicrobial peptides produced by bacteria. Bacteriocins with a narrow spectrum of activity against a target microorganism can offer a therapeutic alternative to traditional antibiotics. Gut-associated bacteriocin producers also have the advantage of producing the antimicrobial *in situ*, and therefore, in these situations, the antimicrobial peptide is not affected by proteolysis during gastric transit or does not need to be encapsulated (Guinane and Cotter 2013). Bacteriocins have been shown to be useful in controlling a number of gastrointestinal pathogens *in vivo*, including *Listeria monocytogenes*, *Salmonella* spp., *Campylobacter jejuni*, and *Clostridium difficile* (Guinane and Cotter 2013, David et al. 2015).

7.5.4 Therapeutic Spectrum of Bacterial Metabolites

Microorganisms are known to produce a large number of secondary metabolites. While these compounds help the producing organisms to survive under unfavorable conditions, they are not essential for growth and reproduction. Many of these metabolites have been known to possess antimicrobial properties that may provide the organism a competition for food and space in its habitat. These metabolites may have significant roles in medicine against human infectious diseases and in the pharmaceutical fields. The therapeutic spectrum and structural classification of the bacterial secondary metabolites is explored in an attempt to somewhat simplify further studies on their structure-function relationships towards establishing the diverse therapeutic potential of these compounds (Singla *et al.* 2014). Exploitation of microorganisms to produce food ingredients has been going on since antiquity. Microorganisms are being employed, since several decades for the large scale production of a variety of bio-chemicals ranging from alcohol to antibiotics and in processing of foods and feeds. Microorganisms have great potential as natural sources of drugs for the treatment and prevention of diseases like cancer, anaemia, diarrhoea, obesity, diabetes, atopic dermatitis, Crohn's disease, etc. (Gupta *et al.* 2014). Microorganisms are also potential sources of natural antioxidants, colours, immuno-suppressants, enzyme inhibitors, hypocholesterolemic agents, vitamins, enzymes, and antibiotics.

7.6 Microbiome and Sustainable Healthcare

The human microbiome is composed of trillions of microorganisms, as well as their genomes, that live in our gut, mouth, skin and elsewhere in our bodies. These microbial communities have numerous beneficial functions relevant to healthcare and supporting life. These resident microbes attribute to human health and well-being, particularly with respect to their roles in stimulating the immune systems, contributing to the digestion of food, and acting as a first line of defense against pathogens. The total number of genes associated with the human microbiome exceeds the total number of human genes (Rodney and Janice 2015). The influence of microbiota reaches from primary metabolites to secondary effects such as substrate competition or the activation of enzymes (Tralau *et al.* 2015). The concept of microbiome and its competent metabolites add enormous complexity to human physiology. Yet, the potential specificity of microbial targets harbors therapeutic promise for diseases such as diabetes, cancer and psychiatric disorders (Tralau *et al.* 2015). A better physiological and biochemical understanding of the microbiome is, thus, of high priority for academia and biomedical research. Management of the microbiome to reduce health risk, spare antibiotic use, and reduce the risk of recurrent infections, may provide a more effective healthcare strategy. Microbiome management, including microbiota seeding and feeding, appears likely to be a core component of a path toward sustainable healthcare. There are potential opportunities for microbiome management at different life stages. A focus on the microbiota particularly during early life might be an effective strategy for disease prevention (Collado *et al.* 2014). In fact, the microbial genetic component of humans is sufficiently significant that it has been referred to as our "second genome" (Rodney and Janice 2015). Our microbiota are the first exposed and the first responders to the food. Microbiome can regulate the function of numerous physiological systems, and thus effective management of the microbiome takes a significantly higher priority.

The bacteria colonizing the human intestinal tract exhibit a high phylogenetic diversity that reflects their immense metabolic potential. Metabolic diversity via the microbiome metabolism is a critical factor for the microbial development and gut microbiota balance. Our microbiota are at the center of: 1) interactions with the environment, 2) nutrition, 3) metabolism, and 3) regulation of our host development and physiology. Therefore, a focused priority on microbiome management should increase the effectiveness of risk assessment, drug discovery and efficacy, prevention, and medical therapies bringing us that much closer to sustainable healthcare. By virtue of their catalytic activity, the human gut microbiota has an impact on gastrointestinal function and host health. All dietary components that escape digestion in the small intestine are potential substrates of the bacteria in the colon (Blaut and Clavel 2007). The bacterial conversion of carbohydrates, proteins and nonnutritive compounds such as polyphenolic substances leads to the formation of a large number of compounds that may have beneficial or adverse effects on human health. A wide variety of factors has been reported to affect the microbiome (Figure 7.4). These include lifestyle choices, diet (probiotic and prebiotic), infections, environmental chemicals, drugs, and stresses (Zhang *et al.* 2015).

7.7 Conclusion and Prospects

It is well known that changes in culture conditions, nutrient, and starvation, all may result in a big alteration in the biosynthesis and diversity of microbial metabolites. The short history, specific features and future prospects in research of bioactive microbial metabolites are reviewed. The microbial origin, diversity of producing species, functions, and various bioactivities of metabolites, unique features of their chemical structures are discussed. However, due to the emergence of multi-resistant strains and the increasing serious problems of chemotherapy, urgent clinical needs and new challenges in the therapy of physiological diseases are highly required. Diversity of microbial metabolites can be exploited for human health and well-being. Diet and lifestyle has an important influence on the composition of human gut microbiota. To investigate the influence of long-term dietary impact on gut microbiota, it is important to carefully assess the habitual food intake. Gut microbiota plays a pivotal role in human health and disease. The resident microbiota influence host metabolism, physiology and immune system development, while trouble of the microbial community can result in chronic gastrointestinal diseases. While the revolution in molecular technologies has provided the tools necessary to more accurately study of gut microbiota, the need to more accurately elucidate for the relationships between the gut microbiota and several intestinal pathologies is required. Understanding the part that microbial populations play in gastrointestinal disease is fundamental to the ultimate development of appropriate therapeutic approaches. Most frequent test organisms such as *Bacillus subtilis*, *Staphylococcus aureus*, *Sarcina lutea*, *Escherichia coli*, *Pseudomonas aeruginosa*, and *Candida albicans* are effectively inhibited by microbial metabolites. However, the infectious diseases are mainly treated with natural antibiotics and their derivatives, while still the majority of drugs applied in the so called physiological diseases are synthetic products. The challenge for natural medical products to treat these diseases is huge.

Tab. 7.3: Diversity of bacterial metabolites (Biopolymers) and their functional attributes (Nwodo *et al.* 2012, Patil and Prajapati 2013).

Biopolymer	Bacterial strain	Monomer Units	Applications
Dextran	*Leuconostoc mesenteriodes*	Glucose	Confectionary products to improve moisture retention,viscosity and inhibit sugar crystallization, In human medicine as blood flow improving agent and as cholesterol-lowering agent, in separation technology and as micro-carrier in tissue/cell culture

Tab. 7.3: (continued)

Biopolymer	Bacterial strain	Monomer Units	Applications
Alginate	*P. aeruginosa*, *Azotobacter vinelandii*	Guluronic acid and mannuronic acid	Food hydrocolloid and medicine: surgical dressings, hypo-allergic wound-healing tissue, and controlled drug release. As an immobilization matrix for viable cells and enzymes, coating of roots of seedlings and plants to prevent desiccation.
Xanthan	*Xanthomonas* spp.	Glucose, mannose, and glucuronic acid	Foods, pharmaceuticals, cosmetics and personal-care products. As viscosifier, stabilizer, emulsifier and suspending agent in food industries, in food as thickening and stabilizing agent- often used in combination with guar gum, in secondary and tertiary crude-oil recovery.
Curdlan	*Rhizobium meliloti*, *Agrobacterium radiobacter*	Glucose	Foods, pharmaceutical industry, heavy metal removal and concrete additive. As a gelling agent, immobilization matrix, displays promising high antiretroviral activity (anti AIDS-drug)
Succino-glycan	*Alcaligenes faecalis* var. myxogenes	Glucose and Galactose	Foods and oil recovery, as a gelling agent and matrix for immobilization.
Glucuronan	*Gluconaceto-bacter hansenii*	Glucuronic acid	Foods and cosmetics products
Colanic acid	*E. coli*, *Shigella* spp., *Salmo-nella* spp. and *Enterobacter* spp.	Fucose, glucose, glucoronate, and galactose	Cosmetics and personal care products
Cellulose	*Acetobacter* spp.	Glucose	Foods (indigestible fiber which can be impregnated with amino acids, vitamins and minerals), biomedical (wound healing, tissue engineered blood vessels), in separation technology, as acoustic membranes in audio-visual equipment
Levan	*Bacillus*, *Streptococcus*, *Pseudomonas*, *Zymomonas*	Fructose	For production of sweet confectionary, ice cream as viscosifer and stabilizer
Emulsan	*Acinetobacter calcoaceticus*	Sugar & fatty acids	Crude-oil recovery and other applications are similar as for alginate

7.8 References

[1] Okasha S. (2010). Does diversity always grow? Nature, 466, 318. Doi:10.1038/466318a.
[2] Hibbing M, Clay F, Parsek M. and Peterson S. (2010). Bacterial competition: surviving and thriving in the microbial jungle. Nat. Rev. Microbiol., 8, 15–25.
[3] Singla R, Harsh D. and Dubey A. (2014). Therapeutic Spectrum of Bacterial Metabolites. Indo Global J. Pharm. Sci., 4, 52–64.
[4] Kratochwil A. (1999). *Biodiversity in ecosystems: some principles*. Pages 5–38 In: A. Kratochwil, (ed.). Biodiversity in Ecosystems: principles and case studies of different complexity levels. Kluwer Academic Publishers. The Netherlands.
[5] Gaston K. (2000). Global patterns in biodiversity. Nature, 405, 220–227.
[6] Das S, Lyla P. and Khan S. (2006). Marine microbial biodiversity: Present status and advanced statistical paradigms. Published by the Society for Conservation Biology Asia.
[7] Sahney S, Benton M, and Ferry P. (2010). Links between global taxonomic diversity, ecological diversity and the expansion of vertebrates on land. Biology Let., 6, 544–7.
[8] Staudt A. 2009. Identification and environmental factors critical to the production of exopolysaccharides by *Rhizobium tropici*. University of Notre Dame. India.
[9] Abdel-Aziz SM, Mouafi F. and Keera A. (2013). Amazing Abiotic Stress Response of a Novel Strain *Bacillus alvei* NRC-14 Isolated from Egyptian Soil. J Basic Appl Sci Res., 3, 299–309.
[10] Abdel-Aziz SM, Hamed H, and Keera A. (2014). Cold-Shock Stress Induces Formation of a Biopolymer Flocculant by *Bacillus alvei* NRC-14 under Static Incubation Conditions. SAJ Biotechnol., 1, 1–11.
[11] Abdel-Aziz SM, Hamed H, Mouafi F, Abdelwahed N. (2011). Extracellular metabolites produced by a novel strain, *Bacillus alvei* NRC-14: 3. Synthesis of a bioflocculant that has chitosan-like structure. Life Sci J., 8, 883–890.
[12] Abdel-Aziz SM. and Mouafi F. (2005). Production and some Properties of an inulinase produced by *Bacillus alvei*. New Egyp J Microbiol., 10, 200–211.
[13] Mouafi F, Abdel-Aziz SM, Moustafa Y, Keera A. (2012*a*). Extracellular Metabolites Produced by a Novel Strain, Bacillus alvei NRC-14: 4. Antimicrobial Activity of Chitosan Oligosaccharides. J Basic Appl Sci Res., 2, 11434–11442.
[14] Abdel-Aziz, SM, M. Moharam, H. Hamed and F. Mouafi, 2012*b*. Extracellular metabolites produced by a novel strain, *Bacillus alvei* NRC-14: 1. Some properties of the chitinolytic system. New York Sci J., 5, 53–62.
[15] Abdel-Aziz SM, Mouafi F, Keera A. (2012*c*). Extracellular Metabolites Produced by a Novel Strain Bacillus alvei NRC-14: 2. Molasses-Based Growth for High Production of N-acetylglucosamine. Aust J Basic Appl Sci., 6, 251–261.
[16] Abdel-Aziz SM, Mouafi F. and Keera A. (2012*d*). Extracellular Metabolites Produced by a Novel Strain Bacillus alveiNRC-14: 6. Heat-Shock Stress Induces Production of Antimicrobial Metabolites. J Basic Appl Sci Res., 3, 410–420.
[17] Bardgett R. (2005). The Biology of Soil: A Community and Ecosystem Approach, Oxford University Press.
[18] Nannipieri P, Ascher J, Ceccherini M, Landi L, Pietramellara G. and Renella G. (2003). Microbial diversity and soil functions. Eur J Soil Sci., 54, 655–670.
[19] Ellingsen K. (2002). Soft-sediment benthic biodiversity on the continental shelf in relation to environmental variability. Mar Ecology Prog Series, 232, 15–27.
[20] Vincx M, Westawski J. and Nash R. (2009). Marine Biodiversity and Ecosystem Functioning. Printbase, Dublin, Ireland: ISSN 2009–2539.

[21] Eva Ortega-Retuerta. (2015). The black boxes have just been opened: Linking organic matter composition and microbial diversity in aquatic environments. Institut de Ciencies del Mar, CSIC, Spain.

[22] Bourque A, Rebecca V. and James W. (2015). Microbial community structure and dynamics in restored subtropical sea-grass sediments. Aquat Microb Ecol., 74, 43–57.

[23] Olalla L, Maria E. and Maria I. (2013). Hot spring metagenomics. Life, *2*, 308–320.

[24] Ward, D, Bateson M, Ferris M. and Nold S. (1998). A natural view of microbial biodiversity within hot spring cyanobacterial mat communities. Microbiol Mol Biol Rev., 62, 1353–1170.

[25] Kemp P. and Aller J. (2004). Bacterial diversity in aquatic and other environments: What 16S rDNA libraries can tell us. FEMS Microbiol. Ecol., 47, 161–177.

[26] Chan C, Chan K, Tay Y, Chua Y. and Goh K. 2015. Diversity of thermophiles in a Malaysian hot spring determined using16S rRNA and shotgun metagenome sequencing. Front. Microbiol., 6, 1–15.

[27] Mangrola A, Pravin D, Prakash K, Joshi C. and Pate R. 2015. Shotgun metagenomic sequencing based microbial diversity assessment of Lasundra hot spring, India. Genimics Data, 4, 73–75.

[28] Suman A, Verma B, Ajar N. and Anil S. (2015). Bioprospecting for extracellular hydrolytic enzymes from culturable thermotolerant bacteria isolated from Manikaran thermal springs. Res J Biotechnol., 10, 33–42.

[29] Miller S. and Richard W. (2000). Evolution of thermotolerance in hot spring cyanobacteria of the genus *Synechococcus*. Appl Env Microbial., 66, 4222–4229.

[30] Klatt C, Jason M, Douglas B Rusch D, Bateson M, *et al.* (2011). Community ecology of hot spring cyanobacterial mats: predominant populations and their functional potential. ISME J., 5, 1262–1278.

[31] Mandal S. and Rath J. (2015). *Extremophilic cyanobacteria for novel drug development.* Springer Briefs In: Pharmaceutical Science & Drug Development. Springer International Publishing Switzerland. DOI 10.1007/978-3-319-12009-6_1.

[32] Guinane C. and Cotter P. (2013). Role of the gut microbiota in health and chronic gastro-intestinal disease: understanding a hidden metabolic organ. Therap Adv Gastroenterol., 6, 295–308.

[33] Wang W, Xu S, Ren Z, Tao L, Jiang J. and Zheng S. 2015. Application of metagenomics in the human gut microbiome. World J. Gastroenterol, 21, 803–814.

[34] Fukuda S, Toh H, Taylor T, Ohno H, Hattori M. 2012. Acetate-producing bifidobacteria protect the host from enteropathogenic infection via carbohydrate transporters. Gut Microbes, 3, 449–454.

[35] Tremaroli V. and Backhed F. (2012). Functional interactions between the gut microbiota and host metabolism. Nature, 489, 242–249.

[36] Cani PD. (2014). Metabolism in 2013: The gut microbiota manages host metabolism. Nat Rev Endocrinol, 10, 74–76.

[37] Clarke G, Stilling R, Kennedy P, Stanton C, *et al.* (2014). Gut microbiota: the neglected endocrine organ. Mol Endocrinol, 28, 1221–1238.

[38] David R, Hevia A, Rafael B, Lopez P, *et al.* (2015). Ranking the impact of human health disorders on gut metabolism: Systemic lupus erythematosus and obesity as study cases. Sci Reports, 5, 1–9.

[39] Duffy L, Daniel J, Van S. and Pamela S. (2015). Progress and Challenges in Developing Metabolic Footprints from Diet in Human Gut Microbial Cometabolism. J Nut., 145, 1123–1130.

[40] John Innes. (2015). Microbial Development. Molecular Microbiology Department, Kenyon College.

[41] Joseph E, Steven K, Ryan C, John L, *et al.* (2014). Nutrient addition dramatically accelerates microbial community succession. PLOS ONE, 9: www.plosone.org.

[42] Todar K. (2012). *Regulation and Control of Metabolism in Bacteria*. The American Society for Microbiology.
[43] Kuner J. and Kaiser D. (1982). Fruiting body morphogenesis in submerged cultures of *Myxococcus xanthus*. J Bacteriol 151, 458–461.
[44] Steven S, Jose E, Sigal B, Richard L. and Roberto K. (2001). Fruiting body formation by *Bacillus subtilis*. Proc Natl Acad Sci U S A., 98, 11621–11626.
[45] Heukelekiann H. and A. Heller A. (1940). Relation between food concentration and surface for bacterial growth. J Bacteriol., 40, 547–558.
[46] Henrici AT. (1933). Studies of freshwater bacteria. I. Adirect microscopic technique. J. Bacteriol. 25, 277–287.
[47] ZoBell C. Allen E. (1933). Attachment of marine bacteria to submerged slides. Proc Soc Exp Biol Med., 30, 1409–1411.
[48] Costerton W, Geesey G. and Cheng K. (1978). How bacteria stick. Sci Amr., 238, 86–95.
[49] Prakash B, Veeregowda B. and G. Krishnappa G. (2003). Biofilms: A survival strategy of bacteria. Cur Sci., 85, 1299–1307.
[50] Yoko A, Jiro M, Tsuda T, Kuwabata S, *et al.* (2015). Simple observation of *Streptococcus mutans* biofilm by scanning electron microscopy using ionic liquids. AMB Express, 5, 6–14.
[51] O'Toole G, Kaplan H. and Kolter R. 2000. Biofilm formation as microbial development. Ann. Rev. Microbiol., 54, 49–79.
[52] Nwodo U, Ezeikel G. and Okoh A. (2012). Bacterial Exopolysaccharides: Functionality and Prospects. Int J Mol Sci., 13, 14002–14015.
[53] Henning B, Mack D. and Holger R. (2015). Structural basis of *Staphylococcus epidermidis* biofilm formation: Mechanisms and molecular interactions. Front Cel Inf Microbiol., 5, 1–15.
[54] Hobley L, Harkins C, Cait E. and Nicola R. 2015. Giving structure to the biofilm matrix: an overview of individual strategies and emerging common themes. FEMS Microbiol Rev., 39, 649–669.
[55] Otto M. (2008). Staphylococcal Biofilm. Curr Top Microbiol Immunol., 322, 207–228.
[56] Patel R. (2005). Biofilms and antimicrobial resistance. Clin. Orthop. Relat. Res., 437: 41–47.
[57] Boundless. (2014). https://www.boundless.com/microbiology/textbooks/boundless-microbiology-textbook/industrial-microbiology–17/industrial-microbiology–198/primary-and-secondary-metabolites–999–5345.
[58] Berdy J. (2005). Bioactive Microbial Metabolites. J Antibiotics, 58, 1–26.
[59] Gupta C, Prakash D. and Gupta S. (2014). Natural Useful Therapeutic Products from Microbes. J Microbiol Exp., 1, 1–8.
[60] Xue C. (2015). Finding the Sweet Spot: How Human Fungal Pathogens Acquire and Turn the Sugar Inositol against Their Hosts. mBio., 6, 1–3.
[61] Zhang Y, Li S, Gan R, Tong Z, Xu D. and Li H. (2015). Impacts of gut bacteria on human health and diseases. Int J Mol Sci., 16, 7493–7519.
[62] Panizzon J, Luiz H, Knaak N, Ramos R, *et al.* (2015). Microbial diversity: Relevance and relationship between environmental conservation and human health. Braz. Arc. Biol. Biotechn., 58, 137–145.
[63] Conlon M. and Bird A. 2015. The impact of diet and lifestyle on gut microbiota and human health. Nutr., 7, 17–44.
[64] Gibson G. and Roberfroid M. 1995. Dietary modulation of the human colonie microbiota: Introducing the concept of prebiotics. J. Nutr., 125, 1401–1412, 1995.
[65] Conlon M. and Bird A. 2015. The impact of diet and lifestyle on gut microbiota and human health. Nutr., 7, 17–44.
[66] Neis E, Dejong C. and Rensen S. 2015. The Role of Microbial Amino Acid Metabolism in Host Metabolism. Nutr., 7, 2930–2946.

[67] Brody T. (1999). Nutritional Biochemistry (2nd ed.). Academic Press. p. 320.
[68] Mayu K, Sae H, Takero H, Atsuhiko I. and Ikuo K. Dietary gut microbial metabolites, short-chain fatty acids, and host metabolic regulation. Nutr., 7, 2839–2849.
[69] Hijova E. and chmelarova A. 2007. Short chain fatty acids and colonic health. Bratisl Lek Listy, 108, 354–358.
[70] Rodney R. and Janice M. (2015). The Microbiome and Sustainable Healthcare. Healthcare, 3, 100–129.
[71] Tralau T, Sowada J. and Luch A. (2015). Insights on the human microbiome and its xenobiotic metabolism: what is known about its effects on human physiology? Expert Opin Drug Metab Toxicology, 11, 411–425.
[72] Collado, M, Rautava, S, Isolauri, E. and Salminen S. 2014. Gut microbiota: Source of novel tools to reduce the risk of human disease? *Pediatr. Res.*, 77, 182–188.
[73] Blaut M. and Clavel T. (2007). Metabolic diversity of the intestinal microbiota: Implications for health and disease. J Nutr., 137: 751S–755S.
[74] Patel A. and Prajapati J. 2013. Food and Health Applications of Exopolysaccharides produced by Lactic acid Bacteria. Adv. Dairy Res., 1, 1–7.

Nádia S. Parachin, and Grace F. Ghesti

8 Microbes in wine and beer industries

8.1 Abstract

Wine and Beer are alcoholic beverages produced trough microbial fermentation of grapes and malt respectively. Although those are produced since the Egyptian period, it is only with advance of microbiology studies that it begins to be optimized. This chapter describes the principles of wine and beer making with emphasis in the microbial role during their production. Initially the principal roles of the microorganisms are described in relation do alcohol and flavor formation in both beverages. Then the main desired characteristics for the microorganism to be used in wine and beer industries are listed. Moreover the main spoilage microorganisms and its effect in both beverages are discussed. Finally the main perspectives in wine and beer industries related to microorganism engineering and process design are given.

8.2 Introduction

Although the production of fermented beverages dates to about 7,000 years ago, it wasn't until the 17th century when Antonie van Leeuwenhoek observed microorganisms in the fermentation process that it began to be studied. Overall, beer and wine fermentation have common steps (Figure 8.1). The substrate, must (wine) or wort (beer) is prepared to release the sugars present either in the seed (beer) or grape (wine). Then, fermenting microorganisms are responsible for catabolizing those sugars, producing ethanol, glycerol and acetic acid. From the point of view of the microorganism, this process is fundamental for the synthesis of ATP in oxygen-limited conditions. Since the microorganisms have to hydrolyze different sugars and saccharides such as glucose, fructose, sucrose, maltose, and maltotriose, the fermentation rate will depend on the capability of the microorganism to convert all sugars present in the must. After this primary fermentation, most of sediments are removed and a sec-

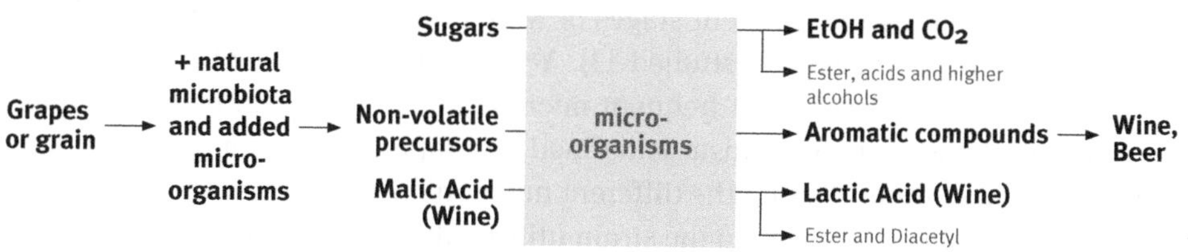

Fig. 8.1: Main steps during wine and beer production. Initially, substrate is prepared so sugars are released for microbial fermentation. During fermentation, other than ethanol (EtOH) formation other, aromatic compounds are also produced.

ondary fermentation is carried out where precursors are biotransformed into some of the wine and beer flavors. After this stage, maturation takes place where flavors are also formed. Therefore, both beer and wine are result of type of the must, fermenting microorganism and maturation process used.

Microorganisms play an essential role during beverage production since they are responsible for ethanol and flavor formation. Thus, the type of microorganisms present in beer and wine must be one of the first aspects to be studied by microbiologists. In both processes, the yeast Saccharomyces cerevisiae was ultimately identified as the principal microorganism responsible for wine and beer production. Nevertheless, whereas the utilization of a pure culture technique is most common in the beer industry, other yeast and bacteria will be essential for flavor creation in the wine industry [1]. For example, the bacterium Oenococcus oeni was shown to be the main microorganism responsible for malolactic fermentation, which is very important for the reduction of wine acidity [2].

Finally, there are also types of microorganisms that are often encountered in wine and beer spoilage. They enter the process either at the beginning of the fermentation step or during secondary fermentation or maturation. Since wine is not pasteurized like beer, there is also a contamination source during bottling.

Although there are many aspects to be considered when producing an alcoholic beverage, this chapter will focus on the role of microorganisms in wine and beer production. Furthermore, the main aspects to be considered in terms of avoiding contamination will also be pointed out. Finally, critical conclusions and perspectives will be offered about how genetic engineering techniques could be applied for the improvement of wine and beer as well as the policies of different governments for the use of genetically modified organisms (GMO) in beverage production.

8.3 Microbes in Wine

It is well known among wine specialists that final flavor is dependent on a series of factors, including the soil where the grape is grown, the grape type and its sugar content, the microorganisms present during fermentation and the maturation process. Moreover, in contrast to brewing beer, wine fermentation has different microorganisms dominating the different stages of wine production. Thus, the ecology of vineyards has been previously studied [3]. Yeasts, filamentous molds, lactic and acetic acid bacteria are the most frequent microorganisms found in mature grapes. The amount of each microorganism is defined by grape content, temperature, and humidity among others. Among the different microorganisms, yeasts are considered to influence the wine according to the strain utilized during production. Aspects that are exclusively dependent of yeast strains have been summarized as fermentation performance, downstream wine processing, concentration of ethanol, and levels of flavors and off-flavors [4].

Microbial influence on wine production starts right after must preparation. The so-called apiculate yeasts (yeast species present in the skin of grapes) initiate sugar consumption. Some of the microorganisms that inhabit the healthy skin of grapes are *Aureobasidium pullulans, Metschnikowia, Hanseniaspora (Kloeckera), Kloeckera apiculata, Hanseniaspora uvarum, Cryptococcus* and *Rhodotorula species* (depending on stage of grape maturity) [5]. The non-Saccharomyces yeasts contribute to fermentation since they can reach populations of about up 106–107 cells/ml, affecting both growth kinetics and metabolism of Saccharomyces sp. These yeasts are capable of anaerobic as well as aerobic growth and may persist during fermentation, competing with Saccharomyces sp. for nutrients, and may produce secondary compounds affecting the aroma of the final product [11].

These species are overtaken by the *S. cerevisiae* species responsible for primary fermentation mainly due to its resistance to ethanol. Nowadays in established vineyards, selected Saccharomyces sp. are utilized as starters introduced into the must. Nevertheless, in smaller vineyards fermentation might also occur spontaneously. This is reported to confer a more unique flavor into the wine, although it is more difficult to control fermentation [6]. By the end of fermentation, bacterial species will be responsible for malic acid fermentation during secondary fermentation. Finally, flavors will be finalized during maturation processes. Research reviews of the microorganism's role during wine production have already been published [6–10].

In fact, there are many microorganisms involved in wine production. Most starters of fermentation are either yeasts in pure *S. cerevisiae* cultures or in co-cultures with other species such as *S. bayanus* and *S. paradoxus* [11]. Since these Saccharomyces strains dominate wine fermentation, some aspects of its ideal behavior could be summarized as [5, 7]:

- Fast and robust fermentation of grape juice.
- Tolerance towards sulfur dioxide, frequently added to the juice as an antioxidant and antimicrobial agent.
- Capability of mixing uniformly throughout the fermenting juice and dispersing sediment quickly from the wine at the end of fermentation.
- Production of moderate ethanol levels (48 % v/v) since an excess is described to overtake flavor. Therefore, genetic engineering strategies have been used to reduce ethanol formation during vinification such as the redirection of sugar into the glycerol biosynthetic pathway. Other strategies for production of reduced alcoholic wines have recently been reviewed [12].
- Increased capacity of flavor formation during fermentation. When the physiology of fermented yeasts is well studied, a proper strain may be chosen in order to achieve the target wine aroma.
- Reduced capacity to produce sulfur aroma compounds, such as hydrogen sulfide (which confers a taste know as rotten eggs), acetic acid and ethyl acetate.
- Hyperproduce extracellular enzymes that improve wine clarification such as proteases, glucanases and pectinases.

8.4 Fermenting microorganisms undergoing primary fermentation

After must preparation, the wine undergoes primary fermentation. During this stage, microorganisms influence wine quality depending on the types that are present in the grapes before harvest, the performance of alcoholic fermentation and the synthesis of aromatic compounds. During primary fermentation, the most common yeasts present in the must are the Saccharomyces species. However, if for any reason the yeast delays growth, various species of lactic acid bacteria and acetic acid bacteria may overtake the must and inhibit yeast growth, resulting in stagnant or stuck fermentation.

Microorganisms present during primary fermentation are submitted to osmotic stress and the presence of inhibitory compounds such as sulfuric molecules, which are frequently used to control microbial contamination. Nevertheless, at this stage the sugars present in the must are consumed concomitantly to ethanol, glycerol and acetic acid formation in higher amounts, to the order of grams per liter (Figure 8.2) [9]. Among these, glycerol is considered to confer a sweet flavor into the wine.

Moreover, acetaldehyde and diacetyl are also produced during this fermentation stage (Figure 8.2). The first one is described to confer a fruity flavor while diacetyl at concentrations below 8 mg/L is described to confer yeasty, nutty, and toasty aromas to wine. Nevertheless, in higher concentrations (> 10 mg/L) it turns into an unappreciated buttery flavor. Moreover, due to their high reactivity, these molecules might bind to proteins or amino acids, generating a range of flavors. The *S. cerevisiae* species is known to produce diacetyl during alcoholic fermentation. Nevertheless, it is commonly metabolized by the yeast into acetoin and 2,3-butanediol (Figure 8.2) [10]. The amount of diacetyl, acetoin and 2,3-butanediol is strain dependent as opposed to ethanol, glycerol, acetic acid and ethyl acetate, which present little variation among

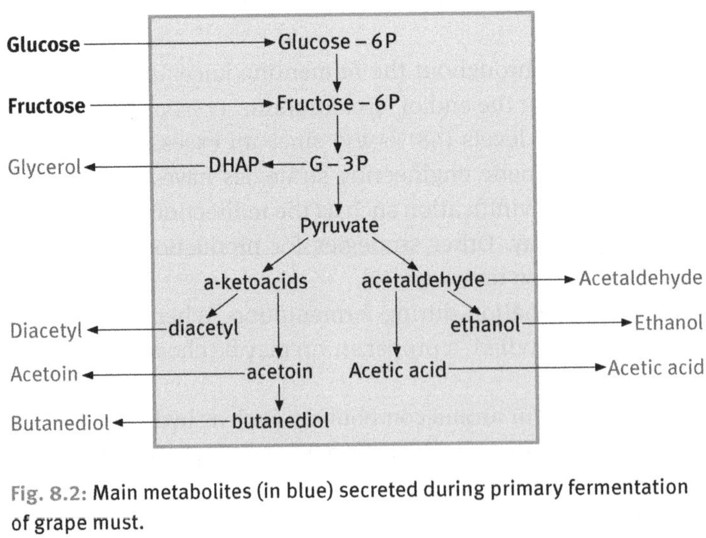

Fig. 8.2: Main metabolites (in blue) secreted during primary fermentation of grape must.

yeast strains. Moreover, it has been previously reported that acetoin and 2,3-butanediol are not present in high concentrations. When acetoin is highly produced, only low levels of 2,3-butanediol are encountered and vice-versa [9]. This is explained by the activity of a NAD dependent Butanediol dehydrogenase which reduces acetoin to 2,3-butanedio by the concomitant oxidation of NADH. During wine fermentation, the enzyme level has been shown to be limited which impairs 2,3-Butanediol formation and consequently acetoin accumulation [13].

Finally, yeast metabolites can also reduce wine quality, as *Saccharomyces* is responsible for the production of several off-flavors in wine, such as the sulfur-containing volatiles, higher alcohols, and aldehydes [14]. By the end of alcoholic fermentation, there is also high biomass concomitant to low sugar concentration. Usually, most of the yeast flocculates into the bottom of the fermenter and eventually die, releasing their intracellular contents. Other microorganisms might use these during the maturation process for growth.

8.5 Flavoring microorganisms undergo secondary fermentation

Final wine flavor is influenced by a wide variety of compounds with different aromatic properties. This includes aromatic compounds originating from grapes, compounds formed during operations of extraction and conditioning of must, and other compounds produced by yeasts and bacteria during alcoholic and malolactic fermentation. Finally, there are compounds that appear during the aging process. Aromatic compounds in wine are determined by their odor activity value (OAV), which is the concentration of a specific aroma molecule divided by its detected threshold concentration.

Nevertheless, it is during secondary fermentation where most of the wine aroma is formed. Yeasts will contribute to wine flavor by producing new aromatic compounds from sugars present in grapes and by transforming the precursors formed during must preparation. For instance, the non-saccharomyces species are known to secrete several hydrolases in the must that produced active aromatic compounds from grape precursor compounds [9].

During secondary fermentation, malic acid, citric acid and tartrate are the main organic acids originating from grapes left over in the must. Malic acid can be decarboxylated into lactic acid and CO_2 in a process known as malolactic fermentation (MLF). An excess of malic acid (more than 10 g/L) in the wine confers high acidity and undesirable flavor. When malic acid is converted into lactic acid, the increase in pH results in a softer wine. The lactic acid bacterium *Oneococcus oeni* is the most frequent species found during MFL, although *Lactobacillus* spp. and *Pediococcus* spp are also encountered at this stage [10]. A *S. cerevisiae* strain has also been engineered to produce lactic acid from malic acid. The gene from *Lactococcus lactis* (mleS) encoding for the malic enzyme was expressed in *Saccharomyces cerevisiae*, allowing for the production of Lactic acid from malic acid in high yields [15]. Nevertheless, there are no reports of this strain being utilized for industrial wine production.

Citric acid and glucose are co-utilized by *O. oeni* and can generate an excess of acetic acid that negatively influences the wine aroma (Figure 3). Besides lactic acid and acetic acid, lactic acid bacteria can also produce other aromatic compounds via their own metabolism such as esters that are described to enhance wine's fruitiness [9]. Moreover, lactic acid bacteria might produce high levels of diacetyl (> 8 mg/L) which bestows a buttery taste to the wine. A previous review has summarized the main aspects that influence diacetyl formation during MLF such as [14]: bacterial strain, inoculum size, oxygenation, presence of citric acid, lower temperature (18 °C), acid pH and presence of sulfur compounds.

Also during secondary fermentation, ester and higher alcohols are synthesized and interfere with wine flavor. The main aromatic compounds produced during secondary fermentation are summarized in Figure 8.3. After MLF, increased diacetyl and acetoin concentrations, mainly derived from citric acid catabolism, are frequently obtained.

At least during the aging of wine, the enzymes released from lysed yeasts are responsible for decreasing the concentration of volatile compounds. These are usually converted to long chain alcohols and fatty acids. Furthermore, when the yeast cells lyse, they also release cell wall proteins such as mannoproteins, which are known to stabilized wine color and clarification [10].

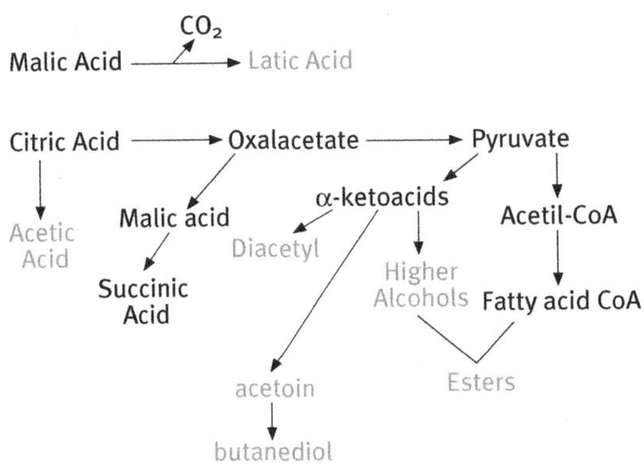

Fig. 8.3: Main aroma compounds produced (in green) during secondary fermentation.

8.6 Spoilage microorganisms

Up to this point, the ways different yeast and bacterial strains influence wine flavor and ethanol formation have been discussed. Nevertheless, it is also relevant to discuss the most common contaminants responsible for wine spoilage. After primary fermentation, wine is not a susceptible environment for microbial contamination since it has a low concentration of carbon sources and oxygen, high ethanol concentrations, acidic pH, presence of sulfur compounds and it is stored at cool temperatures during maturation. Therefore, during wine production, most concerns regarding microbial contamination lie after must preparation (since it is rich in sugar content), during bottling (where new microorganisms might be added into the wine), and finally predicting its shelf life since wine is not pasteurized (as this would change its characteristics). Instead, wine is often filtered in order to be stabilized [3].

To be considered contamination-free from the inspection point of view, the bottled wine should have a clear aspect without any haze. That means a microbial load of less than 10^4–10^5 CFU/ml for microorganisms producing powdery sediments. Another aspect that is taken into account in order to determine whether the wine is contaminated is its sensorial characteristics. Considering both aspects, spoilage microorganisms are rarely encountered during wine fermentation, but during storage or aging and during the bottling process [3].

The yeasts from the genera *Zygosaccharomyces* and *Brettanomyces* are the cause of the most common microbial contaminations in the wine industry [14]. The latter may also be referred to as Dekkera, the spore-forming variety of *Brettanomyces*. Together, these two genera have been described as the biggest microbiological problem for preventing spoilage. *Zygosaccharomyces* bailii contamination is described to cause refermentation and CO_2 production in sweet wines or grape juice concentrate. This happens since Zygosaccharomyces bailli prefers to utilize fructose instead of glucose while *S. cerevisiae* does the opposite. Spoilage activity manifested as carbon dioxide production within sealed containers of grape juice concentrate or sweetened wine is a major concern with *Zygosaccharomyces* contamination. Excessive gas formation carbonates still wines or even causes containers to burst or explode.

Brettanomyces bruxellensis spoilage often contributes off-odors and flavors to red wines [14]. Volatile phenols produced by *B. bruxellensis* include 4-ethylphenol and 4-ethylguaiacol. Oxygen is not required for the growth of *B. bruxellensis*, a yeast that is considered to be a facultative anaerobe. Under certain conditions, however, *Brettanomyces* spp. may metabolize carbohydrates to acetic acid rather than ethanol in glycolysis.

Grapes infected by the necrotrophic fungus *Botrytis cinerea* are dehydrated, leading to noble rot, which allows the production of sweet wines. Nevertheless, this fungus might also produce the so-called grey rot in wet environments, resulting in the lost of the affected bunches. Nevertheless, it is not clear nowadays whether *B.cinerea* infection alters the chemical composition of grapes, impairs natural microbiota growth and fermentation, or both [3].

Beside yeasts and filamentous fungi, some bacterial groups such as lactic acid bacteria (LAB) and acetic acid bacteria (AAB) are also considered spoilage wine microorganisms. The main off flavors produced by spoilage bacteria have been previously summarized [7]. Although some lactic acid bacteria play a key role in the malolactic fermentation of wine, others may cause serious faults. Excessive volatile acidity, mannitol taint, ropiness, mousiness, acrolein formation and bitterness, tartaric acid degradation, diacetyl overproduction and rancidness, as well as the very unpleasant geranium off flavor, are often the consequence of uncontrolled growth of some species of *Lactobacillus* [7]. Within the lactic acid bacteria, species of *Lactobacillus* and *Pediococcus* are most relevant, along with the aforementioned *Oenococcus oeni*, which conducts malolactic fermentation [5].

Regarding contamination by AAB, this group usually confers a vinegary flavor into the wine due to acetic acid production [7, 16]. Although acetic acid is produced during alcoholic and malolactic fermentation, it is usually in such low concentrations that it can hardly be detected. Nevertheless, the presence of AAB results in a surplus of acetic acid formation. Essentially, it is fundamental that wine be stored under anaerobic conditions in order to avoid contamination by AAB which is strictly aerobic. The presence of oxygen leads to the oxidation of ethanol and acetaldehyde into acetic acid. Moreover, acetaldehyde and ethyl acetate might also be produced, bestowing a solvent-like aroma into the wine. AAB contamination is more frequent during barrel maturation, although lack of control in the bottling process and storage might also favor the growth of these microorganisms. Within the acetic acid bacteria, *Acetobacter aceti, Acetobacter pasteurianus* and *Gluconobacter oxydans* are commonly found [5]. Finally, it is also important to emphasize that AAB may also contaminate damaged grapes, resulting in arrested primary fermentation [16].

8.7 Microbes in Beer

The main steps utilized for beer production are summarized in Figure 8.4. The type of beer to be produced will depend on the raw material utilized and the fermenting yeast strain. In general, beer can be classified into two main categories: ales and lagers. Both ales and lagers run the entire course of distinct flavors for each kind of beer. They both vary in a wide range of colors, bitterness, aroma, and maltiness. The main difference between lagers and ales lies in the fermentation step [17].

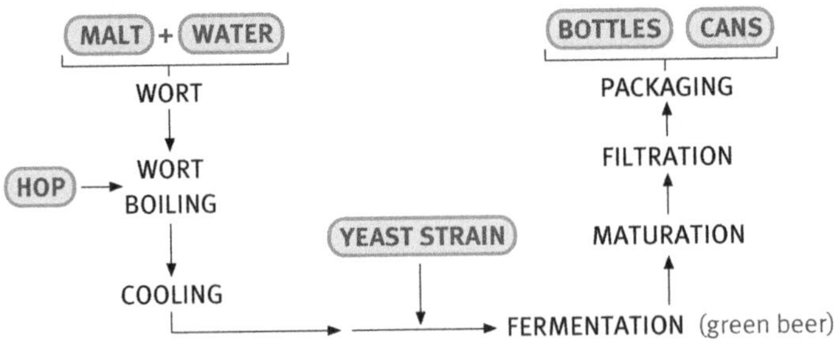

Fig. 8.4: Main steps during beer production.

It is during this stage that a different yeast strain and fermenting temperature are used. Ales are fermented with top fermenting yeasts that hang around at the top of the fermentation tank. Ales are generally fermented at warmer temperatures, about 13–25 °C, by *Saccharomyces cerevisiae* strains, during short time periods (3 to 5 fermentation days). Lagers, on the other hand, are fermented at cooler temperatures, about 7–15 °C, by *Saccharomyces uvarum*, formerly known as *Saccharomyces carlsbergensis*, for longer periods of time (6 to 10 fermentation days).

Lagers use bottom-fermenting yeast which hang out at the bottom of the fermentation tank at the end of fermentation. The result of this cooler, longer fermentation and bottom fermenting yeast is generally a crisper and clearer appearance typically associated with lagers. The final flavor of the lager will depend mainly on the Saccharomyces strain as well as the temperature utilized during the fermentation. Finally, since lager yeast stay at the bottom of the fermenter by the end of fermentation, it can be re-utilized in a subsequent beer batches [18].

The rate of sugar consumption in the fermentation step and consequently the time needed for beer production will depend on many aspects that directly or indirectly affect the fermenting yeast. For instance, during wort preparation, sugars and Free Nitrogen amino acids are measured as a control of wort quality. After this, the coagulated protein is boiled (Figure 8.4) and hop resins are removed since they give a bitter taste to the beer and lead to the production of yeast inhibitors.

During the fermentation step, increased temperature increases ethanol formation rate up to the optimum temperature of utilized yeast strain. Finally, the surface of contact between the yeast and wort has an important influence on sugar utilization. With an increase in yeast concentration, the contact surface is increased. As a rule, yeast is inoculated into the wort at a rate of 20 million cells/mL, approximately = 0.6 to 0.7 L of thick yeast slurry per 100 L. The yeast dosage must be homogeneous during tank filling to avoid stratification of beer fermentation and consequently, different flavor formation in the same batch [19].

In order to produce a high quality beer with desirable sensory evaluation, the important aspects to be considered are [17, 20]:

- the right yeast – One of the main differences between the various beer styles is the yeast strain that was used. Different yeast strains produce beers with different characteristics. Actually, there are many different yeast strains for brewing production, and each produces a different flavor profile. Some Belgian strains produce fruity aromas that resemble banana and cherry. Some German strains produce phenols, highlighting the clove aroma.
- the right nutrients – Nutritional deficiencies can have an adverse effect on yeast performance which can include deficiencies in metals (eg zinc), vitamins, oxygen, sterols and/or unsaturated fatty acids.
- consistent behavior when cultivating the yeast – Various traits can vary within a given yeast culture including: flocculation, fermentation rate, ability to reduce diacetyl and the ability to produce sulfur dioxide. In the case that the yeast is not

cultivated properly, one or more traits might be lost and the yeast strain may have to be 're-selected' to restore its behavior;
- contaminant free – Wild yeasts, other brewing yeasts, and contaminant bacteria may be present; controling levels of contaminants through attention to plant hygiene and yeast acid treatments must be emphasized. Beer cannot be easily contaminated because in its composition there is ethanol, CO_2, O_2, low sugar and nutrient content, and reduced pH values. However, some contaminants can survive in these conditions such as *Lactobacillus*, *Pediococcus*, *Streptococcus* and *Pectinatus*. These bacteria irreversibly spoil the beer. As a consequence, it is possible to observe an acid taste, sulfury flavors, aldehydic flavors, phenolic or spicy flavors, butter-like flavor, indole, haze or turbidity, fining problems, rope, gushing, nitrosamines content, biogenic amines content, mycotoxins and aflatoxins;
- pure culture – DNA typing methods can reveal to what extent the yeast is genetically homogeneous. In contrast to wine fermentations, beer fermentations result in a better a quality product if a single microorganism is utilized during the entire batch;
- safe storage – Yeast is an integral part of a brewer's brand. Therefore, storing it under conditions that assure its preservation, both in the short term and long term, is essential. The best practice is liquid nitrogen storage (–196 °C) at a minimum of two independent sites.

8.8 Yeast Management

Yeast management is one of the most important steps for beer production. After strain selection, it is important to ensure the appropriate yeast growth during all propagation steps. Figure 8.5 summarizes the main steps utilized for yeast propagation until its inoculation (pitching) into the wort. Initially, yeast preserved into a slant is inoculated to a tube containing 10 mL of wort. This is considered a suitable volume since it reaches saturation in a short amount of time and therefore minimizes potential contamination.

Inoculation or pitching rates have been shown to have a profound effect on fermentation performance and are one of the main factors which contribute to consistency between beer batches. It is usually expressed as the amount of yeast found in 1 ml of wort after pitching. The recommended optimal pitching rates have been described to be 6–10 million cells/ml for ales and 10–15 million cells/ml for lagers according to the formula presented bellow [21].
- Ale pitching rate = 6–10 million cells/ml × (1 – Original Gravity)/48
- Lager Pitching rate = 10–15 million cells/ml × (1 – Original Gravity)/48

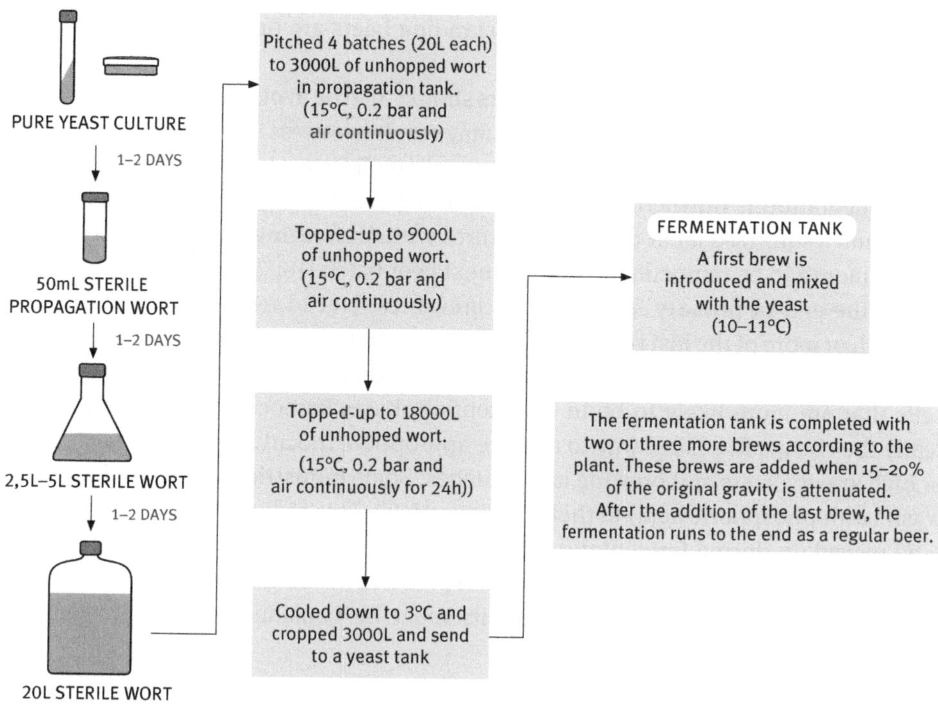

Fig. 8.5: Yeast propagation in a large brewery.

For example, 1.096 of original gravity ale wort should be pitched with 14–20 million cells/ml. A fermentation tank contains approximately 20,000 ml. This means that for 1.050 gravity ale, you would need to add between 120 billion (20,000 ml × 6 million cells/ml) and 200 billion cells (20,000 ml × 10 million cells/ml) [21].

During yeast propagation, there should be no more than 3 or 4 cell divisions taking place during the first few hours of fermentation. In case a low volume of yeast is added into the wort, the yeast will enter a lag phase and consequently spend much more time starting the fermentation process. Furthermore, during this extended growth period the yeast tends to secrete more esters and fusel alcohols that will negatively affect the beer flavor. Moreover, they may not have a sufficient number to adequately metabolize all fermentable sugars [18, 20]. Therefore, the yeast should not be too diluted when passing from one propagation step to another. Thus, 10 ml should not be inoculated into more than 2 L. The rationale behind this principle is to always keep the yeast growing in the exponential phase. Diluting the yeast out too far will slow down their growth and give bacteria a chance to overtake the culture. Since bacteria can grow as much as 6 times faster than yeast, it is important to keep yeast growing rapidly. If more than 1 or 2 L of starter are required, it is best to do more than one step-up [18, 22]. In general, outstanding beers are produced using a clean, healthy, well-aerated 2 L yeast starter.

Inoculation of larger amounts of yeasts should also be avoided. An excess of yeast inoculum will increase yeast growth and may result in excess ester formation, diacetyl, hydrogen sulfide, or stuck fermentations, where the yeast becomes inactive before the fermentation is finished. Thus, the remaining yeast of a given inoculum may be saved into a sanitized jar at cool temperatures without freezing it. This may be stored up to 1 month to be reinoculated into the next beer batch [18, 22].

At the end of primary fermentation, breweries tend to reinoculate one-third to one-fifth or more of the last batch into a new batch. Note that breweries usually collect only the middle third out of the conical fermentation tank, since these are the yeast cells that are more likely to be in exponential phase. Reinoculation of the yeast is really the cheapest, easiest way to achieve appropriate inoculum rates for lager and/or high gravity beers or if brewing large batches. Some industries usually add the acid wash (with phosphoric acid) to their yeast at pH 2.2. This reduces the amount of bacteria picked up during fermentation, but will also reduce yeast viability. It also does not get rid of wild yeast, which is a common problem in some breweries [20].

During yeast propagation, several details must be considered carefully such as [23–25]:

- During all stages of yeast growth, all yeast management must be performed under sterile conditions until the yeast propagation vessel stage. Spoiled microorganisms cannot be removed during propagation because most of them grow under the same conditions as S. cerevisiae.
- Intensive sterile aeration of the yeast is necessary for rapid yeast growth and consequently for activating the fermenting yeast. In the presence of oxygen, the carbon present in the wort is preferentially directed to biomass formation.
- Yeast multiplies much faster at temperatures of 10 to 37 °C, depending on the yeast strain, than at lower temperatures. However, it is essential to bring yeast to the brewery at operating temperature conditions in the course of propagation for it to achieve full fermenting power in the brewery.
- Yeast propagation is commonly done in the wort without waiting since it contains bittering and inhibitory compounds that affect the growth of contaminants.

Besides oxygen addition, it is also necessary to make sure that the yeast is uniformly distributed into the wort with as few yeast clumps as possible, a process known by the brewers as rousing the yeast. Rousing allows the individual yeast cells to rapidly come into contact with the nutrients in the wort. This is important for the intensity of the initial fermentation. Consequently in modern plants, the yeast is dosed uniformly into the flowing stream of wort. Poor wort aeration can have very undesirable consequences such as defective initial fermentation and consequently longer fermentation time. Moreover, when the yeast inoculum is older, the fermentation becomes more sensitive to cooling and tends to flocculate faster. When the yeast floculates, secondary fermentation stops, and the resulting beer frequently has a higher off-flavor content [26, 27].

8.9 Primary fermentation

Primary fermentation is started by adding the yeast to the fermenter after its propagation, a process known as pitching the yeast among brewers. The yeast is added uniformly to the wort while it is intensively aerated so that fermentation can begin rapidly and vigorously. As soon as the yeast is added, the wort is referred to as young beer. The fermentation temperature and fermentation time are very important during fermentation management. During primary fermentation, according to the yeast strain, the young beer passes through various fermentation stages: initial, young or low Krausen, high Krausen, Krausen collapsing and collapsed foam [20, 21, 23, 24]. All these stages are related to the amount of foam produced during the fermentation. Foam formation will rely on the amount of protein released during wort preparation and the amount of CO_2 released during fermentation. Initial and Low Krausen are characterized by the beginning of the formation of a white foam and correspond to the yeast leaving the lag phase of growth and beginning exponential growth. High Krausen is characterized by high foam formation with coarse bubbles and corresponds to yeast exponential growth. Finally, Krausen corresponds to the entrance of the yeast into the stationary phase where the foam looks more brown and decreases in intensity.

Figure 6 shows a typical fermentation profile during the production of lager beer. Reduction of specific beer gravity is associated to concomitant yeast growth and ethanol production (Figure 8.6A). Due to rapid oxygen consumption (Figure 8.6B) the ethanol produced is much higher than the yeast biomass formed. Lager beers are generally of better quality because of reduced formation of by-products, in particular, higher alcohols and esters [17].

Essentially two criteria are utilized in order to determine fermentation performance: the time required for consumption of sugars and flavor production. In order to determine whether the fermentation process is complete the apparent attenuation is determined. This is frequently calculated by the percent reduction in gravity or:

Apparent Attenuation = (Original Gravity – Final Gravity)/ Original Gravity

For a traditional lager, the wort original gravity is about 1.050. The final gravity when the fermentation is done should be near 70–85 % of initial value for most utilized yeast strains. Moreover, this value should be achieved within 3–5 fermentation days. The final degree of attenuation is usually determined before beer transfer to maturation [20].

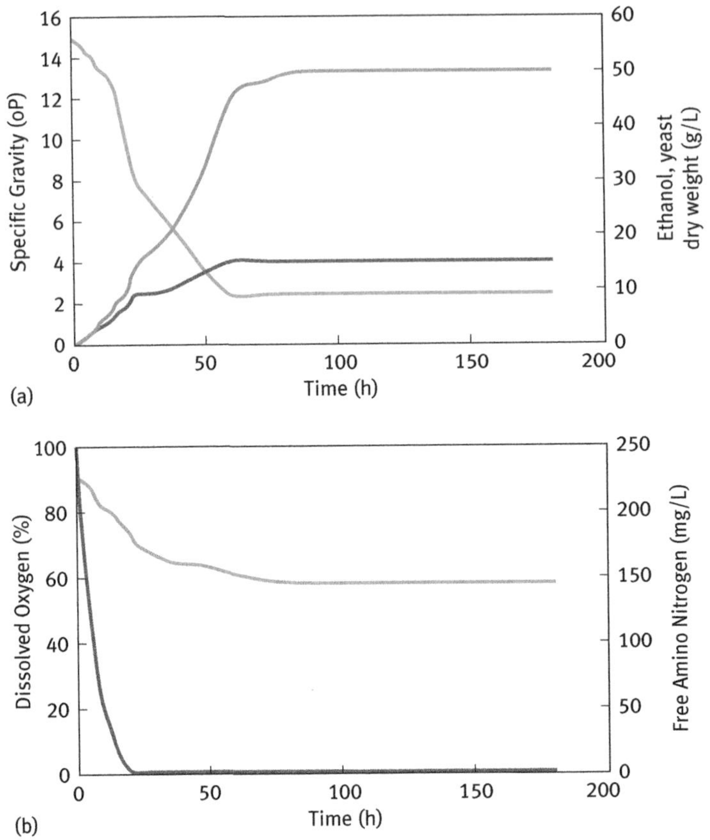

Fig. 8.6: Typical fermentation profile during the production of a lager beer.
(a) Specific gravity-blue, ethanol-green and yeast dry weight-red.
(b) Dissolved Oxygen-red and Free amino nitrogen-blue.

8.10 Maturation and Flavor Formation

The reactions that take place during the conditioning phase, or maturation phase, are primarily done by the yeast metabolism. When the vigorous primary stage is over the majority of the wort sugars were catabolized into ethanol and biomass. Then the majority of the yeast cells are going dormant – but some are metabolically active. Maturation allows for the slow reduction of the remaining fermentable sugars. The yeast has consumed all of the easily fermentable sugars and now starts to metabolize other carbohydrates such as maltotriose. Moreover, the yeast is able to metabolize some of the byproducts that are produced during the primary fermentation phase [27, 28].

When lysed, the dormant yeast on the bottom of the fermented tank releases amino and fatty acids into the beer. In the case that the beer is not separated from its sediments, dormant yeast and remaining proteins, the beer will be rich in soapy flavors. For these reasons, it can be important to get the beer off of the trub and dormant yeast during the maturation stage [27].

During the secondary fermentation and maturation stages, the most commonly produced byproducts that will affect beer flavor are esters, aldehydes, organic acids and sulfur compounds (Figure 8.7). They contribute to the full-bodied taste of the beer but can also have a very detrimental effect on beer taste and aroma besides determining foam stability [27, 29, 30].

With the introduction of new processes to shorten the fermentation and maturation time, fermentation by-products and the factors that influence their formation and removal become increasingly important to the brewing industry. This applies in particular to the following fermentation by-products, divided in two groups [31]:

- Green beer aroma compounds – such as diacetyl, aldehyde and sulfur molecules.
- Maturated beer aroma – such as fusel alcohols and esters. These essentially determine the aroma of the beer and their presence, in defined concentration ranges, are a necessary requirement for a high quality beer.

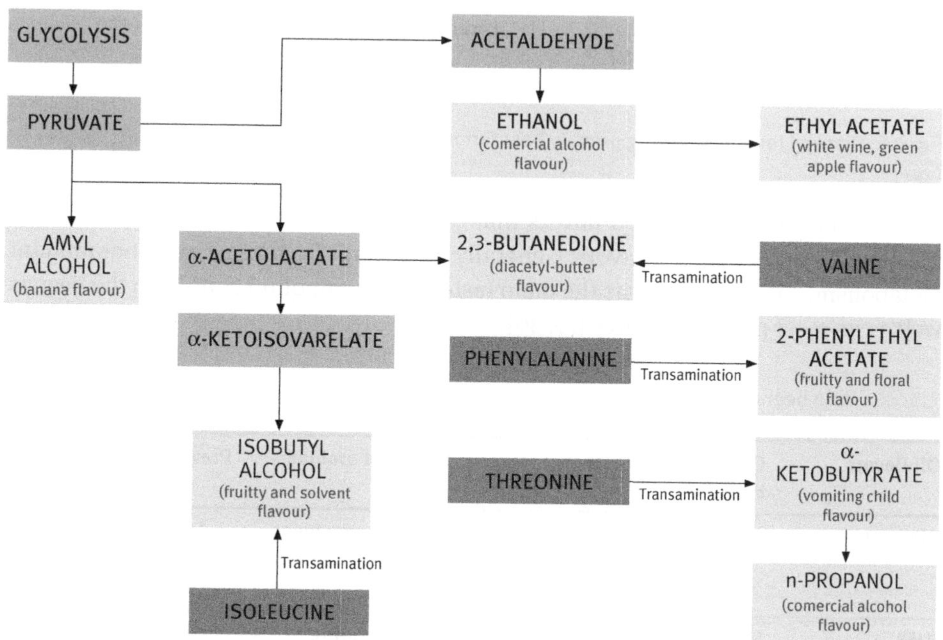

Fig. 8.7: Saccharomyces cerevisiae biosynthetic pathways for production of volatile compounds, in green. Products derived from glycolysis are highlighted in blue whereas products of amino acid metabolism are marked in red.

The beer aromas produced during primary fermentation result in an unclean, immature, unripe, unbalanced taste and aroma and, at higher concentrations, more than 20 mg/L, impairs beer quality. During the course of secondary fermentation and maturation, off- flavors can be biochemically removed again from the beer. This is essentially the main purpose of beer maturation, to reduce levels of diacetyl and fusel alcohols still using yeast metabolism [31, 32].

One of the main beer off-flavors, especially in pale lager beers, is diacetyl. This is formed from alpha-acetolactate that is produced during primary fermentation (Figure 8.8). High diacetyl levels, > 10 mg/L, may also be encountered in contaminated beers. Thus, in large breweries, the aging or maturation time is associated with the reduction of diacetyl concentration. When it reaches concentrations below 10 mg/L, the maturation process is interrupted and the maturated beer can be sent to the next step of the process, filtration or packaging (Figure 8.4) [29, 30].

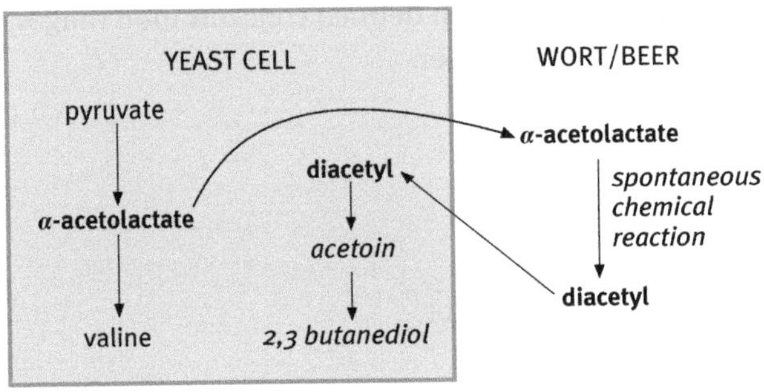

Fig. 8.8: Diacetyl formation during primary beer fermentation.

Besides diacetyl, the main off-flavors that can be produced during beer production are listed in main beer off-flavors. Generally they result from yeast or a contaminant metabolism. Table 8.1 also lists the main tastes each compound confers to the beer as well as its most probable source [29, 30].

Tab. 8.1: Main beer off-flavors.

Off flavors	Chemical class	Taste	Source of problem	Prevention
Diacetyl	Carbonyls	Buttery, butterscotch	Yeast management	– Avoid contamination – Proper yeast – cultivation
DMS	Sulfurs	Cooked vegetable	– Yeast during fermentation – Slow fermentation – Early capping	– Yeast type – Strong fermentation – Later capping

Tab. 8.1: (continued)

Off flavors	Chemical class	Taste	Source of problem	Prevention
Suphury Other	Sulfurs	Meaty, rotten vegetable, burnt matches, burnt grain	– Yeast during fermentation – Slow fermentation – Early capping	– Strong fermentation – Later capping
4-Vinyl guaiacol	Phenolics	Herbal, clove-like, spicy phenolic	Cross contamination Lager – Ale yeast	– Proper yeast strain separation – Cleaning
Acetaldehyde	Aldehydes	Green apple	Yeast stress – long time at 20 °C or beer stored with excess yeast	– Fast diacetyl reduction and fast cooling – Cone cooling
Sour	Organic acids	Acidic	– rapid growth of yeast – pH drop	– Right pitching, dissolved oxygen, temperature
Lactic	Organic acids	Sour, acidic	– Yeast infection	– Proper hygiene
Yeast negative	Sulfurs	Rotten yeast	Autolysis – high dead cells	– Fast diacetyl reduction and fast cooling

8.11 Main strategies for avoiding microorganism spoilage in wine and beer

In order to prevent wine spoilage, sulfur dioxide has been traditionally used to control unwanted microorganisms during winemaking. It chemical is frequently added to bins of machine-harvested grapes and after malolactic fermentation. Sulfur dioxide acts as both an antimicrobial agent and an antioxidant in wine.

Sorbic acid (2,4-hexadienoic acid) can also be used as a chemical preservative in sweetened wines. This is commonly added during bottling to prevent yeast fermentation after packaging. However, several LAB species, such as O. oeni strains, are able to utilized sorbic acid, resulting in the formation of 2-eth- oxyhexa-3,5-diene, which has an odor of crushed geranium leaves. Another compound that can be added to wine is dimethyl dicarbonate. This is a chemical inhibitor of microorganisms that inactivates cellular enzymes. It is usually hydrolyzed to methanol and carbon dioxide, natural constituents of grape juice and wine. Therefore, the addition of dimethyl dicarbonate does not affect wine flavor or color. Nevertheless, it does not effectively inhibit LAB or AAB [33].

In the brewing industry, brew vessels, yeast and fermentation tanks as well as filtration systems must be cleaned during the brewing process to ensure sterility. To avoid disassembly, the cleaning of the entire pipeline system is regularly done in place. The treatment agents for the flushing processes include caustic soda and various acids, usually phosphoric acid and nitric acid. Moreover, additives such as hydrogen peroxide and peracetic acid may also be utilized. The use of tenside free detergents in modern flushing processes can help saving time and water [34].

8.12 Perspectives in wine and beer industry

Today, with the development of next-generation DNA sequencing strategies, up to 80 genome sequences of Saccharomyces cerevisiae strains are available all over the world [15]. This allows for a deeper understanding of the genetics of the major strains utilized for wine and beer production. For instance, a comparative genome analysis has concluded that wine *Saccharomyces cerevisiae* strains had insertions along its genome related to the improvement of sugar consumption, either fructose or xylose. Moreover, this analysis revealed the presence of five genes hypothesized to be the result of horizontal transfer between *S. cerevisiae* and *Zygosaccharomyces* sp [15]. All this information together allows for the construction of databases that can be utilized in genetic engineering strategies.

Nevertheless, there is great resistance against implementing genetically modified organisms (GMOs) into wine and beer industry. Generally, the entire European community as well as Australia and New Zealand have forbidden the utilization of GMOs for commercial beverage production. Only the USA and Canada have allowed for the utilization of transgenic wine yeast [8]. Still, most of decisions against GMOs are based on the negative feedback of consumers. Therefore, it is necessary to instruct people regarding the harmlessness of utilizing transgenic yeast and or grapes/grains for beverage production.

In order not to generate transgenic microorganisms, most researchers utilized classical breeding and random mutation in order to achieve desirable traits. Among different studies, genetic modification has aimed to improve the flavor profile in a given strain. Perspectives relating yeast metabolism to flavor and off-flavor formation have been previously summarized as [14]: the discovery of novel yeast biosynthetic pathways for flavor formation, gene manipulation for the enhancement of desired flavors without the need of GMOs, and the addition of flavors produced by other microorganisms into a unique yeast strain in order to get a reproducible beverage profile.

Specifically regarding beer production, high gravity fermentations are implemented in large breweries in order to increase beer productivity. It consists in the production of wort containing a higher sugar concentration. Therefore, fermentation results in higher ethanol production and allows beer dilution and blending. Initial tests have shown the techno-economical feasibility of high-gravity beer production. Nevertheless, most challenges in overcoming scaling-up high gravity beer have been listed as poor extract yields that result in poor free amino nitrogen, poor trub compaction and consequently reduced wort clarity and poor or incomplete yeast fermentation.

Another trend in the brewing industry is to reduce the maturation time and consequently increase beer productivity. Therefore, different sets of enzymes and genetically modified organisms have been considered to maximize diacetyl reduction. Probably, in the near future, maturation time will not exist according to market tendency and brewing technology development.

8.13 Bibliography

[1] Amerine, M.A. and R.E. Kunkee, Microbiology of Winemaking. Annual Review of Microbiology, 1968. 22: p. 323–&.

[2] Borneman, A.R., S.A. Schmidt, and I.S. Pretorius, At the cutting-edge of grape and wine biotechnology. Trends Genet, 2013. 29(4): p. 263–71.

[3] Loureiro, V. and M. Malfeito-Ferreira, Spoilage yeasts in the wine industry. Int J Food Microbiol, 2003. 86(1–2): p. 23–50.

[4] Borneman, A.R., P.J. Chambers, and I.S. Pretorius, Yeast systems biology: modelling the winemaker's art. Trends Biotechnol, 2007. 25(8): p. 349–55.

[5] Fleet, G.H., Yeast interactions and wine flavour. Int J Food Microbiol, 2003. 86(1–2): p. 11–22.

[6] Fleet, G.H., Wine yeasts for the future. FEMS Yeast Res, 2008. 8(7): p. 979–95.

[7] Pretorius, I.S., Tailoring wine yeast for the new millennium: novel approaches to the ancient art of winemaking. Yeast, 2000. 16(8): p. 675–729.

[8] Pretorius, I.S., C.D. Curtin, and P.J. Chambers, The winemaker's bug: From ancient wisdom to opening new vistas with frontier yeast science. Bioeng Bugs, 2012. 3(3): p. 147–56.

[9] Romano, P., et al., Function of yeast species and strains in wine flavour. Int J Food Microbiol, 2003. 86(1–2): p. 169–80.

[10] Styger, G., B. Prior, and F.F. Bauer, Wine flavor and aroma. J Ind Microbiol Biotechnol, 2011. 38(9): p. 1145–59.

[11] Querol, A., et al., Adaptive evolution of wine yeast. Int J Food Microbiol, 2003. 86(1–2): p. 3–10.

[12] Schmidtke, L.M., J.W. Blackman, and S.O. Agboola, Production technologies for reduced alcoholic wines. J Food Sci, 2012. 77(1): p. R25–41.

[13] Ehsani, M., et al., Engineering of 2,3-butanediol dehydrogenase to reduce acetoin formation by glycerol-overproducing, low-alcohol Saccharomyces cerevisiae. Appl Environ Microbiol, 2009. 75(10): p. 3196–205.

[14] Bisson, L.F. and J.E. Karpel, Genetics of yeast impacting wine quality. Annu Rev Food Sci Technol, 2010. 1: p. 139–62.

[15] Ansanay, V., et al., Malolactic fermentation by engineered Saccharomyces cerevisiae as compared with engineered Schizosaccharomyces pombe. Yeast, 1996. 12(3): p. 215–225.

[16] Bartowsky, E.J. and P.A. Henschke, Acetic acid bacteria spoilage of bottled red wine – a review. Int J Food Microbiol, 2008. 125(1): p. 60–70.

[17] Kunze, W., Technology Brewing and Malting. 4th ed, ed. O. Hendel. 2010: VLB Berlin, Germany.

[18] Hough, J.S., et al., Malting and Brewing Science: Hopped Wort and Beer. 2nd ed. Vol. 2. 1982: Aspen Publication.

[19] Kuhbeck, F., W. Back, and M. Krottenthaler, Influence of lauter turbidity on wort composition, fermentation performance and beer quality – A review. Journal of the Institute of Brewing, 2006. 112(3): p. 215–221.
[20] Briggs, D.E., et al., Brewing Science and practice. 2004, Cambridge, England: Woodhead Publishing Limited and CRC Press, LLC.
[21] Goldammer, T., The brewer's handbook. 2008: Apex Publishers.
[22] Hough, J.S., Biotechnologia de la cerveza y de la malta. 1990, Zaragoza: Ed. Acribia.
[23] Boulton, C.A., Applications of Yeast Physiology for Improvements in the Brewing Process. Journal of Chemical Technology and Biotechnology, 1991. 50(1): p. 135–137.
[24] Leedham, P.A., Control of the Fermentation in the Brewery Via Yeast Growth. Monatsschrift Fur Brauwissenschaft, 1983. 36(5): p. 199–200.
[25] Stewart, G.G. and I. Russell, 100 Years of Yeast Research-and-Development in the Brewing Industry. Journal of the Institute of Brewing, 1986. 92(6): p. 537–558.
[26] Verstrepen, K.J., et al., Flavor-active esters: Adding fruitiness to beer. Journal of Bioscience and Bioengineering, 2003. 96(2): p. 110–118.
[27] Pires, E.J., et al., Yeast: the soul of beer's aroma-a review of flavour-active esters and higher alcohols produced by the brewing yeast. Applied Microbiology and Biotechnology, 2014. 98(5): p. 1937–1949.
[28] Linko, M., et al., Recent advances in the malting and brewing industry. Journal of Biotechnology, 1998. 65(2–3): p. 85–98.
[29] Bellido-Milla, D., J.M. Moreno-Perez, and M.P. Hernandez-Artiga, Differentiation and classification of beers with flame atomic spectrometry and molecular absorption spectrometry and sample preparation assisted by microwaves. Spectrochimica Acta Part B-Atomic Spectroscopy, 2000. 55(7): p. 855–864.
[30] Coghe, S., et al., Sensory and instrumental flavour analysis of wort brewed with dark specialty malts. Journal of the Institute of Brewing, 2004. 110(2): p. 94–103.
[31] Hutter, K.J., Flow cytometry – A new tool for direct control of fermentation processes. Journal of the Institute of Brewing, 2002. 108(1): p. 48–51.
[32] Weiss, A., et al., Sensory and analytical characterisation of reduced, isomerised hop extracts and their influence and use in beer. Journal of the Institute of Brewing, 2002. 108(2): p. 236–242.
[33] Bartowsky, E.J., Bacterial spoilage of wine and approaches to minimize it. Lett Appl Microbiol, 2009. 48(2): p. 149–56.
[34] Ault, R.G. and R. Newton, Brewing Hygiene and Biological Stability of Beers, in Modern Brewing Technology, W.P.K. Findlay, Editor. 1991, Cleveland, Ohio: The Macmillan Press.

Massimiliano Marvasi, George A. O'Connor, and Max Teplitski

9 Use of tetracyclines and β-lactams in agriculture: Fate in the environment and occurrence of antibiotic-resistance determinants

9.1 Introduction

Antibiotics have been widely used in animal production for decades. Although the European Union in late 2000 banned the use of antibiotics for livestock production, other countries use massive amounts of the chemicals in veterinary therapies and at sub-therapeutic doses in feed supplements to promote animal growth. Tetracyclines and β-lactams (penicillins) are two classes of medically important antibiotics commonly used in animal production and for treating human infections (FDA, 2011). In the USA, 5,591,000 kg of chlortetracycline, oxytetracycline, and tetracycline and 870,000 kg of penicillins, amoxicillin, ampicillin, cloxacillin and penicillin were used in animal production in 2011 (FDA, 2011). Approximately 61 % of these antibiotics sold in the USA were approved for use in food-producing animals. Only 5 % of the antibiotics were administered via injection or topical administration to animals while the rest were administered via water or feed either to systematically prevent diseases and for food production (FDA, 2011). For example, in the USA, 200 g of antibiotics per ton of feed is commonly approved as a 14-day treatment for healthy animals (Boyd, 2001).

In countries such as USA and China, antibiotics are still considered an un-avoidable component of animal production. There are several reasons for the use of antibiotics as feed supplements. For example, tetracyclines and ampicillins are thought to reduce mortality, increase feed utilization efficiency and enhance growth (Jay, Boland, & Silbergeld, 2007). Indeed, during the last 50 years, the wide use of antibiotics in animal production has been associated with a significant increase of average market weight of commercial broilers (from 3.5 lb in 1965 to 4.7 lb in 1995) (Boyd, 2001). However, improved agricultural technologies, including introduction of new crosses, were likely responsible for this increase of growth (Boyd, 2001; Cahaner & Leenstra, 1992). Some evidence supports the notion that the administration of antibiotics at sub-therapeutic doses improves feed utilization efficiency and growth promotion, as recently highlighted by Hao et al 2014 and other scientists (Boyd, 2001; Hao et al., 2014; Jay et al., 2007). Hao and collaborators analyzed several recent studies (from 1985 to 2011) demonstrating the role of tetracyclines and β-lactams in improving feed utilization, growth promotion, reproductive performances and carcass quality (Hao et al., 2014) (Table 9.1, from Hao, H. 2014 – with permission). The mechanisms by which antibiotics administered at sub-therapeutic doses improve animal growth are not clear. Experiments with germ-free chicks led to hypotheses that: i) administration of antibiotics may protect nutrients from bacterial degradation in the gut; ii) absorption of nutrients may be improved because of a thinning of the intestinal barrier; iii) sub-therapeutic doses of antibiotics decrease production of toxins by bacteria; iv) general reduction of intestinal infections is beneficial to animal growth (Hao et al., 2014; Jay et al., 2007). It is also possible that antibiotics at sub-therapeutic doses alter the composition or functionality of host microbiota in livestock (Allen et al., 2013).

Tab. 9.1: Evidences for the role of antimicrobials on feed utilization, growth promotion, reproductive performance, and carcass quality as reported by Hao and collaborators (2014); (*) ↑ Shows increase, while ↓ shows decrease (Hao et al., 2014).

Reference	Drugs and animals	Parameters of animal production (growth rate)
1 (Hao et al., 2014; Van Lunen, 2003)	Antimicrobials to growing and fattening pigs	Digestion of energy (5.10 %↑)* Digestion of nitrogen (1.80 %↑) Digestion of phosphorus (3.40 %↑)
2 (Hao et al., 2014; Van Lunen, 2003)	Antimicrobials to swine	Feed utilization (7 %↑) Average weight gain (3.3–8.8 %↑)
3 (Hao et al., 2014; Van Lunen, 2003)	Antimicrobials to young pigs Antimicrobials to grower pigs	Feed utilization (4.6 %↑) Average weight gain (6.8 %↑) Feed utilization (1.7 %↑) Average weight gain (1.9 %↓)
4 (Hao et al., 2014; Van Lunen, 2003)	Antimicrobials to piglet	Feed utilization (6.9 %↑)
5 (Hao et al., 2014)	Antimicrobials to food animal	Average weight gain (4–5 %↑)
6 (Hao et al., 2014; Van Lunen, 2003)	Antimicrobials to piglet	Average weight gain (16.4 %↑)
7 (Nagaraja & Taylor, 1987)	Salinomycin to weanling piglets	Average weight gain (15.82 %↑)
8 (He et al., 2011; Zhou et al., 2011)	(Bacitracin zinc, colistin sulfate, Flavomycin and florfenicol) to (Carassius, Carp or hybrid tilapia)	Average weight gain (25.5~40.87 %↑)
9 (Soma & Speer, 1975)	Chlortetracycline to sow	Conception rate (4.1 %↑) Farrowing rate (5.8 %↑)
10 (Lindemann et al., 1985)	Folic acid to sow	Gestation gain (18 %↑)
11 (Cromwell et al., 1984; Cromwell, Stahly, Speer, & O'Kelly, 1984; Hao et al., 2014; Lindemann et al., 1985)	Tiamulin, nosiheptide, and salinomycin to pig	Thickness of backfat (9.7 %↓) Thickness of total fat (8 %↓) Eye muscle area (9.8 %↑) Lean meat (4.4 %↑)

Whether or not the use of antibiotics in animal feed supplementation is commercially justified is a part of the broader discussion, which focuses on the public and environmental health consequences of this practice. It is commonly accepted that the antibiotic use in general contributes to the rise and spread of drug-resistance genes and microbes carrying them. A quick search in PubMed reveals over 9,000 papers that focus on surveys and characterization of antibiotic-resistant bacteria in a variety of environment, ranging in the degree of anthropogenic impact (e.g. (Allen et al., 2013; Aubry-Damon et al., 2004; Chee-Sanford et al., 2001, 2009; Hao et al., 2014). Discoveries of genetic determinants of antibiotic resistance and microbes carrying them in relatively pristine habitats (Allen et al., 2010) may serve as an indication of far-reaching consequences of human impact and antibiotic over-use, or may be a reflection of the fact that antibiotic-resistance genes normally occur in microbial populations that synthesize antibiotics. While establishing causal relationships in these observations is important, there is also little doubt that the spread of the genetic determinants of antibiotic resistance and pathogens carrying them is one of the major threats to public health worldwide (Allen et al., 2013).

It is well known that antibiotic resistant genes are present in the natural environments. Most antibiotics are produced by strains of fungi and bacteria that occur naturally (Allen et al., 2010). Many of the known antibiotic resistant genes are found on transposons, integrons, and plasmids, and can be mobilized and transferred to other bacteria readily (Allen et al., 2010; Nandi et al., 2004). Therefore, resistance genes are harmful even if they are harbored in non-pathogenic bacteria: Studies demonstrate that human pathogens acquire antibiotic resistance genes from environmental bacteria (Finley et al., 2013; Gaze et al., 2013).

9.2 Fates of tetracyclines and β-lactams in waste and soils

Lability of waste-borne antibiotics applied to land is expected to vary with numerous antibiotic soil, and waste factors. Antibiotic chemical characteristics (e.g., solubility, Kow, pKas, hydrolysis tendencies, etc.) are critically important, but soil (or waste-amended soil) characteristics (e.g., pH, charge kind and amount, specific surface area, chemical composition of the organic matter, etc.) can be equally critical. Given the almost infinite number of possible soil-chemical-waste combinations, forecasting waste-borne antibiotics fate, transport, and risk can appear daunting. When the risk of antibiotic resistant bacteria development is considered, the task is further complicated by the likely issues of bioavailability of soil-sorbed antibiotics. For example, antibiotics retained on soil and/or organic particles may nonetheless be bioavailable to microorganisms, and/or bioavailability may change over time depending on a number of biological, chemical and physical processes (Subbiah, Mitchell, Ullman, & Call, 2011). Furthermore, the chemical nature of the organic matter (humus vs plant cell wall polymers not fully digested by animals and present in manures) will interact with antibiotics and their residuals differently, and will have different susceptibility to microbial degradation. Traditional conceptual and mathematical models of organic chemical retention in soils focus on hydrophobic-based interactions (e.g., hydrophobic "bonding", Nernst partitioning) (Higgins et al., 2010; Southworth & Keller, 1986). The approach works best for poorly soluble, non-polar, and moderate to high K_{ow} value organic chemicals. Although their physicochemical properties vary widely, most antibiotics are at least moderately water-soluble and have low K_{ow} values owing to various polar functional groups. For example, different sorption and desorption behaviors of oxytetracycline and ofloxacin are influenced by soil organic matter content and cation exchange capacity by soil pH. Complexation and cation-exchange reactions are the main sorption mechanisms (Peng et al., 2014).

Association/dissociation of protons from the functional groups (e.g., NH2, R-COOH, R-OH) alters the charge on the molecule, leading to cations, anions, or zwitterions in addition to neutral (undissociated) species in proportion to system pH. Chemical speciation strongly influences chemical retention by soils to negatively, positively, or neutral-charged soil components, whose charge can also be a function of pH. Predictions of antibiotics fate based solely on Kow-partitioning, thus, are usually insufficient (Higgins et al., 2010). A prime example of the complexities discussed above is the soil reactions of the tetracyclines, as detailed by Sassman and Lee (2005).

Soil (and waste) solid components other than organic matter, such as cation- and anion-sorbing sites can be important adsorbents. Indeed, cation retention mechanisms dominate tetracycline behavior throughout the environmentally relevant pH range, owing to the dominance of the (positive portion of the) zwitterion. Antibiotic retention across soils is highly variable, but normalizing the distribution coefficients (Kd values) to the cation exchange capacities (CEC values) of the soils yields a parameter (Kd/CEC) that greatly reduces the variability. Adjustments to account for the fraction of antibiotic species occurring at specific soil pH values (Sassman & Lee, 2005) and their individual Kd values further reduces variability in the parameter used to account for retention. Retention of the tetracycline class of antibiotics is well researched, and theoretical descriptions are available (Figueroa et al., 2004). β-lactams are less well studied, but the chemistry (pKas) of β-lactams suggests that similar polar-based consideration of compound speciation should complicate soil/waste retention. The biological half-life of β-lactams, however, is much lessthan for the tetracycline class of antibiotics (Subbiah et al., 2011) and tends to dominate soil behavior.

Retention mechanisms, based on compound speciation and system pH, can also be expected to influence compound fate and bioavailability. Strong chemical retention/association with soils and wastes typically reduces bioavailability and prolongs chemical persistence. Thus, while the total soil concentration of the chemical may remain elevated following contamination, impacts of the chemical (e.g., plant or animal uptake, promotion of antibiotic resistant bacteria development) may decrease significantly. Subbiah et al. (2011) showed dramatic reductions in antibiotic activity of some, but not all, of 10 antibiotics added to soil-water slurries, even at high (≥ 200 ppm) antibiotics concentrations after only 24 h reaction. Antibiotic "dissipation" (attributed to soil adsorption – as biological degradation was dismissed) was essentially complete for tetracycline, but was minimal for ampicillin, a β-lactam. Tetracycline adsorbed onto soil particles was deemed biologically unavailable when spread onto a lawn of antibiotic-sensitive *E. coli* K-12 (Subbiah et al., 2011). Soil impacts on antibiotics retention and bioavailability were evaluated in three soils varying in texture (sand to silt loam), total organic carbon content (0.24 to 2.8 %), and pH (4.7 to 7.8) (Subbiah et al., 2011). Effects of antibiotic bioavailability on *E. coli* were qualitatively consistent with expectations of chemical retention. No attempt, however, was made to quantify expectations, or bioavailability issues, that could be generalized to other soils and waste-amended soils. Further, no attempt was made to quantify possible changes in bioavailability (e.g., formation of bound residues) with time or the nature of the antibiotics source, e.g., waste that is land applied.

Tetracyclines and β-lactams can persist in manures and soils, but it is not clear how and to what extent residual antibiotics are available to microbes and whether these antibiotics exert selective pressure to maintain and amplify antibiotic resistance genes within bacterial populations in soil, manure and compost.

9.3 Mechanisms of resistance to tetracyclines and ß-lactams and spread of the genetic determinants of antibiotic resistance in soil and manure

9.3.1 Tetracycline resistance genes

Tetracyclines are a family of antibiotics that inhibit protein synthesis by preventing attachment of aminoacyl-tRNA to the ribosomal acceptor site (Chopra & Roberts, 2001). Tetracyclines traverse the outer membrane of gram-negative bacteria through the OmpF and OmpC porin channels, as positively charged cation-tetracycline coordination complexes (Chopra & Roberts, 2001). Resistance to tetracyclines has now emerged in many commensal and pathogenic bacteria due to the acquisition of tet genes. More than thirty tetracycline resistance (tet) genes and three oxytetracycline resistance (otr) genes have been characterized (Roberts, 2005). Eighteen of the tet genes and one of the otr genes code for efflux pumps, and seven of the tet genes and one of the otr genes, otrA, code for ribosomal protection proteins (Chopra & Roberts, 2001). The most frequent mechanism of resistance isolated from the animal waste and the environment involves a membrane-associated protein (TetA) that exports the antibiotic out of the bacterial cell before it inhibits polypeptide elongation (Chee-Sanford et al., 2001, 2009; Chen et al., 2010; Gueimonde et al., 2006; Heuer et al., 2011; Karczmarczyk et al., 2011[a]; Karczmarczyk et al., 2011[b]). Adjacent to tetA and divergently oriented is tetR, which encodes the tetracycline receptor and a repressor of both tetA and tetR expression (Ramos et al., 2005). In tetR-A expression systems the expression of tetA is proportional to the levels of tetracycline (Georgi, Buerger, Hillen, & Berens, 2012; Poole, 2004). Tetracycline resistance dispersion is also facilitated by transferable plasmids and transposons (Karczmarczyk et al., 2011; Karczmarczyk, Walsh et al., 2011).

9.3.2 β-lactam resistance genes

A high rate of isolation of β-lactam resistant pathogens from the manure and environment has been reported (Heuer et al., 2011; Keen & Montforts, 2012; Lachmayr et al., 2009). β-lactams target transpeptidases involved in bacterial cell-wall biosynthesis. In the presence of the antibiotic, transpeptidases form a lethal covalent penicilloyl – enzyme complex that blocks the normal trans-peptidation reaction. β-lactams are rendered inactive through three primary mechanisms of resistance: i) degradation of the antibiotics, ii) alteration of the antibiotic target site, and iii) prevention of access of the antibiotic through altered permeability or forced efflux (Poole, 2004; Wilke et al., 2005). Four classes of β-lactamases (which hydrolyze the antibiotic) confer resistance to the antibiotics: A, B, C and D. Classes A, C and D share a common serine hydrolase mechanism, while cass B β-lactamases are metalloenzymes (Szarecka et al., 2011). Plasmid-mediated class A and class C enzymes are frequently reported, whereas class B and class D β-lactamases are relatively rare (Rayamajhi et al., 2010). The class A, extended spectrum β-lactamases are derivatives of enzymes called TEM (named after Temoneira, a Greek patient from whom it was first isolated (Datta & Kontomichalou, 1965) and SHV (sulfydryl variable; (Gupta, 2007)), both responsible for ampicillin degradation (Bradford, 2001). There are more than 90 TEM-type β-lactamases and more than 25 SHV-type enzymes (Bradford, 2001). TEM-1 is the most commonly encountered β-lactamase in gram-negative bacteria isolated from manure (Rayamajhi et al., 2010). The *blaTEM* gene is carried and transferred by large-plasmids and bacteriophages.

The spread of the resistant plasmid and TEM genes was high in manure and within a wide variety of clinical and opportunistic pathogens (Binh et al., 2008). Of 228 transconjugant plasmids isolated from 15 different manures using selective media supplemented with amoxicillin, sulfadiazine or tetracycline, Binh et al. (2008) recovered 81 plasmids representing different antibiotic resistance patterns. bla-TEM was detected on 44 plasmids, while 77 genes conferred tetracycline resistances. Class C resistance genes have also been isolated from manure (Bradford, 2001; Colomer-Lluch et al., 2011; Ingram et al., 2011; Karczmarczyk et al., 2011[a]; Karczmarczyk et al., 2011[b]; Rayamajhi et al., 2010). It is not clear how and to what extent residual β-lactams exert selective pressure on the spread and maintenance of the β-lactam resistance determinants in bacterial populations. Unlike tetracyclines, which are typically pumped out of the resistant cells (thus increasing local concentrations in the vicinity of the cells carrying tetAR), β-lactamases can be extracellular and reduce local concentrations of the antibiotic and potentially favor proliferation of susceptible cells.

9.3.3 Spread of tetracyclines and β-lactams resistant bacteria in the environment

Several studies have demonstrated a correlation between application of manures to soils and prevalence of antibiotic-resistant bacteria. DNA extracted from soil from different locations from 1940 to 2008, showed that all classes of antimicrobial resistance genes significantly increased over time. Resistance to tetracyclines increased dramatically, with some samples being >15-fold more resistant to tetracyclines in 2010 than in the 1970s (Knapp et al., 2010). Similarly, laboratory studies showed that the frequency of bacteria carrying antibacterial resistance determinants is correlated with the use of antibiotics in livestock production (Chee-Sanford et al., 2001, 2009; Fessler et al., 2011; Heuer et al., 2011; Johnsen et al., 2005; Peak et al., 2007). Chee-Sanford and collaborators demonstrated that tet genes occur in the environment as a direct result of antibiotic use in livestock production and suggested that groundwater was a potential source of antibiotic resistance in the food chain. The presence of tetracycline resistance determinants in waste lagoons and in groundwater underlying two swine farms was analyzed. Eight classes of genes encoding tetracycline mechanism of resistance [*tetO*, *tetQ*, *tetW*, *tetM*, *tetBP*, *tetS*, *tetT*, and *otrA*] were found in total DNA extracted from water of two lagoons. In particular *tetM*, which was not predominant in the environmental samples, was identical to *tetM* isolated from the lagoons. In addition, the presence of the resistance genes in typical soil bacteria suggests that antibiotic resistance gene dissemination is not limited to strains of gastrointestinal origin but it can be mobilized into the indigenous soil microbiota (Allen et al., 2010; Chee-Sanford et al., 2001, 2009). Further studies were focused on determining the spread of tetracycline resistant bacteria among animals. Newborn piglets were colonized with antibiotic resistant strains by infected sows. To prevent the spread of antibiotic resistant bacteria, newborn piglets were also colonized with autoctonal antibiotic-sensitive strains. The new microbial community delayed but did not prevent colonization by maternal resistant strains (Allen et al., 2010; Chee-Sanford et al., 2009; Stanton & Humphrey, 2011).

Conversely, reducing the use of antibiotics during food production apparently decreased occurrence of resistance (Aarestrup & Frimodt-Møller, 2011; Klare, et al., 2001; Wegener, 2003). For example, Del Grosso and collaborators studied fecal carriage of avoparcin- (vancomycin-like molecule) resistant enterococci in broiler chickens and slaughter pigs in Italy before the avoparcin ban and after (Del Grosso et al., 2000; Wegener, 2003). After the avoparcin ban, the rate of avoparcin resistant enterococci contamination in meat products decreased. The decrease was statistically significant in poultry (from 18.8 % to 9.6 %), but not in pork products (from

9.7 % to 6.9 %). However, pertinent studies pre-post antibiotic ban are limited (Del Grosso et al., 2000; Hao et al., 2014; Wegener, 2003) but results as showed by Del Grosso et al. (2000) allow a better comparison of the data about the fate of antibiotic resistant determinants in soil before and after the ban.

The increase in selective pressure and spread of antibiotic resistance within exposed microbial populations makes good evolutionary sense. Nevertheless some studies found antibiotic-resistant bacteria isolated from humans and wild animals without a recent history of antibiotic exposure (Gueimonde et al., 2006; Lachmayr et al., 2009; Luo et al., 2005; Zhang et al., 2011). Bacteria carrying antibiotic resistance genes are present in native microbial populations. Indeed, there are several lines of evidence that human pathogens have acquired resistance genes from environmental bacteria (Finley et al., 2013; Gaze et al., 2013). Commercial use of antimicrobials at therapeutic and sub-therapeutic levels is assumed to exert selective pressure on native microbial populations, but the extent of this pressure and the evolutionary and ecological consequences are not yet completely clear.

Antibiotic resistance genes are common in natural environments and existed, even on plasmids, before the wide-spread commercial use of antibiotics (Allen et al., 2010). From an ecological point of view, resistance genes are proposed to contribute to growth and survival of bacteria in ecological niches (Groh et al., 2007). Groh and collaborators identified membrane efflux components (such as *mexF* gene) required for sediment fitness of nonpathogenic bacteria which exhibited homology to multidrug resistance genes (Groh et al., 2007). Chloramphenicol-adapted *Shewanella oneidensis* MR-1 cultures exhibited mutations in the gene encoding a TetR family regulatory protein, indicating a role for this protein in regulating expression of the *mexEF* operon. The authors concluded that membrane efflux components are important for sediment fitness and suggested that antibiotic efflux may be a required process for bacteria living in sediment systems (Groh et al., 2007). However, a complete picture of the ecology of the resistance determinants is still missing.

9.4 Current Developments and future challenges

All antimicrobial agents belonging to substance classes used for therapeutic treatments in humans (including β-lactams and tetracyclines) have been forbidden as antimicrobial growth promoters (AGPs) since 1972 in Switzerland. In Europe, the last phase of the wide ban on tetracyclines and β-lactams usage in animal feed took effect in 2007 (EC Regulation No. 1831/2003) (Millet & Maertens, 2011). During this time, the reduction in the use of antibiotics in food production appeared to have been linked to a decrease in the spread of resistances (Aarestrup & Frimodt-Møller, 2011; Boerlin et al., 2001). However, the issue is still controversial: when tetracycline was banned as a growth promoter in Europe, studies showed reduced levels of resistance, whereas other studies showed no changes in the levels of spread of resistances (Hao et al.,

2014; Wegener, 2003). The difficulty in properly assessing the reduction of antibiotic resistances is primarily due to the heterogeneity of the data compared. The main reason is because many studies observed only the resistance phenotype rather than the typology of resistance determinants and the bacterial genotype. A mere counting of resistant bacteria cannot be an appropriate approach because it can include naturally occurring resistant bacteria and not strictly related to the kind of antibiotics amended to the feed.

In terms of sustainability of livestock production, the abolition of antibiotics seems not to have had a detrimental effect. Antimicrobial growth promoters were phased out in Denmark, Norway and Sweden in 1998–1999, 1995 and 1986, respectively. Despite the reduction of antibiotic usage, there was no change in production efficiency of livestock and chickens illustrating that it is possible to cope with the ban on β-lactams and tetracyclines (Grave et al., 2006). Furthermore, in recent years, several studies have tested a number of innovative strategies and alternative products that were developed to sustain livestock growth by replacing antibiotics.

9.4.1 Probiotics

In terms of reduced mortality and feed efficiency, a balanced gut microbiota is certainly a primary factor (Montagne et al., 2003). It is well know that an appropriate gut microbiota constitutes an efficient barrier against pathogen colonization, produces metabolic substrates (e.g. vitamins and short-chain fatty acids) and stimulates immune system in a non-inflammatory manner (Gaggia et al., 2010). The European Communion has recently released comprehensive documentation listing several "Qualified Presumption of Safety" (QPS) organisms used as additive in food or feed. Among these macroorganisms several species of *Bifidobacterium* sp., *Bacillus* sp. and *Lactobacillus* sp. (EFSA, 2008) are already commonly used as probiotic supplements in animals and humans.

9.4.2 Prebiotics and other molecules

Current research also focuses on providing prebiotics defined as "non-digestible food ingredients that beneficially affect the host by selectively stimulating the growth and/or activity of one or a limited number of bacteria in the colon" such as fibers (Gaggia et al., 2010; Gibson & Roberfroid, 1995). Other alternative chemical products have been reviewed by Huyghebaert and coworkers (Huyghebaert et al., 2011). Some alternatives, that in some case can partially compensate the antibiotic treatments, are the utilization of exogenous enzymes, herbs, etheric oils and organic acids (Huyghebaert et al., 2011).

9.4.3 New generation of antibiotics

The generation of antibiotics unable to lead to the development of resistance is also currently studied and it seems to open very interesting scenarios. For example, antibiotic teixobactin was discovered in a screen of uncultured bacteria. Teixobactin inhibits cell wall synthesis by binding to a highly conserved motif of a precursor of peptidoglycan and cell wall component teichoic acid. Ling and coworkers (Ling et al., 2015) did not obtain any mutants of *Staphylococcus aureus* or *Mycobacterium tuberculosis* resistant to teixobactin. Properties of this compound suggest a path towards developing antibiotics that are less likely to lead to the development of resistance (Ling et al., 2015). Further studies need to be carried out on *Enterobaceraiceae* (*Escherichia coli* or *Salmonella enterica*). It is almost certain, however, that such potent antibiotics will have a detrimental effect on the host's normal gut microbiota.

9.4.4 Manure management

Recently, manure management practices, such as composting, have been tested as potential routes for reducing antibiotic concentrations in manure before land application, thereby minimizing the potential for environmental contamination (Dolliver et al., 2008). To that end, Dolliver and coworkers measured the presence of chlortetracycline in three manure composting treatments. Despite significant differences in temperature, mass, and nutrient losses between the composting treatments and the control, chlortetracycline concentrations declined rapidly during composting with a first-order decay per day with no difference in antibiotic degradation among the treatments. (H. K. Allen et al., 2010; Chee-Sanford et al., 2009; Dolliver et al., 2008). However further studies need to be addressed in further different treatment conditions by acting on the manure pile with different mixing, moisture content adjustments, vessel composting, typology of manure used and enrichment with chemicals.

9.5 Conclusions

Even though some nations have banned the use of therapeutic antibiotics for livestock production, the use of these chemicals continue to support agriculture in many other countries (e.g. USA and China). The antibiotics are deposited in the environment with organic wastes, and their fate in amended soils and in crop production environment remains controversial. Several studies have demonstrated correlations between antibiotic detection in manure-amended soils and the prevalence of antibiotic-resistant bacteria (Knapp et al., 2010), but the correlation is not perfect. For example, after antibiotics were banned, studies showed a reduction or no changes in the levels of spread of resistances (Wegener, 2003). In addition, the minimum concentrations of antibiotics needed to maintain this pressure are not known. Research is complicated by the difficulty of measuring the biological and chemical availability of antibiotics in soil due to the adsorption and chelation properties by soil particles.

9.6 References

[1] Aarestrup, F. M., & Frimodt-Møller, N. (2011). Increasing transmission of antibiotic resistance from animals to humans. Ugeskr Laeger, 173, 2180–2183.

[2] Allen, H. K., Donato, J., Wang, H. H., Cloud-Hansen, K. A., Davies, J., & Handelsman, J. (2010). Call of the wild: Antibiotic resistance genes in natural environments. Nature Reviews. Microbiology, 8(4), 251–259.

[3] Allen, H. K., Levine, U. Y., Looft, T., Bandrick, M., & Casey, T. A. (2013). Treatment, promotion, commotion: Antibiotic alternatives in food-producing animals. Trends in Microbiology, 21(3), 114–119.

[4] Aubry-Damon, H., Grenet, K., Sall-Ndiaye, P., Che, D., Cordeiro, E., Bougnoux, M. E., et al. (2004). Antimicrobial resistance in commensal flora of pig farmers. Emerging Infectious Diseases, 10(5), 873–879.

[5] Binh, C. T., Heuer, H., Kaupenjohann, M., & Smalla, K. (2008). Piggery manure used for soil fertilization is a reservoir for transferable antibiotic resistance plasmids. FEMS Microbiology Ecology, 66, 25–37.

[6] Boerlin, P., Wissing, A., Aarestrup, F. M., Frey, J., & Nicolet, J. (2001). Antimicrobial growth promoter ban and resistance to macrolides and vancomycin in enterococci from pigs. Journal of Clinical Microbiology, 39(11), 4193–4195.

[7] Boyd, W. (2001). Making meat: Science, technology, and american poultry production. The Johns Hopkins University Press. 51(1), 631–664.

[8] Bradford, P. A. (2001a). Extended-spectrum beta-lactamases in the 21st century: Characterization, epidemiology, and detection of this important resistance threat. Clinical Microbiology Review, 14, 933–51.

[9] Bradford, P. A. (2001b). Extended-spectrum β-lactamases in the 21st century: Characterization, epidemiology, and detection of this important resistance threat. Clinical Microbiology Reviews, 14(4), 933–951.

[10] Cahaner, A., & Leenstra, F. (1992). Effects of high temperature on growth and efficiency of male and female broilers from lines selected for high weight gain, favorable feed conversion, and high or low fat content. Poultry Science, 71(8), 1237–1250.

[11] Chee-Sanford, J., Aminov, R. I., Krapac, I. J., Garrigues-Jeanjean, N., & Mackie, R. I. (2001). Occurrence and diversity of tetracycline resistance genes in lagoons and groundwater underlying two swine production facilities. Applied Environmental Microbiology, 67, 1494–1502.

[12] Chee-Sanford, J., Mackie, R. I., Koike, S., Krapac, I. G., Lin, Y. F., Yannarell, A. C., et al. (2009). Fate and transport of antibiotic residues and antibiotic resistance genes following land application of manure waste. Journal of Environmental Quality, 38, 1086–1108.

[13] Chen, J., Michel, F. C., Sreevatsan, S., Morrison, M., & Yu, Z. (2010). Occurrence and persistence of erythromycin resistance genes (erm) and tetracycline resistance genes (tet) in waste treatment systems on swine farms. Microbial Ecolology, 60, 479–486.

[14] Chopra, I., & Roberts, M. (2001). Tetracycline antibiotics: Mode of action, applications, molecular biology, and epidemiology of bacterial resistance. Microbiology and Molecular Biology Review, 65, 232–60.

[15] Colomer-Lluch, M., Imamovic, L., Jofre, J., & Muniesa, M. (2011). Bacteriophages carrying antibiotic resistance genes in fecal waste from cattle, pigs, and poultry. Antimicrobial Agents and Chemotherapy, 55, 4908–4911.
[16] Cromwell, G. L., Stahly, T. S., Jensen, A. H., Plumlee, M. P., Krider, J. L., & Russett, J. C. (1984). Efficacy of thiopeptin as a growth promotant for growing barrows and gilts – a cooperative study. Journal of Animal Science, 59(4), 892–895.
[17] Cromwell, G. L., Stahly, T. S., Speer, V. C., & O'Kelly, R. (1984). Efficacy of nosiheptide as a growth promotant for growing-finishing swine – a cooperative study. Journal of Animal Science, 59(5), 1125–1128.
[18] Datta, N., & Kontomichalou, P. (1965). Penicillinase synthesis controlled by infectious R factors in Enterobacteriaceae. Nature, 208, 239–241.
[19] Del Grosso, M., Caprioli, A., Chinzari, P., Fontana, M. C., Pezzotti, G., Manfrin, A., et al. (2000). Detection and characterization of vancomycin-resistant enterococci in farm animals and raw meat products in Italy. Microbial Drug Resistance, 6(4), 313–318.
[20] Dolliver, H., Gupta, S. F., & Noll, S. (2008). Antibiotic degradation during manure composting. Journal of Environmental Quality, 37(3), 1245–53.
[21] EFSA. (2008). The maintenance of the list of QPS microorganisms intentionally added to food or feed. No. EFSA-Q-2008-006, European Food Safety Authority.
[22] FDA. (2011). Summary report on antimicrobials sold or distributed for use in food-producing animals. http://www.fda.gov/downloads/ForIndustry/UserFees/AnimalDrugUserFee-ActADUFA/ucm277657.pdf.
[23] Fessler, A. T., Kadlec, K., Hassel, M., Hauschild, T., Eidam, C., Ehricht, R., et al. (2011). Characterization of methicillin-resistant Staphylococcus aureus isolates from food and food products of poultry origin in Germany. Applied Environmental Microbiology, 77, 7151–7157.
[24] Figueroa, R. A., Leonard, A., MacKay, Allison., A. (2004). Modeling tetracycline antibiotic sorption to clays. Environmental Science & Technology. 38(2), 476–483.
[25] Finley, R. L., Collignon, P., Larsson, D. G., McEwen, S. A., Li, X. Z., Gaze, W. H., et al. (2013). The scourge of antibiotic resistance: The important role of the environment. Clinical Infectious Diseases : An Official Publication of the Infectious Diseases Society of America, 57(5), 704–710.
[26] Gaggia, F., Mattarelli, P., & Biavati, B. (2010). Probiotics and prebiotics in animal feeding for safe food production. International Journal of Food Microbiology, 141 Suppl 1, S15–28.
[27] Gaze, W. H., Krone, S. M., Larsson, D. J., Li, X., Robinson, J. A., Simonet, P., et al. (2013). Influence of humans on evolution and mobilization of environmental antibiotic resistome. Emerging Infectious Diseases, 19(7), e120871.
[28] Georgi, C., Buerger, J., Hillen, W., & Berens, C. (2012). Promoter strength driving TetR determines the regulatory properties of tet-controlled expression systems. PloS One, 7(7), e41620.
[29] Gibson, G. R., & Roberfroid, M. B. (1995). Dietary modulation of the human colonic microbiota: Introducing the concept of prebiotics. The Journal of Nutrition, 125(6), 1401–1412.
[30] Grave, K., Jensen, V. F., Odensvik, K., Wierup, M., & Bangen, M. (2006). Usage of veterinary therapeutic antimicrobials in Denmark, Norway and Sweden following termination of antimicrobial growth promoter use. Preventive Veterinary Medicine, 75(1–2), 123–132.
[31] Groh, J. L., Luo, Q., Ballard, J. D., & Krumholz, L. R. (2007). Genes that enhance the ecological fitness of Shewanella oneidensis MR-1 in sediments reveal the value of antibiotic resistance. Applied and Environmental Microbiology, 73(2), 492–498.
[32] Gueimonde, M., Salminen, S., & Isolauri, E. (2006). Presence of specific antibiotic (tet) resistance genes in infant faecal microbiota. FEMS Immunology and Medical Microbiology, 48, 21–25.

[33] Gupta, V. (2007). An update on newer beta-lactamases. The Indian Journal of Medical Research, 126(5), 417–427.
[34] Hao, H., Cheng, G., Iqbal, Z., Ai, X., Hussain, H. I., Huang, L., et al. (2014). Benefits and risks of antimicrobial use in food-producing animals. Frontiers in Microbiology, 5, 288.
[35] He, S., Zhou, Z., Meng, K., Zhao, H., Yao, B., Ringo, E., et al. (2011). Effects of dietary antibiotic growth promoter and saccharomyces cerevisiae fermentation product on production, intestinal bacterial community, and nonspecific immunity of hybrid tilapia (Oreochromis niloticus female x Oreochromis aureus male). Journal of Animal Science, 89(1), 84–92.
[36] Heuer, H., Schmitt, H., & Smalla, K. (2011). Antibiotic resistance gene spread due to manure application on agricultural fields. Current Opinion in Microbiology, 14, 236–243.
[37] Higgins, C., Sharp, J., Sepulvado, J., Littrell, B., O'Connor, G., Snyder, E., et al. (2010). Trace organic chemicals in biosolids-amended soils: State-of-the-science review. IWA Publishing.
[38] Huyghebaert, G., Ducatelle, R., & Van Immerseel, F. (2011). An update on alternatives to antimicrobial growth promoters for broilers. Veterinary Journal, 187(2), 182–188.
[39] Ingram, P. R., Inglis, T. J., Vanzetti, T. R., Henderson, B. A., Harnett, G. B., & Murray, R. J. (2011). Comparison of methods for AmpC β-lactamase detection in Enterobacteriaceae. Journal of Medical Microbiology, 60, 715–721.
[40] Jay, G. P., Boland, J. J., & Silbergeld, E. (2007). Growth promoting antibiotics in food animal production: An economic analysis. Public Health Reports, 122(1), 79–87.
[41] Johnsen, P. J., Østerhus, J. I., Sletvold, H., Sørum, M., Kruse, H., Nielsen, K., et al. (2005). PMC544227; persistence of animal and human glycopeptide-resistant enterococci on two Norwegian poultry farms formerly exposed to avoparcin is associated with a widespread plasmid-mediated vanA element within a polyclonal Enterococcus faecium population. Applied and Environmental Microbiology, 71, 159–168.
[42] Karczmarczyk, M., Abbott, Y., Walsh, C., Leonard, N., & Fanning, S. (2011). Characterization of multidrug-resistant Escherichia coli isolates from animals presenting at a university veterinary hospital. Applied and Environmental Microbiology, 77, 7104–7112.
[43] Karczmarczyk, M., Walsh, C., Slowey, R., Leonard, N., & Fanning, S. (2011). Molecular characterization of multidrug-resistant Escherichia coli isolates from irish cattle farms. Applied and Environmental Microbiology, 77, 7121–7127.
[44] Keen, P. L., & Montforts, M. H. M. M. (2012). Antimicrobial resistance in the environment. (Ed. Patricia L. Keen, Mark H. M. M. Montforts). Hoboken, N.J. Wiley.
[45] Klare, I., Werner, G., & Witte, W. (2001). Enterococci. habitats, infections, virulence factors, resistances to antibiotics, transfer of resistance determinants. Contribution to Microbiology, 8, 108–122.
[46] Knapp, C. W., Dolfing, J., Ehlert, P. A., & Graham, D. W. (2010). Evidence of increasing antibiotic resistance gene abundances in archived soils since 1940. Environmental Science & Technology, 44, 580–587.
[47] Lachmayr, K. L., Kerkhof, L. J., Dirienzo, A. G., Cavanaugh, C. M., & Ford, T. E. (2009). Quantifying nonspecific TEM beta-lactamase (blaTEM) genes in a wastewater stream. Applied and Environmental Microbiology, 75(1), 203–211.
[48] Lindemann, M. D., Kornegay, E. T., Stahly, T. S., Cromwell, G. L., Easter, R. A., Kerr, B. J., et al. (1985). The efficacy of salinomycin as a growth promotant for swine from 9 to 97 kg. Journal of Animal Science, 61(4), 782–788.
[49] Ling, L. L., Schneider, T., Peoples, A. J., Spoering, A. L., Engels, I., Conlon, B. P., et al. (2015). A new antibiotic kills pathogens without detectable resistance. Nature, 517(7535), 455–459.
[50] Luo, N., Pereira, S., Sahin, O., Lin, J., Huang, S., Michel, L., et al. (2005). Enhanced in vivo fitness of fluoroquinolone-resistant Campylobacter jejuni in the absence of antibiotic selection

pressure. Proceedings of the National Academy of Sciences of the United States of America, 102(3), 541–546.
[51] Millet, S., & Maertens, L. (2011). The European ban on antibiotic growth promoters in animal feed: From challenges to opportunities. Veterinary Journal, 187(2), 143–144.
[52] Montagne, L., Pluske, J. R., & Hampson, D. J. (2003). A review of interactions between dietary fibre and the intestinal mucosa, and their consequences on digestive health in young non-ruminant animals. Animal Feed Science and Technology, 108(1–4), 95–117.
[53] Nagaraja, T. G., & Taylor, M. B. (1987). Susceptibility and resistance of ruminal bacteria to antimicrobial feed additives. Applied and Environmental Microbiology, 53(7), 1620–1625.
[54] Nandi, S., Maurer, J. J., Hofacre, C., & Summers, A. O. (2004). Gram-positive bacteria are a major reservoir of class 1 antibiotic resistance integrons in poultry litter. Proceedings of the National Academy of Sciences of the United States of America, 101(18), 7118–7122.
[55] Peak, N., Knapp, C. W., Yang, R. K., Hanfelt, M. M., Smith, M. S., Aga, D. S., et al. (2007). Abundance of six tetracycline resistance genes in wastewater lagoons at cattle feedlots with different antibiotic use strategies. Environmental Microbiology, 9, 143–151.
[56] Peng, F. J., Zhou, L. J., Ying, G. G., Liu, Y. S., & Zhao, J. L. (2014). Antibacterial activity of the soil-bound antimicrobials oxytetracycline and ofloxacin. Environmental Toxicology and Chemistry, 33(4), 776–783.
[57] Poole, K. (2004). Uninhibited antibiotic target discovery via chemical genetics. Nature Biotechnology, 22, 1528–1529.
[58] Ramos, J. L., Martínez-Bueno, M., Molina-Henares, A., Terán, W., Watanabe, K., Zhang, X., et al. (2005). PMC1197418; the TetR family of transcriptional repressors. Microbiology and Molecular Biology Reviews, 69, 326–356.
[59] Rayamajhi, N., Jung, B. Y., Cha, S. B., Shin, M. K., Kim, A., Kang, M. S., et al. (2010). Antibiotic resistance patterns and detection of blaDHA-1 in Salmonella species isolates from chicken farms in South Korea. Applied and Environmental Microbiolology, 76, 4760–4764.
[60] Roberts, M. C. (2005). Update on acquired tetracycline resistance genes. FEMS Microbiology Letter, 245, 195–203.
[61] Sassman, S. A., & Lee, L. S. (2005). Sorption of three tetracyclines by several soils: Assessing the role of pH and cation exchange. Environmental Science and Technology, 39, 7452–7459.
[62] Soma, J. A., & Speer, V. C. (1975). Effects of pregnant mare serum and chlortetracycline on the reproductive efficiency of sows. Journal of Animal Science, 41(1), 100–105.
[63] Southworth, G. R., & Keller, J. L. (1986). Hydrophobic sorption of polar organics by low organic carbon soils. Water, Air, and Soil Pollution, 28(3–4), 239–248.
[64] Stanton, T. B., & Humphrey, S. B. (2011). Persistence of antibiotic resistance: Evaluation of a probiotic approach using antibiotic-sensitive Megasphaera elsdenii strains to prevent colonization of swine by antibiotic-resistant strains. Applied and Environmental Microbiology. 77(20): 7158–7166.
[65] Subbiah, M., Mitchell, S. M., Ullman, J. L., & Call, D. R. (2011). β-lactams and florfenicol antibiotics remain bioactive in soils while ciprofloxacin, neomycin, and tetracycline are neutralized. Applied and Environmental Microbiology, 77, 7255–7260.
[66] Szarecka, A., Lesnock, K. R., Ramirez-Mondragon, C. A., Nicholas, H. B., & Wymore, T. (2011). The class D β-lactamase family: Residues governing the maintenance and diversity of function. Protein Engineering, Design and Selection. 24(10), 801–9.
[67] Van Lunen, T. A. (2003). Growth performance of pigs fed diets with and without tylosin phosphate supplementation and reared in a biosecure all-in all-out housing system. The Canadian Veterinary Journal. La Revue Veterinaire Canadienne, 44(7), 571–576.
[68] Wegener, H. C. (2003). Antibiotics in animal feed and their role in resistance development. Current Opinion in Microbiology 6(5), 439–45.

[69] Wilke, M. S., Lovering, A. L., & Strynadka, N. C. (2005). Beta-lactam antibiotic resistance: A current structural perspective. Current Opinion in Microbiology, 8, 525–533.
[70] Zhang, L., Kinkelaar, D., Huang, Y., Li, Y., Li, X., & Wang, H. H. (2011). Acquired antibiotic resistance: Are we born with it? Applied and Environmental Microbiology, 77(20), 7134–7141.
[71] Zhou, Z., He, S., Liu, Y., Cao, Y., Meng, K., Yao, B., et al. (2011). Gut microbial status induced by antibiotic growth promoter alters the prebiotic effects of dietary DVAQUA(R) on aeromonas hydrophila-infected tilapia: Production, intestinal bacterial community and non-specific immunity. Veterinary Microbiology, 149(3–4), 399–405.

Mattia Pia Arena, Pasquale Russo, Daniela Fiocco, and Giuseppe Spano

10 Industrial microorganisms: tolerance to antibiotics and application of antimicrobial agents

10.1 Abstract

Lactic acid bacteria (LAB) are a group of microorganisms generally associated to food industry as they are traditionally applied in the production of several fermented food. Beside the aim to optimize the quality of the end product determining the organoleptic and nutritional profile, LAB can also contribute to the safety assessment of food by producing different compounds with antimicrobial activity such as bacteriocins, antimicrobial peptides and low molecular weight compounds. The production of antimicrobial compounds could be considered a suitable trait to be sought in the selection of starter, non-starter and probiotic strains due to the realistic use of these bacteria in the food industry in order to reduce spoilage and pathogenic microorganisms, increase the shelf life, moderate or eliminate the chemical additives. Moreover, the antagonistic activity can also aim to contrast the alarming problem of antibiotic resistance that is a crucial safety pre-assessment. Here, the potential use of LAB as biopreservatives and biotherapeutic agents in the food industry, and an overview about their resistance to antibiotics are discussed.

10.2 Introduction

Several microorganisms are traditionally associated to the production of many food products in which the microbial aptitude to ferment food matrices is the main desirable characteristic required to obtain end product with specific features. Lactic acid bacteria (LAB), a group of microorganisms generally recognized as safe (GRAS) (FAO/WHO, 2002), are largely applied in food manufacturing having a long history as both microbial starter and probiotic microorganisms due to their multipurpose strain-specific characteristics (Caplice and Fitzgerald, 1999; Wood and Holzapfel, 1995; Altay et al., 2013; Georgieva et al. 2009; Cebeci and Gürakan, 2003; Guidone et al., 2014). LAB are Gram-positive, usually non-spore forming and catalase-negative bacteria and their strictly fermentative metabolism is exploited for industrial processing of dairy, meat, vegetable and cereal products (Šušković et al., 2010).

Commonly, the practice to inoculate selected starter cultures has the aim to optimize the quality of the end product, to abbreviate the manufacture process and to prevent the development of spontaneous and uncompleted fermentation (Leroy and De Vuyst, 2004). Although industries are still exploiting the autochthonous microflora to ferment food matrix, at present, there is a growing demand for improved starter cultures (Law, 2001; Leroy et al., 2002). Starter cultures of LAB steer the rapid fermentation process of food raw material and determine the organoleptic and nutritional profile of the end product by producing organic acids, sugar polymers, sweeteners, aromatic compounds and vitamins (Leroy and De Vuyst, 2004). Moreover, LAB can also contribute to the safety assessment of food product due to the production of different compounds with antimicrobial nature such as bacteriocins and low molecular weight compounds. This latter feature fully meets the remarkable attention of the consumers towards food products in which the addition of chemicals for the storage and the increase of shelf life are limited or absent (Šušković et al., 2010). Furthermore, the demand for minimally processed food is expanding. Indeed, some physical and physicochemical technologies that preserve food from spoilage and avoid the development of undesirable microorganisms, including heat and cold treatment and modified atmosphere storage, can lead to a depletion of flavour and nutritional value (Law, 2001; Thomas et al., 2002). Thus, the implementation of selection criteria for starter cultures is focusing on microorganisms that possess, in addition to the ability to complete an optimal fermentation, extra properties, including probiotic activities and the capacity to produce useful enzymes, nutraceutical compounds and antimicrobials (Capozzi et al., 2012; Pidcock et al., 2002). Moreover, taking advantage of the intrinsic antimicrobial activity of microbial starters could be a promising approach to reduce the employment of antibiotics in the food chain.

It is well known that the antibiotic resistance pattern of starter and probiotic cultures is an essential part of a crucial safety pre-assessment (Snydman, 2008). The main concern is the potential exchange of tolerance factors with other microorganisms, with the risk of transferring antibiotic resistance traits to pathogenic bacteria. Recent genomic and metagenomic studies in humans, animals, food and environment have led to the discovery of a huge reservoir of antibiotic resistance genes called the resistome that could be mobilized and transferred from these sources to human pathogens (Rolain, 2013). This threat is even more insidious when considering that probiotics act directly in the human gut (Gueimonde et al., 2013). Indeed the human gut microbiota is suggested to be a reservoir of antibiotic resistance genes, which are potentially exchangeable between transient colonizers and residing bacteria (Haug et al., 2011). Moreover, lactobacilli have been reported as sporadic etiological agents of endocarditis that can be only controlled by antibiotic therapy (Salvana and Frank, 2006).

Here, we focus on the antimicrobial activity of LAB as a potential strategy to be valued for the biopreservation of food. Moreover, we give an overview on the question of antibiotic resistance in non-pathogenic bacteria throughout the food chain.

10.3 Antimicrobial agents

The achievement of food preservation by novel biological methods, e.g. profiting from the microbiota or the use of antimicrobial metabolites, is a central topic for the scientific field and the food industry. The antagonistic capability of LAB against foodborne pathogenic and spoilage microorganisms has attracted considerable attention (Reis et al., 2012). LAB are able to produce several antimicrobial metabolites, both protein and non-protein substances. Such molecules have a small impact on the organoleptic properties of the end product, contrarily to chemical agents and physicochemical treatments (Gálvez et al., 2010). Moreover, the inoculation of LAB and/or their metabolites into the food matrix is feasible and does not require large manufacturing costs (Kos et al., 2008).

The antimicrobial metabolites produced by LAB can be included into a non-bacteriocin group, comprising low molecular mass substances (<1 kDa), or in a bacteriocin and bacteriocin-like group including high molecular mass substances (>1 kDa) (Šušković et al., 2010; Collins et al., 2010; Cleveland et al., 2001).

10.3.1 Non-bacteriocin substances

LAB are able to obtain metabolic energy through oxidation of carbohydrates. Some LAB show homofermentative metabolism, which consists in the conversion of hexoses into lactic acid as a principal end-product, via the Embden-Meyerhof-Parnas pathway. Conversely, heterofermentative bacteria are able to degrade hexoses producing different compounds, such as lactic acid, ethanol, acetate, carbon dioxide and also formic, propionic, butyric acids, diacetyl and acetaldehyde, via the 6-phosphogluconate/phosfoketonate pathway (Prückler et al., 2015). Such compounds (<1 kDa) (Fig. 1.1) are essential to give sensorial attributes to the fermented product; on the other hand, they could have additional intrinsic properties, including inhibiting and inactivating activities against other microorganisms that are placed within the same food matrix. The total amount of these molecules is correlated to the chemical composition of the raw material and is strain-dependent.

Lactic acid and acetic acid are the most important microbial products in the fermented food. The undissociated forms of these organic acids act as antimicrobial molecules because they are able to pass through the bacterial cell membrane, thereby altering the intracellular pH and membrane potential of target microorganisms (Lorca and de Valdez, 2009). Additionally, lactic acid can be metabolized to other antimicrobial products such as acetic acid, formic acid and CO_2 (Doelle, 2014). Among these, carbon dioxide can contribute to food preservation as it creates an anaerobic environment, which is hostile to the growth of most spoilage microbes, and shows the ability to interfere with cell membrane, leading to dysfunction of cell permeability (Šušković et al., 2010).

LAB possessing flavoprotein oxidases are also able to produce hydrogen peroxide by reduction of oxygen. Hydrogen peroxide holds oxidizing activity against cellular components and can destabilize the molecular structures of proteins, enzymes, membrane constituents and also nucleic acids of microbial cells (Finnegan et al., 2010). The antimicrobial activity of hydrogen peroxide is predominantly against anaerobic and gram-negative bacteria (Reid, 2008).

Heterofermentative LAB can also produce acetoin, acetaldehyde and diacetyl. All these substances, besides having a very characteristic flavor, also show a considerable inhibitory activity (Vanderbergh, 1993). Moreover, many LAB can produce an organic acid named 3-phenyllactic acid (2-hydroxy-3-phenylpropanoic acid or β-phenyllactic acid, PLA) which shows specific antifungal and antibacterial properties (Mu et al., 2012). The chemical formula of PLA is $C_9H_{10}O_3$ and it can occur as two chiral isomers, i.e. D- and L-PLA, with D-PLA having a slightly higher antagonistic activity (Dieuleveux et al., 1998a). The mode of action of PLA has not been completely elucidated, although PLA probably damage the cell wall structure leading to a breakage which determines the leakage of cellular contents (Dieuleveux et al., 1998b; Lavermicocca et al., 2003). Numerous LAB, mainly belonging to *Lactobacillus plantarum* species, have been characterized as PLA producers (Lavermicocca et al., 2000; Ohhira et al., 2004; Wang et al., 2012; Prema et al., 2010).

10.3.2 Bacteriocin and bacteriocin-like substances

LAB also produce proteinaceous compounds (>1 kDa) (Fig. 10.1) presenting narrow or broad spectrum of growth inhibitory action for related species and across genera, respectively (Cotter et al., 2005). These compounds are ribosomally synthetized and include bacteriocins and bacteriocin-like substances.

Presently, the classification of bacteriocins identifies three or four classes of such molecules (De Vuyst and Leroy, 2007; Calo-Mata et al., 2008). Class I includes the so-called lantibiotics, which are small peptides (<5 kDa) containing post-translationally modified amino acids, such as lanthionine, dehydroalanine, α-dehydroalanine, and dehydrobutirine. Such bacteriocins have been shown to be heat-stable and can be further divided into two sub-classes, i.e. class Ia and class Ib. Class Ia contains net positive charge lantibiotics with linear structure, while class Ib includes negative or no net charge lantibiotics showing globular structure. Class Ia bacteriocins, such as nisin, affect the cell membranes of target microorganisms by interacting with the anionic lipids and through the formation of pores, with negative consequences on membrane potential, whereas class Ib antibacterial proteins are able to inhibit cellular enzymes (Guder et al., 2000).

Class II comprises non-lanthionine-containing bacteriocins, small (<10 kDa) and heat-stable proteins principally active against *Listeria* genera. Bacteriocins of class II are able to dissipate the proton motive force (PMF) forming pores in the cell mem-

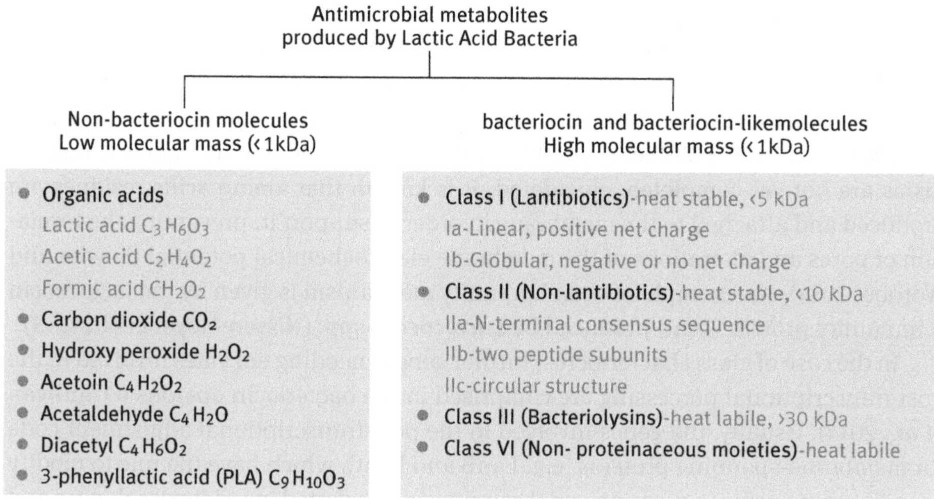

Fig. 10.1: Antimicrobial substances produced by lactic acid bacteria (LAB).

brane. This class can be further categorized into three subclasses. Class IIa (pediocin-like bacteriocins) comprises bacteriocins having a chemical structure similar to the bacteriocin pediocin produced by *Pediococcus* spp., presenting an N-terminal consensus sequence (–Tyr-Gly-Asn-Gly-Val-Xaa-Cys). The antimicrobial activity of this subclass is closely correlated to a mannose permease of the phosphotransferase system (PTS) (Papagianni and Anastasiadou, 2009). Class IIb bacteriocins have the particular characteristic to be constituted by a two-peptide structure which is essential for the antimicrobial action based on the dissipation of PMF. Proteins displaying a circular structure are comprised in class IIc and can induce the permeabilization of cell membrane.

Class III-bacteriocins includes bacteriolysins which are large (>30 kDa), heat-labile and lytic molecules; thus, the peculiar characteristic of this group is the cell wall-degrading activity of these proteins that lead to hydrolysis of cell wall of target microorganisms (Nilsen et al., 2003). Additionally, class IV, not yet recognized by the entire scientific community, consists of bacteriocins acting as antimicrobial compounds with non-proteinaceous moieties, such as lipids or carbohydrates (Šušković et al., 2010).

Usually, the genes coding for bacteriocins are organized in operons and can be located both in the chromosomal and plasmid DNA. The clusters for lantibiotic production are the most described operons. Typically, the operon includes genes coding for the structural peptide, a bacteriocin transporter, regulatory proteins and an immunity protein (Saeed et al., 2014). The translocation of bacteriocins to the extracellular environment can be performed by the so-called ABC transporter or sec-dependent systems (Nishie et al., 2012; George and Jones 2014). Both the ABC and sec-dependent transporter proteins are able to recognize distinctive leader sequences, i.e. several class II bacteriocins possess a double glycine-leader that is specific for the ABC translocators. The immunity genes are essential to protect the bacteriocin-producing microorganism from the effects of its own bacteriocins (Klaenhammer, 1993; Nes et al., 1996; Moll et al., 1999; Quadri, 2002). Although the immune mechanisms are not yet completely elucidated, it is known that amino acids residues are produced and attached to the membrane in order to support it, preventing the formation of pores and alterations of the membrane electrochemical potential (Snyder and Worobo, 2014). An example of such immunity mechanism is given by the lactococcin A immunity protein (LcnI) produced by *Lactococcus* spp. (Nissen-Meyer et al., 1993).

In the case of class I bacteriocins, further genes encoding enzymes involved in the post-transcriptional processing are comprised in the bacteriocin operons (O'Sullivan et al., 2014). Usually, the genes involved in the post-transcriptional adjustment code for membrane-spanning proteins, e.g. LanB and LanC, which have the role to modify specific amino acids (e.g. serine and threonine are converted into dehydroalanine and dehydrobutyrine precursors for lanthionine and methyl-lanthionin), and/or to turn inactive pre-peptides into active antimicrobial compounds (Ehrmann et al., 2000).

10.3.3 Application of antimicrobials agents in Food Industry

The production of antimicrobial compounds, including bacteriocins, is a suitable feature for microbiological starters in order to inhibit the development of pathogenic and spoilage microflora (Chen and Hoover, 2003). Thus, bacteriocins and antimicrobial compound-producing microorganisms recruited for industrial application could represent an interesting opportunity to achieve improvements in the storage and preservation (biopreservation) processes through the use of natural compounds (Fig 10.2). Actually, the biopreservation purpose could be reached using appropriately selected starters able to create a disadvantaged environment for spoilage and pathogenic bacteria. Thus, the starter culture could carry out both the fermentation process and the preservation maintenance (De Vuyst et al., 2004). Moreover, in fermented food in which the fermentation process has been already standardized and the technological quality of the end product is well-established, it is recommended not to further inoculate with microbial starters in order to preserve the organoleptic qualities of the product. However, non-starter cultures could be added synergistically with appropriated starter microorganisms. In these cases, starter bacteria are involved in the fermentation process, while non-starter flora meets the necessity to produce antimicrobial compounds and achieve a longer shelf life and a higher safety (Bravo et al., 2009). Nevertheless, the non-starters could be selected on the basis of the ability to realize seasoning and secondary transformation of the food product such as the ability to lyse the starter cells which leads to a release of enzymes capable to contribute to the flavour (O'Sullivan et al., 2003). Moreover, preservative cultures

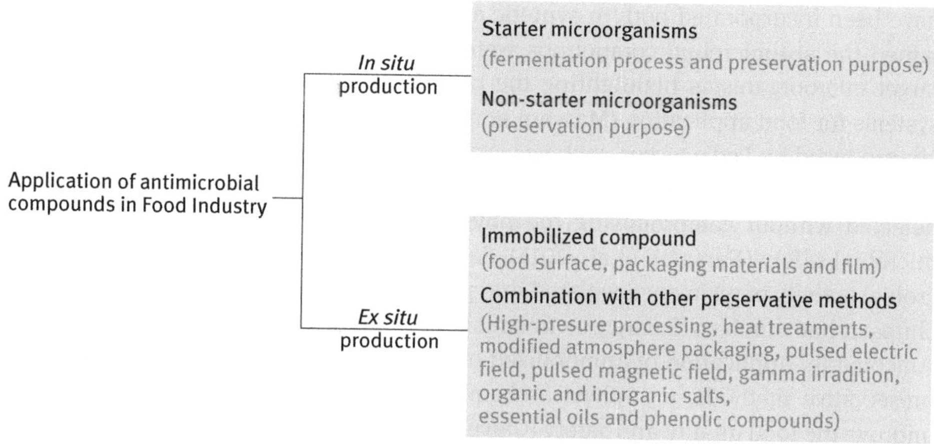

Fig. 10.2: Application of antimicrobial metabolites produced by lactic acid bacteria in the food industry.

could be supplemented afterwards to fermentation with the aim to extend the shelf-life by *in situ* production of antimicrobial compounds. Anyhow, the addition of the antagonistic strain is the lowest cost strategy, although the accuracy in the selection of the microorganism to be used is the crucial factor for the success of the preservative purpose. The addition of purified or semi-purified antimicrobial compounds, e.g. bacteriocin *ex situ* produced and harvested from microbial cultures, have been suggested in recent years, although several problems related to the yield process and chemical stability of bacteriocins have to be overcome (Campos et al., 2013). In fact, extraction, concentration and purification processes at industrial scale are needed with the aim to obtain a sufficient content of bacteriocins to be used as food additive, but these procedures require time and economic resources. In addition, the use of bacteriocins as additives in food preparation is regulated by legislative approbation conversely to the addition of bacteriocin-producing strain; the latter possesses the 'Generally Recognised As Safe' (GRAS) status when the producer microorganisms is recognized as GRAS (Sobrino-López and Martín-Belloso, 2008). Furthermore, bacteriocins are proteinaceous compound thus they can be degraded by enzymatic activity present in food matrix (Aasen et al., 2003). Consequently, the composition of food and the possibility to apply strategic methods to improve the persistence and the preservative effect of bacteriocins need to be valued. The use of immobilized bacteriocins seems to give good results leading to a slow release of antimicrobial compound during the shelf life and enhance the effects at the surface of food products as the bacteriocins are combined with food particles, packaging materials or films (Aasen et al., 2003; Bali, et al., 2014a, b).

Several studies on numerous bacteriocins have been carried out. Lactocin 705 and lactocin AL705, two bacteriocins produced by *Lactobacillus curvatus* CRL705, have been incorporated both in syntetic and gluten films. The active packages contained the antimicrobial compounds were able to inhibit efficiently the growth of target microorganisms highlighting the possibility to use antimicrobial packaging systems for food application (Massani et al., 2014). Moreover, the bacteriocin pediocin produced by *Pediococcus pentosaceus* was incorporated in alginate coatings and tested on minimally processed papaya. A more persistent antimicrobial activity was detected without compromising the physico-chemical qualities of papaya or the microbial safety (Narsaiah et al., 2015). da Silva et al. (2010) proved that the antimicrobial activity of nisin encapsulated into phosphatidylcholine liposomes containing chitosan had similar efficiency as free nisin in inhibiting strains of *Listeria* spp. The concomitant application of immobilized or free antimicrobial compounds and other preservative methods, e.g. physico-chemical treatments, has been ascertained to improve the food quality and safety (Gâlvez et al., 2014). For example, synergic effect on antimicrobial activity has been demonstrated using antimicrobial compounds and high-pressure processing (HPP), heat treatments, modified atmosphere packaging, pulsed electric field, pulsed magnetic field and gamma irradiation (Deegan et al., 2006). In this way, the spectrum of action of bacteriocins could be enhanced; indeed, it has been shown a greater sensibility for Gram-negative bacteria upon exposure to bacteriocins when combined with chelating agents, e.g. EDTA. Additionally, the ability of bacteriocins to inhibit the growth of spoilage or pathogenic microorganisms can also be increased in combination with organic and inorganic salts, organic acids, essential oils, phenolic compounds (Fang and Tsai, 2003).

Presently, even though several bacteriocins have been identified, nisin, a bacteriocins firstly isolated by *Lactococcus lactis*, is the only antimicrobial protein approved as food preservative (E234) in over 50 countries (FAO/WHO, 2002; de Arauz et al., 2009; WHO, 1969). Several studies showed the ability of nisin to inhibit Gram-positive bacteria, including *Listeria*, *Staphylococcus*, *Mycobacterium* and the formation of spores in *Bacillus* and *Clostridium* spp. (Gharsallaoui et al., 2016; Scherer et al., 2015). Nisin, whose Nisaplin™ is the most known commercial formulation, and six other variants are used in the manufacturing of dairy products, fermented beverages, bacon and fish (Field et al., 2015, 2008).

Beside the tangible potential of using starter LAB in the food industry as bio-preservative agents, the antagonistic ability of these valuable microorganisms could be further exploited in the production of functional food containing probiotic LAB strains. In fact, probiotics showing antimicrobial activity can be more competitive in the gastrointestinal tract and can discourage the adhesion and colonization of undesirable pathogenic bacteria (Chen and Hoover, 2003). Probiotics are defined as live microorganisms which when administered in adequate amounts confer a health benefit on the host (Arena et al., 2014a); they could be an alternative and/or adjuvant in the treatment and prevention of gastrointestinal diseases as biotherapeutics (Periti and Tonelli, 2002). Once beneficial microorganisms have reached and colonized the intestine, they can reduce the availability of vital nutrients to pathogens and, additionally, produce various molecules, including vitamins and short-chain fatty acids (Arena et al., 2014b). Among these compounds, bacteriocins and non-proteinaceous substances can also originate from microbial metabolism (Dobson et al., 2012; Gillor et al., 2008). These aforementioned molecules could lead to an unfavorable modification of redox potential and pH values resulting in an inhibition of pathogen growth (Dobson et al., 2012). A number of studies on functional food claimed that the consumption of products containing probiotics can promote the health of the consumer (Marsh et al., 2014; Miklavec et al., 2015). Possible therapeutic uses of LAB strains have been widely shown by several authors. Liévin-Le Moal and Servin (2014) emphasized the antimicrobial ability of probiotic *Lactobacillus* strains suggesting their use as innovative anti-infectious agents both in luminal and intracellular gastrointestinal tract. Similary, Shokryazdan et al. (2014) characterized nine LAB strains from different sources for their antimicrobial activity and found a good ability to inhibit the growth of target pathogenic strains. Strains of *L. plantarum* and *Lactobacillus fermentum* applied during processing of fresh-cut pineapples were shown to inhibit the growth of *Listeria monocytogenes* and *Escherichia coli* O157:H7 (Russo et al., 2014). Moreover, purified nisin were also shown to inhibit MRSA (methicillin-resistant *Staphylococcus aureus*), VRE (vancomycin-resistant enterococci) and *Pseudomonas aeruginosa* (Brumfitt et al., 2002; Giacometti et al., 1999).

10.4 Antibiotic resistance of microbial food cultures

10.4.1 Strategies for microbial antibiotic resistance

It is known that LAB can develop antibiotic resistance through different ways, which can be summarized in three mainly types of resistance: intrinsic, acquired and mutational. However, multiresistant strains may adopt different resistant mechanisms depending on the antibiotic. Intrinsic resistance is specific for a bacterial species or genus and includes four mechanisms: enzymatic degradation of antibiotics, modification of the target, changes in the cell wall permeability or developing of alternative metabolic pathways (Verraes et al., 2013). Examples of enzymatic degradation are the beta-lactamase enzymes hydrolyzing the beta-lactam ring of beta-lactam antibiotics such as, cephalosporin, carbapenems, and monobactams. In general, this ability is carried out by plasmids that are responsible for transferring the antibiotic resistance genes (Deasy, 2009). Some microorganisms are able to contrast the bactericidal effect by changing the cell wall permeability to reduce the antibiotic intake or remove them from the cell lumen by efflux pump mediated systems (Li and Nikadio, 2009). Moreover, some antibiotics can bind proteins on the cell wall of sensitive microorganism e.g. penicillin binding protein, whose alteration can confer resistance to the microorganism. In the case of methicillin resistant *S. aureus*, the *mecA* gene coding for a variant penicillin binding protein PBP2A with very low affinity for β-lactams. Finally, bacteria can develop antibiotic resistance by modification of its targeted metabolic pathways resulting in internal changes to prevent the antibiotic in reaching its target site (Sharma et al., 2014).

Acquired resistance is generally attributed to either gaining of exogenous DNA or to the mutation of indigenous genes. From a safety point of view, bacterial strains carrying an acquired resistance, that is shown to be due to chromosomal mutation, present a low risk of horizontal dissemination (EFSA, 2012). More hazardously, antibiotic resistance may be horizontally transferred by conjugative plasmids, transposons, integrons and insertion elements, and lytic and temperate bacteriophages (Sharma et al., 2014). In particular, lateral transmission of genetic material represents a serious risk to transfer resistance genes among different species which share the same environment (FEEDAP, 2008).

10.4.2 The European regulatory framework

The industrial employment of microbial food cultures is strictly ruled by competent authorities to prevent the introduction of antibiotic resistance determinants into the food chain. In Europe, the question of antibiotic resistance in non-pathogenic bacteria has been contemplated by the Scientific Panel on Additives and Products or substances used in the Animal Feed (FEEDAP) of the European Commission that established rigorous breakpoint values to determine the antimicrobial resistance to antibiotics of human and veterinary clinical significance (FEEDAP, 2008; EFSA, 2012). The cut-off values identified by the FEEDAP Panel should be seen as a pragmatic response intended to introduce consistency in the separation of strains with acquired resistance from susceptible strains (EFSA, 2012). Actually, on the basis of the European concept of Qualified Presumption of Safety (QPS), starter cultures are always subject to a case-by-case assessment before their authorization for industrial purposes (EFSA, 2013). This does not mean in absolute terms that microorganisms demonstrating higher resistance to a specific antimicrobial respect to the other strains of the same taxonomical unit cannot be authorized for introduction into the food chain. However, if antibiotic resistance is not an intrinsic trait, and the genetic basis of the antimicrobial resistances have not been clearly established, the extent of the risk of horizontal gene transfer to other bacteria in the food chain and in the environment cannot be established. Building on these considerations, the absence of known antimicrobial resistance genes is not sufficient to explain the nature of the detected resistance (EFSA, 2012). With a concrete example, the expert FEEDAP panel expressed a negative Scientific Opinion on the safety of a *Lactobacillus pentosus* strain resistant to tetracycline, ampicillin and clindamycin as a silage additive for all animal species due to the absence of information on the genetic nature of the demonstrated resistances (EFSA, 2011).

10.4.3 Antibiotic resistance in traditional fermented foods

The employment of microbial starter cultures is strongly recommended in the food industry to ensure standard and safe products to the consumers. However, in several typical artisanal foods a spontaneous fermentation of raw matrices relies upon naturally occurring wild microorganisms whose ability to transmit antibiotic resistance is unknown. Therefore, in the last years, several studies focused on the safety of traditional fermented foods with the aim to determine if indigenous lactic acid bacteria could be considered a potential reservoir of antibiotic resistance (Tab. 10.1).

Tab. 10.1: Antibiotic resistance determinants in microbial food cultures. (a) Numbers in bracket correspond to the strains tested. In bracket is reported the percentage of the resistant strains (b), and of the main antibiotic resistance found (c). (d) Number in bracket is the percentage of the identified resistance genes found in the resistant population.

Microbial Food Cultures[(a)]	Source	Main resistant strains[(b)]	Main Antibiotic Resistance[(c)]	Identified resistance genes[(d)]	Reference
Coagulase-negative *Staphylococcus* (390)	Salame Piacentino (Italy)	*S. xylosus*	Erythromycin (68.4 %) Tetracycline (83.7 %) Kanamycin (40.8 %)	*ermC* (37.8 %) *tet*(L) (31.6 %) *tet*(K) (34.7 %)	Rebecchi et al., 2015
Lactic Acid Bacteria (65)	Sucuk (Turkey)	*Enterococcus* spp. (68 %) 2–8 antibiotics			Yüceer and Özden Tuncer, 2015
Lactic Acid Bacteria (72)	Green table olives (Spain)	*L. pentosus* *L. pseudomesenteroides*	Ciprofloxacin Tetracycline Clindamycin	*norA*, *mepA*, *mdeA*	Casado Muñoz et al., 2014
Lactic Acid Bacteria (409)	Cheeses (Japan)	*E. faecalis* *E. faecium* *L. lactis* *L. casei.* *L. rhamnosus*	Dihydrostreptomycin Erythromycin Oxytetracycline	*erm*(B) *tet*(L), *tet*(M), *tet*(S)	Ishihara et al., 2013
Coagulase-negative *Staphylococcus* (93)	Raw milk Cheeses (Italy)	*S. xylosus*		*tet*(K) (10 %)	Ruaro et al., 2013

Tab. 10.1: (continued)

Microbial Food Cultures[a]	Source	Main resistant strains[b]	Main Antibiotic Resistance[c]	Identified resistance genes[d]	Reference
Enterococcus spp. (337)	Cheeses (French)	*E. faecalis* (81 %) *E. faecium* (13 %) *E. durans* (6 %)	Tetracycline Minocycline Erythromycin Kanamycin Chlor-amphenicol	*tet*(M) (90 %) *erm*(B) (96 %) *aph3'IIIa* (100 %) *cat*(A) (80 %)	Jamet et al., 2012
Coagulase-negative *Staphylococcus* (116)	Traditionally fermented sausages Meat products (Switzerland)	*S. xylosus*	Tetracycline Penicillin	*tet*(K) *bla*(Z)	Marty et al., 2012
Lactic Acid Bacteria (43)	Yogurts (China)	*S. thermophilus* *L. bulgaricus*	Kanamycin Aminoglycosides Tetracyclin	*aph3'IIIa* *ant(6)* *tet*(M)	Zhou et al., 2012
Lactic Acid Bacteria (84)	Fermented foods (China)	Lactic Acid Bacteria	Penicillin (2.3 %) Erythromycin (10.7 %) Clindamycin (5.9 %) Tetracycline (16.6 %)	 *erm*(B) *tet*(M), *tet*(S)	Nawaz et al., 2011
Coagulase-negative *Staphylococcus* (84)	Dairy products (Portugal)	*Staphylococcus* spp.	(57.1 %) (28.6 %) multiresistant		Soares et al., 2011
Starter cultures (30)	Meat starter cultures	*P. pentosaceus* *S. carnosus*	100 % at least three antimicrobials	Lincosamide, macrolide, and tetracycline not found	Cordeiro et al., 2010
Coagulase-negative staphylococci (330)	Feremented foods Starter cultures	*S. xylosus*	Tetracycline (22 %) Penicillin (69 %) Lincomycin	*tet*(K) *bla*(Z), *lnu*(A)	Resch et al., 2008
Lactic Acid Bacteria (473)	Including probiotic	Probiotic (1.6 %)	Streptomycin	*erm*(B) and/or *tet*(W), *tet*(M)	Klare et al., 2007

Tab. 10.1: (continued)

Microbial Food Cultures[(a)]	Source	Main resistant strains[(b)]	Main Antibiotic Resistance[(c)]	Identified resistance genes[(d)]	Reference
Lactobacillus *Staphylococcus* *Bifidobacterium* *Pediococcus* (200)	Starter cultures Probiotics	*Staphylococcus* (2.5 %) *B. lactis* *L. reuteri*		*tet*(K) *tet*(W) *lnu*(A)	Kastner et al., 2006
Bifidobacterium spp. (100)	Humans, animals, probiotic products	Probiotic (7 %)	Tetracycline	*tet*(K)	Masco et al., 2006

Spontaneous fermented meat products have often been linked to a potential hazard, due to the occurrence of coagulase-negative staphylococci (CNS) (Zdolec et al., 2013). In order to provide new meat starter strains, Marty et al. (2012) explored the antibiotic susceptibility in CNS isolates from traditionally fermented sausages manufactured with meat from wildlife or animals grown under low- or no-antibiotic pressure. Less than 50 % of these CNS isolates showed phenotypic resistances to at least one antibiotic where resistances to penicillins and tetracycline are the most reported (Marty et al., 2012). In a recent study, Rebecchi and coauthors (2015), suggest that, although Salame Piacentino was a source of antibiotic resistant CNS, the salami manufacturing process could lead to a strong reduction of the antibiotic resistant strains. In this work, eight antibiotic resistant strains, isolated from skin and salami samples, were detected in different steps along the manufacturing process, with high frequency of phenotypic tetracycline and erythromycin and a lower kanamycin resistant. However, only one *Staphylococcus xylosus* strain that harboured *tet(K)* gene was detected during the salami ripening (Rebecchi et al., 2015).

Recently, LAB isolated from sucuk (a dry, spicy sausage) samples were found to be resistant to clinically important antibiotics, suggesting that LAB in sucuk may play a role to spread the antibiotic resistance between other bacteria including pathogens. In particular, about 70 % of the *Enterococcus* strains showed multiple antibiotic resistance patterns, while all *Pediococcus* and *Lactobacillus* strains were found completely sensitive to ampicillin, cephalothin, chloramphenicol, clindamycin, erythromycin, rifampicin and tetracycline (Yüceer and Özden Tuncer, 2015).

Coagulase-negative staphylococci showed a low incidence of potential pathogenicity traits in raw milk and cheese of North Italy, since despite the presence of ten transferable antibiotic resistance genes, multiple antibiotic resistance patterns were not detected (Ruaro et al., 2013). In contrast, 57.1 % of the isolates in the environment and dairy products of a small manufacturing dairy plant showed at least resistance against one antibiotic, and 28.6 % were multi-resistant, which might accomplish resistance for up to 6 antibiotics simultaneously (Soares et al., 2011).

The widespread antibiotic resistance trends among *Enterococcus* spp. have been extensively investigated since they are a critical concern for the safety of the cheese ecosystem (Ogier and Serror, 2008). Some recent surveys have tried to answer the question if marketing cheeses could be considered a potential reservoir of antibiotic resistance. The antimicrobial resistance profiles of enterococci in French cheeses from retail stores showed a total of 337 antibiotic resistant strains isolated in 29 % and 60 % of pasteurized-milk and raw-milk cheeses, respectively (Jamet et al., 2012). In the same study, *Enterococcus faecalis* was the predominant (81 %) antibiotic resistant species recovered, followed by *Enterococcus faecium*, and *Enterococcus durans*, while the most prevalent antibiotic resistances were tetracycline and minocycline, followed by erythromycin, kanamycin and chloramphenicol. However, although multiple antibiotic resistance phenotypes were identified, the absence of vancomycin- or ampicillin- resistant enterococci indicate that cheeses cannot be considered as a major reservoir for nosocomial multi-drug resistant enterococci (Jamet et al., 2012).

A similar assay performed on cheeses commercialized in Japan revealed that imported cheeses were less safe than cheeses produced in Japan since a higher percentage of LAB resistant to dihydrostreptomycin, erythromycin, and/or oxytetracycline was found in strains isolated from imported cheeses (Ishihara et al., 2013). In particular, antimicrobial resistance genes were detected in *E. faecalis* (*tet*L, *tet*M, and *erm*B; *tet*L and *erm*B; *tet*M), *E. faecium* (*tet*M), *L. lactis* (*tet*S), *Lactobacillus casei* and *Lactobacillus paracasei* (*tet*M or *tet*W), and *Lactobacillus rhamnosus* (*erm*B) isolated from seven imported cheeses. Moreover, *E. faecalis* isolates were able to transfer antimicrobial resistance genes increasing the risk of antibiotic lateral transmission into the food chain (Ishihara et al., 2013).

Nawaz et al., (2011) reported that acquired antibiotic resistance is well dispersed in Chinese fermented foods and represent a potential risk since both *erm*(B) gene and *tet*(M) gene from resistant LAB had been successfully transferred to *E. faecalis*. Widespread resistance to ampicillin, chloramphenicol, chlortetracycline, tetracyclines, lincomycin, streptomycin, neomycin, and gentamycin was found among *Lactobacillus bulgaricus* and *Streptococcus thermophilus* isolated from Chinese yogurt (Zhou et al., 2012). Similarly, the resistance incidences of Chinese fermented sausages were much higher than that of fermented vegetables, and oft associated to multi-resistance phenotypes. Moreover, genes *tet*M and *erm*B were detected on both plasmid and chromosome; gene *aph*A3 was only detected on plasmid while gene *mef*A was associated to chromosome (Pan et al., 2011).

Several acquired genes encoding for tetracycline and erythromycin were found among LAB isolated from naturally Indian fermented foods, mainly belonging to *Lactobacillus*, *Pediococcus*, *Enterococcus*, and *Leuconostoc* spp. Moreover, the prevalence of tetracycline resistance genes in several erythromycin resistance strains indicated that resistance genes encoding both antibiotics could be linked (Thumu and Halami, 2012). The same authors found different macrolides-lincosamides and streptogramin B phenotypes, multiple resistance genes and transferable plasmids among enterococci isolated from naturally fermented Idli batter (Thumu and Halami, 2014).

In another study, 121 strains of *L. paracasei*, isolated from Italian fermented foods were tested for their tetracycline and erythromycin susceptibility, and *tet*M and *erm*B genes were the most frequently detected in the resistant strains (Comunian et al., 2010). Similarly, phenotypically erythromycin-resistant and tetracycline-resistant were widely distributed among *Lactobacillus* spp. from fermented dry sausages, with a high incidence of microorganisms resistant to both antibiotics (Zonenschain et al., 2009).

In a screening on strains isolated from naturally fermented green table olives, most *L. pentosus* and all *Leuconostoc pseudomesenteroides* were resistant to at least three antibiotics (Casado Muñoz et al., 2014). However, such resistances resulted from an intrinsic mechanism, non-transferable or non-acquired resistance determinants which may in part be due to chromosomally encoded efflux pumps, thus suggesting that these strains could be considered safe for future application as starter cultures or as probiotics (Casado Muñoz et al., 2014).

10.4.4 Antibiotic resistance: safety of starter and probiotic microorganisms

Starter cultures and probiotic microorganisms are live bacteria, yeasts or moulds used in food production contributing to the one or multiple unique properties of foodstuff especially about taste, flavour, colour, texture, safety, conservation, nutritional value, wholesomeness and/or health benefits (EFFCA, 2013). Although microbial strains deliberately added to the food chain are submitted to a rigorous assessment, several studies reported on the occurrence of antimicrobial resistance traits (Tab. 10.1). Interesting data are recovered in the field of the meat starter cultures, for which the high incidence of antibiotic resistance is probably a reflection of the close phylogenetic relationship of food-associated CNS with the pathogenic *Staphylococcus* species (Resch et al., 2008).

Cordeiro and coauthors (2010) found that all the isolates of meat starter cultures exhibited resistance to at least three antimicrobials regardless of antimicrobial class, with an incidence higher among *P. pentosaceus* than *Staphylococcus carnosus* strains. These strains were related to high risk of lateral transmission, since the absence of known resistance genes suggests that genes or mechanisms other than intrinsic resistance might have contributed to the resistant phenotypes (Cordeiro et al., 2010). A comprehensive insight into the incidence of antibiotic resistances in food-associated CNS showed that meat starter cultures of *S. xylosus* strains revealed a high percentage (93 %) of antibiotic resistance (Resch et al., 2008). In this case, phenotypic resistances could be traced back to the genes such as *bla*Z, *lnu*A and *tet*K, encoding for the ß-lactam antibiotics, lincomycin and tetracycline, respectively (Resch et al., 2008).

In contrast, microbial food cultures exhibiting resistances that were not an intrinsic feature of the respective genera were isolates from Switzerland meat starter cultures and probiotics (Kastner et al., 2006). In particular, tetracycline resistance gene *tet*(K) was detected in five *Staphylococcus* isolates (Kastner et al., 2006), while the tetracycline resistance gene *tet*(W) residing on a plasmid was found in the probiotic cultures *Bifidobacterium lactis* DSM 10140 and *Lactobacillus reuteri* SD 2112, and the lincosamide resistance gene *lnu*(A) in *L. reuteri* SD 2112 (Kastner et al., 2006). A resistance phenotype linked to the presence of an acquired non-plasmid located *tet*(W) gene was detected in seven probiotic isolates belonging to the taxa *Bifidobacterium animalis* subsp. *lactis* and *Bifidobacterium bifidum* (Masco et al., 2006), emphasizing the need for a careful safety evaluation during the selection of *Bifidobacterium* strains for probiotic use. However, resistant probiotic *Bifidobacterium* strains were also found in the commercial formulation Yakult, although unassociated with a high risk of horizontal transmission. Indeed, molecular analysis revealed that a mutation of the *rpsL* gene for ribosomal protein S12 was responsible for the streptomycin resistance *Bifidobacterium breve* strain Yakult (Kiwaki and Sato, 2009). Similarly, the antibiotic resistances erythromycin, neomycin, and streptomycin of *B. bifidum* YIT 4007 were acquired by genes chromosome mutations and not associated with mobile genetic elements like insertion sequences, phages, and plasmids (Sato and Iino, 2010).

In contrast, the occurrence of acquired resistance genes was found in about 10 % of the isolates intended for both probiotic or as starter cultures, with a wide spread of the resistance to the clinically important antibiotics streptomycin, erythromycin and tetracycline (Klare et al., 2007).

10.5 Concluding Remarks

Lactic acid bacteria showing the ability to inhibit spoilage and pathogenic microorganisms could profitably provide an advantage in the food industry. The potential of lactic acid bacteria as biopreservative agents in fermented and non-fermented food encounter the growing consumer request for healthy and natural food products in order to eliminate or reduce chemical additives and contrast the alarming problem of antibiotic resistance. Although strictly ruled by the competent authorities, the antibiotic resistance of microbial cultures deliberately added into the food chain is still an open question posing the insidious risk of horizontal gene transfer of resistance determinants directly in the gut environment. Therefore, the antagonistic ability of LAB is a suitable probiotic feature to be exploited with biotherapeutic purpose in order to accomplish a prophylaxis and/or therapy of different gastrointestinal and urogenital infections. Appropriate selection of LAB and adequate description of their antimicrobial strength could increase the quality, the safety and the functional trait of foods.

10.6 References

[1] Aasen, I. M., Markussen, S., Moretro, T., Katla, T., Axelsson, L. & Naterstad, L. (2003). Interactions of the bacteriocin sakacin P and nisin with food constituents. International Journal of Food Microbiology, 87, 35–43.

[2] Altay, F., Karbancıoglu-Güler, F., Daskaya-Dikmen, C. & Heperkan, D. (2013). A review on traditional Turkish fermented non-alcoholic beverages: Microbiota, fermentation process and quality characteristics. International Journal of Food Microbiology, 167, 44–56.

[3] Arena, M.P., Fiocco, D., Massa, S., Capozzi, V., Russo, P. & Spano, G. (2014a) Lactobacillus plantarum as a strategy for an in situ production of vitamin B2. Journal of Food & Nutritional Disorder, S1–004.

[4] Arena, M.P., Russo, P., Capozzi, V., López, P., Fiocco, D. & Spano, G. (2014b). Probiotic abilities of riboflavin-overproducing *Lactobacillus* strains: a novel promising application of probiotics. Applied Microbiology & Biotechnology, 98, 7569–7581.

[5] Bali, V., Panesar, P.S. & Bera, M.B. (2014a) Potential of immobilization technology in bacteriocin production and antimicrobial packaging. Food Reviews International, 30, 244–263.

[6] Bali, V., Panesar, P.S. & Bera, M.B. (2014b) Trends in utilization of agro-industrial byproducts for production of bacteriocins and their biopreservative applications. Critical Reviews of Biotechnology, 28, 1–11.

[7] Bravo, D., Rodríguez, E. & Medina, M. (2009) Nisin and lacticin 481 coproduction by *Lactococcus lactis* strains isolated from raw ewes' milk. Journal of Dairy Science, 92, 4805–4811.

[8] Brumfitt, W., Salton, M.R.J. & Hamilton-Miller, J.M.T. (2002). Nisin alone and combined with peptidoglycan-moduling antibiotics: Activity against methicillin-resistant *Staphylococcus aureus* and vancomycin-resistant enterococci. Journal of Antimicrobial Chemotherapy, 50, 731–734.

[9] Calo-Mata, P., Arlindo, S., Boehme, K., de Miguel, T., Pascoal, A. & Barros Velazquez, J. (2008). Current applications and future trends of lactic acid bacteria and their bacteriocins for the biopreservation of aquatic food products. Food Bioprocess Technology, 1, 43–63.

[10] Campos, C.A., Castro, M.P., Rivas, F.P. & Schelegueda, L.I. (2013). Bacteriocins in food: evaluation of the factors affecting their effectiveness. In A. Méndez-Vilas (Ed.) Microbial pathogens and strategies for combating them: science, technology and education. (pp. 994–1004). Badajoz: Formatex Research Centre.

[11] Caplice, E. & Fitzgerald, G.F. (1999). Food fermentations: role of microorganisms in food production and preservation. International Journal of Food Microbiology, 50, 131–149.

[12] Capozzi, V., Russo, P., Dueñas, M.T., López, P. & Spano, G. (2012). Lactic Acid Bacteria producing B-group vitamins: a great potential for functional cereals products. Applied Microbiology & Biotechnology, 96, 1383–1394.

[13] Casado Muñoz, M.C., Benomar, N., Lerma, L.L., Gálvez, A. & Abriouel, H. (2014). Antibiotic resistance of *Lactobacillus pentosus* and *Leuconostoc pseudomesenteroides* isolated from naturally-fermented Aloreña table olives throughout fermentation process. International Journal of Food Microbiology, 172, 110–118.

[14] Cebeci, A. & Gürakan, C. (2003). Properties of potential probiotic *Lactobacillus plantarum* strains. Food Microbiology, 20, 511–518.

[15] Chen, H. & Hoover, D.G. (2003) Bacteriocins and their food applications. Comprehensive Reviews in Food Science & Food Safety, 2, 82–100.

[16] Cleveland, J., Montville, T.J., Nes, I.F. & Chikindas, M.L. (2001). Bacteriocins: safe, natural antimicrobials for food preservation. International Journal of Food Microbiology, 71, 1–20.

[17] Collins, B., Curtis, N., Cotter, P.D., Hill, C. & Ross, R.P. (2010). The ABC transporter AnrAB contributes to the innate resistance of *Listeria monocytogenes* to nisin, bacitracin, and various beta-lactam antibiotics. Antimicrobial Agents & Chemotherapy, 54, 4416–4423.

[18] Comunian, R., Daga, E., Dupré, I., Paba, A., Devirgiliis, C., Piccioni, V., Perozzi, G., Zonenschain, D., Rebecchi, A., Morelli, L., De Lorentiis, A. & Giraffa, G. (2010). Susceptibility to tetracycline and erythromycin of *Lactobacillus paracasei* strains isolated from traditional Italian fermented foods. International Journal of Food Microbiology, 138, 151–156.

[19] Cordeiro, R.P., Du, T., Mulvey, M.R., Krause, D.O. & Holley, R.A. (2010). Susceptibility of meat starter cultures to antimicrobials used in food animals in Canada. Journal of Food Protection, 73, 916–922.

[20] Cotter, P.D., Hill, C. & Ross, R.P. (2005). Bacteriocins: developing innate immunity for food. Nature Reviews Microbiology, 3, 777–788.

[21] da Silva, I.M., Boelter, J.F., da Silveira, N.P. & Brandelli, A. (2014). Phosphatidylcholine nanovesicles coated with chitosan or chondroitin sulfate as novel devices for bacteriocin delivery. Journal of Nanoparticle Research, 16, 1–10.

[22] de Arauz, L.J., Jozala, A.F., Mazzola, P.G. & Vessoni Penna, T.C. (2009). Nisin biotechnological production and application: A review. Trends in Food Science & Technology, 20, 146–154.

[23] Deegan, H., Cotter, P.D., Hill, C. & Ross, P. (2006). Bacteriocins: Biological tools for bio-preservation and shelf-life extension. International Dairy Journal, 16, 1058–1071.

[24] Deasy, J. (2009). Antibiotic resistance: The ongoing challenge for effective drug therapy. Journal of the American Academy of Physician Assistants, 22, 18–22.

[25] De Vuyst, L., Avonts, L., Hoste, B., Vancanneyl, M., Neysens, P. & Callewaert, R. (2004). The lactobin A and amylovorin L471 genes are identical, and their distribution seems to be restricted to the species *Lactobacillus amylovorus* that is of interest for cereal fermentations. International Journal of Food Microbiology, 90, 93–106.

[26] De Vuyst, L. & Leroy, F. (2007). Bacteriocins from lactic acid bacteria: Production, purification, and food applications. Journal of Molecular Microbiology & Biotechnology, 13, 194–199.

[27] Dieuleveux, V., Van Der Pyl, D., Chataud, J. & Gueguen, M. (1998a) Purification and characterization of anti-*Listeria* compounds produced by *Geotrichum candidum*. Applied & Environmental Microbiology, 64, 800–803.

[28] Dieuleveux, V., Lemarinier, S. & Gueguen, M. (1998b). Antimicrobial spectrum and target site of D-3-phenyllactic acid. International Journal of Food Microbiology, 40, 177–183.

[29] Dobson, A., Cotter, P.D., Ross, R.P. & Hill, C. (2012). Bacteriocin production: a probiotic trait? Applied & Environmental Microbiology, 78, 1–6.

[30] Doelle, H.W. (2014). Bacterial metabolism. New York: Academic Press.

[31] EFFCA (European Food and Feed Cultures Association). (2013). Microbial Food Culture, Available: http://www.effca.org/content/microbial-food-culture [Access: 01 Nov 2014].

[32] EFSA (European Food Safety Authority). (2011). Scientific Opinion on the safety and efficacy of *Lactobacillus pentosus* (DSM 14025) as a silage additive for all animal species. European Food Safety Authority Journal, 9, 2449.

[33] EFSA (European Food Safety Authority). (2012). Guidance on the assessment of bacterial susceptibility to antimicrobials of human and veterinary importance. European Food Safety Authority Journal, 10, 2740–2750.

[34] EFSA (European Food Safety Authority). (2013). Scientific opinion on the maintenance of the list of QPS biological agents intentionally added to food and feed. European Food Safety Authority Journal, 11, 3449–3559.

[35] Ehrmann, M.A., Remiger, A., Eijsink, V.G. & Vogel, R.F. (2000). A gene cluster encoding plantaricin 1.25beta and other bacteriocin-like peptides in *Lactobacillus plantarum* TMW1.25. Biochimica et Biophysica Acta, 1490, 355–361.

[36] Fang, T.J. & Tsai, H.C. (2003). Growth patterns of *Escherichia coli* O157:H7 in ground beef treated with nisin, chelators, organic acids and their combinations immobilized in calcium alginate gels. Food Microbiology, 20, 243–253.

[37] FAO/WHO. (Food and Agriculture Organization of the United Nations, FAO, World Health Organization, WHO). (2002). Guidelines for the evaluation of probiotics in food. Canada: FAO/ WHO.

[38] FEEDAP (The Panel on Additives and Products or Substances used in Animal Feed). (2008). Technical guidance-update of the criteria used in the assessment of bacterial resistance to antibiotics of human or veterinary importance. European Food Safety Authority Journal, 732, 1–15.

[39] Field, D., O'Connor, P.M., Cotter, P.D., Hill, C. & Ross, R.P. (2008). The generation of nisin derivatives with enhanced activity against specific Gram-positive pathogens. Molecular Microbiology, 69, 218–230.

[40] Field, D., Daly, K., O'Connor, P.M., Cotter, P.D., Hill, C. & Ross, R.P. (2015). Efficacies of nisin A and nisin V semipurified preparations alone and in combination with plant essential oils for controlling *Listeria monocytogenes*. Applied and Environmental Microbiology, 81, 2762–2769.

[41] Finnegan, M., Linley, E., Denyer, S.P., McDonnell, G., Simons, C. & Maillard, J.Y. (2010). Mode of action of hydrogen peroxide and other oxidizing agents: differences between liquid and gas forms. Journal of Antimicrobial Chemotherapy, 2108–2115.

[42] Gálvez, A., Abriouel, H., Benomar, N., & Lucas, R. (2010) Microbial antagonists to food-borne pathogens and biocontrol. Current opinion in biotechnology, 21, 142–148.

[43] Gálvez, A., López, R.L., Pulido, R.P. & Burgos, M.J.G. (2014). Application of Lactic Acid Bacteria and Their Bacteriocins for Food Biopreservation. In: Food Biopreservation. (pp. 15–22). New York: Springer.

[44] George, A.M. & Jones, P.M. (2014). Bacterial ABC transporters: structure and function. Bacterial Membranes: Structural and Molecular Biology. (pp. 227–248). Norfolk: Caister Academic Press.

[45] Georgieva, R., Iliev, I., Haertlé, T., Chobert, J.M., Ivanova, I. & Danova, S. (2009). Technological properties of candidate probiotic *Lactobacillus plantarum* strains. International Dairy Journal, 19, 696–702.

[46] Gharsallaoui, A., Oulahal, N., Joly, C. & Degraeve, P. (2016). Nisin as a food preservative: Part 1: physicochemical properties, antimicrobial activity, and main uses. Critical Reviews in Food Science & Nutrition, 56, 1262–1274.

[47] Giacometti, A., Cirioni, O., Barchiesi, F., Fortuna, M. & Scalise, G. (1999). In vitro activity of cationic peptides alone and in combination with clinically used antimicrobial agents against *Pseudomonas aeruginosa*. Journal of Antimicrobial Chemotherapy, 44, 641–645.

[48] Gillor, O., Etzion, A. & Riley, M.A. (2008). The dual role of bacteriocins as anti-and probiotics. Applied Microbiology & Biotechnology, 81, 591–606.

[49] Guder, A., Wiedemann, I. & Sahl, H.G. (2000). Postranslationally modified bacteriocins – The lantibiotics. Biopolymers, 55, 62–73.

[50] Gueimonde, M., Sánchez, B., de los Reyes-Gavilán, C.G. & Margolles, A. (2013). Antibiotic resistance in probiotic bacteria. Frontiers in Microbiology, 4, 202.

[51] Guidone, A., Zotta, T., Ross, R.P., Stanton, C., Rea, M.C., Parente, E. & Ricciardi, A. (2014). Functional properties of *Lactobacillus plantarum* strains: A multivariate screening study. LWT – Food Science & Technology, 56, 69–76.

[52] Haug, M.C., Tanner, S.A., Lacroix, C., Stevens, M.J. & Meile, L. (2011). Monitoring horizontal antibiotic resistance gene transfer in a colonic fermentation model. FEMS Microbiology Ecology, 78, 210–219.

[53] Ishihara, K., Nakajima, K., Kishimoto, S., Atarashi, F., Muramatsu, Y., Hotta, A., Ishii, S., Takeda, Y., Kikuchi, M. & Tamura, Y. (2013). Distribution of antimicrobial-resistant lactic acid bacteria in natural cheese in Japan. Microbiology & Immunology, 57, 684–691.

[54] Jamet, E., Akary, E., Poisson, M., Chamba, J., Bertrand, X. & Serror, P. (2012). Prevalence and characterization of antibiotic resistant *Enterococcus faecalis* in French cheeses. Food Microbiology, 31, 191–198.

[55] Kastner, S., Perreten, V., Bleuler, H., Hugenschmidt, G., Lacroix, C. & Meile, L. (2006). Antibiotic susceptibility patterns and resistance genes of starter cultures and probiotic bacteria used in food. Systematic & Applied Microbiology, 29, 145–155.

[56] Kiwaki, M. & Sato, T. (2009). Antimicrobial susceptibility of *Bifidobacterium breve* strains and genetic analysis of streptomycin resistance of probiotic *B. breve* strain Yakult. International Journal of Food Microbiology, 134, 211–215.

[57] Klaenhammer, T.R. (1993). Genetics of bacteriocins produced by lactic acid bacteria. FEMS Microbiology Reviews, 12, 39–85.

[58] Klare, I., Konstabel, C., Werner, G., Huys, G, Vankerckhoven, V., Kahlmeter, G., Hildebrandt, B., Müller-Bertling, S., Witte, W. & Goossens, H. (2007). Antimicrobial susceptibilities of *Lactobacillus*, *Pediococcus* and *Lactococcus* human isolates and cultures intended for probiotic or nutritional use. Journal of Antimicrobial Chemotherapy, 59, 900–912.

[59] Kos, B., Suskovic, J., Beganovic, J., Gjuracic, K., Frece, J., Iannaccone, C. & Canganella, F. (2008). Characterization of the three selected probiotic strains for the application in food industry. World Journal of Microbiology Biotechnology, 24, 699–707.

[60] Lavermicocca, P., Valerio, F., Evidente, A., Lazzaroni, S., Corsetti, A. & Gobbetti, M. (2000). Purification and characterization of novel antifungal compounds from the sourdough *Lactobacillus plantarum* strain 21B. Applied & Environmental Microbiology, 66, 4084–4090.

[61] Lavermicocca, P., Valerio, F. & Visconti, A. (2003). Antifungal activity of phenyllactic acid against molds isolated from bakery products. Applied & Environmental Microbiology, 69, 634–640.

[62] Law, B.A. (2001). Controlled and accelerated cheese ripening: the research base for new technologies. International Dairy Journal, 11, 383–398.

[63] Leroy, F., Verluyten, J., Messens, W. & De Vuyst, L. (2002). Modelling contributes to the understanding of the different behavior of bacteriocin-producing strains in a meat environment. International Dairy Journal, 12, 247–253.

[64] Leroy, F. & De Vuyst, L. (2004). Lactic acid bacteria as functional starter cultures for the food fermentation industry. Trends in Food Science Technology, 15, 67–78.

[65] Li, X. & Nikadio, H. (2009). Efflux-mediated drug resistance in bacteria: An update. Drugs, 69, 1555–1623.

[66] Liévin-Le Moal, V. & Servin, A.L. (2014). Anti-infective activities of lactobacillus strains in the human intestinal microbiota: from probiotics to gastrointestinal anti-infectious biotherapeutic agents. Clinical Microbiology Review, 27, 167–199.

[67] Lorca, G.L. & de Valdez, G.F. (2009). *Lactobacillus* stress responses. In Å. Ljungh & T. Wadström (Eds.) Lactobacillus Molecular Biology: From Genomics to Probiotics. (pp. 115–138). Norfolk: Caister Academic Press.

[68] Marsh, A.J., Hill, C., Ross, R.P. & Cotter, P.D. (2014). Fermented beverages with health-promoting potential: Past and future perspectives. Trends in Food Science Technology, 38, 113–124.

[69] Marty, E., Bodenmann, C., Buchs, J., Hadorn, R., Eugster-Meier, E., Lacroix, C. & Meile, L. (2012). Prevalence of antibiotic resistance in coagulase-negative staphylococci from spontaneously fermented meat products and safety assessment for new starters. International Journal of Food Microbiology, 159, 74–83.

[70] Masco, L., Van Hoorde, K., De Brandt, E., Swings, J. & Huys, G. (2006) Antimicrobial susceptibility of *Bifidobacterium* strains from humans, animals and probiotic products. Journal of Antimicrobial Chemotherapy, 58, 85–94.

[71] Massani, M.B., Molina, V., Sanchez, M., Renaud, V., Eisenberg, P. & Vignolo, G. (2014). Active polymers containing *Lactobacillus curvatus* CRL705 bacteriocins: Effectiveness assessment in Wieners. International Journal of Food Microbiology, 178, 7–12.

[72] Miklavec, K., Pravst, I., Grunert, K.G., Klopčič, M. & Pohar, J. (2015). The influence of health claims and nutritional composition on consumers' yoghurt preferences. Food Quality & Preference, 43, 26–33.

[73] Moll, G.N., Konings, W.N. & Driessen, A.J.M. (1999). Bacteriocins: Mechanism of membrane insertion and pore formation. Antonie van Leeuwenhoek, 76, 185–189.

[74] Mu, W., Yu, S., Zhu, L., Zhang, T. & Jiang, B. (2012). Recent research on 3-phenyllactic acid, a broad-spectrum antimicrobial compound. Applied Microbiology & Biotechnology, 95, 1155–1163.

[75] Narsaiah, K., Wilson, R.A., Gokul, K., Mandge, H.M., Jha, S.N., Bhadwal, S., Anurag, R.K., Malik, R.K. & Vij, S. (2015). Effect of bacteriocin-incorporated alginate coating on shelf-life of minimally rocessed papaya (Carica papaya L.). Postharvest Biology & Technology, 100, 212–218.

[76] Nawaz, M., Wang, J., Zhou, A., Ma, C., Wu, X., Moore, J.E., Millar, B.C. & Xu, J. (2011). Characterization and transfer of antibiotic resistance in lactic acid bacteria from fermented food products. Current Microbiology, 62, 1081–1089.

[77] Nes, I.F., Diep, D.B., Havarsteuin, L.S. & Brurberg, M.B. (1996). Biosynthesis of bacteriocins in lactic acid bacteria. Antonie van Leeuwenhoek, 70, 113–128.

[78] Nilsen, T., Nes I.F. & Holo, H. (2003). Enterolysin A, a cell wall-degrading bacteriocin from *Enterococcus faecalis* LMG 2333. Applied & Environmental Microbiology, 69, 2975–2984.

[79] Nishie, M., Nagao, J.I. & Sonomoto, K. (2012). Antibacterial peptides "bacteriocins": an overview of their diverse characteristics and applications. Biocontrol Science, 17, 1–16.

[80] Nissen-Meyer, J., Havarstein, L.S., Holo, H., Sletten, K. & Nes, I.F. (1993). Association of the lactococcin A immunity factor with the cell membrane: purification and characterization of the immunity factor. Journal of General Microbiology, 139, 1503–1509.

[81] Ohhira, I., Kuwaki, S., Morita, H., Suzuki, T., Tomita, S., Hisamatsu, S., Sonoki, S. & Shinoda, S. (2004). Identification of 3-phenyllactic acid as a possible antibacterial substance produced by *Enterococcus faecalis* TH10. Biocontrol Science, 9, 77–81.

[82] Ogier, J.C. & Serror, P. (2008). Safety assessment of dairy microorganisms: the *Enterococcus* genus. International Journal of Food Microbiology, 126, 291–301.

[83] O'Sullivan, L., Ross, R.P. & Hill, C. (2003). A lacticin 481-producing adjunct culture increases starter lysis while inhibiting nonstarter lactic acid bacteria proliferation during cheddar cheese ripening. Journal of Applied Microbiology, 95, 1235–1241.

[84] O'Sullivan, O., Hill, C., Ross, P. & Cotter, P. (2014). Bacteriocin Mining in Metagenomes. In Encyclopedia of Metagenomics. (pp. 1–8). New York: Springer.

[85] Pan, L., Hu, X., & Wang, X. (2011). Assessment of antibiotic resistance of lactic acid bacteria in Chinese fermented foods. Food Control, 22, 1316–1321.

[86] Papagianni, M. & Anastasiadou, S. (2009). Pediocins: The bacteriocins of *Pediococci*. Sources, production, properties and applications. Microbial Cell Factories, 8, 1–3.

[87] Periti, P. & Tonelli, F. (2002). Biotherapeutics and biotherapy of surgical enteropathies. Digestive & Liver Disease, 34, 87–97.

[88] Pidcock, K., Heard, G.M. & Henriksson, A. (2002). Application of nontraditional meat starter cultures in production of Hungarian salami. International Journal of Food Microbiology, 76, 75–81.

[89] Prema, P., Smila, D., Palavesam, A. & Immanuel, G. (2010). Production and characterization of an antifungal compound (3-phenyllactic acid) produced by *Lactobacillus plantarum* strain. Food & Bioprocess Technology, 3, 379–386.
[90] Prückler, M., Lorenz, C., Endo, A., Kraler, M., Dürrschmid, K., Hendriks, K., da Silva, F.S., Auterith, E., Kneifel, W. & Michlmayr, H. (2015). Comparison of homo-and heterofermentative lactic acid bacteria for implementation of fermented wheat bran in bread. Food Microbiology, 49, 211–219.
[91] Quadri, L.E.N. (2002). Regulation of antimicrobial peptide production by autoinducer-mediated quorum sensing in lactic acid bacteria. Antonie van Leeuwenhoek, 82, 133–145.
[92] Reid, G. (2008). Probiotic lactobacilli for urogenital health in women. Journal of Clinical Gastroenterology, 42, 234–236.
[93] Rebecchi, A., Pisacane, V. & Callegari, M.L. (2015). Ecology of antibiotic resistant coagulase-negative staphylococci isolated from the production chain of a typical Italian salami. Food Control, 53, 14–22.
[94] Reis, J.A., Paula, A.T., Casarotti, S.N. & Penna, A.L.B. (2012). Lactic acid bacteria antimicrobial compounds: characteristics and applications. Food Engineering Reviews, 4, 124–140.
[95] Resch, M., Nagel, V. & Hertel, C. (2008). Antibiotic resistance of coagulase-negative staphylococci associated with food and used in starter cultures. International Journal of Food Microbiology, 127, 99–104.
[96] Rolain, J.M. (2013). Food and human gut as reservoirs of transferable antibiotic resistance encoding genes. Frontiers in Microbiology, 4, 173.
[97] Ruaro, A., Andrighetto, C., Torriani, S. & Lombardi, A. (2013). Biodiversity and characterization of indigenous coagulase-negative staphylococci isolated from raw milk and cheese of North Italy. Food Microbiology, 34, 106–111.
[98] Russo, P., de Chiara, M.L.V., Vernile, A., Amodio, M.L., Arena, M.P., Capozzi, V., Massa, S. & Spano, G. (2014). Fresh-cut pineapple as a new carrier of probiotic lactic acid bacteria. BioMed Research International, 309183.
[99] Saeed, M., Khan, W.A., Shabbir, M.A., Khan, M.I., Randhawa, M.A. & Yasmin, I. (2014). Bacteriocins as a natural antimicrobial agent in food preservation: a review. Pakistan Journal of Food Science, 24, 244–255.
[100] Salvana, E.M.T. & Frank, M. (2006). *Lactobacillus endocarditis*: case report and review of cases reported since 1992. Journal of Infection, 53, 5–10.
[101] Sato, T. & Iino, T. (2010). Genetic analyses of the antibiotic resistance of *Bifidobacterium bifidum* strain Yakult YIT 4007. International Journal of Food Microbiology, 137, 254–258.
[102] Scherer, K.M., Spille, J.H., Sahl, H.G., Grein, F. & Kubitscheck, U. (2015). The Lantibiotic Nisin induces lipid II aggregation, causing membrane instability and vesicle budding. Biophysical Journal, 108, 1114–1124.
[103] Sharma, P., Tomar, S.K., Goswami, P., Sangwan, V. & Singh, R. (2014). Antibiotic resistance among commercially available probiotics. Food Research International, 57, 176–195.
[104] Shokryazdan, P., Sieo, C.C., Kalavathy, R., Liang, J.B., Alitheen, N.B., Faseleh Jahromi, M. & Ho, Y.W. (2014). Probiotic potential of *Lactobacillus* strains with antimicrobial activity against some human pathogenic strains. BioMed Research International, 927268.
[105] Snyder, A.B. & Worobo, R.W. (2014). Chemical and genetic characterization of bacteriocins: antimicrobial peptides for food safety. Journal of the Science of Food and Agriculture, 94, 28–44.
[106] Snydman, D.R. (2008). The safety of probiotics. Clinical Infectious Disease, 46, 104–111.
[107] Soares, J.C., Marques, M.R., Tavaria, F.K., Pereira, J.O., Malcata, F.X. & Pintado M.M. (2011). Biodiversity and characterization of *Staphylococcus* species isolated from a small

manufacturing dairy plant in Portugal. International Journal of Food Microbiology, 146, 123–129.

[108] Sobrino-López, A. & Martín-Belloso, O. (2008). Use of nisin and other bacteriocins for preservation of dairy products. International Dairy Journal, 18, 329–343.

[109] Šušković, J., Kos, B., Beganović, J., Leboš Pavunc, A., Habjanič, K. & Matošić, S. (2010). Antimicrobial activity–the most important property of probiotic and starter lactic acid bacteria. Food Technology & Biotechnology, 48, 296–307.

[110] Thomas, L.V., Ingram, R.E., Bevis, H.E., Davies, E.A., Milne, C.F. & Delves-Broughton, J. (2002). Effective use of nisin to control *Bacillus* and *Clostridium* spoilage of a pasteurized mashed potato product. Journal of Food Protection, 65, 1580–1585.

[111] Thumu, S.C.R. & Halami, P.M. (2014). Phenotypic expression, molecular characterization and transferability of erythromycin resistance genes in *Enterococcus* spp. isolated from naturally fermented food. Journal of Applied Microbiology, 116, 689–699.

[112] Thumu, S.C.R. & Halami, P.M. (2012). Presence of erythromycin and tetracycline resistance genes in lactic acid bacteria from fermented foods of Indian origin. Antonie van Leeuwenhoek, 102, 541–551.

[113] Vanderbergh, P.A. (1993). Lactic acid bacteria, their metabolic products and interference with microbial growth. FEMS Microbiol Reviews, 12, 221–238.

[114] Verraes, C., Van Boxstael, S., Van Meervenne, E., Van Coillie, E., Butaye, P., Catry, B., de Schaetzen, M.A., Van Huffel, X., Imberechts, H., Dierick, K., Daube, G., Saegerman, C., De Block, J., Dewulf, J. & Herman, L. (2013). Antimicrobial resistance in the food chain: a review. International Journal of Environmental Research & Public Health, 10, 2643–2669.

[115] Wang, H., Yan, Y., Wang, J., Zhang, H. & Qi, W. (2012). Production and characterization of antifungal compounds produced by *Lactobacillus plantarum* IMAU10014. PloS one, 7:e29452.

[116] WHO (World Health Organisation). (1969). Specifications for identity and purity of some antibiotics. (pp. 53–67). WHO Food Additives Series 69.34. Geneva: WHO.

[117] Wood, B.J.B., Holzapfel, W.H. (1995). The genera of lactic acid bacteria. London: Blackie Academic & Professional.

[118] Yüceer, O. & Özden Tuncer, B. (2015). Determination of antibiotic resistance and biogenic amine production of lactic acid bacteria isolated from fermented turkish sausage (Sucuk). Journal of Food Safety, 2, 276–285.

[119] Zdolec, N., Rači, I., Vujnović, A., Zdelar-Tuk, M., Matanović, K., Filipović, I., Dobranić, V., Cvetnić, Z. & Špičić, S. (2013). Antimicrobial resistance of coagulase-negative staphylococci isolated from spontaneously fermented sausages. Food Technology & Biotechnology, 51, 240–246.

[120] Zhou, N., Zhang, J.X., Fan, M.T., Wang, J., Guo, G. & Wei, X.Y. (2012). Antibiotic resistance of lactic acid bacteria isolated from Chinese yogurts. Journal of Dairy Science, 95, 4775–4783.

[121] Zonenschain, D., Rebecchi, A. & Morelli, L. (2009). Erythromycin- and tetracycline-resistant lactobacilli in Italian fermented dry sausages. Journal of Applied Microbiology, 107, 1559–1568.

Leonora Rios de Souza Moreira, Caio de Oliveira Gorgulho Silva, Barbara Calheiros Neumann, and Edivaldo Ximenes Ferreira Filho

11 Microbial biofuel production: An overview on recent developments

11.1 Abstract

Currently, the growing energy demand together with the claiming for reduction in environmental pollution has led scientists to search for new energy sources. Ethanol presents advantages of reduction of greenhouse gases and be a renewable fuel. First generation bioethanol is a well-stabilized pathway for this fuel production which is produced mainly in Brazil and USA, from sugarcane and corn, respectively. In past decades the concern about reduction of food feedstock prompted the development of second generation bioethanol that is produced from lignocellulosic feedstock. This production utilizes microbial enzymes to hydrolysate the polysaccharides contained in biomass for subsequent fermentation. Third and four generation bioethanol are the fuel produced from aquatic microbial oxygenic photoautotrophs (AMOPs) such as algaes and cyanobacterias and engineered AMOPs, respectively. They can be cultivated in areas that do not present the competition with the production of feed and food. This chapter focuses on reviewing the main advantages and challenges of bioethanol from first to fourth generation.

11.2 Introduction

The world's economy today is highly dependent on fossil energy sources (coal, oil, natural gas) which are used to produce fuels, electricity, chemicals, and other goods. As a result of increasing population, demand of energy is escalating throughout the world. Utilization of fossil energy sources in the long run is not considered to be sustainable. In this scenario, use of renewable resources is an alternative [1]. Although estimates vary, the economically recoverable fossil fuel reserves include almost 1 trillion metric tons of coal, more than 1 trillion barrels of petroleum, and over 150 trillion cubic meters of natural gas [2].

Extreme consumption of fossil fuels, especially in large urban areas has caused more pollution due to release of greenhouse gases (GHGs) especially during the last few decades. The concentration of GHGs in the biosphere has hugely increased [3–5]. According to the World Energy Council, approximately 82 % of global energy demand is provided by fossil resources such as petroleum, natural gas and coal. Transportation sector alone consumes about 60 % of the petroleum based fuel produced globally and is responsible for production of one fifth of total CO_2 emissions [6].

Incentives for mitigating global climate change further stimulate international communities to invest in development and utilization of renewable energy [5]. To reduce our dependency on fossil fuels, except for nuclear energy, the most suitable alternative to fossil fuels is renewable sources such as hydroelectric, biomass, wind, solar, geothermal heat, and marine tidal for the energy industry [7, 8]. Renewable energy is a basic ingredient for sustainable development. Such sources can supply the energy we need for indefinite periods of time polluting far less overall than fossil or nuclear fuels [9]. Renewable energy is one of the most efficient ways to achieve sustainable development. Increasing its share in the world matrix will help prolong the existence of fossil fuel reserves, address the threats posed by climate change, and enable better security of the energy supply on a global scale [8]. Biofuels refer to plant biomass and the refined products to be combusted for energy (heat and light). Similar to fossil fuels, biofuels exist in solid, liquid, and gaseous forms [5]. Microbial biofuel synthesis is a cost effective and environmentally sustainable way of producing replacements for gasoline, diesel, and jet fuel from lignocellulosic biomass [10]. Microorganisms are a rich source of enzyme systems displaying glycoside hydrolase and other activities, involved in the breakdown of plant cell wall polysaccharides [11]. Microbial enzymes are known to play a crucial role as metabolic catalysts, leading to their use in various industries and applications. Polymer-degrading activities are suitable for saccharification of lignocellulosic biomass for the production of second generation biofuels [12]. Fungi belonging to the genus *Trichoderma* have high levels of production of holocellulases. Particularly, cellulolytic system of *T. reesei* has been focus of research for more than 50 years [13]. Cellulolytic fungi (*T. viride, T. longibrachiatum, T. reesei*) have long been considered the most productive and powerful destroyers of crystalline cellulose [14]. Beyond the fungi belonging to the genus *Trichoderma, Aspergillus niger* (and other species of *Aspergillus*) are also commonly used in the industrial production of enzymes. In addition to *Aspergillus* species, various *Penicillium* species are been considered as an alternative to *T. reesei* for production of biomass-degrading enzymes [14].

11.3 Ethanol

Ethanol has been known as a potential alternative fuel for well over one hundred years [15]. Ethanol's unique properties and characteristics make it an extremely valuable component of gasoline. For more than three decades, refiners and blenders have used low levels of ethanol to boost the octane rating and oxygen content of finished gasoline [16]. Ethanol contains 35 % oxygen, which results in a more efficient combustion of fuel and reduced levels of harmful gases. Moreover, ethanol production uses mainly energy from renewable sources, reducing greenhouse gas emissions and the net carbon dioxide added to the environment. Pure ethanol can be burned with greater efficiency, and is thought to produce smaller amounts of ozone precursors, thus decreasing urban air pollution. It is particularly beneficial with respect to low net CO_2 put into the atmosphere [17].

11.4 First generation ethanol

First generation ethanol corresponds to ethanol produced from food or feed feedstocks, namely sucrose containing crops (such as sugarcane, sugar beets and sweet sorghum) and starch containing crops (such as corn, wheat, barley, rice and cassava) [18]. Although produced from several sources, this discussion will focus on ethanol from sugarcane sucrose and corn starch, which are quite well established technologies used in Brazil and the US, respectively.

In Brazil, most part of sugarcane mills are able to produce both ethanol and sugar, consisting of two annexed plants in the same production facility, one specific for each product. Mills with annexed plants usually address 50 % of the sugarcane to sugar production and 50 % to ethanol production, and may change the sugarcane input direction according to market situation. Autonomous distilleries, where all sugarcane is diverted for ethanol production, are minority [19, 20]. Ethanol production from sugarcane is much simpler than that from corn starch and does not require enzymatic steps. After washing and milling the sugarcane stalks, extracted sugarcane juice is screened to clean off sand and fiber debris and goes through a clarification process, where impurities are removed with the addition of a flocculant agent and the pH is adjusted. Juice, initially with 15 wt % solids, is concentrated to around 22 wt % solids prior to the addition of yeast for the fermentation step. Molasses (containing high concentrated sucrose) obtained as residue from the crystallization of sugarcane juice for sugar production in annexed plants are also sent to fermentation.

Brazilian sugarcane mills follow the Melle-Boinot process, where fed batch is employed, yeasts are recycled and nitrogen content in fermentation substrate is low. Fermentation fed batch process in Brazilian mills is characterized by a gradual change in yeast composition along the production season. When ethanol production season begins, barker's yeast (*Saccharomyces cerevisiae*) is traditionally employed in fermentations as the starter fermenting microorganism. Industrial fermenting conditions are not aseptic and contaminations with wild yeasts and bacteria are inevitable. After every fermentation cycle, fermented wine is centrifuged and the separated yeast cells are recycled for the next cycles. Centrifuged yeast cream is treated with sulphuric acid (pH 1.8–2.5 for 1–3 h) prior to new cycle in order to reduce contaminations with bacteria, and this process is repeated during all production season. Yeasts are recycled up to 600 times during the 200–250 days of the harvest seasons. As cells are recycled every fermenting cycle, wild *S. cerevisiae* strains contaminate and gradually take over as the dominant fermenting microorganism, while original starter yeast becomes less abundant. Fermentation is actually performed by a mixture of different wild strains of *S. cerevisiae*. Starter barker's yeast strains cannot compete with wild ones, which are more adapted to the stressful conditions, have higher ethanol yields and do not cause flocculation or foaming. Robust wild strains tolerate harsh conditions that include high glucose osmotic pressure, high temperatures, pH variations and high concentrations of ethanol. Several wild *S. cerevisiae* strains have been isolated from ethanol plants and some (PE-2, CAT-1, SA-1, BG-1, FT858 and JP-1) have already become commercially available. Nevertheless, introducing wild strains as starter and dominant microorganisms in industrial fermentation is difficult, since strain composition changes along time and even dominant strains may disappear [21]. Fed batches in Brazilian mills are usually short, lasting from 6 to 12 hours (a very competitive characteristic of the process) and yielding a final ethanol titer of 7–12 % w/v. Fermentation yield has improved from 75–80 % in 1975–1980 to current 92–93 % of the maximum theoretical ethanol yield from sucrose. The remaining 7–8 % of sucrose is used mainly for yeast cell growth, metabolism and glycerol side production [22]. Fermented wine is then sent to distillation. After the distillation, the remaining material, named vinasse, is considered a valuable co-product used as soil fertilizer. Sugarcane bagasse generated after milling is used to co-generate heat and power by combustion, providing steam and electricity to support the whole production process. Surplus electricity, generated in approximately 12 % of Brazilian mills is sold as valuable co-product [19, 23–25].

In US, corn ethanol may be produced by dry-grind or wet-grind processes, being the former the most widely method used for that purpose [26]. Wet-grind mills have a much larger portfolio of products besides ethanol, including corn oil, corn gluten feed and corn gluten meal (both used as animal feed), sweeteners (such as high fructose corn syrup) and solvents [27, 28]. Meanwhile, dry-grind mills are focused on ethanol production, generating the so called distiller's grains residue with solubles (DGS), a valuable co-product used as animal feed that can be sold in wet (WDGS) or dried (DDGS) forms [28, 29]. Recently, dry-mills incorporated technologies to extract corn oil before fermentation [29] and, in 2014, 85 % of dry-mills also extracted corn oil [16]. Corn distillers oil may also be used as animal feed ingredient or for the production of biodiesel [16].

Considering starch cannot be readily fermented, ethanol production from corn starch involves enzymatic hydrolysis steps. In dry-grind mills, cleaned corn kernels are milled and mixed with water. The resulting mash containing starch is usually treated first with α-amilase (EC 3.2.1.1, glycoside hydrolase family 13), which has an endo-type mechanism and cleaves α-1,4 glucosidic bonds of starch, generating α-dextrins (glucose oligossacharides) of different degrees of polymerization. This step, called liquefaction, rapidly reduces the viscosity of the starch solution and is carried out at high temperatures (70–90 °C) and pH 6.5 using thermostable enzymes. α-Amylases are usually obtained from the thermophilic organisms *Bacillus liqueniformis* or *Bacillus stearothermophilus*, that produce more thermostable amylases then the originally used *Bacillus amyloliquefaciens*. *Aspergillus oryzae* and *Aspergillus niger* are also sources of less thermostable α-amylases. *Bacillus* enzymes, although stable at temperatures as high as 105 °C, do not tolerate low pH, requiring pH adjustment of the starch mash prior to liquefaction. α-Amylases with higher stability at low pH and low Ca^{2+} levels (stabilizing agent) are preferable, as they reduce costs by avoiding pH adjustments, calcium addition and posterior desalting steps. *Bacillus* α-amylases have been engineered to fit those requirements [27,30]. The second enzymatic step is called saccharification and consists of converting soluble oligosaccharides to glucose through treatment with glucoamylase (EC 3.2.1.3, glycoside hydrolase family 15), an exo-type acting enzyme that cleaves α-1,4 bonds and α-1,6 bonds from chain ends, producing glucose from branched and unbranched dextrins [28, 30, 31]. Glucoamylases are usually prepared from *A. niger*, *A. awamori* and *A. oryzae* and are stable at 60 °C and low pH (4.2–4.5) [27]. Pullulanases, isopullulanases and isoamylases, which cleaves α-1,6 bonds, are also important enzymes that may be used to speed the debranching reaction and improve the yield of glucose [27]. Saccharification step is usually carried out at pH 4.5 and 60 °C. Yeast, usually *Saccharomyces cerevisiae*, is added to the hydrolysate after saccharification or in a simultaneous saccharification and fermentation (SSF) process. In separate hydrolysis and fermentation (SHF) processes, saccharified starch is cooled to 32 °C prior to addition of yeast. Fermentation decreases pH due to liberation of CO_2, causing increase in glucoamylase activity, which continues saccharification, and inhibits contamination. SHF reduces the amount of glucoamylase required to full saccharification. On the other hand, in SSF processes, saccharification and fermentation occurs in the same temperature (32 °C). Ethanol inhibits contamination of the saccharified starch, and the lower glucose concentrations reduce osmotic stress in yeast. SSF is generally more energy-efficient. Fermentation usually lasts 48–72 hours, yielding final ethanol concentrations of 10–12 % [28].

After fermentation, in both sugarcane and corn mills, fermented liquor containing 10–12 % ethanol concentration is distilled to 95.4 % purity ethanol (containing 4.6 % water). Ethanol and water form an azeotrope, meaning that water cannot be further separated from ethanol at 95.4 % by conventional distillation. Further purification of ethanol (dehydration) is then achieved by either using solvents, usually benzene, to break the azeotrope and allow purification of ethanol by distillation, or using molecular sieves that absorb water [28].

11.4.1 Energy balance and greenhouse emissions

Ethanol, although a renewable biofuel, requires fossil fuels in its production process, directly or indirectly. Fossil fuels are required in the agriculture, industrial and transportation stages of the production process. Considering that, a positive energy balance and net reductions in GHG emissions can only be achieved if the amount of fossil fuels employed during production of ethanol are smaller than the amount of those avoided by the use of ethanol [32]. Ethanol production from sugarcane is much less

GHG intensive than ethanol from corn. Combustion of sugarcane bagasse to provide energy for ethanol mills is the main factor contributing the lower GHG emission, in contrast to American corn ethanol mills that use mostly natural gas [29, 32]. The use of pesticides and fertilizers in both crops is an important limiting factor in energy balance of these processes and in GHG emission reduction [37]. It is noteworthy that the use of corn ethanol co-products as animal feed also offers CO_2 credits to the corn ethanol production process. Although 36.8 % of corn production in US is currently addressed to ethanol production [33], the co-products DGS, corn gluten feed, corn gluten meal and corn distillers oil are able to partially compensate the large amount of corn addressed to ethanol as they are used as animal feed [29].

According to Crago et al. [25], Brazilian ethanol production from sugarcane emits 470 kg of CO_2 per m^3 anhydrous ethanol produced, which sums up to 550 kg/m^3 when transportation to the US is included. Meanwhile, American corn ethanol production, including agriculture and industrial stages, reaches 1173 kg CO_2 per m^3 of ethanol produced, 113 % higher than sugarcane ethanol emissions. Macedo et al. [32] estimated that anhydrous sugarcane ethanol production in Brazil emitted 436 kg of CO_2 per m^3 in the 2005/2006 harvest, which represented net avoided emissions of 1886 kg of CO_2 per m^3 of anhydrous ethanol blended with gasoline.

According to Goldemberg [8], the best example of a large growth in the use of renewables is given by the sugarcane ethanol program in Brazil. The production of ethanol from sugarcane can be replicated in other countries without serious damage to natural ecosystems. Still according to Goldemgerg [8], expanding the Brazilian ethanol program by a factor of 10 would supply enough ethanol to replace 10 % of the gasoline used in the world. This land area is a small fraction compared to the crops already harvested on the planet [8]. Today the production of ethanol in Brazil is 27.96 billion of liters in the crop of 2013/2014 [34]. In the same period, the United States production was 14.3 billion of gallons of ethanol (approximately 54.34 billion liters) [16]. These 2 nations represent almost the totality of global production of 87,2 billion liters [4].

Much debate on whether US corn ethanol results in negative or positive energy balance has been taking place for decades. Although many studies from the 1980's and before estimated negative energy balances for the production of ethanol, more recent research, from 90's until today, show positive energy balances. That may be due to current improved corn productivity and industrial efficiency [29] or due to incorrectly ignored co-products and obsolete data [35].

According to Crago et al [25], on an energy-equivalent basis, use of US corn ethanol decreases GHG emissions by 44 % when compared to gasoline, while Wang et al [29] estimated 24 % GHG emission reductions. With better results, Brazilian sugarcane ethanol reduces emission of GHG by 74 %, according to Crago et al [25] or 63 % as estimated by EPA [36]. Considering sugarcane ethanol reduces emission of GHG by more than 50 %, it fits the concept of advanced biofuels of the Renewable Fuel Standard (RFS) program (discussed below). Thus, Brazilian sugarcane ethanol can compete with cellulosic ethanol in the US, representing a great opportunity for Brazil to become, once again, an exporter of ethanol do the US [25].

11.4.2 First generation ethanol in Brazil

Brazil's interest in replacing imported fossil fuels by domestically produced ethanol for powering cars, motors and lamps dates back to the beginning of the 20th century. Increase in oil prices during World War I, coupled with low sugar prices due to Brazilian overproduction, influenced the government to consider producing ethanol from sugarcane as a way to increase profitability of the sugarcane industry and having a domestic alternative fuel to oil. In 1931 federal government established for the first time at national level that gasoline should contain 5 % of ethanol [37, 38] and this governmental influence stimulated the sugarcane industry. Sugar and Alcohol Institute (Instituto do Açúcar e do Álcool – IAA) was created in 1933 to regulate this growing sugarcane sector. This public institution executed the government interventions on the sector for six decades, until its deactivation in 1990 [39]. In the 1930's, the IAA supported the construction of new ethanol mills so the sector could meet national demands. Northeastern Brazilian states, especially Alagoas and Pernambuco, were the main sugarcane producers, which headquartered important ethanol brands of the time such as Usina Serra Grande and Azulina [37]. Difficult transportation of sugar and ethanol from northeast to southeast region, especially during World War II, led the implantation of sugarcane sector in São Paulo state. Superior transportation and energy infrastructure, land of good quality and proximity to industrial complexes and to research centers consolidated sugarcane in the region, overtaking northeastern production in the 1950's [40]. Creation of the Brazilian state oil company Petrobras in 1953 and pressures from the oil and car industries halted the plan of a domestic ethanol fuel for the following decades [37].

Brazilian robust and well established sugarcane ethanol industry existing today is a consequence of the national interest on ethanol that re-emerged during the 1970's. A timeline of recent progress of ethanol in Brazil is illustrated in figure 11.1. In November 1975, the National Alcohol Program (Programa Nacional do Álcool), or ProAlcohol, was launched in response to the first oil shock caused by the 1973 OPEC oil embargo, which prompted oil price to quadruplicate in 1973/74. Brazil imported more than 80 % of the gasoline national demand at that time, and elevated oil prices contributed to inflation, gasoline shortages and rationing [39]. Meanwhile, the international sugar prices in the second half of 70's were low due to world overproduction. ProAlcohol was a strategy to overcome Brazil's high dependency on expensive imported oil and replace gasoline by domestically produced sugarcane ethanol. Originally, other biomass sources such as cassava were also considered, but sugarcane was chosen by political and economic reasons [41]. ProAlcohol was also a means to guarantee profitability of the sugarcane sector during low sugar price period. The government associ-

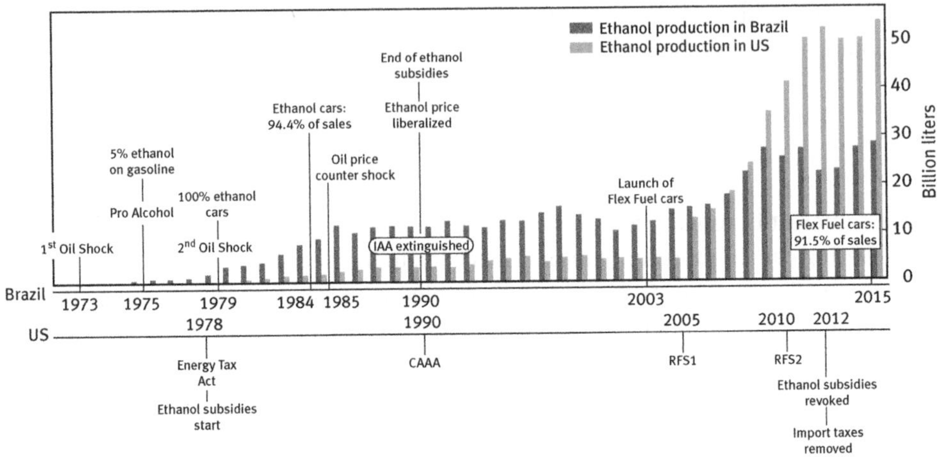

Fig. 11.1: Timeline of ethanol production in Brazil and USA.

ated the potential of growth of the sugarcane industry with the opportunity to lessen the dependence on imported oil after the oil shock [37, 40, 42].

Among ProAlcohol initial measures was the offering of financial incentives at low interest rates for the construction and expansion of ethanol mills. From 1975 to 1979, a mandate to blend 5 % of ethanol into gasoline was established. During ProAlcohol's first phase, ethanol production went from 664 million liters in 1976/77 to 3.7 billion liters in the 1980/81 crop, largely contributed by governmental incentives [37, 40, 42].

From 1979 to 1985, boosted by the second oil crisis after Iranian Revolution, the ProAlcohol entered a new phase, focusing on the use of neat ethanol as a full substituent of gasoline. The government, by reducing taxes on alcohol cars, encouraged automobile manufacturers to develop and commercialize vehicles powered by 100 % hydrated ethanol [37]. During that period, the sales of cars fueled exclusively by gasoline dropped, accompanied by sharp increase in ethanol fueled cars sales, which came to represent 94.4 % of automobile production in 1984 [43]. Ethanol production in Brazil reached 11.9 billion liters in 1985. Petrobras was responsible for purchasing and distributing ethanol and regulating prices to keep ethanol more cost favorable than gasoline, even if artificially [39, 42].

Success of the ProAlcohol program and substitution of gasoline by ethanol was heavily dependent on government subsidies and tax incentives, necessary for the construction and expansion of mills, sales of ethanol fueled cars and keeping the price of ethanol fix and advantageous in relation to gasoline, among others. Decrease in the international oil price in 1985, the so-called oil counter-shock, and the severe Brazilian economic crisis in the mid and late 1980's jeopardized the government's financial capacity to support the program. Recovery of international sugar prices and liberalization of the sugar export market re-directed sugarcane to sugar production.

Ethanol production no longer could meet national demand and Brazil faced shortages during final 1980's. The 1990's was a transition phase for the ethanol sector. IAA was extinguished, subsidies were revoked, fixed ethanol prices were abolished and ethanol market was liberalized. Hydrated ethanol was not available at the pumps during the 1990's [39, 40, 42, 44].

A re-emergence of global interest on biofuels, especially ethanol, has been occurring since the 2000's driven by increase in oil prices and concerns about greenhouse gases emission, global warming and sustainability. In 2003, the development and commercialization of flex fuel vehicles that can run on ethanol, gasoline or any volumetric combination of these two fuels, was a breakthrough and encouraged consumption of ethanol once again in Brazil. In 2014, 91.5 % of the cars produced in Brazil were flex-fuel, and only 8.3 % were gasoline fueled [43].

Although Brazilian ethanol is not subsidized since early 1990's, mandatory blending still guarantees internal consumption of ethanol. Since 1993, law 8.723 determines that gasoline must contain ethanol [45]. Today, the government is allowed to change the ethanol concentration between 18 and 27.5 % is accordance to ethanol supply, which may vary due to climate variations. Poor sugarcane harvests usually leads to temporary decrease in ethanol percentage [44]. Today ethanol blend is fixed at 27 % [46].

The existing project of an ethanol pipeline for transportation of ethanol from the main producing areas to distribution centers and the port of Santos (Brazil's largest port) in expected to increase Brazilian ethanol competitiveness by reducing its GHG intensity and its cost for both domestic and international consumers.

Various research centers have played key role in the improvement of the sugarcane sector in Brazil, which has been achieved through incremental cumulative non-disruptive contributions. Agronomic Institute of Campinas (IAC) pioneered sugarcane breeding before ProAlcohol. Planalsucar, public national sugarcane breeding program, and Copersucar Sugarcane Technology Center (CTC), a private research center funded by mills, brought significant improvements in sugarcane yields during ProAlcohol. Planalsucar and CTC's varieties covered large proportions of sugarcane fields in Brazil, as well as they contributed to industrial process melioration. Dedini is the main manufacturer of industrial equipment for sugarcane mills and also added contributions. Today, Ridesa (former Planalsucar), CTC, CTBE (Brazilian Bioethanol Science and Techonology Laboratory) and EMBRAPA (Brazilian Agricultural Research Corporation) Agroenergia are the major research centers focusing on the sugarcane and ethanol sector [40].

11.4.3 First generation ethanol in USA

The idea to use ethanol as fuel is dated to before the discovery of petroleum. The first internal combustion engines created independently by Samuel Morey in the US and by Nicholas Otto in Germany were designed to run on ethanol containing fuels [47]. Henry Ford's first automobile of 1896 was powered by ethanol and the Ford Model T of 1908, the first popular automobile marketed in the US, was designed as a flexible fuel vehicle that could run on gasoline, ethanol or a combination of both. Sustainable production of ethanol was already discussed in the beginning of the twentieth century in US. Henry Ford, a great supporter of ethanol fuel, predicted the development of cellulosic ethanol: "The fuel of the future is going to come from fruit like that sumac out by the road, or from apples, weeds, sawdust – almost anything". In 1917, Alexander Graham Bell also pointed the feasibility of using several types of feedstock, such as crop residues, farm waste, grasses and city garbage for ethanol production, and declared, "We need never fear the exhaustion of our present fuel supplies so long as we can produce an annual crop of alcohol to any extent desired" [47–49].

Government policies, taxes and subsidies have played key roles in determining the price of ethanol and curbing or launching the use of ethanol as alternative fuel in the US. In 1861, a tax of 0.55 cents per liter of ethanol ($2.08 per gallon – equivalent to approximately $11 per liter today) was established to cellect funds for the Civil War. Although originally directed to alcoholic beverages, fuel ethanol was not spared. Ethanol tax was not revoked until 1906, drastically reducing ethanol competitiveness with fossil fuels such as kerosene during that period. By the time ethanol tax was removed, new ethanol plants were built but gasoline industry was better established and gas had become the favored fuel for automobiles [50]. Moreover, discovery of new oil spots in Texas made ethanol prices soar. Ethanol demand increased to 227 million liters (60 million gallons) during World War I due to rationing of gasoline. In the mid 1930's, the emergence of the Farm Chermugy movement encouraged the production of bio-based materials from renewable agricultural sources. Ethanol production increased and blends of 5–17 % ethanol in gasoline under the names of Agrol or Alcolene were commercialized in the Midwest during that period. During World War II, ethanol demand also increased, but primarily for the production of synthetic rubber and not for use as fuel. Ethanol consumption reached 2.27 billion liters (600 million gallons) per year. However, after World War II, the use of tetraethyl lead as an antiknock agent and octane booster contributed to the prevalence of leaded gasoline over ethanol, considering the former was cheaper and easier to produce [47–49].

The ethanol fuel production in the US was effectively set in motion by the Energy Tax Act (ETA) of 1978, which implemented a subsidy of 10.6 cents per liter (40 cents per gallon) of ethanol produced [51]. The world was facing an oil crisis scenario, when the oil prices rose sharply in 1973 after the OPEC Arab oil embargo [51]. Subsidy on ethanol, which was modified a few times in terms of values and mechanisms, lasted until 2012 [52] and was of major importance for prompting ethanol industry. Major milestones of ethanol progess in US are shown in figure 11.1.

The Clean Air Act Amendments (CAAA) of 1990 created new market perspectives for ethanol. The CAAA acknowledged the contribution of vehicle fuels to air pollution problems, namely high levels of carbon monoxide and ground-level ozone, and established the use of oxygenated fuels blended to gasoline in many areas of the US that did not attain federal air quality standards. Higher content of oxygenate fuels improves the complete burning of gasoline, reducing the release of carbon monoxide. Furthermore, evaporated compounds from gasoline may increase formation of ground-level ozone, or smog, which is harmful to the human respiratory system. The amount of evaporative emissions of gasoline was also regulated by the US government, imposing a limit to its Reid Vapor Pressure (RVP). Ethanol fitted the RVP requirements for reducing the volatility of the blended fuel. Although ethanol was already a viable option as a gasoline substitute, methyl tertiary butyl ether (MTBE), an oxygenated fuel produced from natural gas, was the fuel additive of choice during the 90's until it was found to be toxic and associated to contamination of groundwater supplies, posing risks to public health and the environment. MTBE use was reduced and eventually banned in many states and replaced mainly by ethanol or the ethanol derived ETBE (ethyl tertiary butyl ether). Corn ethanol, for being renewable and more environmentally friendly than MTBE, reducing the volatility of gasoline and having minimum health concerns finally gained popularity and took off as the main oxygenate fuel to be blended into gasoline [53].

The Energy Policy Act (EPAct) of 2005 eliminated the requirements of blending oxygenate fuels to gasoline, but on the other hand implemented the Renewable Fuel Standard (RFS) program, created to ensure a minimum value of renewable fuels added to gasoline. Originally, the RFS1 required 28.4 billion liters of ethanol to be blended into gasoline by 2012, which was increased to 136 billion liters by 2022 under the Energy Independence and Security Act (EISA) of 2007 and reinforced by the EISA of 2010 [54]. The so called RFS2 also establishes that conventional biofuels (e.g. corn starch ethanol) should make up to 56.7 billion liters (15 billion gallons) from 2015 on, and that advanced biofuels should make up to 79.5 billion liters (21 billion gallons) by 2022. Advanced biofuels are those with life-cycle GHG emissions reduced to at least 50 % in comparison to gasoline, in which cellulosic ethanol and Brazilian sugarcane first generation ethanol are included. RFS2 also requires that corn starch ethanol should achieve a GHG emission reduction of 20 % relative to gasoline [55].

Until 2012, domestic ethanol in the US was protected from competing with lower-cost Brazilian ethanol by importation tariffs [51]. Tariffs summed up to 30 % of the imported ethanol, while fossil fuels had very low import tariffs [56]. In 2012, the import tariffs were withdrawn, providing free commercialization of ethanol between Brazil and US. This was after Brazil removed taxes on imported American ethanol in 2010. American import tax removal in 2012 was accompanied by the elimination of the $0.45 per gallon tax credit to blenders. That alleviated around $6 billion per year on subsidies for the American government [52]. Guaranteed demand on ethanol was maintained by RFS [57], though. Free commercialization of ethanol between Brazil and US is a step forward for ethanol becoming a global energy commodity, which is an interest of both parts [56].

11.5 Second generation bioethanol

Fossil fuels have always been the main energy source for transportation and industry. Current scarcity of petroleum reserves, together with necessity of reducing environmental impact caused by fossil fuels, leads to utilization of renewable fuels sources, like ethanol. However, ethanol produced from sugarcane, sugar beet and corn has a problem since they have value as food and feed. In this context arises the need for utilization of new sources to produce ethanol, like lignocellulosic feedstock, which does not hold debate between food and fuel. Considering that plant biomass is an attractive source for production of 2nd generation bioethanol owing to its large availability, low cost and it does not bring about the problematic of competition with food supplying.

Plant biomass presents a promising renewable energy that could suppress at least in part our demand for fossil fuels. In addition to reducing our dependence on fossil fuels, biomass utilization will also impact positively on many environmental issues and help to minimize the net production of greenhouse gases [58]. Plant biomass corresponds to the majority of cheap and abundant nonfood materials available from plants. Employment of biomass is a cost effective and sustainable source for commercial production of bioenergy as bioethanol. Second generation biofuels are produced from biomass in a more ecologically friendly way, which is truly carbon neutral or even carbon negative in terms of its impact on CO_2 concentrations [4, 59–61]. For large-scale bioethanol, it is worthy to use cheaper and abundant feedstocks in order to lower its high price which has hindered the use of fuel ethanol in the energy industries [62].

Agricultural residues are used as animal feed, domestic fuel and as fuel to run boilers. Utilization fraction of each feedstock is low and varies according to the geographic region [63]. Generally, only a small portion of the produced agricultural residues is used as animal feed, the rest is commonly burned, a common practice all over the world that increases the air pollution [64]. A large amount of agricultural feedstock is available for bioethanol production. In 2004, approximately 1.55×10^9 tons of agricultural residues was produced around the world. The four major agrowastes were rice straw (7.3×10^8 tons), wheat straw (3.5×10^8 tons), corn stover (2.0×10^8 tons) and sugarcane bagasse (1.8×10^8 tons) [63].

The plant cell wall is composed of three major components, cellulose, hemicellulose and lignin. Typically, most of the agricultural lignocellulosic biomass is comprised of about 10– 25 % lignin, 20–30 % hemicellulose, and 40–50% cellulose [65]. Cellulose is the major structural component of plant cell wall, important for mechanical strength, it is a linear polymer of β-1,4-linked glucose units. Hemicelluloses are often composed of pentoses and hexoses, which include a variety of polysaccharides with linear or branched polymers derived from sugars such as D-xylose, D-galactose, D-mannose, D-glucose and L-arabinose. Pectin is a linear polymer of α-1,4 galacturonic acid units, some of which are methylated at C-6, some acetylated at C-2 and some with more extensive substitution [11,65]. Lignin contains three aromatic alcohols (coniferyl alcohol, sinapyl alcohol and p-coumaryl alcohol) and it forms a physical barrier against the attack on cellulose and hemicellulose. Lignocellulosic feedstock includes agricultural wastes, forestry residues, grasses and woody materials [65]. From a biotechnological point of view, a broad diversity of lignocellulosic feedstock is available to be converted into high value bioproducts such as bioethanol/biofuels. Hypothetically, all plant materials can be used to generate bioethanol. Nevertheless, it is fairly challenging to deconstruct these compound polymers into simple sugars [5, 65].

11.5.1 The role of pretreatment

One of the greatest obstacles for the production of 2nd generation bioethanol is lignin.

Many reasons have been presented as responsible for inhibition of hydrolysis by lignin, among them:

1. It acts as a physical barrier that prevents the access of cellulases and hemicellulases to their substrate [66,67];
2. It is covalently linked to hemicellulose, conferring mechanical strength to the cell wall [67];
3. Not only the presence of lignin, but the distribution and the kind of lignin also affect the enzymatic hydrolysis [68];
4. Lignin-derivable phenolic compounds have an inhibitory effect on the enzymatic activity [69,70];
5. Lignin inhibits the action of cellulases by non-specific adsorption [71].

Due to these obstacles, pretreatment is an important tool for achieving effective hydrolysis of biomass. Pretreatment is one of the most expensive steps in the biochemical conversion of biomass, accounting for up to 40 % of the total processing cost [72], however, non-pretreated lignocellulosic biomass has low yield, which increases the total fixed capital per annual gallon of capacity. Pretreated lignocellulosic biomass has high yield and lower capital requirements per annual gallon of capacity as compared to the no pretreatment case [73]. Generally, pretreatment aims to break down the lignin barrier, allowing a greater access to cellulose, which will be subjected to enzymatic hydrolysis to convert into fermentable sugars [65]. Different lignocellulosic biomass subjected to different pretreatment will have a variety of types and concentrations of sugars [74]. A variety of results will be achieved according to the pretreatment used, like the interruption of the structure of lignin and its connections to the rest of the biomass [75]; removal of a part or all of the lignin which leads to a greater porosity in the substrate [76]; removal of hemicellulose to increase the access of cellulases to cellulose [75]; disruption in the hemicellulose structure [75]; reducing the degree of polymerization and the crystallinity of cellulose [75]; reducing the size of the particle [66].

11.5.2 Enzymatic hydrolysis of the plant cell wall

Enzymatic hydrolysis is an advantageous method to achieve fermentable sugars in a eco-friendly reaction from the pretreated biomass [65], so the majority of strategies for converting lignocellulosic biomass to ethanol or other liquid fuels involve the enzyme-catalyzed depolymerization of polysaccharides [77]. Enzymes have benefits over the traditional chemical treatments due to the higher conversion efficiency, no substrate loss because of chemical modifications and the use of moderate physical-chemical conditions (lower reaction temperatures, almost neutral pH, and the use of biodegradable and nontoxic reagents), which makes the process more environmentally friendly [78]. Nevertheless, enzymes are costly since they must be produced by living organism. In view of the recalcitrance of lignocellulose, high enzymes loadings are necessary to obtain valid degradation rates. The price of production of enzymes and the pretreatment costs are hindering the development of an economically possible second generation bioethanol industry [77]. The rate and the productivity of conversion of biomass to fermentable sugars depends on the starting biomass material, the pretreatment and action of a wide amount of enzymes [74].

A great variety of enzymes are required to deconstruct the cell wall structure because of its recalcitrant nature and heterogeneity. These enzymes are produced by a wide range of fungi and bacteria, *Cellulomonas fimi* and *Thermomonospora fusca* have been extensively studied for cellulase bioethanol production [79]. Among fungi, the most used species are *Trichoderma reesei, Aspergillus niger, Trichoderma longibrachiatum* [80], *Sclerotium rolfisii, Phanerochaete chrysosporium* beyond species from *Schizophyllum* and *Penicilium* [79]. Mutant strains of *Trichoderma sp.* (*T. viride, T. reesei, T. longibrachiatum*) have long been considered to be the most productive and powerful destroyers of crystalline cellulose [81].

Hydrolytic enzymes like cellulases, xylanases, manannases contributes to the degradation of holocelulose structure by hydrolysis of glyosidic bounds [82]. Swollenins facilitate the hydrolysis by breaking down hydrogen bonds between cellulose microfibrils or cellulose and other polysaccharides [82]. Oxidative enzymes like lytic polysaccharide monooxygenases helps in the degradation of cellulose chains by an oxidative mechanism of intrachain cellobiosyl moiety [83]. For the complete hydrolysis of holocellulase, the synergistic action of a broad range of endo and exo enzymes with different specificities is necessary. Xylan-degrading enzymes are produced by a variety of fungi and bacteria, like *Trichoderma* spp., *Penicillium* spp., *Talaromices* spp., *Aspergillus* spp. and *Bacillus* spp. [18].

11.5.2.1 A brief overview on enzymatic degradation of the plant cell wall

As mentioned in the last section, the complete degradation of lignocellulosic biomass requires the coordinate action of enzymes with different specificities.

As mentioned before, lignocellulose contains significant amount of cellulose which can be used as a valuable carbon source for production of value-added chemicals [84]. Cellulose is degraded by the synergistic action of endoglucanases (EC 3.2.1.4), that randomly act on soluble and insoluble cellulose chains; cellobiohydrolases (EC 3.2.1.91) which release cellobiose from the nonreducing (CBH I) and reducing (CBH II) ends of the cellulose chain; and β-glucosidase (EC 3.2.1.21) release glucose from cellobiose [82].

Hemicellulases are frequently classified according to their action on distinct substrates. Considering the heterogeneous structure and complex chemical nature of xylan, its fully hydrolysis requires a large amount of enzymes, including endo-β-1,4 xylanase (EC 3.2.1.8), which hydrolyze the glycosidic bonds in the xylan backbone releasing oligosaccharides with small degree of polymerization. β-xylosidases (EC 3.2.1.37) that catalyze the hydrolysis of small xylooligosaccharides and xylobiose, releasing xylose from the non-reducing terminus [85]. Enzymes that acts on side chain include acetylxylan esterases (EC 3.1.1.6) that remove the O-acetyl groups of acetyl xylan [85], α-L-arabinofuranosidases (EC 3.2.1.55) that cleave the terminal α -L-1,2-, α -L-1,3- and α -L-1,5-arabinofuranosyl residues [82], α-glucuronidases (EC 3.2.1.-) cleave the α-1,2 bonds between glucuronic acid residues and the main chain in glucuronoxylan [85]. Moreover, esterases activities are observed in feruloyl esterases (E.C. 3.1.1.73) and p-coumaloyl esterases (E.C. 3.1.1.73) which are capable of hydrolysate the ester-link between arabinose and monomeric or dimeric ferulic acid and the linkages between arabinose and p-coumaric acid, respectively [82, 85]. Other enzymes are also involved in the modification of xyloglucans, like endo-transglycosylase/hydrolases (2.4.1.2070), a xyloglucan-active β-D-galactosidase (3.2.1.23), a xyloglucan specific α-L-fucosidase (3.2.1.51/3.2.1.63) and a xyloglucan oligosaccharide-specific α-D-xylosidase (3.2.1.37) [86].

The mannan-degrading enzymes system is composed of several enzymes including β-mannanase (EC 3.2.178), that randomly hydrolyses the β-1,4 linkages of the mannan backbone, β-mannosidase (EC 3.2.1.25), an exo-type enzyme that hydrolyzes the β-1,4-linked mannosides, releasing mannose from the nonreducing end of mannans and mannooligosaccharides. β-glucosidase (EC 3.2.1.21) an exo-type enzyme, cleaves the β-D- 1,4 glucopyranose at the nonreducing end of the oligosaccharides released from glucomannan and galactoglucomannan by β-mannanase [87]. Other additional enzymes are necessary, such as acetyl mannan esterase (EC 3.1.1.6) and α-galactosidase (EC 3.2.1.22), which are debranching enzymes. The first one cleaves the linkage in the acetyl groups from galactoglucomannan, while the second one cleaves the α-1,6 linked D-galactopyranosyl side chains of galactomannan and galactoglucomannan [87].

With regards to the degradation of pectic substances, the enzymes involved in this process may be divided in three major groups. The hydrolytic enzymes, which includes polygalaturonases, like endopolygalaturonase (EC 3.2.1.15) and exopolygalacturonase (3.2.1.67) cleaves polygalacturonic acid chain forming oligogalacturonates and monogalacturonates, respectively. Pectinesterases (EC 3.1.1.11) catalyze the deesterification of methyl ester linkages of galacturonan backbone of pectic substances releasing acidic pectins and methanol. Lyases carry out the breakdown of pectinases through trans-elimination. Some exemples of these third group include endopolygalacturonase lyase (EC 4.2.2.2) and exopolygalacturonase lyase (EC 4.2.2.9) that act on pectic acids releasing unsaturated oligogalacturonates and digalacturonates, respectively [88].

Beyond the holocellulolytic enzymes, laccases, lignin peroxidase and manganese peroxidase are fundamental for delignification process. The presence of laccase, alone or together with lignin peroxidase and manganese peroxidase, has been detected in a wide variety of white rot fungi, making clear its fundamental role during delignification [82]. Due to the complex and random phenylpropanoic polymer structure of lignin, enzymes involved in this process must have broad substrate specificity [89]. Laccases (EC 1.10.3.2) oxidize phenolic compounds and aromatic amines using the molecular oxygen as a terminal electron acceptor. Manganese peroxidases (EC 1.11.1.13) and lignin peroxidases (EC 1.11.1.14) are low substrate specificity enzymes capable of oxidizing lignin at the phenolic and non-phenolic aryl-ether positions, respectively [90].

Besides these hydrolytic enzymes, another kind of enzymes also has important roles in lignocellulosic biomass degradation. Oxidative enzymes, classified as auxiliary activities such as cellobiose dehydrogenase (EC 1.1.99.18) are secreted by cellulose degrading fungi. This enzyme is capable of oxidizing a variety of sugars including cellobiose, lactose and in some cases, glucose. During substrate oxidation the cofactor FAD is fully reduced [91, 92]. Other suggested roles for that enzyme includes the activation of hydrolytic cellulases by relief of cellobiose product inhibition [93]. Currently, cellobiose dehydrogenase are classified into the AA3 family [92]. Lytic polysaccharide monooxygenases (LPMO) are classified into the AA9 family (formerly known as GH61) and are mononuclear type II copper enzymes that require an external electron donor, a small molecule, or a partner redox- protein for activity. The breakdown of the glucosidic bounds occurs with the oxidation of C-1, C-4 and C-6. LPMO degrades cellulose using C-H activation followed by O_2-dependent chain cleavage. [83, 92].

Other proteins are also associated with the plant cell wall degradation. Swollenins, proteins with sequence similarity to the plant expansins, break down hydrogen bonds between cellulose microfibrils [82]. Swollenins contain carbohydrate binding domains (CBM) and are thought to interrupt cellulose structure via non-hydrolytic mechanisms, however, the biochemichal action of these proteins needs to be completely elucidated [94].

11.5.2.2 Second generation bioethanol global production

Commercial bioethanol is currently produced from starch/sugar-based crops like sugarcane, sugar beet, sweet sorghum, corn, wheat, barley, potato, yam and cassava. A huge increase in ethanol production was observed since 1990 when the commercial bioethanol production was 4.0 billion gallons in that year. It had a small increase in 2000 (4.5 billion gallons) and then rapidly increased to 23.3 billion gallons in 2010 [5]. In 2014 the global production of bioethanol was 24,565 million of gallons, United States and Brazil were the largest producers of bioethanol with 14,300 (58 %) and 6,190 (25 %) millions of gallons, respectively. Europe had a share of 6 %, China and Canada 3 and 2 % respectively, while Thailand, Argentina and India produced 1 % each. The rest of the world together represented 3 % of the total [16].

Developing new industrial technologies requires several test stages proportional to the production perspectives: the first step in bioethanol is assessing technology through lab-scale processes, in which reactor volumes are usually less than 1000 mL and involves manual intervention at varying degrees, and experiments a large range of parameters; pilot-scale then incorporates continuous processing under distinct operating conditions, with reactors extending between 1 and 100 L; demonstration-scale is implemented when most variables are established and the technology is substantially developed, in order to mitigate risks and prove to investors that the process is efficient and satisfies market expectations using up to 40,000 L capacity. This course of action is decisive towards accomplishing success in commercial-scale technologies considering factors such as solids handling, particles size and nature, flowability of feedstock delivering, among all things that interfere on plant engineering [95, 96]. A recent production of second-generation ethanol targeting the general public has been possible as a result of decades of research in several stages. Pilot and demonstration plants have produced smaller amounts with testing goals since early 2000's. Government support and subsides were vital for every commercial-scale plant established thus far; USA and Brazil are yet again leader countries in this subject, along with Italy [97]. The first commercial-scale cellulosic ethanol plant (Crescentino Bio-refinery, Crescentino, Vercelli, Italy) entered into full operation on October, 2013. The plant is running by the Italian company Beta Renewables to annually produce 20 million gallons of ethanol from wheat straw and giant reed using the patented "Proesa" technology in which plant biomass is pre-treated with steam (high temperature and pressure), followed by enzymatic hydrolysis [5].

Studies indicate that the total international production of ethanol from lignocellulosic biomass could reach 442 billion liters [63] based on agricultural waste data. This corresponds to 5 times current 1st generation ethanol production of 87,2 billions [4]. Nevertheless, this production is only starting to emerge and 2014 has been a fruitful year, with at least three plants inaugurated and other four under construction (Tab. 11.1). Projects involve multiple companies from different countries uniting for technology development, process optimization and providing resources such as enzymes, area, and funding, physical structure and biomass material. This is a crucial moment to analyze results, strategies and reassess perspectives for the next few years.

Tab. 11.1: Examples of second generation bioethanol producing plants.

Location		Company	Scale	Ethanol production (Kl/year)	Feedstocks used	Ref
USA	City of Nevada Iowa	Dupont	C	114,000	corn stover	[98–100]
USA	Vonore Tenessee	Dupont, Genecor International	D	950	stalks, switchgrass,	[98, 99, 101]
USA	Hugoton City, Texas	Abengoa	C	95,000	corn stover, wheat straw, milo stubble and prairie grasses	[99, 102]
USA	Emmetsburg, Iowa	POET-DSM Advanced Biofuels	C	95	corncobs, leaves and husks	[99, 103]
USA	Vero Beach, Florida	INEOS Bio and New Plant Energy Florida LLC	C	30,400	yard, wood and vegetative wastes	[104, 105]
USA	Fayetteville, Ark	Bioenergy Resources Inc and INEOS	P	ND	wide variety	[106]
BRA	São Miguel dos Campos, Alagoas	Granbio	C	311,600	sugarcane bagasse and straw	[107]
BRA	Piracicaba, São Paulo	Raízen and Iogen Corporation	C	152,000	sugarcane bagasse and straw	[108–110]
CAN	Ottawa, Ontario	Iogen Corporation	D	700	cereal straws, corn stover, sugarcane bagasse, sugarcane tops and leaves	[109]
ITA	Crescentino	Beta Renewables	C	285,000	wheat straw, rice straw, giant cane	[111–113]
SWE	Örnsköldsvik	SEKAB	D	ND	ND	[114]
DNK	Kalundborg	DONG Energy	D	5,400	wheat straw	[115–117]
ESP	Babilafuente, Salamanca	Abengoa	D	5,000	grain, urban solid waste	[118]
RUS	Kirov	Biochemical Plant	C	40,000	wood waste	[119]

11.5.3 Fermenting microorganisms

Unlike the saccharification of starch and sugar-based feedstock that results mostly in hexoses, lignocellulosic feedstock is composed of cellulose and hemicellulose which results in hexoses (C6) and pentoses (C5). So, beyond an efficient pretreatment, is necessary the fermentation of C6 and C5 sugars to bring second generation bioethanol to commercial reality [120]. Principal choice for second generation ethanol production from lignocellulosic biomass by biological pathways is enzymatic hydrolysis followed by fermentation – separate hydrolysis and fermentation or simultaneous saccharification and fermentation (SSF). SSF may include the conversion of C5 and C6 sugars [121].

Generally, a microorganism to be used in bioethanol production should give a high ethanol yield, high productivity and be capable of withstand elevated ethanol concentrations. Moreover, the capacity of using different sugars are essential for SSF [121].

Commonly, the most used yeast for industrial ethanol fermentation is *Saccharomyces cerevisiae*, a facultative anaerobic yeast, only capable of glucose fermentation, which is used in starch or sucrose based ethanol production [120, 121]. *Zymomonas mobilis*, a gram-negative bacterium are also commonly used to convert C6 sugars. Both, *S. cerevisiae* and Z. mobilis, are adapted to ethanol fermentation, with a high ethanol tolerance, and amenability to genetic modifications. However, both are not capable of fermenting C5 sugars [122]. Since C5 sugars are not naturally fermentable by the most common used yeast (*S. cerevisiae*), a genetic engineered microorganisms are used to convert xylose, beyond glucose [123]. Other options of engineered microorganisms include *Escherichia coli* [124] and *Zymomonas mobilis* [125]. Comparative performance trials on glucose showed that *Z. mobilis* can achieve 5 % higher bioethanol yields and up to 5-fold higher bioethanol volumetric productivity compared to traditional *S. cerevisiae* yeast [126].

Naturally fermenting yeasts are also an option for fermentation of pentoses, *Pichia stipitis* is able to ferment a wide range of sugars like glucose, xylose, mannose, galactose and cellobiose, has a high ethanol yield and relatively little xylitol [127, 128]. *Candida shehatae* also offers a good potential, despite a low tolerance to ethanol, a low ethanol yield and inactivity at low pH. Thermophilic bateria, e.g. *Thermoanaerobacterium sacchaarolyticum, Thermoanaerobacter ethanolicus, Clostridium thermocellum*, are extreme anaerobic bacteria, resistant to an extremely high temperature. These bacteria can ferment a variety of sugars, display cellulolytic activity, but exhibit a low tolerance to ethanol [122]. Bacteria, like *Z. mobilis, E. coli, Klebsiella oxytoca* have attracted attention due to their rapid fermentation capacity, which can be minutes, compared to hours for yeasts [129].

11.6 Third generation bioethanol

The major discussion over biofuels production has been focused in the problems associated to the use of higher plants as its source, like the reduction of ecosystems and the increase in the food prices [130]. Moreover, the recalcitrant nature of plant biomass imposes some challenges to the utilization of this material to the production of bioethanol [131]. In this context, another potential source of production of biofuel to replace petroleum is using aquatic microbial oxygenic photoautotrophs (AMOPs), like cyanobacteria, algae and diatoms for this propose [132, 133]. The application of marine biomass as an energy source was investigate in USA and Japan in 1970s after the oil crises, but when the prices stabilized, the studies were discontinued [134].

Marine biomass significantly differs from terrestrial biomass in terms of their chemical composition. Some features of algaes and cyanobacterias have been attracting attention for the production of third generation bioethanol, among them:

1. AMOPs are buoyant aquatic microcells and for this reason structural biopolymers are not necessary. So, they present low percentage of lignin and hemicellulose compared to lignocellulosic plants, which reduces their recalcitrance [132, 133]. Due to the reduced amount of lignin and hemicellulose, enzymatic hydrolysis of the biomass do not require pretreatment and bypass the expensive pretreatment processes required for hydrolysis of higher plants [133];
2. This kind of microorganism presents rapid growth [131] and have high productivity (in comparison to typical terrestrial crops), therefore the biomass cultivation for commercial production is efficient [130];
3. They can be cultivated in areas that do not present the competition with the production of feed and food [131];
4. They do not need freshwater for their cultivation [135];
5. They use a broad variety of water sources (fresh, seawater, wastewater, brackish) [130, 134], which allows the selection of species adapted for cultivation in locally available aquifers, or may even possess anaerobic metabolism aiming the production of hydrogen, ethanol and/or organic acids by autofermentation [133];
6. They are capable of execute photosynthesis using water as electron donor [130];
7. AMOPs are non-food based feedstocks [130]. The only exception is macroalgae which are used in a few East Asia countries for food, hydrocolloids, fertilizer, and animal feed, so it presents a low risk of competition for food in this countries [135];
8. They produce co-products of high value beyond biofuels [130];
9. Seaweeds do not require an internal transport system of nutrient and water, which saves energy and enhances the mass productivity in relation to land plants [136].

11.6.1 Algae

Based on morphology and size, algae can be classified as microalgae and macroalgae. Microalgae are microscopic organisms and macroalgae are composed of multi-cellular plant like structure. Macroalgae is composed of high amount of sugars (at least 50 %) that can be used in fermentation for bioethanol production [136]. Chemical composition beyond pigment and growth are affected by their habitats conditions such as light, temperature, salinity, nutrient, pollution, and even water motion [135].

The main carbohydrates contained in macroalgae are mannan; ulvan a polysaccharide mainly built on disaccharides repeating sequences composed of sulfated rhamnose and glucuronic acid, iduronic acid or xylose [137]; agar, which is made up of alternating β-D-galactose and α-L-galactose with scarce sulfations [135]; carrageenan, that consists of repeating D-galactose unit and anhydrogalactose, which may or may not be sulfated [135] and starch [135]. This carbohydrates are not present in microalgae [135]. As mentioned before, macroalgae almost do not contain lignin because they float in the water, so the cell wall of this organisms are structurally flexible [132, 135]. Microalgae are composed of starch and monosaccharides [135]. The residual algal biomass generated in the lipid extraction for biodiesel can be appropriately utilized for the production of bioethanol or biomethane [138].

Green algae such as *Spirogyra* and *Chlorococum* are microalgaes that have being shown to accumulate high amounts of polysaccharides which make them good candidates for bioethanol production [132]. Moreover, saccharification of red algae *Gelidium elegans* consisting of acid and enzymatic hydrolysis that converted galactan and glucan to galactose and glucose, increased the ethanol concentration (5.5 %) which is higher than economically feasible concentration (4–5 %) for distillation (Lobban and Wynne, 1981, cited by Jung et al [135]).

11.6.2 Cyanobacteria

Concerning to biomass utilization, cyanobacteria presents some convenience in relation to eukaryotic algae. Cyanobacterial cell wall contains peptidoglycan layer that more closely resembles that of Gram-positive bacteria, consequently it can be degraded by lysozyme [131, 139]. Its cell wall is less complex and less diverse than that of algaes. Moreover, cyanobacteria possess glycogen as a storage carbohydrate which is preferred over starch as a fermentation feedstock due to the fact that in vivo glycogen mobilization by heating and enzymatic treatment requires less energy than starch mobilization [131]. Furthermore, cyanobacteria presents superior photosynthesis ability, converting up to 10 % of the sun's energy into biomass, which is much more than 5 % converted by algaes and 1 % archived by conventional crops like corn and sugarcane [130].

11.6.3 Bioethanol production

Polysaccharides can be converted to simple sugars and then fermented to produce bioethanol. Marine biomass contains unique different carbohydrates from those observed in terrestrial biomass, so the technology applied for terrestrial biomass cannot be applied to marine biomass. Thus, the conversion to sugars and the choice of the appropriate microorganism is important for successful sugar fermentation [135].

Hydrolysis of marine biomass has been performed either with physicochemical methods or enzymatic methods. In the acid hydrolysis, acid concentration and hydrolysis time influence the total yield of reducing sugars and consequently the ethanol production. Enzymatic hydrolysis is a soft approach to the saccharification of marine biomass. However, since marine organisms contain several types of polysaccharides, enzymatic hydrolysis of such materials for ethanol production has still been underdeveloped [135]. Chemical and enzymatic hydrolyses have been combined to more effectively obtain mono-sugars from seaweeds [140, 141]. Ge et al., [140] demonstrated that acid hydrolysis (dilute sulfuric acid pretreatment) expanded the reaction surface area and for this reason, the enzymatic hydrolysis (cellulases and cellobiases) were increased. The fermentation by *S. cerevisiae* resulted in a 41.2 % of ethanol conversion which corresponds to 80.8 % of the theoretical yield.

Pretreatment is used in an attempt to solubilize some polysaccharides, as laminarin. However pretreatment can inhibit microbial fermentation and for this reason a detoxification step is required to effective utilization of this biomass [135]. Heat and pH pretreatments reduced bioethanol yields from the brown *algae Saccharina latissima* [142].

Besides pretreatment, contamination is also a problem for the cultivation of AMOPs. NREL (National Renewable Energy Laboratory) showed that non-native AMOPs species were contaminated by native AMOP species, creating a serious problem facing large-scale cultivation in open reactors [143]. Some improvements suggested include the selection of robust AMOPs that can be cultivated in environments were contamination is lower (high salt concentrations and high pH) [133].

Algal and cyanobacterial presents advantages over traditional energy crops. Nevertheless, to make its production economically feasible, environmental and economical obstacles must be overcome, with the advent of new technologies [130]. Improvements in efficiency, cost structure and ability to scale up AMOPs growth and biofuels production must be done to produce economically viable biofuels. For this purpose a defined set of new technology breakthroughs will be necessary to develop the optimum utilization of algal biomass for the commercial production of biofuel [138]. The main goal is to enhance the productivity of AMOPs to supply the demand of a rapid growing market. The inherent capacity of those organisms can be improved by genetic and metabolic engineering. The sequencing of AMOPs genomes brings facilities to cloning and manipulation of genes [144].

Metabolic engineered algae for production of biofuel is considered as fourth generation of biofuel and has great potential in providing sustainable and clean energy [145]. This new technology increases carbon entrapment ability, cultivation, harvesting, fermentation or oil extraction transesterification, or thermochemical process. This technique presents the advantages of improved yield with high lipid containing algae and more CO_2 capture ability, beyond high production rate. These crops are genetically engineered to consume more amounts of CO_2 from the atmosphere than they will produce later during combustion. Some fourth generation technology pathways include: pyrolysis, gasification, upgrading, solar-to-fuel, and genetic manipulation of organisms to secrete hydrocarbons [146]. However, the initial investment for four generation bioethanol is still high [147]. Figure 11.2 (a, b) summarizes the main differences and similarities among first to fourth generation bioethanol.

11.7 Conclusion and perspectives

Over the past one hundred years with the world's population growth and industrial development, the annual global primary energy consumption has increased nearly 10 times [5]. A wide range of microorganisms are currently used for biofuel production. Utilization of biofuels reduces the GHG emission which minimizes the greenhouse effect. Furthermore, crops used for production of bioethanol of second, third and fourth generations do not have the issue of competition with food crops.

Currently, a broad range of fungi and bacteria such as *Trichoderma* spp., *Aspergillus* spp., *Penicillium* spp., *Bacillus* spp., *Thermoanaerobacterium* spp., *Thermoanaerobacter* spp., and *Clostridium* spp., among others, have been used to enzymes production aiming to degrade lignocellulosic feedstock to bioethanol production. Traditionally, fermentation has been performed by *S. cerevisiae*, however, new possibilities have arisen for fermentation of second and third bioethanol like *Z. mobilis, P. stipilis, C. shehatae*.

Over the next years biofuel technologies are expected to change dramatically, with increasingly sophisticated industrial processing and a much more biotechnology-based agronomy [58]. In future, biofuels will present an important part in fulfilling the world's energy demands. To meet the expanding energy requirement, an uninterrupted energy supply is necessary [147]. In this context, lignocellulosic raw materials play a significant role, since they are cheap and abundant.

In terms of economic performance, today, first generation biofuel is the most cost effective fuel, however its production is limited to certain countries only as its production is highly land intensive. On the other hand, second and third generation biofuel presents elevated costs, mainly because of the high investment prices and low conversion efficiencies of feedstock into biofuel [147]. In this scenario, is necessary to improve the processes involving the production of second and third generation biofuels, aiming to reduce their costs and enhance their conversion efficiency.

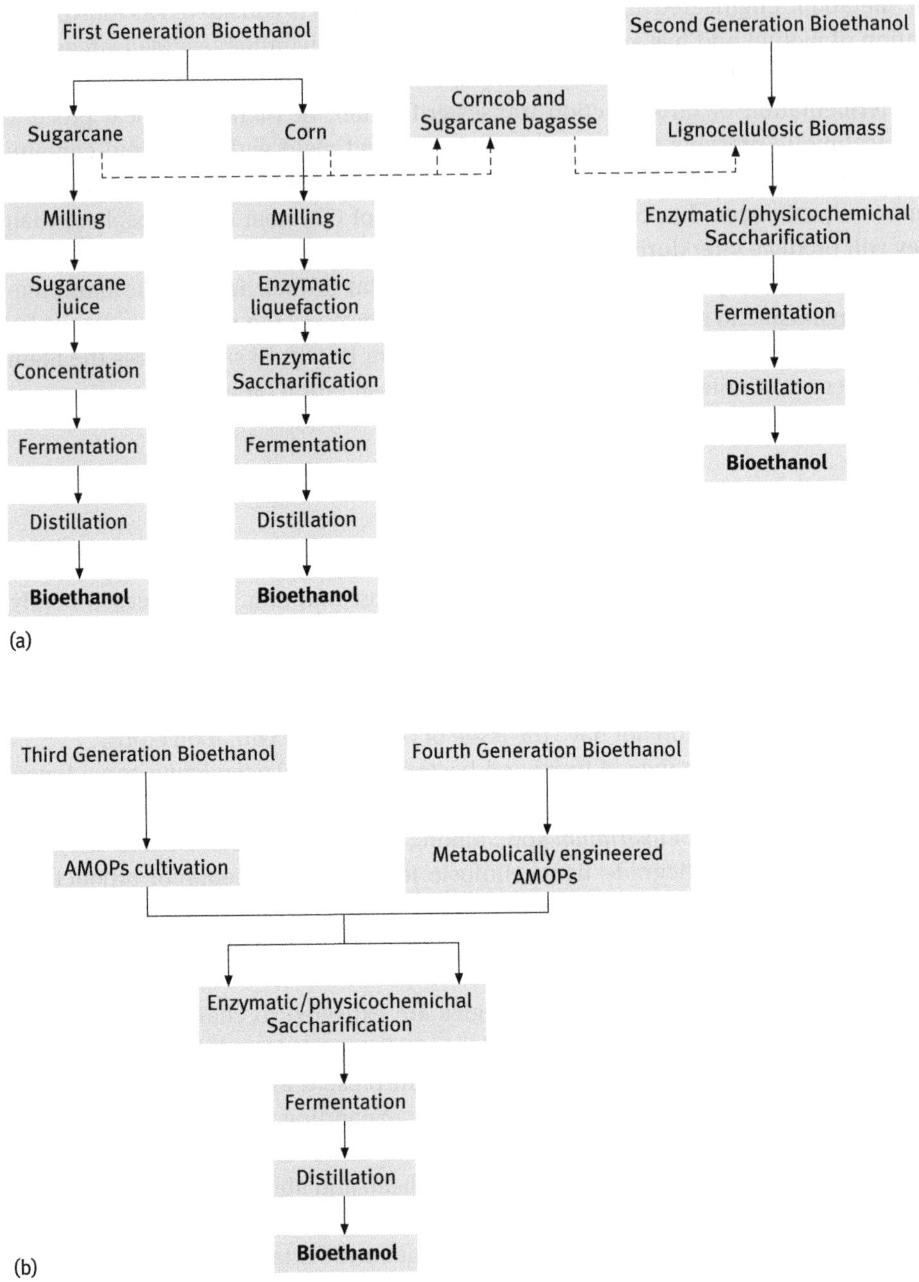

Fig. 11.2: Summary of the pathways for ethanol production from first and second generation (a), and third and fourth generation (b).

Acknowledgements
The authors acknowledge the receipt of financial support from the Brazilian Research Council (CNPq), Coordination for the Improvement of Higher Level Personnel (CAPES), Foundation for Research Support of Federal District (FAPDF) and the National Institute for Science and Technology of Bioethanol.

11.8 References

[1] Uihlein A, Schebek L. Environmental impacts of a lignocellulose feedstock biorefinery system: An assessment. Biomass and Bioenergy 2009,33,793–802.
[2] Chow J, Kopp RJ, Portney PR. Energy Resources and Global Development. Science 2003,302,1528–31.
[3] Ballesteros I, Negro M, Oliva J, Cabañas A, Manzanares P, Ballesteros M. Ethanol production from steam-explosion pretreated wheat straw. Appl Biochem Biotechnol 2006,130,496–508.
[4] Gupta A, Verma JP. Sustainable bio-ethanol production from agro-residues: A review. Renewable and Sustainable Energy Reviews 2015,41,550–67.
[5] Guo M, Song W, Buhain J. Bioenergy and biofuels: History, status, and perspective. Renewable and Sustainable Energy Reviews 2015,42,712–25.
[6] Shaheen M, Choi M, Ang W, et al. Application of low-intensity pulsed ultrasound to increase bio-ethanol production. Renewable Energy 2013,57,462–8.
[7] Karp A, Shield I. Bioenergy from plants and the sustainable yield challenge. New Phytologist 2008,179,15–32.
[8] Goldemberg J. Ethanol for a Sustainable Energy Future. Science 2007,315,808–10.
[9] Goldemberg J, Coelho ST. Renewable energy—traditional biomass vs. modern biomass. Energy Policy 2004,32,711–4.
[10] Ruegg TL, Kim E-M, Simmons BA, et al. An auto-inducible mechanism for ionic liquid resistance in microbial biofuel production. Nature communications 2014,5.
[11] Siqueira FG, Filho EXF. Plant Cell Wall as a substrate for the production of enzymes with industrial applications. Bentham Science Publishers 2010,7,55–60.
[12] Adrio JL, Demain AL. Microbial Enzymes: Tools for Biotechnological Processes. Biomolecules 2014,4,117–39.
[13] Hasunuma T, Okazaki F, Okai N, Hara KY, Ishii J, Kondo A. A review of enzymes and microbes for lignocellulosic biorefinery and the possibility of their application to consolidated bioprocessing technology. Bioresource Technology 2013,135,513–22.
[14] Gusakov AV. Alternatives to Trichoderma reesei in biofuel production. Trends in Biotechnology 2011,29,419–25.
[15] Ahuja SK, Ferreira GM, Moreira AR. Utilization of enzymes for environmental applications. Critical reviews in biotechnology 2004,24,125–54.
[16] 2015 Ethanol industry outlook. 2015. (Accessed 03/01/2015, at http://ethanolrfa.3cdn.net/c5088b8e8e6b427bb3_cwm626ws2.pdf.)
[17] Bhatia L, Johri S, Ahmad R. An economic and ecological perspective of ethanol production from renewable agro waste: a review. AMB Express 2012,2,1–19.
[18] Balat M. Production of bioethanol from lignocellulosic materials via the biochemical pathway: a review. Energy Conversion and Management 2011,52,858–75.
[19] Dias MO, da Cunha MP, Maciel Filho R, Bonomi A, Jesus CD, Rossell CE. Simulation of integrated first and second generation bioethanol production from sugarcane: comparison between

different biomass pretreatment methods. Journal of industrial microbiology & biotechnology 2011,38,955–66.
[20] Cavalett O, Junqueira TL, Dias MO, et al. Environmental and economic assessment of sugarcane first generation biorefineries in Brazil. Clean Technologies and Environmental Policy 2012,14,399–410.
[21] Della-Bianca BE, Basso TO, Stambuk BU, Basso LC, Gombert AK. What do we know about the yeast strains from the Brazilian fuel ethanol industry? Applied microbiology and biotechnology 2013,97,979–91.
[22] Amorim H, Lopes M, de Castro Oliveira J, Buckeridge M, Goldman G. Scientific challenges of bioethanol production in Brazil. Applied microbiology and biotechnology 2011,91,1267–75.
[23] Quintero J, Montoya M, Sánchez O, Giraldo O, Cardona C. Fuel ethanol production from sugarcane and corn: comparative analysis for a Colombian case. Energy 2008,33,385–99.
[24] Junqueira T, Dias M, Jesus CD, et al. Simulation and evaluation of autonomous and annexed sugarcane distilleries. Chemical Engineering Transactions 2011,25,941–6.
[25] Crago CL, Khanna M, Barton J, Giuliani E, Amaral W. Competitiveness of Brazilian sugarcane ethanol compared to US corn ethanol. Energy Policy 2010,38,7404–15.
[26] Kwiatkowski JR, McAloon AJ, Taylor F, Johnston DB. Modeling the process and costs of fuel ethanol production by the corn dry-grind process. Industrial crops and products 2006,23,288–96.
[27] Nigam P, Singh D. Enzyme and microbial systems involved in starch processing. Enzyme and Microbial Technology 1995,17,770–8.
[28] Bothast R, Schlicher M. Biotechnological processes for conversion of corn into ethanol. Applied microbiology and biotechnology 2005,67,19–25.
[29] Wang MQ, Han J, Haq Z, Tyner WE, Wu M, Elgowainy A. Energy and greenhouse gas emission effects of corn and cellulosic ethanol with technology improvements and land use changes. Biomass and Bioenergy 2011,35,1885–96.
[30] Crabb WD, Mitchinson C. Enzymes involved in the processing of starch to sugars. Trends in Biotechnology 1997,15,349–52.
[31] Bon EPS, Gírio F, P. JN. Enzimas na produção de etanol. In: Bon EPS, Ferrara MA, Corvo ML, eds. Enzimas em Biotecnologia: Produção aplicação e mercado. Rio de Janeiro: Editora Interciência, 2008,241–71.
[32] Macedo I, Seabra J, da Silva J. Greenhouse gases emissions in the production and use of ethanol from sugarcane in Brazil: The 2005/2006 averages and a prediction for 2020. Biomass and Bioenergy 2008.
[33] Ag Decision Marker: An Agricutural economics and business website. 2015. (Accessed 03/23/2015, 2015, at https://www.extension.iastate.edu/agdm/crops/outlook/cornbal-ancesheet.pdf.)
[34] Acompanhamento Brasileiro da safra de cana de açúcar. 2014. (Accessed 03/03/2015, at http://www.conab.gov.br/OlalaCMS/uploads/arquivos/14_04_10_09_00_57_boletim_cana_portugues_-_4o_lev_-_13.pdf.)
[35] Farrell AE, Plevin RJ, Turner BT, Jones AD, O'Hare M, Kammen DM. Ethanol Can Contribute to Energy and Environmental Goals. Science 2006,311,506–8.
[36] EPA USEPA-. Lifecycle GHG analysis, in: Renewable Fuel Standard Program (RFS2) Regulatory Impact Analysis . EPA-420-R-10-006. In; 2015:299–515.
[37] Matsuoka S, Ferro J, Arruda P. The Brazilian experience of sugarcane ethanol industry. In: Biofuels: Springer, 2009,157–72.
[38] Oliveira V. Intervencionismo governamental no mercado do etanol. In. Piracicaba: Universidade de São Paulo, escola Superior de Agricultura Luiz de Queiroz, Departamento de Economia,

Adiministração e Sociologia, Grupo de Pesquisa e Extensão em Logística Agroindustrial – ESALQ-LOG; 2014.
[39] Martines-Filho J, Burnquist HL, Vian CE. Bioenergy and the rise of sugarcane-based ethanol in Brazil. Choices 2006,21,91–6.
[40] Furtado AT, Scandiffio MIG, Cortez LAB. The Brazilian sugarcane innovation system. Energy Policy 2011,39,156–66.
[41] Goldemberg J. The Brazilian biofuels industry. Biotechnol Biofuels 2008,1,6.
[42] Xavier MR. The Brazilian sugarcane ethanol experience: Competitive Enterprise Institute Washington, DC, 2007.
[43] Brazilian Automotive Industry Yearbook 2015. (Accessed 03/17/2015, 2015, at http://www.anfavea.com.br/anuario.html.)
[44] Torres D, Martha G, Pray C, et al. The Brazilian ethanol industry: an overview of its production, technology, location, land use, regulations, and future prospects. In: 15th ICABR Conference "SUSTAINABILITY AND THE BIOECONOMY". Rome, Italy; 2011.
[45] República Pd. Lei Nº 8723. In. Brazil; 1993.
[46] Mistura de etanol na gasolina sobe hoje. 2015. (Accessed 03/18/2015, 2015, at http://g1.globo.com/economia/noticia/2015/03/mistura-de-etanol-na-gasolina-sobe-hoje.html)
[47] Songstad DD, Lakshmanan P, Chen J, Gibbons W, Hughes S, Nelson R. Historical perspective of biofuels: learning from the past to rediscover the future. In Vitro CellDevBiol-Plant 2009,45,189–92.
[48] Solomon BD, Barnes JR, Halvorsen KE. Grain and cellulosic ethanol: History, economics, and energy policy. Biomass and Bioenergy 2007, 31, 416–25.
[49] It's still (essentially) your great-grandfather's great-grandfather's ethanol., 2015. (Accessed 03/05/2015, 2015, at http://icminc.com/timeline/start.html.)
[50] Dimitri C, Effland A. Fueling the automobile: an economic exploration of early adoption of gasoline over ethanol. Journal of Agricultural & Food Industrial Organization 2007,5.
[51] Tyner WE. The US ethanol and biofuels boom: Its origins, current status, and future prospects. BioScience 2008,58,646–53.
[52] Spreading the world about clean solutions from sugarcane. 2015. (Accessed 05/03/2015, 2015, at http://sugarcane.org/global-policies/policies-in-the-united-states/us-biofuel-policy.)
[53] Lee S. Ethanol from Corn. In: Lee S, Speight JG, Loyalka SK, eds. Handbook of alternative fuel technologies: crc Press, 2014.
[54] Renewable Fuel Standard (RFS). 2015. (Accessed 03/05/2015, 2015, at
[55] Sorda G, Banse M, Kemfert C. An overview of biofuel policies across the world. Energy Policy 2010,38,6977–88.
[56] Etanol. 2010. (Accessed 13/17/2015, 2015, at http://www.unica.com.br/noticia/2401714692036406485/unica-parabeniza-decisao-do-governo-brasileiro-de-zerar-tarifa-de-importacao-de-etanol/.)
[57] Ethanol Subsidies: Not Gone, Just Hidden a Little Better. 2015. (Accessed 03/05/2015, 2015, at http://www.motherjones.com/kevin-drum/2012/01/ethanol-subsidies-not-gone-just-hidden-little-better.)
[58] Souza A, Grandis A, Leite DC, Buckeridge M. Sugarcane as a Bioenergy Source: History, Performance, and Perspectives for Second-Generation Bioethanol. BioEnergy Research 2014,7,24–35.
[59] Gomez LD, Steele-King CG, McQueen-Mason SJ. Sustainable liquid biofuels from biomass: the writing's on the walls. New Phytologist 2008,178,473–85.
[60] Naik SN, Goud VV, Rout PK, Dalai AK. Production of first and second generation biofuels: A comprehensive review. Renewable and Sustainable Energy Reviews 2010,14,578–97.

[61] Zabaniotou A, Ioannidou O, Skoulou V. Rapeseed residues utilization for energy and 2nd generation biofuels. Fuel 2008,87,1492–502.

[62] Cheng K-K, Cai B-Y, Zhang J-A, et al. Sugarcane bagasse hemicellulose hydrolysate for ethanol production by acid recovery process. Biochemical Engineering Journal 2008,38,105–9.

[63] Kim S, Dale BE. Global potential bioethanol production from wasted crops and crop residues. Biomass and Bioenergy 2004,26,361–75.

[64] Sarkar N, Ghosh SK, Bannerjee S, Aikat K. Bioethanol production from agricultural wastes: An overview. Renewable Energy 2012,37,19–27.

[65] Anwar Z, Gulfraz M, Irshad M. Agro-industrial lignocellulosic biomass a key to unlock the future bio-energy: A brief review. Journal of Radiation Research and Applied Sciences 2014,7,163–73.

[66] Várnai A, Siika-aho M, Viikari L. Restriction of the enzymatic hydrolysis of steam-pretreated spruce by lignin and hemicellulose. Enzyme and Microbial Technology 2010,46,185–93.

[67] Zhao X, Zhang L, Liu D. Biomass recalcitrance. Part I: the chemical compositions and physical structures affecting the enzymatic hydrolysis of lignocellulose. Biofuels, Bioproducts and Biorefining 2012,6,465–82.

[68] Merino S, Cherry J. Progress and Challenges in Enzyme Development for Biomass Utilization. In: Olsson L, ed. Biofuels: Springer Berlin Heidelberg, 2007,95–120.

[69] Duarte G, Moreira LRS, Jaramillo P, Filho E. Biomass-Derived Inhibitors of Holocellulases. BioEnergy Research 2012,5,768–77.

[70] Tejirian A, Xu F. Inhibition of enzymatic cellulolysis by phenolic compounds. Enzyme and Microbial Technology 2011,48,239–47.

[71] Qi B, Chen X, Su Y, Wan Y. Enzyme adsorption and recycling during hydrolysis of wheat straw lignocellulose. Bioresource Technology 2011,102,2881–9.

[72] Zhang Y-HP, Berson E, Sarkanen S, Dale BE. Sessions 3 and 8: pretreatment and biomass recalcitrance: fundamentals and progress. Appl Biochem Biotechnol 2009,153,80–3.

[73] Eggeman T, Elander RT. Process and economic analysis of pretreatment technologies. Bioresource Technology 2005,96,2019–25.

[74] Banerjee G, Scott-Craig JS, Walton JD. Improving enzymes for biomass conversion: a basic research perspective. BioEnergy Research 2010,3,82–92.

[75] Van Dyk J, Pletschke B. A review of lignocellulose bioconversion using enzymatic hydrolysis and synergistic cooperation between enzymes—factors affecting enzymes, conversion and synergy. Biotechnology Advances 2012,30,1458–80.

[76] Mansfield S, Mooney C, Saddler J. Substrate and enzyme characteristics that limit cellulose hydrolysis. Biotechnol Prog 1999,15,804–16.

[77] Merino S, Cherry J. Progress and challenges in enzyme development for biomass utilization. Adv Biochem Eng Biotechnol 2007,108,95–120.

[78] Andreaus J, Filho EXF, Bon EPdS. Biotechnology of Holocellulose-Degrading Enzymes. In: Hou CT, Shaw J, eds. Biocatalysis and Bioenergy: John Wiley & Sons, Inc., 2008,195–229.

[79] Sun Y, Cheng J. Hydrolysis of lignocellulosic materials for ethanol production: a review. Bioresour Technol 2002,83,1–11.

[80] Talebnia F, Karakashev D, Angelidaki I. Production of bioethanol from wheat straw: An overview on pretreatment, hydrolysis and fermentation. Bioresource Technology 2010,101,4744–53.

[81] Gusakov AV, Salanovich TN, Antonov AI, et al. Design of highly efficient cellulase mixtures for enzymatic hydrolysis of cellulose. Biotechnology and Bioengineering 2007,97,1028–38.

[82] Moreira LRS, Milanezi NvG, Filho EXF. Enzymology of Plant Cell Wall Breakdown: An Update. In: Buckeridge MS, Goldman GH, eds. Routes to Cellulosic Ethanol: Springer New York, 2011,73–96.

[83] Eibinger M, Ganner T, Bubner P, et al. Cellulose surface degradation by a lytic polysaccharide monooxygenase and its effect on cellulase hydrolytic efficiency. Journal of Biological Chemistry 2014,289,35929–38.

[84] Juturu V, Wu JC. Microbial cellulases: engineering, production and applications. Renewable and Sustainable Energy Reviews 2014,33,188–203.
[85] Polizeli ML, Rizzatti AC, Monti R, Terenzi HF, Jorge JA, Amorim DS. Xylanases from fungi: properties and industrial applications. Applied microbiology and biotechnology 2005,67,577–91.
[86] Minic Z, Jouanin L. Plant glycoside hydrolases involved in cell wall polysaccharide degradation. Plant Physiology and Biochemistry 2006,44,435–49.
[87] Moreira LRS, Filho EXF. An overview of mannan structure and mannan-degrading enzyme systems. Applied microbiology and biotechnology 2008,79,165–78.
[88] Jayani RS, Saxena S, Gupta R. Microbial pectinolytic enzymes: a review. Process Biochemistry 2005,40,2931–44.
[89] Janusz G, Kucharzyk KH, Pawlik A, Staszczak M, Paszczynski AJ. Fungal laccase, manganese peroxidase and lignin peroxidase: Gene expression and regulation. Enzyme and Microbial Technology 2013,52,1–12.
[90] Crestini C, Melone F, Saladino R. Novel multienzyme oxidative biocatalyst for lignin bioprocessing. Bioorganic & Medicinal Chemistry 2011,19,5071–8.
[91] Knöös P, Schulz C, Piculell L, Ludwig R, Gorton L, Wahlgren M. Quantifying the release of lactose from polymer matrix tablets with an amperometric biosensor utilizing cellobiose dehydrogenase. International Journal of Pharmaceutics 2014,468,121–32.
[92] Carbohydrate-Active Enzyme database. 2015. (Accessed 01/15/2015, at http://www.cazy.org/.)
[93] Langston J, Shaghasi T, Abbate E, Xu F, Vlasenko E, Sweeney M. Oxidoreductive cellulose depolymerization by the enzymes cellobiose dehydrogenase and glycoside hydrolase 61. Applied and environmental microbiology 2011,77,7007–15.
[94] Glass NL, Schmoll M, Cate JHD, Coradetti S. Plant Cell Wall Deconstruction by Ascomycete Fungi. Annual Review of Microbiology 2013,67,477–98.
[95] Edwards D, Eng P. Scaling Up Bioenergy Technologies. CHEMICAL ENGINEERING PROGRESS 2015,111,58–61.
[96] Canadian cellulosic ethanol project attracts government funding. 2015. (Accessed 04/24/2015, 2015, at http://www.ethanolproducer.com/articles/10309/canadian-cellulosic-ethanol-project-attracts-government-funding.)
[97] Cellulosic ethanol: Pros & Cons. 2015. (Accessed 04/24/2015, 2015, at http://greenthefuture.com/CELLETHANOL_PROSCONS.html.)
[98] Dupont biofuels solution. 2015. (Accessed 03/17/2015, 2015, at http://biofuels.dupont.com)
[99] Four Commercial Scale Cellulosic Ethanol Biorefineries to Enter Production This Year. 2015. (Accessed 03/12/2015, 2015, at http://www.fuelsamerica.org/blog/entry/four-commercial-scale-cellulosic-ethanol-biorefineries-to-enter-production.)
[100] Update from DuPont on its Nevada project – labor shortages, new ethanol marketer deal. 2015. (Accessed 03/16/2015, 2015, at http://www.biofuelsdigest.com/bdigest/2014/10/28/update-from-dupont-on-its-nevada-project-labor-shortages-new-ethanol-marketer-deal/.)
[101] Dupont Industrial Biosciences. 2015. (Accessed 03/17/2015, 2015, at http://biosciences.dupont.com/.)
[102] Abengoa. 2015. (Accessed 03/17/2015, 2015, at http://www.abengoa.com/web/en/index3.html.)
[103] POET DSM Advanced Biofuels. 2015. (Accessed 03/17/2015, 2015, at http://poet-dsm.com/.)
[104] New Planet Energy. 2015. (Accessed 03/17/2015, 2015, at http://www.npeholdings.com/NewPlanetEnergy/Welcome.html.)
[105] INEOS. 2015. (Accessed 15/03/2015, 2015, at http://www.ineos.com/.)

[106] Ineos declares commercial cellulosic ethanol online in Florida. 2015. (Accessed 03/17/2015, 2015, at http://www.ethanolproducer.com/articles/10096/ineos-declares-commercial-cellulosic-ethanol-online-in-florida.)
[107] GranBio. 2015. (Accessed 03/11/2015, 2015, at http://www.granbio.com.br/en/conteudos/biofuels/.)
[108] Raizen. 2015. (Accessed 03/11/2015, at http://www.raizen.com/en/.)
[109] IOGEN Corporation. 2015. (Accessed 03/11/2015, 2015, at http://www.iogen.ca/.)
[110] Raizen, Iogen commence cellulosic ethanol production in Brazil. 2015. (Accessed 03/17/2015, 2015, at http://www.biofuelsdigest.com/bdigest/2014/12/17/raizen-iogen-commence-cellulosic-ethanol-production-in-brazil/.)
[111] Crescentino. 2015. (Accessed 03/17/2015, 2015, at http://www.biochemtex.com/references/1/crescentino.)
[112] European Biofuels Technology Plataform - Cellulosic Ethanol 2015. (Accessed 03/11/2015, 2015, at http://www.biofuelstp.eu/cellulosic-ethanol.html.)
[113] Biofuels and biochemicals from agricultural residues and non-food plants. 2015. (Accessed 03/11/2015, 2015, at http://www.betarenewables.com/.)
[114] FROM BIOETHANOL INTO FOSSIL-FREE GREEN CHEMICALS AND BIOFUELS, 2015. (Accessed 03/21/2015, 2015, at http://www.sekab.com/.)
[115] DONG Energy's climate partnerships ensure cost savings and CO_2 reduction. 2015. (Accessed 03/20/2015, 2015, at http://www.dongenergy.com/en.)
[116] Inbicon at the World Bio Markets in Amsterdam. 2015. (Accessed 03/20/2015, 2015, at http://www.inbicon.com/en.)
[117] Kalundborg Bioethanol Demonstration Plant, Denmark. 2015. (Accessed 03/20/2015, 2015, at http://www.chemicals-technology.com/projects/kalundborg_bioethano.)
[118] Babilafuente demonstration plant. 2015. (Accessed 03/22/2015, 2015, at http://www.abengoabioenergy.com/web/en/nuevas_tecnologias/tecnologias/planta_babilafuente.)
[119] Rabinovich ML. Wood hydrolysis industry in the Soviet Union and Russia: A minireview. Cellulose Chemistry & Technology 2010,44,173.
[120] Biswas R, Uellendahl H, Ahring BK. Conversion of C(6) and C(5) sugars in undetoxified wet exploded bagasse hydrolysates using Scheffersomyces (Pichia) stipitis CBS6054. AMB Express 2013,3,42–.
[121] Olofsson K, Bertilsson M, Lidén G. A short review on SSF-an interesting process option for ethanol production from lignocellulosic feedstocks. Biotechnol Biofuels 2008,1,1–14.
[122] Baeyens J, Kang Q, Appels L, Dewil R, Lv Y, Tan T. Challenges and opportunities in improving the production of bio-ethanol. Progress in Energy and combustion science 2015,47,60–88.
[123] Carrasco C, Baudel H, Peñarrieta M, et al. Steam pretreatment and fermentation of the straw material "Paja Brava" using simultaneous saccharification and co-fermentation. Journal of bioscience and bioengineering 2011,111,167–74.
[124] Asghari A, Bothast R, Doran J, Ingram L. Ethanol production from hemicellulose hydrolysates of agricultural residues using genetically engineeredEscherichia coli strain KO11. Journal of Industrial Microbiology 1996,16,42–7.
[125] Zhang M, Eddy C, Deanda K, Finkelstein M, Picataggio S. Metabolic engineering of a pentose metabolism pathway in ethanologenic Zymomonas mobilis. Science 1995,267,240–3.
[126] Sáez-Miranda JC, Saliceti-Piazza L, McMillan JD. Measurement and analysis of intracellular ATP levels in metabolically engineered Zymomonas mobilis fermenting glucose and xylose mixtures. Biotechnology progress 2006,22,359–68.
[127] Carrasco C, Baudel H, Roslander C, Galbe M, Lidén G. Fermentation of the Straw Material Paja Brava by the Yeast Pichia stipitis in a Simultaneous Saccharification and Fermentation Process. 2013.

[128] Agbogbo FK, Coward-Kelly G. Cellulosic ethanol production using the naturally occurring xylose-fermenting yeast, Pichia stipitis. Biotechnology Letters 2008,30,1515–24.
[129] Hayes DJ. An examination of biorefining processes, catalysts and challenges. Catalysis Today 2009,145,138–51.
[130] Parmar A, Singh NK, Pandey A, Gnansounou E, Madamwar D. Cyanobacteria and microalgae: a positive prospect for biofuels. Bioresource Technology 2011,102,10163–72.
[131] Möllers KB, Cannella D, Jørgensen H, Frigaard N-U. Cyanobacterial biomass as carbohydrate and nutrient feedstock for bioethanol production by yeast fermentation. Biotechnol Biofuels 2014,7,64.
[132] Jones CS, Mayfield SP. Algae biofuels: versatility for the future of bioenergy. Current Opinion in Biotechnology 2012,23,346–51.
[133] Dismukes GC, Carrieri D, Bennette N, Ananyev GM, Posewitz MC. Aquatic phototrophs: efficient alternatives to land-based crops for biofuels. Current Opinion in Biotechnology 2008,19,235–40.
[134] Kumar S, Gupta R, Kumar G, Sahoo D, Kuhad RC. Bioethanol production from Gracilaria verrucosa, a red alga, in a biorefinery approach. Bioresource Technology 2013,135,150–6.
[135] Jung KA, Lim S-R, Kim Y, Park JM. Potentials of macroalgae as feedstocks for biorefinery. Bioresource Technology 2013,135,182–90.
[136] Wi SG, Kim HJ, Mahadevan SA, Yang D-J, Bae H-J. The potential value of the seaweed Ceylon moss (Gelidium amansii) as an alternative bioenergy resource. Bioresource Technology 2009,100,6658–60.
[137] Robic A, Rondeau-Mouro C, Sassi JF, Lerat Y, Lahaye M. Structure and interactions of ulvan in the cell wall of the marine green algae Ulva rotundata (Ulvales, Chlorophyceae). Carbohydrate Polymers 2009,77,206–16.
[138] Singh A, Olsen SI. A critical review of biochemical conversion, sustainability and life cycle assessment of algal biofuels. Applied Energy 2011,88,3548–55.
[139] Hoiczyk E, Hansel A. Cyanobacterial Cell Walls: News from an Unusual Prokaryotic Envelope. Journal of Bacteriology 2000,182,1191–9.
[140] Ge L, Wang P, Mou H. Study on saccharification techniques of seaweed wastes for the transformation of ethanol. Renewable Energy 2011,36,84–9.
[141] Jang J-S, Cho Y, Jeong G-T, Kim S-K. Optimization of saccharification and ethanol production by simultaneous saccharification and fermentation (SSF) from seaweed, Saccharina japonica. Bioprocess and biosystems engineering 2012,35,11–8.
[142] Adams JM, Gallagher JA, Donnison IS. Fermentation study on Saccharina latissima for bioethanol production considering variable pre-treatments. Journal of applied Phycology 2009,21,569–74.
[143] Sheehan J, Dunahay T, Benemann J, Roessler P. A look back at the US Department of Energy's aquatic species program: biodiesel from algae: National Renewable Energy Laboratory Golden, 1998.
[144] Rosenberg JN, Oyler GA, Wilkinson L, Betenbaugh MJ. A green light for engineered algae: redirecting metabolism to fuel a biotechnology revolution. Current Opinion in Biotechnology 2008,19,430–6.
[145] Lu J, Sheahan C, Fu P. Metabolic engineering of algae for fourth generation biofuels production. Energy & Environmental Science 2011,4,2451–66.
[146] Demirbas MF. Biofuels from algae for sustainable development. Applied Energy 2011,88,3473–80.
[147] Dutta K, Daverey A, Lin J-G. Evolution retrospective for alternative fuels: First to fourth generation. Renewable Energy 2014,69,114–22.

Shafiquzzaman Siddiquee

12 Fungal cell factories and their applications

12.1 Introduction

Fungal cell factories are widely exploited for the controlled production of substances for food, agriculture, pharmaceutical, enzyme and industrial biotechnology. Recent increasing food and energy demand, the worldwide warning of macro-and micro-climate changes are the major challenges for humans. Biofuels are the main alternative sources of energy; its have unlimited potential to harmonize the food–energy–environment trilemma (Tilman et al., 2009; Harvey and Pilgrim, 2010). Photosynthetic organisms including plants, algae, and cyanobacteria have ability to capture solar energy and store it as chemical energy of their biomass (bioenergy). Thus, agriculture might serve as a source of food and bioenergy. Bioenergy mostly occurred by photosynthetic CO_2 fixation; therefore, carbon-containing biofuels have diverse affinity to neutralize carbon after combustion.

However, human-controlled fungal production and transformation are much older; the cell factory concept was fully established in the 80's through the intensive public and private investment. The promising DNA recombinant technologies are strongly recommended and supported by the approval of recombinant insulin (ANONIMOUS, 1982; Johnson 1983). The principle of controlled biological production as convenient source of difficult-to-obtain molecules (especially those of high added value) intensely penetrated the industrial tissue, so it become a widespread platform targeting at cost effective for large-scale industrially production (Marston, 1986). First generation cell factories (composed by the bacterium *Escherichia coli* and the yeast *Saccharomyces cerevisae*) are replaced by engineered variants.

Based on the biotechnology aspect, enzymes production from fungi should develop when as usually apply their traditional technique. This has led to correctly identify the potential strains and to allow them large amounts of the production of variety of industrial enzymes. The developed molecular genetic approaches have used in bacteria such as *Escherichia* coli and *Bacillus* spp, and afterwards in fungi; have recognized a new stepping stone towards the uses of fungi in protein production.

In recent industrial development, fungal cell factories have used for the expression of specific-target gene products which have high protein-secreting mutant made by traditional random mutagenesis, afterwards the characteristics have modified further by genetic engineering (Lee et al., 2012). In an overall comparison with other available expression systems for heterologous proteins, filamentous fungi always performed very well and produced high yield and relatively cheap when using in liquid culture in Table 12.1.

Tab. 12.1: Salient characteristics of liquid culture-based expression systems used for the production of heterologous proteins (Nevalainen et al., 2005).

Characteristics	*Escherichia coli*	Filamentous fungi	Yeast	Insect cells	Mammalian cells	Plant cell culture
Cell growth[a]	Hours to days	Days to 1 week	Days to 1 week	Days to 1 week	Weeks	Months
Cost of growth medium	Low to medium	Low to medium	Low to medium	High	High	Medium to High
Expression level	Low to high	Low to high	Low to high	Low to high	Low to high	Low
Secretion capability	Secretion to periplasm	Secretion to medium	Secretion to Medium	Secretion to medium	Secretion to medium	Secretion to medium
Post-translational modifications						
Protein folding	Refolding usually required	Refolding might be Required	Refolding might be required	Proper folding	Proper folding	Proper folding
N-linked glycosylation	None	Mammalian-type core, no sialic acid, non-human sugars added	High mannose, no sialic acid, non-human sugars added	Complex, no sialic acid, non-human sugars added	Complex, non-human sugars added, e.g. by murine cells	Complex, no sialic acid, non-human sugars added
O-linked glycosylation, phosphorylation, acetylation, acylation	No	Yes	Yes	Yes	Yes	Yes

[a]Cultivation time to product harvesting.

Filamentous fungi are dominated in the markets when production hosts asexually reproducing *Aspergillus niger*, *A. oryzae*, *Trichoderma reesei*, *T. harzianum* and so on. Subsequently, the maximum information on heterologous protein expression has received when using specific fungi and genetically well-characterized of *Aspergillus nidulans*. Fungi could prove to be valuable as alternative eukaryotic hosts for foreign gene expression. In contrast to *S. cerevisiae*, many species particularly the genus of *Aspergillus*, can secrete copious quantities of wide range of enzymes. In addition, they have several numbers of other attractive features (shared with yeast) as production of hosts (Table 12.2).

Tab. 12.2: Advantage of fungi as hosts (Saunders *et al.*, 1989).

- Naturally secrete a wide range of proteins
- Glycosylate in a similar fashion to higher eukaryotes
- Many species generally regarded as safe (GRAS) by regulatory authorities
- Extensive fermentation and down-stream processing experience already established

Fungi have included a diverse group of eukaryotes with compact genomes and widely distributed in nature. Many species are directly relevant to humans and human activities, with some species are playing important roles in industry, agriculture, and medicine, and also involved in the production of primary and secondary metabolites, including organic acids, enzymes, exopolysaccharides, antibiotics, and biofilms for industrial uses (Siddiquee et al., 2012; Gutiérrez-Correa et al., 2012; Meyer, 2008).

Fungi have ability to grow on rather simple and inexpensive substrates as well as their capacity to produce wide range of commercially potential metabolites that are significantly involved in biotechnology industry. To date, fungi are broadly applied in biotechnology as cell factories for large scale of products. Diverse compounds ranging are produced from simple organic acids to complex secondary metabolites for the use in several market segments (Table 12.3).

Tab. 12.3: Diverse compounds for the use in various market segments. Sources: Bennett, 1998; Archer, 2000; Ooi and Liu, 2000; Willke and Vorlop, 2001; Adrio and Demain, 2003; Leathers, 2003; Linder et al., 2005; Polizeli et al., 2005; Olempska-Beer et al., 2006). PUFA: Polyunsaturated fatty acids.

Compound	Organism	Main application areas
Acids		
Citric acid	*Aspergillus niger*	Food and beverage industry
Itaconic acid	*A. terreus*	Polymer industry
Kojic acid	*A. oryzae*	Food industry
Enzymes		
α-Amylase	*A. niger, A. oryzae*	Starch processing and food industry
Chymosin	*A. niger*	Food industry
Cellulase	*Trichoderma viride, T. reesei*	Textile, pulp and paper industry
Cellobiohydrolase	*T. viride, T. reesei*	Textile, pulp and paper Industry

Tab. 12.3: (continued)

Compound	Organism	Main application areas
Glucoamylase	*A. phoenicis, Rhizopus delemar*	Starch processing industry
Glucose oxidase	*A. niger, A. oryzae*	Textile industry, Biosensor
Laccase	*Trametes versicolor*	Textile, pulp and paper industry
Lipases	*A. niger, A. oryzae*	Food and detergent industry
Pectin lyase	*T. reesei*	Food industry
Proteases	*A. niger, A. oryzae, R. delemar*	Food and detergent industry
Phytase	*A. niger, A. oryzae*	Food industry
Rennin	*Mucor miehei*	Food industry
Xylanases	*T. reesei, T. konignii, A. niger*	Textile, pulp, paper and bakery industry
Exopolysaccharides		
Scleroglucan	*Sclerotium rolfsii*	Oil production, cosmetics industry
Pullulan	*Aureobasidium pullulans*	Food and pharmaceutical industry
Schizophyllan	*Schizophyllum commune*	Pharmaceutical industry
PSK, PSP	*Tr. versicolor*	Pharmaceutical industry
Secondary metabolites		
Cephalosporin	*Acremonium chrysogenum*	Pharmaceutical industry
Cyclosporin	*Tolypocladium nivenum*	Pharmaceutical industry
Ergot alkaloids	*Claviceps purpurea*	Pharmaceutical industry
Griseofulvin	*P. griseofulvum*	Pharmaceutical industry
Lovastatin	*Monascus rubber, A. terreus*	Pharmaceutical industry
Penicillin	*P. chrysogenum*	Pharmaceutical industry
Taxol	*Taxomyces andrenae*	Pharmaceutical industry
Zeranol	*Fusarium graminearum*	Livestock farming
Others		
PUFA	*Mucor circinelloides*	Food, feed and pharmaceutical industry
Panthothenic acid	*F. oxysporum*	Food, feed and pharmaceutical industry
Hydrophobin	*T. reesei*	Tissue engineering, nanotechnology
Biomass	*Agaricus bisporus,*	Food industry
	F. venentatum (Quorn™)	Food industry

The enzymes productions from fungal and non-fungal origin have outstanding ability to express and secrete proteins. In order to native or recombinant enzymes are commonly produced by *Aspergillus niger*, *A. oryzae* and *Trichoderma reesei* and also other strains are currently under development (Punt et al., 2002). In fungi are naturally produced the surprising wealth of secondary metabolites and few of them are applied as biotechnologically and clinically. The β-lactam group of antibiotics, including penicillin and cephalosporin, is the first group that developed by molecular techniques (Brakhage and Caruso, 2004). Moreover, bioactive compounds have produced by microbes that are more important for human welfare, for example, cyclosporine A (immunosuppressive agent), lovastatin (cholesterol-lowering agent), taxol (anticancer agent) and griseofulvin (antifungal agent).

Other fungal metabolites already produced commercially or potentially valuable in biotechnology involving exopolysaccharides such as pullulan suitable for coating foods or for improved delivery of therapeutic agents (Leathers, 2003) and hydrophobins useful as surface modifiers for (nano)technical and medical applications due to their ability to self-assemble at hydrophilic–hydrophobic interfaces into amphipathic films (Scholtmeijer *et al.*, 2001; Hektor and Scholtmeijer, 2005). In addition, several species have become model organisms for understanding basic biological processes. The objective of this book chapter is to summarize and discuss the currently available options for genetic manipulation of potential fungal strains, with special attention to those aspects relevant for the biotechnological applications such as food, agriculture, medicine, enzymes and bioenergy.

12.2 Genetic transformation of the potential specific gene

Genetic engineering is a powerful approach for fungi in order to increase productivity and to minimize the unwanted by-product formation. However, before engineering become a routine work for introducing the wanted genetic manipulation of the fungus, often represents a challenge. Foremost, the developed transformation method is not trivial for many fungi. Besides, the mode and frequency of individual integration events resulting from homologous or illegitimate recombination is not only dependent on the transformation host itself but also on the applied transformation technique (Meyer, 2008). However, an engineering design firstly required the consideration of the most suitable transformation method.

Protoplast-mediated transformation (PMT) of the yeast *Saccharomyces cerevisiae* is the first reported by Hinnen *et al.* (1978), the use of protoplasts for transformation has comprehensively applied to several fungal species (Fincham, 1989; Ruiz-Diez, 2002). However, the frequency of transformation is very low as comparing to yields obtained with *S. cerevisiae*. In order to improve transformation of fungi species strains, the progress made over the past years that resulted in the establishment of alternative methods for fungal transformation such as electroporation, biolistic transformation (Ruiz-Diez, 2002) and Agrobacteriummediated transformation (AMT) (Michielse et al., 2005b). These methods have proven to be valuable for fungal strains that do not form sufficient numbers of protoplasts or whose protoplasts do not regenerate sufficiently (MacKenzie et al., 2004). In Table 12.4 summarizes main characteristics as well as advantages and disadvantages of these transformation techniques.

Tab. 12.4: Four common strategies for transformation of fungi. Source: Fincham, 1989; Ruiz-Diez, 2002; Michielse *et al.*, 2005b.

Method	Principle	Advantage	Disadvantage
PMT	Preparation of protoplasts using cell wall-degrading enzymes	Different cell types (spores, germlings, hyphal tissue) can be used	Particular batch of lytic enzyme alters TR
	Uptake of DNA is achieved by the addition of PEG and CaCl2		Requires regeneration procedure copy number of DNA insertions is often high
AMT	*A. tumefaciens* carries two vectors (the binary vector containing the DNA of interest between the 24 bp border repeat and the T-vector containing the virulence region important for DNA transfer)	Different cell types (spores, germlings, hyphal tissue) can be used Copy number of DNA insertions is low	Various parameters during co-cultivation affect TR
	DNA transfer is achieved during co-cultivation of *A. tumefaciens* with the fungus	Improves targeted integration	More time-consuming than the other methods
EP	Reversible membrane permeabilisation induced by local application of electric pulses mediates DNA uptake	Different cell types (spores, germlings) can be used Simple and cheap method	Often requires protoplast formation to render cells competent
BT	Particles (tungsten, gold) are coated with DNA and become accelerated at high velocity into cells	Recipient cells can retain their cell walls (no pre-treatment necessary)	Requires special equipment

PMT: Protoplast-mediated transformation; AMT: Agrobacterium-mediated transformation; EP: Electroporation; BT: Biolistic transformation; TR: transformation rate.

All four techniques are commonly essential to optimize every method for the interest fungal strains, only one or two of these methods can be applied to a at varying particular fungal species (Meyer *et al.*, 2003). Importantly, AMT is an efficient transformation method for some fungi that are recalcitrant to the other methods previously described by Michielse *et al.* (2005b). However, AMT has reported to be less successful or even fails to produce transformants, e.g. in *A. niger* (Michielse *et al.*, 2005b). Thus, no general rule can be applied to predict the usefulness of the particular transformation technique for ungal strains. Individual species have to be considered independently and identify the most appropriate method for each fungal strains of interest.

Single-copy integration events are discovered when AMT is used for the introduction of DNA into fungi such as *A. awamori, A. giganteus, Calonectria morganii, Fusarium oxysporum* and *Suillus bovines*. In contrast, when these fungi are transformed by PMT, preferentially multicopy integration events are observed (de Groot *et al.*, 1998; Malonek and Meinhardt, 2001; Mullins and Kang, 2001; Meyer *et al.*, 2003). The impact of transformation technique on the transforming DNA is significantly influence on the design of metabolic engineering strategy. For example, when targeted integration or gene deletion is envisaged, AMT would be the method of choice.

PMT could be the optimal method when multiple copies of gene of interest can integrate at random sites in the genome. It has been reported that such strategies (multiple integration of a gene encoding the desired protein) have improved the protein production in *Aspergillus* and *Trichoderma* (Verdoes *et al.*, 1995; Lee *et al.*, 1998; Askolin *et al.*, 2001).

Remarkably, AMT has showed the improvement of homologous recombination (HR) frequency in *A. awamori* (Michielse *et al.*, 2005a). As discussed by Michielse *et al.* (2005b), one possible explanation for the increase in HR frequency obtained by AMT is the fact that *A. tumefaciens* delivers its DNA to the host as single-stranded DNA (ssDNA), whereas double-stranded DNA (dsDNA) has is transferred in the case of the other transformation methods. In *S. cerevisiae*, it has showed the ssDNA transforms cells at greater HR efficiency than that of the dsDNA, thus suggesting that the recombination machinery has a preferential affinity for the ssDNA (Simon and Moore, 1987).

12.3 Re-engineering gene targeting

Targeted integration of genes to a genomic locus known to improve expression of the protein of interest and thereby to maximize the ability of the particular metabolic process. In addition, deletion of genes is often envisaged as the strategy for metabolic engineering due to reduce metabolic fluxes of side pathways and to redirect fluxes to the product-forming pathway. However, gene targeting in fungi is often hampered by very low frequencies of HR and it is not trivial to target gene of interest to the particular genomic locus or to delete an endogenous gene. While in *S. cerevisiae* a minimal homologous fragment length of 30–50 bp is sufficient to ensure high yield of HR (Hua *et al.*, 1997), several hundred base-pairs up to several kilo-bases are necessary to achieve HR in fungi (Chevalet *et al.*, 1992; Wu and Linz, 1993; Shiotani and Tsuge, 1995; Weidner *et al.*, 1998; Young *et al.*, 1998; Fernandez-Martin *et al.*, 2000;

Yu *et al.*, 2004; Meyer *et al.*, 2007). The frequency of HR in fungi is very low (usually between 0 and 30 % (Asch and Kinsey, 1990; Bird and Bradshaw, 1997; Meyer *et al.*, 2002; Meyer *et al.*, 2007), is significantly species-dependent (Kwon-Chung *et al.*, 1998) and its occurrence is strongly dependent on the transcriptional status of the targeted genomic locus (Bird and Bradshaw, 1997), multitude of transformants have commonly analyzed to identify the correct one, this process which makes fungal transformation tedious and laborious.

The above limitation is possible to overcome the use of specific fungal strains which are defective for the nonhomologous end joining pathway (NHEJ). In eukaryotes, integration of the DNA fragment into the genome requires the action of a double-strand break repair mechanism. The two major pathways, HR and NHEJ, are conserved through evolution and have described to mediate double-strand break repair (Dudasova *et al.*, 2004; Krogh and Symington, 2004). HR involves the interaction between homologous sequences and thus leads to the targeted integration. In contrast, NHEJ mediates ligation of DNA strands sharing no homology and hence results in random integration. The HR pathway depends on the Rad52 epistasis group, whereas, the NHEJ pathway depends on the Ku heterodimer (Ku70/Ku80-protein complex) and the DNA ligase IV-Xrcc4 complex (Dudasova *et al.*, 2004; Krogh and Symington, 2004).

According to the 'gatekeeper' model (Haber, 1999), combine pathways always compete with each other. When Rad52p binds to the ends of the introduced DNA, the DNA ends can be processed by HR, however, when Ku binds, the DNA becomes integrated via the NHEJ pathway. In fungi, like in other multicellular organisms and in contrast to *S. cerevisiae*, the NHEJ-pathway seems to be dominant over the HR-pathway. Studies in non-*Saccharomyces* yeasts and fungi have disclosed that by deleting components of the NHEJ-pathway, the random integration of DNA fragments are strongly reduced. For example, deletion of the *Ku70* homologue in *A. niger* (*KusA*) intensely improved homologous integration efficiency and reached more than 80 % when compared to 7 % in the wild-type background, while only 500 bp homologous flanks are used (Fig. 12.1).

In similar, mutants deficient for Ku70, Ku80 or Lig4 homologues gave high HR frequencies in other fungal systems (Table 12.5) such as *A. nidulans* (Nayak *et al.*, 2006), *A. oryzae* (Takahashi *et al.*, 2006), *A. fumigatus* (da Silva Ferreira *et al.*, 2006; Krappmann *et al.*, 2006), *A. sojae* (Takahashi *et al.*, 2006), *Claviceps purpurea* (Haarmann *et al.*, 2008), *Cryptococcus neoformans* (Goins *et al.*, 2006), *Kluyveromyces lactis* (Kooistra *et al.*, 2004), *Magnaporthe grisea* (Villalba *et al.*, 2008), *Neurospora crassa* (Ninomiya *et al.*, 2004; Ishibashi *et al.*, 2006) and *Sordaria macrospora* (Pöggeler and Kück, 2006). Hence, mutants defective for NHEJ offer the opportunity to make targeted genetic alterations in fungi more straightforward and less time-consuming and will serve as a powerful tool for genetic engineering.

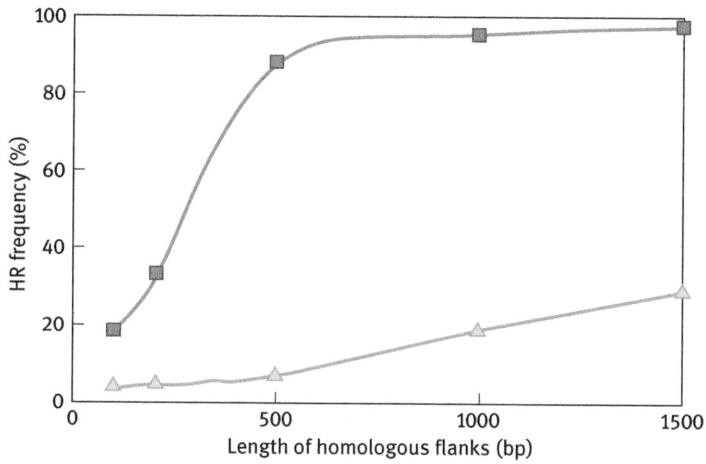

Fig. 12.1: HR-dependent frequency in A. niger wild type (closed diamonds) and ΔkusA (closed squares) recipient strains (Meyer et al., 2007).

Tab. 12.5: Examples of gene targeting frequencies obtained in wild-type and Δku70 strains.

Strain	**Length of homologous flanks (bp)**			**References**
	100	500	1000	
A. fumigatus	wt: 0 %	wt: 2 %	wt: 0 %	Krappmann et al., 2006
	mut: 75 %	mut: 84 %	mut: 96 %	
A. nidulans	wt: n.d.	wt: n.d.	wt: n.d.	Nayak et al., 2006
	mut: n.d.	mut: 89 %	mut: 92 %	
A. niger	wt: 4 %	wt: 7 %	wt: 19 %	Meyer et al., 2007
	mut: 18 %	mut: 88 %	mut: 95 %	
A. sojae	wt: n.d.	wt: n.d.	wt: n.d.	Takahashi et al., 2006
	mut: 0 %	mut: 14 %	mut: 71 %	
N. crassa	wt: 2 %	wt: 9 %	wt: 21 %	Ninomiya et al., 2004
	mut: 10 %	mut: 91 %	mut: 100 %	
S. macrospora	wt: n.d.	wt: n.d.	wt: 4 %	Pöggeler and Kück, 2006
	mut: n.d.	mut: n.d.	mut: 100 %	

wt: wild-type; mut: mutant strain deleted for the Ku70 homologue; n.d.: not determined.

In this context, the Ku proteins are important to maintain telomere length in yeast and plants, and are necessary to ensure chromosome stability in mammals (Boulton and Jackson, 1996; Bailey *et al.*, 1999; Bundock *et al.*, 2002; Riha *et al.*, 2002). Similarly, phenotypic analysis of fungal strains defective in NHEJ shows higher susceptibility to various toxins and irradiation (Müller *et al.*, 2001; Ninomiya *et al.*, 2004; da Silva Ferreira *et al.*, 2006; Meyer *et al.*, 2007). To bypass this problem, fungal strains in which proteins of the NHEJ machinery are transiently silenced could be an attractive option (Nielsen *et al.*, 2008; Ueno *et al.*, 2007).

12.4 RNA technologies for genetic engineering

The possibility to encourage or to silence the given metabolic pathway is not only restricted to DNA-based approaches. An alternative strategy circumventing the need for gene targeting and in particular gene deletion can be the use of RNA-based methods that silence gene expression post-transcriptionally. These tools are mainly valuable when (i) gene targeting approaches fail, (ii) multiple copies of the gene of interest are present in the genome and/or (iii) isogenes might compensate for the knockout of the deleted gene (Akashi *et al.*, 2005). Three RNA-based methods – antisense RNA, hammerhead ribozymes and RNA interference – have showed more valuable tools for gene silencing in eukaryotes (Müller and Stahl, 2004). Table 12.6 outlines the main principles of the tools and their advantages and disadvantages.

Tab. 12.6: RNA-based tools for gene silencing in eukaryotes. (Source: Muller and Stahl, 2004).

Method	Principle	Advantage	Disadvantage
AS	Expressed antisense RNA hybridises by Watson-Crick pairing to the target RNA	Simple and inexpensive	Sometimes low affinity to target RNA due to secondary structures of the antisense molecule
HHR	Translation or processing of the target mRNA becomes inhibited Small catalytic active RNA molecule hybridises with the target RNA (two recognition arms are complementary to the target RNA) Sequence-specific cleavage of the target RNA	High specificity	Laborious
RNAi	Double stranded RNA is diced into siRNAs (21–25 bps long) siRNAs hybridise with the target RNA and guide it to a silencing complex (RISC) that degrades the target RNA	High activity	Requires the activity of endogenous protein complexes

AS: Antisense RNA; HHR: Hammerhead ribozymes; RNAi: RNA interference; siRNA: small interfering RNA. For detailed information see text.

Gene silencing of Fungi have successful constructed with artificial antisense previously reported by several researchers (Zheng *et al.*, 1998; Kitamoto *et al.*, 1999; Ngiam *et al.*, 2000; Moralejo *et al.*, 2002; Lombrana *et al.*, 2004; Blanco and Judelson, 2005). For example, an antisense showing homology to two different genes coding for extracellular carboxypeptidases has transferred by PMT into the genome of *A. oryzae*. Randomly integrated in multiple copies have constructed into the genome and resulted about 70 % of protease activities reduced. This antisense strategy has improved the capacities of *A. oryzae* as host for heterologous protein expression as more stable and higher level of human lysozyme are detected (Zheng *et al.*, 1998). In similar results have reported by Moralejo *et al.* (2002) that antisense silencing of the protease *aspergillopepsin* B found the reduction of 10–70 % of protease levels and resulted in a 30 % increase in heterologous thaumatin production in *A. awamori*.

To date, no zero levels of antisense-mediated reduction of gene expression has been reported, thus suggesting that the efficiency of this strategy is limited. However the complete knockdown of particular gene can be lethal to the specific fungus of interest, attenuation of its expression by the antisense tool can be more powerful approach. For example, the wide-domain transcription factor CreA, the key component of carbon catabolite repression in *Aspergillus* (Dowzer and Kelly, 1991; Ruijter and Visser, 1997; Shroff *et al.*, 1997), negatively regulates number of industrially important enzymes. As shown by Bautista *et al.* (2000), partial suppression of creA expression in *A. nidulans* by its antisense molecule (about 50 % reduced expression has estimated) yielded in partial alleviation of glucose repression and thereby in a substantial increase of the productivities of intra- and extracellular glucose repressible enzymes.

Catalytic RNA molecules are referred to as ribozymes, offer an alternative strategy for gene silencing approaches. The hammerhead ribozyme is the smallest, best-known and most widely used class of RNA-based enzymes (Müller and Stahl, 2004; Akashi *et al.*, 2005). The substrate-recognition arms can be engineered; the arms are complementary to the RNA target, enabling the ribozyme to bind to its target. The cleavage of the substrate RNA occurs immediately adjacent to a "NUX" triplet, where N represents any base and X can be A, U, or C (Shimayama *et al.*, 1995). The applicability of ribozyme based technology as a tool for effective post-transcriptional suppression of gene expression has showed for bacterial, yeast, plant and mammalian systems (Bussiere *et al.*, 2003; Akashi *et al.*, 2005; Isaacs *et al.*, 2006).

Müller *et al.* (2006) have discovered a proof of principle for the use of hammerhead ribozymes to control gene expression in fungi. As reference organism, *A. giganteus* has used as gene deletions, are not possible due to the lack of HR (Meyer *et al.*, 2002). A reporter-based model system using the β-glucuronidase transcript (*uidA*) as the target mRNA has used. This system has validated the activity of seven in silico selected hammerhead ribozymes targeting different sites within the uidA mRNA. All ribozymes tested are able to reduce the reporter activity in *A. giganteus* (up to a maximum of 100 %), demonstrating that ribozyme technology is indeed a useful tool for fungal metabolic engineering and can be used the complete silence gene. First, target sites within the 5'-region of the substrate mRNA preferentially recognized as uidA expression is completely removed in these cases. When ribozymes targeting the 3'-region of uidA is tested, expression of the target gene is reduced only 20–50 %. Secondly, it cannot be guaranteed that the in silico predicted target site is definitely recognized and cleaved by the tailored ribozyme in vivo as other uidA target sites than the predicted ones are accepted under in vivo conditions for two out of seven ribozymes tested. It is difficult to predict the actual in vivo ribozyme binding sites by in silico methods (Akashi *et al.*, 2005).

Finally, RNA interference (RNAi) has been shown remarkably potential to silence fungal genes post-transcriptionally (Nakayashiki, 2005; Nakayashiki *et al.*, 2005). In silencing mechanism, double-stranded RNA with homology gene expressed in the fungal host strain and triggers degradation of its cognate RNA via 21–25 bps long small interfering RNA (siRNA) molecules. A prerequisite for this method is that the fungal strain involves the components of the RNAi silencing machinery, such as RNA-dependent RNA polymerase (RdRP), dicer-like proteins and the RNA-induced silencing complex (RISC) (Nakayashiki, 2005). Referred to some fungal strains, e.g. *Ustilago maydis*, *Candida albicans* and *S. cerevisiae*, lack components of the RNAi silencing machinery, indicating that this tool is not appropriated for these organisms (Nakayashiki, 2005). However, specific inhibition of gene expression by RNAi has showed more suitable for the multitude of fungi such as *A. nidulans* (Hammond and Keller, 2005), *A. fumigatus* (Mouyna *et al.*, 2004; Bromley *et al.*, 2006; Khalaj *et al.*, 2007), *A. oryzae* (Yamada *et al.*, 2007), *Bipolaris oryzae* (Moriwaki *et al.*, 2007), *Colletotrichum lagenarium* (Nakayashiki *et al.*, 2005), *Coprinus cinereus* (Namekawa *et al.*, 2005; Wälti *et al.*, 2006), *Fusarium solani* (Ha *et al.*, 2006), *Magnaporthe oryzae* (Kadotani *et al.*, 2004; Nakayashiki *et al.*, 2005), *Mucor circinelloides* (Nicolas *et al.*, 2007), *Neurospora crassa* (Goldoni *et al.*, 2004) and *Schizophyllum commune* (de Jong *et al.*, 2006).

Similar strategy of antisense, RNAi-induced silencing of fungal gene expression is the most often found incomplete (maximal reduction up to 10 % of wild type level) and full knockout phenotypes are seldom observed. However, the overexpression of RNAi construct either by increased copy numbers or by the use of a strong, inducible promoter can lead to complete repression of gene expression (Khalaj *et al.*, 2007). As shown for *A. fumigatus*, about 50 % of the transformants lost the RNAi construct or parts of it after prolonged cultivation of the transformants (Goldoni *et al.*, 2004; Henry *et al.*, 2007). As suggested by the authors, one possible explanation for this phenomenon might be the loss of one of the inverted repeats after the first mitotic events. These problems might be associated with this tool; however, the molecular mechanisms responsible for RNA-mediated gene silencing in fungi are not yet fully understood. Fungi such as *N. crassa* and *A. nidulans* have provided excellent experimental model systems to unravel the molecular machinery involved in fungal gene silencing (Nakayashiki, 2005; Nakayashiki *et al.*, 2005; Fulci and Macino, 2007).

12.5 Designing gene engineering strategy

The molecular techniques for genetic engineering of fungi have considerably enlarged over the past years. Meyer (2008) suggested that the most appropriate engineering strategy has applied to design using the genetic modification. For example, is the intention to fully silence a gene or is it partially down regulates a gene of interest due to circumvent detrimental effects for the fungal host? Is it possible to silence a group of genes simultaneously by using only a single genetic modification? The overexpression of the particular gene has envisaged and is it thus suitable to integrate multiple copies of the gene into the genome? Each of these considerations will call for different strategy in order to introduce the best genetic modification for metabolic engineering.

In Figure 12.2A exemplarily highlights some genetic engineering tactics that could be considered in overexpressing a gene. In this case, Meyer (2008) suggested that at least three strategies can be followed such as: i) PMT is suggested as transformation method when heterologous integrations of multiple gene copies are envisaged. ii) AMT is the method of choice when the single copy of the gene should integrate via homologous recombination to the specific genomic locus. However, AMT is not applicable for a certain fungus, iii) PMT of NHEJ-deficient recipient strain could be used instead. Figure 12.2B shows two main strategies that could be mapped out either partially or completely down regulates gene of interest. i) Antisense or RNAi constructs cab be introduced by PMT into the fungus of interest, whereby multiple copies of these constructs ectopically integrate into the genome. As the efficiency of both RNA-based methods have limited, partial gene silencing can be achieved and even controlled. This strategy recommended when the target gene is essential for fungal growth. ii) For the complete gene silencing (if the target gene does not have an essential function for the fungus), three alternative possibilities could be considered: either AMT or PMT using the transiently deficient NHEJ recipient strain can be used to knock out the gene of interest. Another option is to make the use of the ribozyme technology which a complete gene knock down can be adjusted.

12.6 Heterologous Proteins Expressed in fungi

The first heterologous protein is used calf chymosin and collected from T. *reesei* in 1989 (Harkki *et al.*, 1989), past decades after accidentally discovered of the fungus during the Second World War in the Solomon Islands (Mandels and Reese, 1957) and two years after genetic transformation system has developed by using fungus (Penttilä *et al.*, 1987). The development of *Trichoderma* expression systems are included: (i) some potential strains found highly protein-secretion (Durand *et al.*, 1988; Nevalainen *et al.*, 1990; Nevalainen and Penttilä, 1995), which developed as hosts for recombinant protein production (Mäntylä and Suominen, 1998); (ii) harnessing the strong cbh1 (cellobiohydrolase 1) promoter for the expression of recombinant gene products (Harkki *et al.*, 1989; Paloheimo *et al.*, 1993); and (iii) production of recombinant proteins as fusions to well-expressed and efficiently secreted native protein such

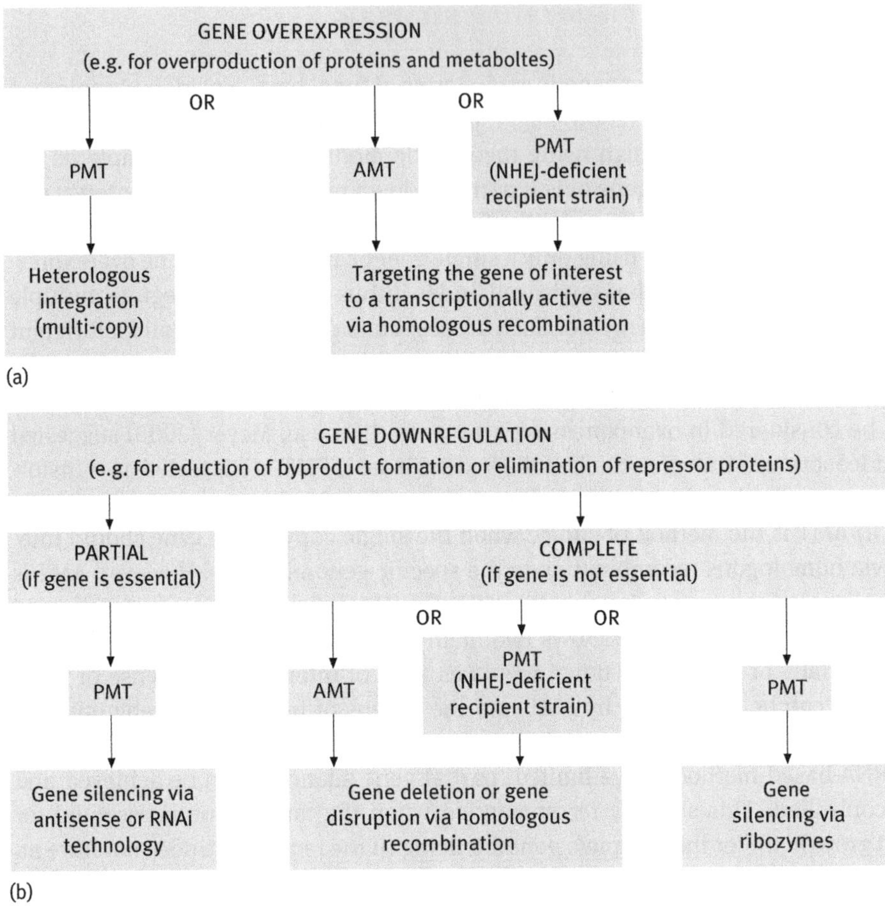

Fig. 12.2: Guidelines for the design of genetic engineering approaches for fungi aiming at overexpression (a) or down regulation (b) of a gene of interest. PMT: Protoplast-mediated transformation; AMT: Agrobacteriummediated transformation; NHEJ: nonhomologous end joining pathway. (Source: Meyer, 2008).

as CBHI (Harkki *et al.*, 1989; Nyyssönen *et al.*, 1993; Bergquist *et al.*, 2002; Zou *et al.*, 2012). Even though several improvements have contributed to the understanding and partial unlocking of the bottlenecks in the expression and secretion of heterologous proteins in *T. reesei*, the majority of approaches still are based on the above three elements.

After calf chymosin, several other heterologous products of mammalian and microbial origin have been expressed and secreted from *T. reesei* (Table 12.7). The highest yields for heterologous proteins expressed in *T. reesei* are those of foreign fungal proteins that have reached gram levels, up to 7.7 g/L for the *Melanocarpus albo-*

myces cellulase Cel5A (Table 12.7; Haakana *et al.*, 2004). The second best yields come from heterologous proteins of bacterial origin nearing and exceeding a gram per liter (Table 12.7; Paloheimo *et al.*, 2003, 2007).

Tab. 12.7: Examples of Heterologous Proteins Expressed in T. reesei under the cbh1 Promoter. Source: Adapted from Peterson and Nevalainen, 2012.

Heterologous Protein	Host Strain	Culture Mode	Yield (g/L)	References
FUNGAL				
Glucoamylase P from *Hormoconis resinae*	Rut-C30 ALKO2221*	Shake flask	0.5 0.7	Joutsjoki *et al.*, 1993a,b
Endochitinase from *Trichoderma harzianum*	Rut-C30	Shake flask	0.13	Margolles-Clark *et al.*, 1996
Acid phosphatase from *Aspergillus niger*	ALKO2221*	Shake flask	0.5	Miettinen-Oinonen *et al.*, 1997
Phytase from *Aspergillus niger*	ALKO2221*	Shake flask	2.0	Paloheimo *et al.*, 1993
Xylanase II from *Humicola grisea*	HEP1*	Shake flask	0.5	de Faria *et al.*, 2002
β-Glucosidase from *Talaromyces emersonii*	Rut-C30	Shake flask	0.0027	Murray *et al.*, 2004
Melanocarpus albomyces Cel5A Cel7A Cel7B	ALKO3620*	Shake flask	7.7 5.4 4.4	Haakana *et al.*, 2004
Laccase from *Melanocarpus albomyces*	Rut-C30	Fermenter (20 L)	0.23	Kiiskinen *et al.*, 2004
XynVI from *Acrophialophora nainiana*	Rut-C30	Shake flask	0.172	Salles *et al.*, 2007
Cinnamoyl esterase EstA from *Piromyces equi*	Rut-C30	Shake flask	0.033	Poidevin *et al.*, 2009
Lipase from *Aspergillus niger*	Tu6 (ATCC MYA-256)*	Shake flask	0.300	Qin *et al.*, 2012
BACTERIAL				
Xyn11A from *Nonomuraea flexuosa*	ALKO3620*	Fermenter (1 L)	0.82	Paloheimo *et al.*, 2003
Truncated Xyn11A from *Nonomuraea flexuosa*	ALKO3620*	Fermenter (2 L)	1.8	Paloheimo *et al.*, 2007

Tab. 12.7: (continued)

Heterologous Protein	Host Strain	Culture Mode	Yield (g/L)	References
Xylanase XynB from *Dictyoglomus thermophilum*	Rut-C30	Shake flask	0.5	Bergquist *et al.*, 2002
PLANT				
Barley endopeptidase B Rut-C30	ALKO2221*	Fermenter (1 L)	0.05 0.5	Saarelainen *et al.*, 1997
MAMMALIAN				
Calf chymosin	Rut-C30	Fermenter (10 L)	0.04	Harkki *et al.*, 1989
CBHI-Fab fusion antibody	Rut-C30	Shake flask Bioreactor	0.04 0.15	Nyyssönen *et al.*, 1993
Human erythropoetin	Rut-C30 M3§ Rut-C30 T108¶	Shake flask	0.097 0.047	Zhong *et al.*, 2011
Human obestatin	Rut-C30	Shake flask	0.000007	Sun *et al.*, 2012

* ALKO2221, ALKO3620, HEP1 and Tu6 (ATCC MYA256) are strains derived from QM9414 following one or more rounds of mutation or transformation.
§ Rut-C30 M3 strain contains a modified *cbh1* promoter (Liu *et al.*, 2008).
¶ Rut-C30 T108 has been transformed with the human *N*-acetylglucosaminyl transferase I gene.

High yields production of mammalian proteins are produced some micrograms to 97 mg/L of human erythropoietin and 150 mg/L of the murine antibody fragment (Zhong *et al.*, 2011; Nyyssönen *et al.*, 1993). Considering these figures is very little improvement for yield production of heterologous mammalian proteins in *T. reesei*. The proteins production of fungal systems spilt out into two categories: industrial enzymes and gene products from higher organisms, which are often of pharmaceutical value (Nevalainen and Peterson, 2014). There are several examples showing that gene products originating from other fungi effectively produced heterologously up to the gram per litre level. These include an *Aspergillus phytase* enzyme produced in *T. reesei* at 2 g/L (Paloheimo *et al.*, 1993), glucoamylase from *Hormoconis resinae* at 0.7 g/L (Joutsjoki *et al.*, 1993) and xylanase 2 originating from *Humicola grisea* var. *thermoidea* at 0.5 g/L (de Faria *et al.*, 2002). Another example of industrially relevant enzymes produced in heterologous fungal host features expression of *Melanocarpus albomyces* cellulase genes in *T. reesei* (Haakana *et al.*, 2004). Yields of some other special types of enzymes such as the thermophile bacterial xylanase XynB (temperature optimum of 85 °C) expressed in *T. reesei* has remained considerably lower (100 mg/L range), even after optimization of the codon usage for fungal expression (Te'o *et al.*, 2000; Bergquist *et al.*, 2004).

Expression of the human interleukin hIL-6 in *Aspergillus* which an elaborate series of host species (*A. nidulans, A. niger, Aspergillus foetidus, A. sojae*) and strains have used in conjunction with different promoters to express a particular product (Punt *et al.*, 2002). As a result, the yields in shake flask cultivations varied from less than 0.1 mg to 100 mg/L. Transformants with the highest yield expressed hIL-6 from the glaA promoter as fusion protein to the endogenous glucoamylase A carrier.

12.7 Fungi as cell factories for fungal enzymes

Fungi have ability to produce large amounts of specific proteins. Both homologous and heterologous enzymes are produced commercially from fungi. Production level of protein from fungal strains are commonly too low for commercial exploitation. The number of heterologous enzymes have approved for food applications, although new heterologous genes successfully expressed in fungi continue including the potential food-use, sweet protein thaumatin (Moralejo *et al.*, 1999). The recombinant enzymes have approved for the use in food and derived from fungal host organism are listed in Table 12.8. It is more valuable information at all although the donor organism is used another fungus. Because the yields of heterologous enzymes from fungi have found very low, especially when the donor organism is not used the same fungus (Archer, 2000).

Tab. 12.8: Recombinant enzymes, derived from a fungal host, approved for use in food (Source; Archer, 2000).

Enzyme	Host	Donor
Catalase	*A. niger*	*Aspergillus* sp.
Cellulase	*A. oryzae*	*Humicola* sp.
Cellulase	*T. reesei (longibrachiatum)*	*Trichoderma* sp.
Chymosin	*A. niger var. awamori*	Calf
α-galactosidase	*A. oryzae*	*Aspergillus* sp.
β-glucanase	*T. reesei (longibrachiatum)*	*Trichoderma* sp.
Glucose oxidase	*A. niger*	*Aspergillus* sp.
Lipase	*A. oryzae*	*Candida* sp.
		Rhizomucor sp. *Thermomyces* sp.
Phytase	*A. niger, A. oryzae*	*Aspergillus* sp.
Protease	*A. oryzae*	*Rhizomucor* sp.

Tab. 12.8: (continued)

Enzyme	Host	Donor
Xylanase	*A. niger (and var. awamori)*	*Aspergillus* sp.
	A. oryzae	*Aspergillus* sp. *Thermomyces* sp.
	T. reesei (longibrachiatum)	*Trichoderma* sp.

However, impressive improvements in protein yield have obtained with traditional strain upgrading strategies based on various mutagenesis approaches. For commercial practices, yields of >30 g/L of the specific protein is not uncommon (Punt *et al.*, 2002). The potential strain always significantly produced high levels of protein.

In protein purification systems, the gene encoding can be cloned by using PCR-based approaches in order to gene characterization and expression in appropriate fungal hosts. Punt *et al.* (2002) suggested that two new gene-cloning approaches are successfully used. One approach is 'expression cloning' which combines simple enzyme assays with the use of *Saccharomyces cerevisiae* expression system for fungal cDNAs (Dalboge, 1997; Dalboge and Lange, 1998). The resulting cDNA clones can easily be characterized and subsequently expressed in more appropriate high-level expression hosts such as *Aspergillus* spp. Similar strategies are applied for development of fungi (Emalfarb, 2001). Another option is the combined use of database mining and molecular screening. Besides, in *S. cerevisiae*, at least two fungal genome sequences are openly available (*Neurospora crassa, neurospora/*; *A. fumigatus*).

Protein targeting and secretion have studied using green fluorescent protein (GFP) tags (Henriksen *et al.*, 1999; Siedenberg *et al.*, 1999). GFP may have proved to be useful in screening for mutant strains with altered secretion properties (Gordon *et al.*, 2000). Only few components of the fungal secretory pathway have been characterized so far the gene level (Whittaker *et al.*, 1999; Kasuya *et al.*, 1999; Saloheimo *et al.*, 1999; Wang and Ward, 2000). Production of heterologous proteins have induced an unfolded protein response (UPR), which leads to the up-regulation of foldases and chaperones (Saloheimo *et al.*, 1999; Wang and Ward, 2000; Ngiam *et al.*, 2000), but the considered up-regulation of those genes have not hitherto proved capable of improving the yields of target heterologous proteins. Down-regulation of *pdiA*, the gene that encodes the major foldase (protein disulfide isomerase), in *A. niger* lowered the secreted yield of glucoamylase (Ngiam *et al.*, 2000), indicating that threshold level of protein disulfide isomerase is necessary for secretion of glucoamylase.

One approach commonly used to enhance yields of secreted heterologous proteins for expression of the gene as fusion contrived with a kexin endoproteolytic cleavage site (Lys–Arg is the preferred recognition and cleavage motif) at the join of the carrier and target proteins. The carrier protein is well-secreted by the fungus, for example, glucoamylase or cellobiohydrolase. Gene encoding of the kexin protease is cloned from *A. niger* (Jalving *et al.*, 2000); its deletion led to the secretion of an unprocessed fusion protein, although other fusion proteins may well be cleaved by other proteases (not necessarily at exactly the same site). Industrial glucoamylase-producing strain of *A. niger* found various effects of mutagenesis, including the transcriptional inactivation of the major protease-encoding gene (MacKenzie *et al.*, 2000). Fungal proteases are an important issue in protein production by fungi; genes encoding proteases and mutant strains that are deficient in proteolytic activity described previously by Mac Kenzie *et al.* (2000). By comparing the endoproteolytic 'nicking' of three lysozymes with variation in sequence, an *A. niger* protease with specificity for Gly–X in exposed positions predicted (Spencer *et al.*, 1999).

Many proteins secreted by fungi are glycosylated. The glycan components have an important in the effectiveness of glycoproteins when used in therapy (Maras *et al.*, 1999). The glycan component can affect the stability and catalytic activity, but not all enzymes. The major impact of the glycan component of fungal enzymes projected for food-use is likely to be allergenicity. Inhibition of glycosylation can reduce secreted protein yields and induce the unfolded protein response (UPR).

Biochemically and genetically characterized components of the cellulolytic system of *T. reesei* has included endoglucanases hydrolyzing internal bonds in the cellulose chains (EGI/Cel7B, EGII/Cel5A, EGIII/Cel12A, EGIV/Cel61A, EGV/Cel45), cellobiohydrolases releasing oligosaccharide units, mostly cellobiose, from the cellulose chain ends (CBHI/Cel7A, CBHII/Cel6A), and β-glucosidases hydrolyzing the produced cello-oligo or -disaccharides further to β-d-glucose or carrying out transglycosylation of the products (BGLI/Cel3A, BGLII/Cel1A). EGIV/Cel61A belongs to the glycoside hydrolase family GH61, which has been included polysaccharide monooxygenases rather than endoglucanases (www.cazy.org). The GH61 members enhanced cellulose hydrolysis in a metal ion dependent way by an oxidative mechanism, and required the presence of a cellobiose dehydrogenase for their action (Saloheimo et al., 2014).

The characterized hemicellulose degrading components of secretome include the xylan, mannan and xyloglucan backbone hydrolyzing enzymes (xylanases XYNI, XYNII, XYNIII and XYNIV; a mannanase MANI; a xyloglucanase Cel74A), as well as a variety of side-chain cleaving enzymes and enzymes hydrolyzing smaller hydrolysis products further (β-xylosidase BXLI, α-l-arabinofuranosidase ABFI, acetyl esterase AESI, acetyl xylan esterase AXEI, α-glucuronidase GLRI, α-galactosidases AGLI-III, β-galactosidase BGAI). Other secreted enzymes characterized from *T. reesei* cultures include enzymes active on chitin (gls93, chi46), other types of cell wall components (glucuronan hydrolase TrGl) or protein glycans (EndoT),as well as proteases, hydrophobins (HFBI-III, small amphipathic proteins with diverse functions e.g. in hyphal development and sporulation), glucoamylase hydrolyzing α-glycoside linkages in starch, and phytase (Saloheimo *et al.*, 2014).

12.8 Fungi as cell factories for non-fungal proteins

High-level production of proteins of non-fungal origin is trailed as fungal proteins approaches. Fungi can produce a variety of compounds with accepted use (e.g. citric acid) or potential use (e.g. vitamins and long-chain PUFAs) as food ingredients. Polyunsaturated fatty acids (PUFAs) are considered as ingredients in baby milk formulations for uses in some countries.

Genes encoding fatty acid desaturases have been described from *Mucor rouxii* (Passorn *et al.*, 1999) and *Mortierella alpine* (Sakuradani *et al.*, 1999). The desaturases have required for synthesis of arachidonic acid (C20:4, ω6); although an ω3 (Δ^{17}) desaturase gene from *M. alpina*, for synthesis of more unsaturated PUFAs, remains elusive. Other organisms which are naturally good sources of ω3 PUFAs, may possibly provide the ω3 desaturase genes. For further analysis of the desaturation system in *M. alpina* is described with the genes encoding cytochrome b_5 and its reductase (Certik *et al.*, 1999; Sakuradani *et al.*, 1999; Kobayashi *et al.*, 1999). A fatty acid elongase gene has not yet been described from *M. alpina*. The challenge will then be to use these genes either within *Mortierella alpina* or in plants to over-produce particular PUFAs. For success, this will probably require more information on the regulation of a complex pathway leading to oils (Wynn *et al.*, 1999).

However, the first use of fungi as protein production hosts, several research groups have addressed the problem of endogenous protease production, so we need to develop and to improve the protease deficient host strains (Mattern *et al.*, 1992; van den Hombergh *et al.*, 1997). Some strains are mostly eased the problem of extracellular degradation. Even in the best protease-deficient strains are residual levels of protease activity, which for some proteins and under certain conditions might still result in serious losses in protein yield.

Another major improvement in the secretion of heterologous proteins is the 'carrier' approach, based on the idea that a secreted fungal protein can act as a carrier for more efficient secretion of heterologous protein. Ward *et al.* (1990) reported the expressed gene-fusion comprising the genes encoding *A. niger* glucoamylase and bovine chymosin, is the first example of this approach in fungi. The yields of chymosin have produced in strains carrying this gene-fusion significantly greater than in strains expressing the unfused chymosin gene. (Gouka *et al.*, 1997). Only the catalytic domain of the glucoamylase has used and a proteolytic processing site is separated the carrier moiety from the protein of interest. The specific amino acid sequence of this processing site (NVISKR) has derived from the pro-peptide of the glucoamylase protein itself. This sequence represents the specific-target site for KEX2-type protein processing protease. These types of proteases are active in the secretion pathway of various eukaryotic organisms (Van de Ven *et al.*, 1993). The gene encoding the corresponding protein in *A. niger* has cloned and characterized by Jalving *et al.* (2000) and found the presence of the proteolytic processing site between the carrier and the target protein is not the main factor for the cleavage of the fusion protein. In similar study done by Punt *et al.* (2002), KEX2-type processing is not vital factor for processing in these cases. The fusion approach is entirely failed for improving the yields of secreted protein (Conesa *et al.*, 2000). However, the gene-fusion approach still remains the first choice in attempts to produce non-fungal proteins in *Aspergillus*.

12.9 Production of bioenergy from microbial biomass

Global climate change, increasing food and energy demand and environment decline are the main challenges for humans. Biofuels, including the photosynthetic organisms such as plants, algae, and cyanobacteria capture solar energy and store it as chemical energy of their biomass. Bioenergy is mostly captured by photosynthetic CO_2 fixation; therefore, carbon-containing biofuels have different tendency to be carbon neutral after combustion. The amount of accomplished largely depends on the nature of feedstock, as like as the agricultural practice, and the industrial process, thus it may have frequently contributed to climate change mitigation (Johnson, 2009).

The use of microbial cell factories for bioenergy has regained attention due to increasing the pressure for higher productivities, novel bioproducts, and environmental protection. It is widely applied of biodiversity in the biosphere. According to the advantages listed in Box 12.1, there is ample interest in developing the technology for the use of photosynthetic microorganisms, such as eukaryotic microlgae or cyanobacteria (Wijffels and Barbosa, 2010; Rasala *et al.*, 2013). Although some startup companies are struggling to commercialize algal fuels (Chisti and Yan, 2011), their actual potential is still debate (Petkov *et al.*, 2012).

Box 12.1: Main advantages of microalgae or cyanobacteria as biofuels feedstock (Ortiz-Marquez *et al.*, 2013).

- At their exponential growth rate they can double their biomass in periods as short as 3.5 h.
- They can produce almost year under favorable weather.
- For oleaginous microalgae, the projected oil productivity per hectare would exceed by tenfold that of the best oilseed crops.
- Despite being aquatic organisms, the culture systems (open ponds or photobioreactors) demand less water than terrestrial crops, minimizing pressure on freshwater resources.
- They can be cultivated in brackish water on nonarable land, not incurring land-use change and associated environmental damage, and competence with the production of food, fodder, and other benefits from crops.
- They can take the most vital nutrients (CO_2, nitrogen, phosphorus, and others) from industrial or municipal waste, helping out to manage waste disposal.
- Microalgae cultivation does not require herbicides or pesticides application.
- They can also produce valuable co-products such as proteins and residual biomass after oil extraction, which may be used as feed or fertilizer, or fermented to produce ethanol or methane.

- Some strains produce biohydrogen either as a result of their metabolism, by means of hydrogenase and/or nitrogenase enzymes or indirectly by biomass fermentation with appropriate microorganisms.

Identifications of the potential microalgae strains are commonly beginning point in the roadmap towards microalgae-based technology development. The most appreciated traits for the 'ideal microalga' are listed in Box 12.2 (Wijffels and Barbosa, 2010; Rasala *et al.*, 2013). There are no available strains excelling in all these traits, which are not surprising, conversely to the development of modern plant crops, no breeding programs for microalgae uses (Georgianna and Mayfield, 2012).

Box 12.2: Most desired characteristics of the ideal photosynthetic microbial cell factory (Ortiz-Marquez et al., 2013).

- High efficiency of light capture and biomass yield (growth rate and final culture density).
- High production of lipid or any other useful energy carrier.
- Production of biomass and energy carrier at the same time.
- Large cells and/or flocculation properties to facilitate harvesting.
- Thin cell walls or with structural characteristics for easy intracellular products extraction.
- Tolerance to high light intensity and oxygen concentration.
- Resistance to contamination with other microorganisms or predators.
- High efficiency of the use of cellular nitrogen and phosphorus and high nutrient-recycling capacity and/or ability to utilize abundant and/or inexpensive alternative sources of nitrogen, phosphorus, and other macroelements.
- Amenability to genetic analysis and manipulation.
- Probably less important aspects such as minimizing respiration rates to minimize carbon dissimilation, decreasing the rate of photoacclimation to low light to slow down self-shading after transition to high light, lowering the carbon:nitrogen ratio of core nitrogenous components and the amount of carbon allocated to cell structure, and the potential to contribute to enhance biofuels yield.

In natural microalgae or cyanobacterial isolates have expected the breeding and/or genetic engineering that would be required to achieve industrial production (Rasala *et al.*, 2013; Georgianna and Mayfield, 2012; Flynn *et al.*, 2013; Larkum *et al.*, 2012) in Box 12.3.

Box 12.3: Multispecies microbial cell factories (Ortiz-Marquez et al., 2013).

- Microbial consortia perform more complex tasks than monocultures and can perform functions that are difficult or even impossible for individual strains or species.
- This approach may alleviate pressure towards increasing yields of heterologous pathways in selected expression platforms, for which overall metabolic coupling is weak and/or mostly poorly understood. In turn, the concept of genetic engineering of multispecies microbial cell factories might rely, as a starting point, on microbes with one or a few outstanding properties

as metabolic/production modules, which can be further optimized for the same or other 'easier-to-accomplish' tasks. This approach takes advantage of the possibility of metabolic compartmentalization between or among cells with different characteristics and/or onto which the burden of overexpressing multiple genes can be split among the partners of the consortium. Communication for trading extracellular metabolites is a key aspect of the principle which on one hand increases the level of complexity and on the other hand might provide opportunities for additional possibilities of regulation of the productive platforms and/or removing toxic byproducts (if any).

- It is expected that better understanding of the natural assemblages of microbial communities and detailed knowledge of naturally occurring symbiosis will represent a lead for biologically inspired design of multispecies microbial cell factories to complement current efforts using more conventional genetic engineering approaches.

12.9.1 Genetic engineering of photosynthetic microbial cell factories

Microalgae and cyanobacteria are designed to produce the specific-target product. There are numerous examples have listed in Table 12.9. One of the most remarkable accomplishments has been the direct conversion of CO_2 into alcohols, partially bypassing the complexity of the formation of biomass. Recombinant ethanol production up to 5.5 g/L has obtained in cyanobacteria (Dexter and Fu, 2009). Also, isobutyraldehyde can be used as a precursor of a variety of chemicals, including isobutanol which is produced a high productivity that could outstrip by five- to six fold estimates for corn and cellulosic ethanol production on the land area basis. Isobutyraldehyde-overproducing is produced an isobutanol as better substitute for gasoline than ethanol by using modified strains (Atsumi *et al.*, 2009). However, the success of importing the heterologous pathway is influenced by variety of drawbacks that may include combinations of different aspects.

Tab. 12.9: Current and potential biofuels from cyanobacteria or modified microalgae (Ortiz-Marquez *et al.*, 2013).

Bioproduct	Strategy	Results/comment	Refs
Cyanobacteria			
	H.E. *pdc* and *adh* from *Zymomonas mobilis* into *Synechocystis* sp. PCC 6803	5.2 mmol ethanol OD_{730}/u.l.d (up to 230 mg/l in 4 weeks)	Dexter and Fu, 2009
	H.E. *pdc* from *Z. mobilis*, overexpressing slr1192, and disruption of the biosynthetic pathway of poly-b-hydroxybutyrate into *Synechocystis* sp. PCC 6803	212 mg/l.d	Gao *et al.*, 2012

Tab. 12.9: (continued)

Bioproduct	Strategy	Results/comment	Refs
Isobutyraldehyde	H.E. *kivD* from *Lactobacillus lactis*, *alsS* from *Bacillus subtilis*, *ilvCD* from *Escherichia coli* and insertion of extra copies of *rbcLS* genes in *Synechococcus elongatus* PCC7942	6230 mg isobutyraldehyde/l.h Higher than cyanobacterial productivities of hydrogen or ethanol.	Atsumi *et al.*, 2009
Isobutanol	H.E. *kivD* from *L. lactis* and *yqhD* from *E. coli* in *S. elongatus* PCC7942	450 mg isobutanol/l in 6 d	Atsumi *et al.*, 2009
1-Butanol	H.E. *hbd, crt,* and *adhE2* genes from *Clostridium acetobutylicum*, *ter* from *Treponema denticola*, and *atoB* from *E. coli* in *S. elongates* PCC7942	14.5 mg 1-butanol /l in 7 d	Lan and Liao, 2011
	H.E. *bldh* from *Clostridium saccharoperbutylacetonicum*, *yqhD* from *E. coli* (among others) in *S. elongates* PCC7942	29.9 mg 1-butanol/l	Lan, and Liao, 2012
Isoprene production	H.E. *ispS* from *Pueraria montana* expressed under regulation of the *psbA2* promoter in *Synechocystis* sp. PCC 6803	50 µg isoprene/g dw.d under high light conditions	Lindberg *et al.*, 2010
Fatty acids	H.E. a thioesterase, among others modifications, in *Synechocystis* sp. PCC 6803	200 mg/L fatty acids secreted into the culture medium	Liu *et al.*, 2011
Fatty alcohols	H.E. far from *Simmondsia chinensis* under the control of P_{rbc} promoter	0.2 mg/L fatty alcohols	Tan *et al.*, 2011
Hydrogen	H.E. [Ni-Fe] hydrogenase *hynSL* from *Alteromonas macleodii*, and *Thiocapsa roseopersicina* and accessory genes in *S. elongatus* PCC7942 [FeFe]-hydrogenase *hydA1* from *Cr* in *Synechocystis* sp. PCC 6803	Accumulation of holo-enzyme that displayed hydrogenase activity *in vitro* in vitro-active hydrogenase	Weyman *et al.*, 2011 Berto *et al.*, 2011
	Insertional disruption of the *hupL* gene in *Nostoc* sp. PCC 7422	100 µmoles H_2/mg chlorophyll *a*.h (three times higher rate than that of the parental strain)	Yoshino *et al.*, 2007
Microalgae			
Triacylglycerol (TAG)	*sta6* and *sta7–10* starchless mutants of *Cr*	2- to 10-fold increase in TAG accumulation	Li *et al.*, 2010; Wang *et al.*, 2009; Work *et al.*, 2010

Tab. 12.9: (continued)

Bioproduct	Strategy	Results/comment	Refs
	sta7–10 (isoamylase mutants) complemented strains	≥400 mg/L starch exceeding in 4 d (significantly higher than those achieved by the parental strains) Higher values in total lipids at 96 h of nitrogen deprivation: 83–118 mg lipid/L	Work *et al.*, 2010
Hydrogen	Replacement of wild type promoter of the Nac2 gene by a copper repressible promoter in *Cr*	20 μmoles H_2/L 1–3.1 μmol/H_2 mol. Chl	Surzycki *et al.*, 2007
	Mutant strain with a double amino acid substitution (L159I-N230Y) in D1 of *Cr*	500 ml H_2/L; 5.77 ml/l.d ≥10-fold higher wild type and fivefold CC124 strain, mostly due to a longer production phase	Torzillo *et al.*, 2009
	Optimization of the e- flux towards hydrogenase state transitions mutant6 (*stm6*) in *Cr*	540 mL H_2/L 4 mL/h (fivefold higher than the wild type)	Kruse *et al.*, 2005

12.9.2 Incompatibility of oxygenic photosynthesis with anaerobic metabolism

1-Butanol, a likely even-better substitute of gasoline than isobutanol that is successfully produced at up to 15 g/L in recombinant *Escherichia coli* cells from glucose (Shen *et al.*, 2011). Conversely, the more challenging task of producing 1-butanol from CO_2 in cyanobacteria remains more elusive because only up to 0.0145 g/L over 7 days could be produced and only under dark, anaerobic conditions at the expense of internal carbohydrate storage (Lan and Liao, 2011).

Production of hydrogen is catalyzed by the combinations of oxygen-sensitive hydrogenases and/or nitrogenases. Hydrogen production is normally low in cyanobacteria and microalgae, and thus genetic engineering approaches have applied for the development of capacity. By means of heterologous expression and/or hydrogenase engineering for oxygen tolerance or increased activity, inactivation of the uptake [NiFe]-hydrogenase in cyanobacteria, fine-tuning of the oxygen-evolving activity of photosystem II (PSII), optimization of the e-flux towards hydrogenase, and others (Srirangan *et al.*, 2011). However, many encouraging proof-of-principle demonstrations, photosynthetic hydrogen production remains low for industrial purposes.

12.9.3 Host tolerance to high concentration of recombinant pathway product and/or precursors

In contrast to ethanologenic microorganisms such as the yeast *Saccharomyces cerevisiae* or the bacterium *Zymomonas mobilis* that tolerate upto 15 % ethanol in the external medium, the cyanobacterium *Synechocystis* sp PCC 6803 tolerates only up to 1 %. Although, such inhibitory concentrations have not been met from ethanol producing cyanobacteria (Dexter and Fu, 2009), indicating that imbalances between the recombinant enzymes activities may result in growth inhibitory concentrations of the toxic intermediate acetaldehyde (Gao *et al.*, 2012). Research study done by Zhang *et al.* (2012) and mentioned that improved recombinant-biodiesel yields in *E. coli* by a dynamic sensor–regulator system that adjusts recombinant enzymes activities.

12.9.4 Assembly of complex enzymes

Enzymes such as hydrogenases or nitrogenases contain at their active sites complex metal centers. Biosynthesis of the iron–molybdenum cofactor and assembly of entirely functional holo-nitrogenase involves at least 15 gene products. Although knowledge on this field has progressively increased, it is still fragmentary (Rubio and Ludden, 2008; Peters *et al.*, 2013) and limits recombinant pathway design and introduction into selected hosts.

12.9.5 General physiological and metabolic adaptations

During the course of evolution, some microbes become exquisitely specialized to carry out some metabolic pathways for which they orchestrate sophisticated arrays of physiological and metabolic adaptations. *Azotobacter vinelandii* displays a concerted set of mechanisms (increased aerobic respiration and exopolysaccharide production and duplication of housekeeping genes) for its particular specialization to run strictly anaerobic pathways (mainly N_2- fixation and H_2 metabolism) together with an obligate aerobic life style (Setubal *et al.*, 2009). Biochemical and genomic analyses have uncovered the unique characteristics for the remarkably high ethanol producer *Z. mobilis*, (Seo *et al.*, 2005). In some cases, the genetic basis of some metabolic switches are not available to design strategies genetic improvement. Subsequently, those targeted bioenergy carriers that constitute the natural carbon reserves of cyanobacteria and microalgae (carbohydrates or lipids) that normally accumulate under unbalanced growth after nutritional or environmental adverse conditions compromise overall productivity (Do Nascimento *et al.*, 2012).

12.9.6 Biotechnological use of microalgal–bacterial consortia as multispecies microbial cell factories

The combined use of algae and bacteria in microbial consortia is gaining renewed interest as biotechnological alternative for the improvement of yields and reducing production costs of algal biomass for bioenergy purposes in integrated bioprocesses together with CO_2 and pollutant removal (Olguin, 2012; Subashchandrabose *et al.*, 2011). The *A. brasilense* stimulation of *Chlorella* spp. growth and lipids and starch accumulation (Choix *et al.*, 2012; de-Bashan *et al.*, 2002) represents a remarkable example of the biotechnological potential of microalgal–bacterial associations applicable to bioenergy. Inoculation with the *Pseudomonas*-related strain GM41 has increased *Synechocystis* sp PCC6803 productivity up to eightfold by assisting to degrade toxic compounds that commonly found in polluted water (Abed, 2010).

Another application of microalgal–bacterial consortia is developed for the microbial solar cells that comprised photoautotrophic microorganisms to harvest solar energy and released organic compounds that are used by electrochemically active microorganisms to generate electricity (Strik *et al.*, 2011).

12.10 Conclusion remarks

The focal application of fungal widely exploited as sources of ingredients in the manufacture of food, agriculture, pharmaceutical, enzyme, bioenergy and industrial biotechnology. The molecular and technical toolboxes into fungal genetics and biology are developed the potential fungal strains for industrial applications. Most of the techniques and designing options have fruitfully proven and validated, although the lack of efficient genetic engineering strategies forms still an obstacle for selected a specific host of fungi commercially producing metabolites. Discover their biotechnological dimensions, these limitations have to be solved. The range of fungal genes encoding proteins with useful enzymatic activities should be extended.

The availability of fungal genomes and their better understanding today open new era to design the specific-potential industrial host strains for industrially large scale production. Genome-wide transcription profiling, proteomics and the reconstruction of the complete metabolic networks will offer valuable in-depth insights into cellular processes of fungi. In consequence, the classical tactic has monitored in fungal biotechnology, i.e. the main focus on individual genes or proteins, will be drawn-out by global approaches and understanding the interaction of the complete set of cellular components. This concept of ‘systems biology’ and will significantly improve industrial fermentation processes and will further increase the value of fungi in the future of industrial biotechnology and other commodities.

12.11 References

[1] Abed, R.M.M. (2010). Interaction between cyanobacteria and aerobic heterotrophic bacteria in the degradation of hydrocarbons. International Biodeterioration and Biodegradation, 64, 58–64.

[2] Adrio, J.L. & Demain, A.L. (2003). Fungal biotechnology. International Microbiology, 6, 191–199.

[3] Akashi, H., Matsumoto, S. & Taira, K. (2005). Gene discovery by ribozyme and siRNA libraries. Nature Reviews Molecular Cell Biology, 6, 413–422.

[4] ANONIMOUS (1982). Human insulin receives FDA approval. FDA Drug Bull, 12, 18–19.

[5] Archer, D.B. (2000). Filamentous fungi as microbial cell factories for food use. Current Opinion in Biotechnology, 11, 478–483.

[6] Asch, D.K. & Kinsey, J.A. (1990). Relationship of vector insert size to homologous integration during transformation of *Neurospora crassa* with the cloned am (GDH) gene. Molecular and General Genetics, 221, 37–43.

[7] Askolin, S., Nakari-Setala, T. & Tenkanen, M. (2001). Overproduction, purification, and characterization of the *Trichoderma reesei* hydrophobin HFBI. Applied Microbiology and Biotechnology, 57, 124–30.

[8] Atsumi, S., Higashide, W. & Liao, J.C. (2009). Direct photosynthetic recycling of carbon dioxide to isobutyraldehyde. Nature Biotechnology, 27, 1177–1180.

[9] Bailey, S.M., Meyne, J., Chen, D.J., Kurimasa, A., Li, G.C., Lehnert. B.E. & Goodwin, E.H. (1999). DNA double-strand break repair proteins are required to cap the ends of mammalian chromosomes. Proceedings of the National Academy of Sciences USA, 96, 14899–148904.

[10] Bautista, L.F., Aleksenko, A., Hentzer, M., Santerre-Henriksen, A. & Nielsen, J. (2000). Antisense silencing of the creA gene in *Aspergillus nidulans*. Applied and Environmental Microbiology, 66, 4579–81.

[11] Beloqui, A., De María, P.D., Golyshin, P.N. & Ferrer, M. (2008). Recent trends in industrial microbiology. Current Opinion in Microbiology, 11, 240–248.

[12] Bennett, J.W. (1998). Mycotechnology: the role of fungi in biotechnology. Journal of Biotechnology, 66, 101–107.

[13] Bergquist, P., Te'o, J., Gibbs, M., Cziferszky, A., de Faria, F.P., Azevedo, M. & Nevalainen, H. (2002). Expression of xylanase enzymes from thermophilic microorganisms in fungal hosts. Extremophiles, 6, 177–184.

[14] Bergquist, P.L., Te'o, V.S.J., Gibbs, M.D., Curach, N.C. & Nevalainen, K.M.H. (2004) Recombinant enzymes from thermophilic microorganisms produced in fungal hosts. Biochemical Society Transactions, 32, 293–297.

[15] Berto, P., D'Adamo, S., Bergantino, E., Vallese, F., Giacometti, G.M. & Costantini, P. (2011). The cyanobacterium *Synechocystis* sp. PCC 6803 is able to express an active [FeFe]-hydrogenase without additional maturation proteins. Biochemical and Biophysical Research Communications, 405, 678–683.

[16] Bird, D. & Bradshaw, R. (1997). Gene targeting is locus dependent in the filamentous fungus *Aspergillus nidulans*. Molecular and General Genetics, 255, 219–25.

[17] Blanco, F.A. & Judelson, H.S. (2005). A bZIP transcription factor from *Phytophthora* interacts with a protein kinase and is required for zoospore motility and plant infection. Molecular Microbiology, 56, 638–648.

[18] Boulton, S.J. & Jackson, S.P. (1996). Identification of a *Saccharomyces cerevisiae* Ku80 homologue: roles in DNA double strand break rejoining and in telomeric maintenance. Nucleic Acids Research, 24, 4639–48.

[19] Brakhage, A.A. & Caruso, M.L. (2004). Biotechnical genetics of antibiotic biosynthesis. In: Esser, K., editor. The Mycota II. Genetics and Biotechnology. Springer-Verlag, p. 317–353.

[20] Bromley, M., Gordon, C., Rovira-Graells, N. & Oliver, J. (2006). The *Aspergillus fumigatus* cellobiohydrolase B (cbhB) promoter is tightly regulated and can be exploited for controlled protein expression and RNAi. FEMS Microbiology Letters, 264, 246–54.

[21] Bundock, P., van Attikum, H. & Hooykaas, P. (20020. Increased telomere length and hypersensitivity to DNA damaging agents in an Arabidopsis KU70 mutant. Nucleic Acids Research, 30, 3395–400.

[22] Bussiere, F., Ledu, S., Girard, M., Heroux, M., Perreault, J.P. & Matton, D.P. (2003). Development of an efficient cis–trans–cis ribozyme cassette to inactivate plant genes. Journal of Plant Biotechnology, 1, 423–35.

[23] Certik, M., Sakuradani, E., Kobayashi, M. & Shimizu, S. (1999). Characterization of the second form of NADH-cytochrome b5 reductase gene from arachidonic acid-producing fungus *Mortierella alpina* 1S-4. Journal of Bioscience and Bioengineering, 88, 667–671.

[24] Chevalet, L, Tiraby, G., Cabane, B. & Loison, G. (1992). Transformation of *Aspergillus flavus*: construction of urate oxidase-deficient mutants by gene disruption. Current Genetics, 21, 447–53.

[25] Chisti, Y. & Yan, J. (2011). Energy from algae: current status and future trends: algal biofuels: a status report. Applied Energy, 88, 3277–3279.

[26] Choix, F.J., de-Bashan, L.E. & Bashan, Y. (2012) Enhanced accumulation of starch and total carbohydrates in alginate-immobilized Chlorella spp. induced by *Azospirillum brasilense*: II. Heterotrophic conditions. Enzyme and Microbial Technology, 51, 300–309.

[27] Conesa, A. Cees, A.M., van den Hondel, J.J. & Punt, P.J. (2000). Studies on the production of fungal peroxidases in *Aspergillus niger*. Applied and Environmental Microbiology, 66, 3016–3023.

[28] da Silva Ferreira, M.E., Kress, M.R., Savoldi, M., Goldman, M.H., Hartl, A., Heinekamp, T., Brakhage, A.A. & Goldman, G.H. (2006). The akuBKU80 mutant deficient for nonhomologous end joining is a powerful tool for analyzing pathogenicity in *Aspergillus fumigatus*. Eukaryot Cell, 5, 207–211.

[29] Dalboge, H. & Lange, L. (1998). Using molecular techniques to identify new microbial biocatalysts. Trends Biotechnology, 16, 265–272.

[30] Dalboge, H. (1997). Expression cloning of fungal enzyme genes; a novel approach for efficient isolation of enzyme genes of industrial relevance. FEMS Microbiology Reviews, 21, 29–42.

[31] de Faria, F.P., Te'o, V.S.J., Bergquist, P.L., Azevedo, M.O. & Nevalainen, K.M.H. (2002). Expression and processing of a major xylanase (XYN2) from the thermophilic fungus *Humicola grisea* var *thermoidea* in *Trichoderma reesei*. Letters in Applied Microbiology, 34, 119–123.

[32] de Groot, M.J., Bundock, P., Hooykaas, P.J. & Beijersbergen, A.G. (1998). Agrobacterium tumefaciens-mediated transformation of filamentous fungi. Nature Biotechnology,16, 839–42.

[33] de Jong, J.F., Deelstra, H.J., Wösten, H.A. & Lugones, L.G. (2006). RNA-mediated gene silencing in monokaryons and dikaryons of *Schizophyllum commune*. Applied and Environmental Microbiology, 72, 1267–9.

[34] de-Bashan, L.E., Moreno, M., Hernandez, J.-P. & Bashan Y. (2002) Removal of ammonium and phosphorus ions from synthetic wastewater by the microalgae *Chlorella vulgaris* coimmobilized in alginate beads with the microalgae growth promoting bacterium *Azospirillum brasilense*. Water Research, 36, 2941–2948.

[35] Dexter, J. & Fu, P. (2009). Metabolic engineering of cyanobacteria for ethanol production. Energy & Environmental Science, 2, 857–864.

[36] Do Nascimento, M., Ortiz-Marquez, J.C., Sanchez-Rizza, L., Echarte, M.M. & Curatti, L. (2012). Bioprospecting for fast growing and biomass characterization of oleaginous microalgae from South-Eastern Buenos Aires, Argentina. Bioresource Technology, 125, 283–290.

[37] Dowzer, C.E. & Kelly, J.M. (1991). Analysis of the creA gene, a regulator of carbon catabolite repression in *Aspergillus nidulans*. Molecular and Cellular Biology, 11, 5701–9.
[38] Dudasova, Z., Dudas, A. & Chovanec, M. (2004). Non-homologous end-joining factors of *Saccharomyces cerevisiae*. FEMS Microbiology Reviews, 28, 581–601.
[39] Durand, H., Baron, M., Calmels, T. & Tiraby, G. (1988). Classical and molecular genetics applied to *Trichoderma reesei* for the selection of improved cellulolytic industrial strains. In: Aubert, J.P., Beguin, P. & Millet, J. (Eds.), Biochemistry and Genetics of Cellulose Degradation, Academic Press, New York, pp. 135–151.
[40] Emalfarb, M.A. (2001). World Patent Application WO 01/25468.
[41] Fernandez-Martin, R., Cerda-Olmedo, E. & Avalos, J. (2000). Homologous recombination and allele replacement in transformants of *Fusarium fujikuroi*. Molecular and General Genetics, 263, 838–45.
[42] Fincham, J.R. (1989). Transformation in fungi. Microbiological reviews, 53,148–170.
[43] Flynn, K.J., Mitra, A., Greenwell, H.C. & Sui, J. (2013). Monster potential meets potential monster: pros and cons of deploying genetically modified microalgae for biofuels production. Interface Focus 3: 20120037.
[44] Fulci, V. & Macino, G. (2007). Quelling: post-transcriptional gene silencing guided by small RNAs in *Neurospora crassa*. Current Opinion in Microbiology, 10, 199–203.
[45] Gao, Z., Zhao, H., Li, Z., Tan, X. & Lu, X. (2012). Photosynthetic production of ethanol from carbon dioxide in genetically engineered cyanobacteria. Energy & Environmental Science, 5, 9857–9865.
[46] Georgianna, D.R. & Mayfield, S.P. (2012). Exploiting diversity and synthetic biology for the production of algal biofuels. Nature, 488, 329–335.
[47] Goins, C.L., Gerik, K.J. & Lodge, J.K. (2006). Improvements to gene deletion in the fungal pathogen *Cryptococcus neoformans*: absence of Ku proteins increases homologous recombination, and co-transformation of independent DNA molecules allows rapid complementation of deletion phenotypes. Fungal Genetics and Biology, 43, 531–44.
[48] Goldoni, M., Azzalin, G., Macino, G. & Cogoni, C. (2004). Efficient gene silencing by expression of double stranded RNA in *Neurospora crassa*. Fungal Genetics and Biology, 41, 1016–24.
[49] Gordon, C.J., Khajal, V., Ram, A.F.J, Archer, D.B., Brookman, J.L., Trinci, A.P.J., Jeenes, D.J., Doonan, J.H., Wells, B., Punt, P.J. van den Hondel, C.A. & Robson, G.D. (2000). Glucoamylase:green fluorescent protein fusion to monitor protein secretion in *Aspergillus niger*. Microbiology, 146, 415–426.
[50] Gouka, R.J., Punt, P.J. & van den Hondel, C.A.M.J.J (1997). Efficient production of secreted proteins by Aspergillus: progress, limitations and prospects. Applied Microbiology and Biotechnology, 47, 1–11.
[51] Gutiérrez-Correa, M., Ludeña, Y., Ramage, G. & Villena, G.K. (2012). Recent advances on filamentous fungal biofilms for industrial uses. Applied Biochemistry and Biotechnology, 167(5), 1235–1253.
[52] Ha et al., 2006.
[53] Ha, Y.S., Covert, S.F. & Momany, M. (2006). FsFKS1, the 1,3-beta-glucan synthase from the caspofungin-resistant fungus *Fusarium solani*. Eukaryot Cell, 5, 1036–42.
[54] Haakana, H., Miettinen-Oinonen, A., Joutsjoki, V., Mäntylä, A., Suominen, P. & Vehmaanperä, J. (2004). Cloning of cellulase genes from *Melanocarpus albomyces* and their efficient expression in *Trichoderma reesei*. Enzyme and Microbial Technology, 34, 159–167.
[55] Haarmann, T., Lorenz, N. & Tudzynski, P. (2008). Use of a nonhomologous end joining deficient strain (Δku70) of the ergot fungus *Claviceps purpurea* for identification of a nonribosomal peptide synthetase gene involved in ergotamine biosynthesis. Fungal Genetics and Biology, 45, 35–44.

[56] Haber, J.E. (1999). DNA repair. Gatekeepers of recombination. Nature, 398, 665–7.
[57] Hammond, T.M. & Keller, N.P. (2005). RNA silencing in *Aspergillus nidulans* is independent of RNA-dependent RNA polymerases. Genetics, 169, 607–17.
[58] Harkki, A., Uusitalo, J., Bailey, M., Penttilä, M. & Knowles, J.K.C. (1989). A novel fungal expression system: secretion of active calf chymosin from the filamentous fungus *Trichoderma reesei*. Nature Biotechnology, 7, 596–603.
[59] Harvey, M. & Pilgrim, S. (2010). The new competition for land: food, energy, and climate change. Food Policy 36 (Suppl. 1), S40–S51.
[60] Hektor, H.J. & Scholtmeijer, K. (2005). Hydrophobins: Proteins with potential. Current Opinion in Biotechnology, 16, 434–439.
[61] Henriksen, A.L.S., Even, S., Muller, C., Punt, P.J., van den Hondel, C.A.M.J.J. & Nielsen, J. (1999) Study of the glucoamylase promoter in *Aspergillus niger* using green fluorescent protein. Microbiology, 145, 729–734.
[62] Henry, C., Mouyna, I. & Latge, J.P. (2007). Testing the efficacy of RNA interference constructs in *Aspergillus fumigatus*. Current Genetics, 51, 277–284.
[63] Hinnen, A., Hicks, J.B. & Fink, G.R. (1978). Transformation of yeast. Proceedings of the National Academy of Sciences USA, 75, 1929–33.
[64] Hua, S.B., Qiu, M., Chan, E., Zhu, L. & Luo, Y. (1997). Minimum length of sequence homology required for in vivo cloning by homologous recombination in yeast. Plasmid, 38, 91–96.
[65] Isaacs, F.J., Dwyer, D.J. & Collins, J.J. (2006). RNA synthetic biology. Nature Biotechnology, 24, 545–54.
[66] Ishibashi, K., Suzuki, K., Ando, Y., Takakura, C. & Inoue, H. (2006). Nonhomologous chromosomal integration of foreign DNA is completely dependent on MUS- 53 (human Lig4 homolog) in Neurospora. Proceedings of the National Academy of Sciences USA, 103, 14871–6.
[67] Jalving, R., van de Vondervoort, P.J.I., Visser, J. & Schaap, P.J. (2000). Characterization of the kexin-like maturase of *Aspergillus niger*. Applied and Environmental Microbiology, 66, 363–368.
[68] Johnson, E. (2009). Goodbye to carbon neutral: getting biomass footprints right. Environmental Impact Assessment Review, 29, 165–168.
[69] Johnson, I.S. (1983). Human insulin from recombinant DNA technology. Science, 219:632–637.
[70] Joutsjoki, V.V., Kuittinen, M., Torkkeli, T.K. & Suominen, P.L., (1993b). Secretion of the *Hormoconis resinae* glucoamylase P enzyme from *Trichoderma reesei* directed by the natural and the cbh1 gene secretion signal. FEMS Microbiology Letters, 112, 281–286.
[71] Joutsjoki, V.V., Torkkeli, T.K. & Nevalainen, K.M.H. (1993a). Transformation of *Trichoderma reesei* with the *Hormoconis resinae* glucoamylase P (gamP) gene: production of a heterologous glucoamylase by *Trichoderma reesei*. Current Genetics, 24, 223–228.
[72] Kadotani, N., Nakayashiki, H., Tosa, Y. & Mayama, S. (2004). One of the two Dicer-like proteins in the filamentous fungi *Magnaporthe oryzae* genome is responsible for hairpin RNA-triggered RNA silencing and related small interfering RNA accumulation. Journal of Biological Chemistry, 279, 44467–74.
[73] Kasuya, T., Nakajima, H. & Kitamoto, K. (1999). Cloning and characterization of the bipA gene encoding ER chaperone BiP from *Aspergillus oryzae*. Journal of Bioscience and Bioengineering, 88, 472–478.
[74] Khalaj, V., Eslami, H., Azizi, M., Rovira-Graells, N. & Bromley, M. (2007). Efficient down regulation of alb1 gene using an AMA1-based episomal expression of RNAi construct in *Aspergillus fumigatus*. FEMS Microbiology Letters, 270, 250–4.
[75] Kiiskinen, L.L., Kruus, K., Bailey, M., Ylösmäki, E., Siika-Aho, M. & Saloheimo, M. (2004). Expression of *Melanocarpus albomyces* laccase in *Trichoderma reesei* and characterization of the purified enzyme. Microbiology, 150, 3065–3074.

[76] Kitamoto, N., Yoshino, S., Ohmiya, K. & Tsukagoshi, N. (1999). Sequence analysis, overexpression, and antisense inhibition of a beta-xylosidase gene, xylA, from *Aspergillus oryzae* KBN616. Applied and Environmental Microbiology, 65, 20–4.
[77] Kobayashi, M., Sakuradani, E. & Shimizu, S. (1999). Genetic analysis of cytochrome b5 from arachidonic acid-producing fungus, *Mortierella alpina* 1S-4: cloning, RNA editing and expression of the gene in *Escherichia coli*, and purification and characterization of the gene product. Journal of Biochemistry, 125, 1094–1103.
[78] Kooistra, R., Hooykaas, P.J. & Steensma, H.Y. (2004). Efficient gene targeting in *Kluyveromyces lactis*. Yeast, 21, 781–92.
[79] Krappmann, S., Sasse, C. & Braus, G.H. (2006). Gene targeting in *Aspergillus fumigatus* by homologous recombination is facilitated in a nonhomologous end-joining deficient genetic background. Eukaryot Cell, 5, 212–5.
[80] Krogh, B.O. & Symington, L.S. (2004). Recombination proteins in yeast. Annual Review of Genetics, 38, 233–271.
[81] Kruse, O., Rupprecht, J., Mussgnug, J. H., Dismukes, G.C., Pandey, D.P. & Gerdes, K. (2005). Improved photobiological H2 production in engineered green algal cells. Journal of Biological Chemistry, 280, 34170–34177.
[82] Kwon-Chung, K.J., Goldman, W.E., Klein, B. & Szaniszlo, P.J. (1998). Fate of transforming DNA in pathogenic fungi. Medical Mycology, 36(Suppl 1), 38–44.
[83] Lan, E.I. & Liao, J.C. (2011). Metabolic engineering of cyanobacteria for 1-butanol production from carbon dioxide. Metabolic Engineering, 13, 353–363.
[84] Lan, E.I. & Liao, J.C. (2012). ATP drives direct photosynthetic production of 1-butanol in cyanobacteria. Proceedings of the National Academy of Sciences USA, 109, 6018–6023.
[85] Larkum, A.W.D., Ross, I.L., Kruse, O. & Hankamer, B. (2012). Selection, breeding and engineering of microalgae for bioenergy and biofuel production. Trends in Biotechnology, 30, 198–205.
[86] Leathers, T.D. (2003). Biotechnological production and applications of pullulan. Applied Microbiology and Biotechnology, 62, 468–473.
[87] Lee ,J.W., Na, D., Park, J.M., Lee, J., Choi, S. & Lee, S.Y. (2012). Systems metabolic engineering of microorganisms for natural and non-natural chemicals. Nature Chemical Biology, 8, 536–546.
[88] Lee, D.G., Nishimura-Masuda, I., Nakamura, A., Hidaka, M., Masaki, H. & Uozumi, T. (1998). Overproduction of alpha-glucosidase in *Aspergillus niger* transformed with the cloned gene aglA. Journal of General and Applied Microbiology, 44, 177–81.
[89] Li, Y., Han, D., Hu, G., Dauvillee, D., Sommerfeld, M., Ball, S. & Hu, Q. (2010) *Chlamydomonas starchless* mutant defective in ADP-glucose pyrophosphorylase hyper-accumulates triacylglycerol. Metabolic Engineering 12, 387–391.
[90] Lindberg, P., Park, S. & Melis, A. (2010) Engineering a platform for photosynthetic isoprene production in cyanobacteria, using Synechocystis as the model organism. Metabolic Engineering, 12, 70–79.
[91] Linder, M.B., Szilvay, G.R., Nakari-Setala, T. & Penttila, M.E. (2005). Hydrophobins: the protein-amphiphiles of filamentous fungi. FEMS Microbiology Reviews, 29, 877–96.
[92] Liu, T., Wang, T., Li, X. & Liu, X. (2008). Improved heterologous gene expression in *Trichoderma reesei* by cellobiohydrolase I gene (cbh1) promoter optimization. Acta Biochimica et Biophysica Sinica, 40, 158–165.
[93] Liu, X.Y., Sheng, J. & Curtiss, R. (2011). Fatty acid production in genetically modified cyanobacteria. Proceedings of the National Academy of Sciences USA, 108, 6899–6904.
[94] Lombrana, M., Moralejo, F.J., Pinto, R. & Martin, J.F. (2004). Modulation of *Aspergillus awamori* thaumatin secretion by modification of bipA gene expression. Applied and Environmental Microbiology, 70, 5145–52.

[95] MacKenzie, D.A., Jeenes, D.J. & Archer, D.B. (2004). Filamentous fungi as expression systems for heterologous proteins. In: Esser, K., editor. The Mycota II. Genetics and Biotechnology (pp. 289–315). Springer-Verlag.

[96] MacKenzie, D.M., Jeenes, D.J., Gou, X. & Archer, D.B. (2000). Molecular basis of glucoamylase overproduction by a mutagenised industrial strain of *Aspergillus niger*. Enzyme and Microbial Technology, 26, 193–200.

[97] Malonek, S. & Meinhardt, F. (2001). Agrobacterium tumefaciens-mediated genetic transformation of the phytopathogenic ascomycete *Calonectria morganii*. Current Genetics, 40, 152–5.

[98] Mandels, M. & Reese, E.T. (1957). Induction of cellulase in *Trichoderma viride* as influenced by carbon source and metals. Journal of Bacteriology, 73, 269–278.

[99] Mäntylä, A. & Suominen, P. (1998). Industrial mutants and recombinant strains of *Trichoderma reesei*. In: Harman, G.E. & Kubicek, C.P. (Eds.), *Trichoderma* and *Gliocladium*, Enzymes, Biological Control and Commercial Applications, Vol 2 (, pp. 291–309.) Taylor and Francis: London.

[100] Maras, M., van Die, I., Contreras, R. & van den Hondel, C.A.M.J.J. (1999). Filamentous fungi as production organisms for glycoproteins of bio-medical interest. Glycoconjugate Journal, 16, 99–107.

[101] Margolles-Clark, E., Hayes, C.K., Harman, G.E. & Penttilä, M. (1996). Improved production of *Trichoderma harzianum* endochitinase by expression in *Trichoderma reesei*. Applied and Environmental Microbiology, 62, 2145–2151.

[102] Marston, F.A. (1986). The purification of eukaryotic polypeptides synthesized in *Escherichia coli*. Biochemical Journal, 240, 1–12.

[103] Mattern, I.E., van Noort, J.M., van den Berg, P., Archer, D.B., Roberts, I.N. & van den Hondel, C.A. (1992). Isolation and characterization of mutants of Aspergillus niger deficient in extracellular proteases. Molecular and General Genetics, 234, 332–336.

[104] Meyer, V. (2008). Genetic engineering of filamentous fungi — Progress, obstacles and future trends. Biotechnology Advances, 26, 177–185.

[105] Meyer, V., Arentshorst, M., El-Ghezal, A., Drews ,A.C., Kooistra, R., van den Hondel, C.A., Ram, A.F. (2007). Highly efficient gene targeting in the *Aspergillus niger* kusA mutant. Journal of Biotechnology, 128, 770–5.

[106] Meyer, V., Müller, D., Strowig, T. & Stahl, U. (2003). Comparison of different transformation methods for *Aspergillus giganteus*. Current Genetics, 43, 371–7.

[107] Meyer, V., Wedde, M. & Stahl, U. (2002). Transcriptional regulation of the antifungal protein in *Aspergillus giganteus*. Molecular Genetics and Genomics, 266:747–57.

[108] Michielse, C.B., Arentshorst, M., Ram, A.F. & van den Hondel, C.A. (2005a). Agrobacterium-mediated transformation leads to improved gene replacement efficiency in *Aspergillus awamori*. Fungal Genetics and Biology, 42, 9–19.

[109] Michielse, C.B., Hooykaas, P.J., van den Hondel, C.A. & Ram, A.F. (2005b). Agrobacterium-mediated transformation as a tool for functional genomics in fungi. Current Genetics, 48, 1–17.

[110] Miettinen-Oinonen, A., Torkkeli, T., Paloheimo, M. & Nevalainen, H. (1997). Overexpression of the Aspergillus niger pH 2.5 acid phosphatase gene in a heterologous host *Trichoderma reesei*. Journal of Biotechnology, 58, 13–20.

[111] Moralejo, F.J., Cardoza, R.E., Gutierrez, S. & Martin, J.F. (1999). Thaumatin production in *Aspergillus awamori* by use of expression cassettes with strong fungal promoters and high gene dosage. Applied and Environmental Microbiology, 65, 1168–1174.

[112] Moralejo, F.J., Cardoza, R.E., Gutierrez, S., Lombrana, M., Fierro, F. & Martin, J.F. (2002). Silencing of the aspergillopepsin B (pepB) gene of *Aspergillus awamori* by antisense RNA

expression or protease removal by gene disruption results in a large increase in thaumatin production. Applied and Environmental Microbiology, 68, 3550–9.

[113] Moriwaki, A., Ueno, M., Arase, S., Kihara, J. (2007). RNA-mediated gene silencing in the phytopathogenic fungus *Bipolaris oryzae*. FEMS Microbiology Letters, 269, 85–9.

[114] Mouyna, I., Henry, C., Doering, T.L. & Latge, J.P. (2004). Gene silencing with RNA interference in the human pathogenic fungus *Aspergillus fumigatus*. FEMS Microbiology Letters, 237, 317–24.

[115] Müller, D., Stahl, U. & Meyer, V. (2006). Application of hammerhead ribozymes in filamentous fungi. Journal of Microbiological Methods, 65,585–95.

[116] Müller, D., Stahl, U. (2004). RNA: a powerful tool for recombination and regulated expression of genes. In: Esser, K., editor. Progress in Botany, Vol. 66 (pp. 31–45). Springer-Verlag.

[117] Müller, E.M., Locke, E.G. & Cunningham, K.W. (2001). Differential regulation of two Ca(2+) influx systems by pheromone signaling in *Saccharomyces cerevisiae*. Genetics, 159, 1527–38.

[118] Mullins, E.D. & Kang, S. (2001). Transformation: a tool for studying fungal pathogens of plants. Cellular and Molecular Life Sciences, 58, 2043–52.

[119] Murray, P., Aro, N., Collins, C., Grassick, A., Penttilä, M., Saloheimo, M. & Tuohy, M. (2004). Expression in *Trichoderma reesei* and characterization of a thermostable family 3 β-glucosidase from the moderately thermophilic fungus *Talaromyces emersonii*. Protein Expression and Purification, 38, 248–257.

[120] Nakayashiki, H. (2005). RNA silencing in fungi: mechanisms and applications. FEBS Letters, 579, 5950–7.

[121] Nakayashiki, H., Hanada, S., Nguyen, B.Q., Kadotani, N., Tosa, Y. & Mayama, S. (2005). RNA silencing as a tool for exploring gene function in ascomycete fungi. Fungal Genetics and Biology, 42, 275–83.

[122] Namekawa, S.H., Iwabata, K., Sugawara, H., Hamada, F.N., Koshiyama, A., Chiku, H., Kamada, T. & Sakaguchi, K. (2005). Knockdown of LIM15/DMC1 in the mushroom Coprinus cinereus by double stranded RNA-mediated gene silencing. Microbiology, 151, 3669–78.

[123] Nayak, T., Szewczyk, E., Oakley, C.E., Osmani, A., Ukil, L., Murray, S.L., Hynes, M.J., Osmani, S.A. & Oakley, B.R. (2006). A versatile and efficient gene-targeting system for *Aspergillus nidulans*. Genetics, 172, 1557–66.

[124] Nevalainen, H. & Penttilä, M. (1995). Molecular biology of cellulolytic fungi. In: Kück, U. (Ed.), The Mycota, Genetics and Biotechnology, Vol. 2 (pp. 303–319). Springer Verlag: Berlin.

[125] Nevalainen K.M.H., Te'o V.S.J. & Bergquist P.L. (2005). Heterologous protein expression in filamentous fungi. Trends in Biotechnology, 23, 468–474.

[126] Nevalainen, H. & Peterson, R. 2014. Heterologous Expression of Proteins in *Trichoderma*. In: Vijay, K.G., Schmoll, M., Herrera-Estrella, A., Upadhyay, R.S., Druzhinina, I. & Tuohy, M.G. (eds), Biotechnology and biology of *Trichoderma*, (pp. 89–101). Elsevier: Netherland.

[127] Nevalainen, H., Saloheimo, M., Teeri, T., Penttilä, M., Harkki, A. & Knowles, J. (1990). *Trichoderma reesei* as production organism for enzymes for the pulp and paper industry. In: Kirk, T.K., Chang, H.-M. (Eds.), Biotechnology in Pulp and Paper Manufacture. Applications and Fundamental Investigations (pp. 593–599). Butterworth-Heinemann: USA.

[128] Ngiam, C., Jeenes, D.J., Punt, P.J., van den Hondel, C.A. & Archer, D.B. (2000). Characterization of a foldase, protein disulfide isomerase A, in the protein secretory pathway of *Aspergillus niger*. Applied and Environmental Microbiology, 66, 775–82.

[129] Nicolas, F.E., de Haro, J.P., Torres-Martinez, S. & Ruiz-Vazquez, R.M. (2007). Mutants defective in a *Mucor circinelloides* dicer-like gene are not compromised in siRNA silencing but display developmental defects. Fungal Genetics and Biology, 44, 504–16.

[130] Nielsen, J.B., Nielsen, M.L. & Mortensen, U.H. (2008). Transient disruption of nonhomologous end-joining facilitates targeted genome manipulations in the filamentous fungus *Aspergillus nidulans*. Fungal Genetics and Biology, 45(3), 165–70.

[131] Ninomiya, Y., Suzuki, K., Ishii, C. & Inoue, H. (2004). Highly efficient gene replacements in Neurospora strains deficient for nonhomologous end-joining. Proceedings of the National Academy of Sciences USA, 101, 12248–53.

[132] Nyyssönen, E., Penttilä, M., Harkki, A., Saloheimo, A., Knowles, J.K.C. & Keränen, S. (1993). Efficient production of antibody fragments by the filamentous fungus *Trichoderma reesei*. Nature Biotechnology, 11, 591–595.

[133] Olempska-Beer, Z.S., Merker, R.I., Ditto, M.D. & DiNovi, M.J. (2006). Food-processing enzymes from recombinant microorganisms — a review. Regulatory Toxicology and Pharmacology, 45, 144–58.

[134] Olguin, E.J. (2012). Dual purpose microalgae-bacteria-based systems that treat wastewater and produce biodiesel and chemical products within a biorefinery. Biotechnology Advances, 30, 1031–1046.

[135] Ooi, V.E. & Liu, F. (2000). Immunomodulation and anti-cancer activity of polysaccharide protein complexes. Current Medicinal Chemistry, 7, 715–29.

[136] Ortiz-Marquez, J.C.F., Do Nascimento, M., Zehr, J.P. & Curatti, L. (2013). Genetic engineering of multispecies microbial cell factories as an alternative for bioenergy production. Trends in Biotechnology, 31(9), 521–529.

[137] Paloheimo, M., Mäntylä, A., Kallio, J. & Suominen, P. (2003). High-yield production of a bacterial xylanase in the filamentous fungus *Trichoderma reesei* requires a carrier polypeptide with an intact domain structure. Applied and Environmental Microbiology, 69, 7073–7082.

[138] Paloheimo, M., Mäntylä, A., Kallio, J., Puranen, T. & Suominen, P. (2007). Increased production of xylanase by expression of a truncated version of the xyn11A gene from *Nonomuraea flexuosa* in *Trichoderma reesei*. Applied and Environmental Microbiology, 73, 3215–3224.

[139] Paloheimo, M., Miettinen-Oinonen, A., Torkkeli, T., Nevalainen, H. & Suominen, P. (1993). Enzyme production by Trichoderma using the cbh1 promoter. In: Suominen, P., Reinikainen, T. (Eds.), Proceedings of the Tricel 93 Symposium Workshop, vol 8 (, pp. 229–238). Foundation for Biotechnical and Industrial Fermentation Research. Espoo: Finland.

[140] Passorn, S., Laoteng, K., Rachadawong, S., Tanticharoen, M. & Cheevadhanarak, S. (1999) Heterologous expression of *Mucor rouxii* D12 desaturase gene in *Saccharomyces cerevisiae*. Biochemical and Biophysical Research Communications, 263, 47–51.

[141] Penttilä, M.E., Nevalainen, H., Salminen, E., Rättö, M. & Knowles, J. (1987). Versatile transformation system for the cellulolytic filamentous fungus *Trichoderma reesei*. Gene 61, 155–164.

[142] Peters, J.W., Boyd, E.S., D'Adamo, S., Mulder, D.W., Therien, J. & Posewitz, M.C. (2013). Hydrogenases, nitrogenases, anoxia, and H2 production in water-oxidizing phototrophs. In Algae for Biofuels and Energy (Borowitzka, M.A. and Moheimani, N.R., eds), (pp. 37–75). Springer

[143] Peterson, R. & Nevalainen, H. (2012). *Trichoderma reesei* RUT-C30—thirty years of strain improvement. Microbiology, 158, 59–68.

[144] Petkov, G., Ivanova, A., Iliev, I. & Vaseva, I. (2012). A critical look at the microalgae biodiesel. European Journal of Lipid Science and Technology, 114, 103–111.

[145] Pöggeler, S. & Kück, U. (2006). Highly efficient generation of signal transduction knockout mutants using a fungal strain deficient in the mammalian ku70 ortholog. Gene, 378, 1–10.

[146] Poidevin, L., Levasseur, A., Paës, G., Navarro, D., Heiss-Blanquet, S., Asther, M. & Record, E. (2009). Heterologous production of the Piromyces equi cinnamoyl esterase in *Trichoderma reesei* for biotechnological applications. Letters in Applied Microbiology, 49, 673–678.

[147] Polizeli, M.L., Rizzatti, A.C., Monti, R., Terenzi, H.F., Jorge, J.A. & Amorim, D.S. (2005). Xylanases from fungi: properties and industrial applications. Applied Microbiology and Biotechnology, 67, 577–91.

[148] Punt, P.J., van Biezen, N., Conesa, A., Albers, A., Mangnus, J. & van den Hondel, C. (2002). Filamentous fungi as cell factories for heterologous protein production. Trends in Biotechnology, 20(5), 200–206.

[149] Qin, L.N., Cai, F.R., Dong, X.R., Huang, Z.B., Tao, Y., Huang, J.Z. & Dong, Z.Y. (2012). Improved production of heterologous lipase in *Trichoderma reesei* by RNAi mediated gene silencing of an endogenic highly expressed gene. Bioresource Technology, 109, 116–122.

[150] Rasala, B.A., Gimpel, J.A., Tran, M., Hannon, M.J., Miyake-Stoner, S.J., Specht, E.A. & Mayfield, S.P. (2013). Genetic engineering to improve algal biofuels production. In: Algae for Biofuels and Energy (Borowitzka, M.A. & Moheimani, N.R., eds), (pp. 99–113). Springer

[151] Riha, K., Watson, J.M., Parkey, J. & Shippen, D.E. (2002). Telomere length deregulation and enhanced sensitivity to genotoxic stress in Arabidopsis mutants deficient in Ku70. EMBO Journal, 21, 2819–26.

[152] Rubio, L.M. & Ludden, P.W. (2008) Biosynthesis of the ironmolybdenum cofactor of nitrogenase. Annual Review of Microbiology, 62, 93–111.

[153] Ruijter, G.J. & Visser, J. (1997). Carbon repression in *Aspergilli*. FEMS Microbiology Letters, 151, 103–14.

[154] Ruiz-Diez, B. (2002). Strategies for the transformation of filamentous fungi. Journal of Applied Microbiology, 92, 189–95.

[155] Saarelainen, R., Mäntylä, A., Nevalainen, H. & Suominen, P. (1997) Expression of barley endopeptidase B in *Trichoderma reesei*. Applied and Environmental Microbiology, 63, 4938–4940.

[156] Sakuradani, E., Kobayashi, M. & Shimizu, S. (1999). Identification of an NADH-cytochrome b5 reductase gene from an arachidonic acid-producing fungus, *Mortierella alpina* 1S-4, by sequencing of the encoding cDNA and heterologous expression in a fungus, *Aspergillus oryzae*. Applied and Environmental Microbiology, 65, 3873–3879.

[157] Salles, B.C., Te'o, V.S., Gibbs, M.D., Bergquist, P.L., Filho, E.X., Ximenes, E.A. & Nevalainen, K.M.H. (2007). Identification of two novel xylanase- encoding genes (xyn5 and xyn6) from *Acrophialophora nainiana* and heterologous expression of xyn6 in *Trichoderma reesei*. Biotechnology Letters, 29, 1195–1201.

[158] Saloheimo, M., Lund, M. & Penttilä, M.E. (1999). The protein disulphide · isomerase gene of the fungus *Trichoderma reesei* is induced by endoplasmic reticulum stress and regulated by the carbon source. Molecular and General Genetics, 262, 35–45.

[159] Saloheimo, M., Pakula, T., Aro, N. & Joensuu, J.J. (2014). Protein Production—Quality Control and Secretion Stress Responses in *Trichoderma reesei*. In: Gupta, V.K., Schmoll, M., Herrera-Estrella, A., Upadhyay, R.S., Druzhinina, I. & Tuohy, M.G. (eds), Biotechnology and biology of *Trichoderma*, (pp. 81–88). Netherland; Elsevier Publisher.

[160] Saunders, G, Picknett, T.M., Tuite, M.F. & Ward, M. (1989). Heterologous gene expression in filamentous fungi. TIBTECH – 7, 283–287.

[161] Scholtmeijer, K., Wessels, J.G. & Wösten, H.A. (2001). Fungal hydrophobins in medical and technical applications. Applied Microbiology and Biotechnology, 56, 1–8.

[162] Seo, J.S., Chong, H., Park, H.S., Yoon, K.O., Jung, C., Kim, J.J., Hong, J.H., Kim, H., Kim, J.H., Kil, J.I., Park, C.J., Oh, H.M., Lee, J.S., Jin, S.J., Um, H.W., Lee, H.J., Oh, S.J., Kim, J.Y., Kang, H.L., Lee, S.Y., Lee, K.J. & Kang, H.S. (2005). The genome sequence of the ethanologenic bacterium *Zymomonas mobilis* ZM4. Nature Biotechnology, 23, 63–68.

[163] Setubal, J.O.C. et al. (2009) Genome sequence of Azotobacter vinelandii, an obligate aerobe specialized to support diverse anaerobic metabolic processes. Journal of Bacteriology, 191, 4534–4545.

[164] Shen, C.R., Lan, T.I., Dekishima, Y., Baez, A., Cho, K.M. & Liao, J.C. (2011) Driving forces enable high-titer anaerobic 1-butanol synthesis in *Escherichia coli*. Applied and Environmental Microbiology, 77, 2905–2915.

[165] Shimayama, T., Nishikawa, S. & Taira, K. (1995). Generality of the NUX rule: kinetic analysis of the results of systematic mutations in the trinucleotide at the cleavage site of hammerhead ribozymes. Biochemistry, 34, 3649–54.

[166] Shiotani, H. & Tsuge, T. (1995). Efficient gene targeting in the filamentous fungus *Alternaria alternata*. Molecular and General Genetics, 248, 142–50.

[167] Shroff, R.A., O'Connor, S.M., Hynes, M.J., Lockington, R.A. & Kelly, J.M. (1997). Null alleles of creA, the regulator of carbon catabolite repression in *Aspergillus nidulans*. Fungal Genetics and Biology, 22, 28–38.

[168] Siddiquee, S., Cheong, B.E., Taslima, K., Hossain, K. & Hasan, M.M. (2012). Separation and identification of volatile compounds from liquid cultures of *Trichoderma harzianum* by GC-MS using three different capillary columns. Journal of Chromatographic Science, 50, 358–367.

[169] Siedenberg, D., Mestric, S., Ganzlin, M., Schmidt, M., Punt, P.J., van den Hondel, C.A.M.J.J. & Rinas, U. (1999). GlaA promoter controlled production of a mutant green fluorescent protein (S65T) by recombinant *Aspergillus niger* during growth on defined medium in batch and fed-batch cultures. Biotechnology Progress, 1, 43–50.

[170] Simon, J.R. & Moore, P.D. (1987). Homologous recombination between single-stranded DNA and chromosomal genes in *Saccharomyces cerevisiae*. Molecular and Cellular Biology, 7, 2329–34.

[171] Spencer, A., Morozov-Roche, L.A., Noppe, W., MacKenzie, D.A., Jeenes, D.J., Joniau, M., Dobson, C.M. & Archer, D.B. (1999). Expression, purification and characterization of the recombinant calcium-binding equine lysozyme secreted by the filamentous fungi *Aspergillus niger*: comparisons with the production of hen and human lysozymes. Protein Expression and Purification, 16, 171–180.

[172] Srirangan, K., Pyne, M.E. & Chou, C.P. (2011) Biochemical and genetic engineering strategies to enhance hydrogen production in photosynthetic algae and cyanobacteria. Bioresource Technology, 102, 8589–8604.

[173] Strik, D.P.B.T.B., Timmers, R.A., Helder, M., Steinbusch, K.J.J., Hamelers, H.V.M. & Buisman C.J.N. (2011). Microbial solar cells: applying photosynthetic and electrochemically active organisms. Trends in Biotechnology, 29, 41–49.

[174] Subashchandrabose, S.R., Ramakrishnan, B., Megharaj, M., Venkateswarlu, K. & Naidu, R. (2011) Consortia of cyanobacteria/microalgae and bacteria: biotechnological potential. Biotechnology Advances, 29, 896–907.

[175] Sun, A., Nevalainen, H., Te'o, J. & Godlewski, M. (2012). Heterologous expression of the human peptide hormone obestatin in *Trichoderma reesei*. Poster, European Conference on Fungal Genetics, 30 March–2 April, Marburg, Germany, 2012.

[176] Surzycki, R., Cournac, L., Peltier, G., Rochaix, J.D. (2007). Potential for hydrogen production with inducible chloroplast gene expression in Chlamydomonas. Proceedings of the National Academy of Sciences USA, 104, 17548–17553.

[177] Takahashi, T., Masuda, T. & Koyama, Y. (2006). Enhanced gene targeting frequency in ku70 and ku80 disruption mutants of *Aspergillus sojae* and *Aspergillus oryzae*. Molecular Genetics and Genomics, 275, 460–70.

[178] Tan, X. Yao, L., Gao, Q., Wang, W., Qi, F. & Lu, X. (2011). Photosynthesis driven conversion of carbon dioxide to fatty alcohols and hydrocarbons in cyanobacteria. Metabolic Engineering, 13, 169–176.

[179] Te'o, V.S.J., Cziferszky1, A.E., Bergquist, P.L. & Nevalainen, K.M.H. (2000). Codon optimisation of the xylanase gene xynB from the thermophilic bacterium *Dictyoglomus thermophilum* for expression in the filamentous fungus *Trichoderma reesei*. FEMS Microbiology Letters, 190, 13–19.

[180] Tilman, D., Socolow, R., Foley, J.A., Hill, J., Larson, E., Lynd, L., Pacala, S., Reilly, J., Searchinger, T., Somerville, C. & Williams, R. (2009). Beneficial biofuels: the food, energy, and environment trilemma. Science, 325, 270–271.

[181] Torzillo, G., Scoma, A., Faraloni, C., Ena, A. & Johanningmeier, U. (2009). Increased hydrogen photoproduction by means of a sulfur-deprived *Chlamydomonas reinhardtii* D1 protein mutant. International Journal of Hydrogen Energy 34, 4529–4536.

[182] Ueno, K., Uno, J., Nakayama, H., Sasamoto, K., Mikami, Y. & Chibana, H. (2007). Development of a highly efficient gene targeting system induced by transient repression of YKU80 expression in *Candida glabrata*. Eukaryotic Cell, 6, 1239–47.

[183] Van de Ven, W.J., Roebroek, A.J. & Van Duijnhoven, H.L. (1993). Structure and function of eukaryotic proprotein processing enzymes of the subtilisin family of serine proteases. Critical reviews in oncogenesis, 4(2), 115–136.

[184] Van den Hombergh, J.P., van de Vondervoort, P.J.I, Fraissinet-Tachet, L., Visser, J. (1997). *Aspergillus* as a host for heterologous protein production: the problem of proteases. Trends in Biotechnology, 15, 256–263.

[185] Verdoes, J.C., Punt, P.J. & van den Hondel, CA. (1995). Molecular genetic strain improvement for the overproduction of fungal proteins by filamentous fungi. Applied Microbiology and Biotechnology, 43, 195–205.

[186] Villalba, F., Collemare, J., Landraud, P., Lambou, K., Brozek, V., Cirer, B., Morin, D., Bruel, C., Beffa, R. & Lebrun, M-H. (2008). Improved gene targeting in *Magnaporthe grisea* by inactivation of MgKU80 required for non-homologous end joining. Fungal Genetics and Biology, 45, 68–75.

[187] Wälti, M.A., Villalba, C., Buser, R.M., Grunler, A., Aebi, M. & Kunzler, M. (2006). Targeted gene silencing in the model mushroom Coprinopsis cinerea (Coprinus cinereus) by expression of homologous hairpin RNAs. Eukaryotic Cell, 5, 732–44.

[188] Wang, H. & Ward, M. (2000). Molecular characterization of a PDI-related gene prpA in *Aspergillus niger var. awamori*. Current Genetics, 37, 57–64.

[189] Wang, Z.T., Ullrich, N., Joo, S., Waffenschmidt, S. & Goodenough, U. (2009). Algal lipid bodies: stress induction, purification, and biochemical characterization in wild-type and starchless *Chlamydomonas reinhardtii*. Eukaryotic Cell, 8, 1856–1868.

[190] Ward, M., Wilson, L.J., Kodama, K.H., Rey, M.W. & Berka, R.M. (1990). Improved production of chymosin in *Aspergillus* by expression as a glucoamylase-chymosin fusion. Biotechnology, 8, 435–440.

[191] Weidner, G., d'Enfert, C., Koch, A., Mol, P.C. & Brakhage, A.A. (1998). Development of a homologous transformation system for the human pathogenic fungus *Aspergillus fumigatus* based on the pyrG gene encoding orotidine 5'- monophosphate decarboxylase. Current Genetics, 33, 378–85.

[192] Weyman, P.D., Vargas, W.A., Tong, Y., Yu, J., Maness, P.-C., Smith, H.O. & Xu, Q. (2011). Heterologous expression of *Alteromonas macleodii* and *Thiocapsa roseopersicina* [NiFe] hydrogenases in *Synechococcus elongatus*. PLoS ONE 6, e20126.

[193] Whittaker, S.L., Lunness, P., Milward, K.J., Doonan, J.H. & Assinder, S.J. (1999). sodIVC is an alpha-COP-related gene which is essential for establishing and maintaining polarized growth in *Aspergillus nidulans*. Fungal Genetics and Biology, 26, 236–252.

[194] Wijffels, R.H. & Barbosa, M.J. (2010). An outlook on microalgal biofuels. Science, 329, 796–799.

[195] Willke, T. & Vorlop, K.D. (2001). Biotechnological production of itaconic acid. Applied Microbiology and Biotechnology, 56, 289–295.

[196] Work, V.H., Radakovits, R., Jinkerson, R.E., Meuser, J.E., Elliott, L.G., Vinyard, D.J., Laurens, L.M.L., Dismukes G.C. & Posewitz, M.C. (2010). Increased lipid accumulation in the *Chlamydomonas reinhardtii* sta7–10 starchless isoamylase mutant and increased carbohydrate synthesis in complemented strains. Eukaryotic Cell, 9, 1251–1261.

[197] Wu, T.S. & Linz, J.E. (1993). Recombinational inactivation of the gene encoding nitrate reductase in *Aspergillus parasiticus*. Applied and Environmental Microbiology, 59, 2998–3002.

[198] Wynn, J.P., Hamid, A.B.A. & Ratledge, C. (1999). The role of malic enzyme in the regulation of lipid accumulation in filamentous fungi. Microbiology, 145, 1911–1917.

[199] Yamada, O., Ikeda, R., Ohkita, Y., Hayashi, R., Sakamoto, K. & Akita, O. (2007). Gene silencing by RNA interference in the koji mold *Aspergillus oryzae*. Bioscience, Biotechnology, and Biochemistry, 71, 138–44.

[200] Yoshino, F., Ikeda, H., Masukawa, H. & Sakurai, H. (2007). High photobiological hydrogen production activity of a *Nostoc* sp. PCC 7422 uptake hydrogenase-deficient mutant with high nitrogenase activity. Marine Biotechnology, 9, 101–112.

[201] Young, C., Itoh, Y., Johnson, R., Garthwaite, I., Miles, C.O., Munday-Finch, S.C. & Scott, B. (1998). Paxilline-negative mutants of *Penicillium paxilli* generated by heterologous and homologous plasmid integration. Current Genetics, 33, 368–377.

[202] Yu, J.H., Hamari, Z., Han, K.H., Seo, J.A., Reyes-Dominguez, Y. & Scazzocchio, C. (2004). Double-joint PCR: a PCR-based molecular tool for gene manipulations in filamentous fungi. Fungal Genetics and Biology, 41, 973–981.

[203] Zhang, F., Carothers, J.M. & Keasling, J.D. (2012). Design of a dynamic sensor-regulator system for production of chemicals and fuels derived from fatty acids. Nature Biotechnology, 30, 354–359.

[204] Zheng, X.F., Kobayashi, Y. & Takeuchi, M. (1998). Construction of a low-serine-typecarboxy-peptidase- producing mutant of Aspergillus oryzae by the expression of antisense RNA and its use as a host for heterologous protein secretion. Applied Microbiology and Biotechnology, 49, 39–44.

[205] Zhong, Y., Liu, X., Xiao, P., Wei, S. & Wang, T. (2011). Expression and secretion of the human erythropoetin using an optimized cbh1 promoter and the native CBHI signal sequence in the industrial fungus *Trichoderma reesei*. Applied Biochemistry and Biotechnology, 165, 1169–1177.

[206] Zou, G., Shi, S.H., Jiang, Y., van den Brink, J., de Vries, R.P., Chen, L., Zhang, J., Ma, L., Wang, C.S. & Zhou, Z.H. (2012). Construction of a cellulase hyper-expression system in *Trichoderma reesei* by promoter and enzyme engineering. Microbial Cell Factories, 11, 21.

Karla Victoria Jiménez-Guevara, Leonel E. Amábilis-Sosa, Landy Irene Ramírez-Burgos, Irina Salgado-Bernal, and María-del-Carmen Durán-Domínguez-de-Bazúa

13 Evaluation of an indirect method for rapid bacterial quantification in wastewaters

13.1 Abstract

The statistical certainty of an indirect method for bacterial quantification in different types of wastewaters was evaluated. A calibration curve was built by a metals tolerant bacterial strains consortium and quantified the number of microorganisms per CFU mL^{-1} with its respective absorbance. Strains were characterized of genus Bacillus, Gram positive and with presence of endospores. With the calibration curve, theoretical CFU mL^{-1} of six wastewater industrial and municipal samples were obtained, and analysis r&R was applied to dataset for statistical validation. The results indicated three samples with a high degree of certainty and certainty to optimize in another sample. Two samples that did not show validity, exhibited pH values below 3.

13.2 Introduction

Among the major problems facing society today, the generation, treatment, and disposal of wastewaters is a serious one, since wastewaters are characterized by their large quantities, pollutants diversity, and overall geographical and environmental site problems (Bernal-González *et al.*, 2015).

In addition to the low efficiency in depuration by many wastewater treatment plants, there exists a large amount of waste liquid effluents that are either untreated or only partially treated and are therefore discharged into surface water bodies and/or soils (WHO, 2001).

The main source of water bodies pollution are wastewaters discharges, which are defined as wastewaters of varied composition, coming from discharges of domestic installations, municipal networks, and industrial and agricultural uses. These facts mean that probably contain pollutants coming from household uses, public services, mining, chemical, paper, textile, pharmaceuticals industries among others, as well as agricultural and farmer wastes, which depending on the source of wastewaters, may contain heavy metals, pesticides, pharmaceuticals compounds, nutriments, organic matter, and microorganisms, viruses, etc., plus the mixture of some of them (Sarkis & Cordeiro, 2001).

Considering the above, surface waters pollution due to uncontrolled wastewaters discharges has become a problem that can cause from minor disturbances to destruction of habitats due to toxic compounds, eutrophication phenomena, and bioaccumulation of heavy metals and organic compounds in trophic chains. In general, these effects do not only cause affectations to the hydrologic cycle, but to all ecosystems (Lüderitz, 2001).

In specific, the microorganisms contained in the wastewaters, regardless of the source, are considered as pollutants, because some species are pathogenic, causing many human diseases. Some particular microorganisms are not harmful to health but they perform process that sustain life and are used for the benefit of humanity, through biotechnological developments ranging from improving agricultural commodities, up to catalysis, particularly of difficult steps in chemical reactions, and naturally including antibiotics (Lui *et al.*, 2007; Korneli *et al.*, 2013; Miethe *et al.*, 2013).

Among the general characteristics of the microorganisms, size and surface/volume ratios are two of the more evident, since the diameter of the majority of the bacteria is lower to one millimeter, meaning about one micrometer (1 μm = 10^{-3} mm). This small size results in a very large ratio between exposed superficial or surface area and its volume, which renders a great interaction with the environment. There results in a high exchange of matter, which considering that bacteria assimilate organic matter and other nutrients by osmosis, is ideal for assimilation of biodegradable organic matter and nutrients, when these are considered as pollutants (Schmidt-Nielson, 1984).

Besides, these microorganisms have as advantages the adaptation to inhospitable environments as are industrial wastewaters (acid or basic pH values, extreme temperatures, and high concentrations of pollutants) and they have an easy propagation in numerous culture media (different sources of carbon, nitrogen and phosphorus compounds) (application of Bergmann's classic paper, 1847).

Moreover, bacteria are also used for the treatment of toxic compounds such as heavy metals, because they have developed tolerance through endospores, although the vast majority of the bacteria, able to develop these mechanisms, are Gram positive *Bacillus* (Cornell University, 2015).

With all the above, the wastewater treatment methods most widely used are the socalled biological systems (traditional secondary stage in a wastewater treatment plant) in which microorganisms are the fundamental component of the bioreactors, due to the fact that they are responsible for the removal efficiency of the contaminants, because they degrade organic matter. Likewise bacteria transform nitrogen and phosphorus compounds in wastewaters. Thus, the removal efficiency of contaminants is in accordance with the presence of the prokaryotic organisms (Metcalf-Eddy, 1972).

In this regard, treatment and accurate quantification of bacteria are procedures of great importance for the treatment of all wastewaters.

Generally, microbial quantification assays are grouped into direct methods and indirect methods. In the first ones, the total amount of microorganisms (living and dead) is determined, while in the second ones only the viable microorganisms (living ones) is quantified (Ramírez-Gama *et al.*, 2012).

One of the most used methods for quantification of microorganisms in wastewater is the most probable number (MPN), classified as a technique for viable cell count of coliforms and based on the theory of probability. The enumeration of coliform organisms can be done by using solid and liquid culture mediums with selective and differential characteristics. The main drawback in using the most probable number is that wastewater does not only contain coliform bacteria, for which it is specific, and thus, the rest of prokaryotic organisms are not quantified. Although it is a good indicator for pathogenic wastewater issues it has this drawback.

With respect to the direct counting of colony-forming unit (CFU), this method is internationally accepted and endorsed. Furthermore, it has the advantage of generating a great certainty in the results to count bacterial cell, due to the fact that it considers at least seven dilutions from an initial sample, and these repetitions are also used as statistics replicas. The disadvantage of this method is that it requires a large number of reagents and materials, which finally implies more analytical process time and higher analytical costs.

Furthermore, there is an indirect quantification of the total number of microorganisms by turbidimetry, wherein in a water sample, the amount of bacterial cells is related to the turbidity or optical density and both are inversely proportional to the amount of light that can penetrate the water column, measured as transmittance.

Thus, the number of microorganisms present in a suspension by measuring turbidity might be an interesting alternative, since transmittance can be measured with a nephelometer or a spectrophotometer (Madigan *et al.*, 2009).

The microbial suspension is deposited into a chamber or translucent tube for leading a constant light source to the suspension (incident light) and by the use of a photoelectric cell transmitted light intensity is recorded. The methodology is useful for suspensions of low and medium microbial densities, where microorganisms are unicellular and with sizes within the scale of few micrometers. These characteristics allow them to remain suspended and evenly or homogenously distributed. In contrast, this type of methodology is not suitable with the presence of larger microorganisms and with those that are polysaccharides producers (Ramírez-Gama *et al.*, 2012).

Based on the above statements, the optical density is proportional (within some limits) to the number of bacterial cells, so turbidity reading can be used as a substitute for direct counting methods. Nonetheless, if turbidity is used for this purpose, it is first necessary to build a calibration curve, statistically relating the indirect measurement (turbidity) with a direct measurement of the number of microbial cells (microscopic or recount of viable cells), previously standardized. Furthermore, turbidity measurements should be precise and have the virtue of being rapid, economic, and easy to perform. Besides, these determinations should be carried out without destroying or modifying the sample.

Here, a rapid method of bacterial quantification is proposed, based on statistical calibrations, with precision and accuracy, considering as reference the results of the colony-forming unit method.

13.3 Methodology

13.3.1 Bacterial isolates

Bacterial strains used in this study, were selected from research conducted by Salgado *et al.* (2012), who isolated microorganisms from hydrophyte plants rhizosphere of the specie *Typha dominguensis*, located in the natural wetlands hydrographic basin Almendares-Vento, City of The Habana, Cuba. In fact, these microorganisms were also evaluated assessing its tolerance to heavy metals of environmental importance, such as Pb, Cr, and Hg (Amábilis-Sosa *et al.*, 2015b).

Of the isolated strains, those identified for the authors as TAN 117, TAN 119, TAN 217, TAN 1113, and TAN 1115 were chosen for this research. All of them were of genus *Bacillus*. These strains were selected due to its high tolerance to chromium, mercury, and lead, since these toxic compounds were used to gradually reduce the bacterial concentration and thus, were suitable for developing a calibration curve of wide range as described later.

The selected strains were also morphologically characterized and evaluated using Gram stain and presence of endospores, using selective staining techniques (Amábilis-Sosa, 2015).

Furthermore, effect of dissolved oxygen over microbial strains growth was determined by thioglycolate culture medium. The purpose of this analysis is to know if the calibration tests and bacterial counts are applicable to all wastewaters, particularly to those with anoxic-anaerobic conditions (Masters & Ela, 2008).

13.3.2 Calibration curve and bacterial quantification

The validity of the indirect quantification method proposed is based on statistical similarity in comparison to the direct method of quantification of colony forming units (CFU), which is endorsed and standardized by international entities. In this manner, first, a calibration curve was built by plotting concentrations as units (abscissas) versus the absorbance data measured for each concentration (ordinates).

Secondly, the microbial consortia with the five selected strains were incubated at 33 °C during 48 hours in five 250 mL Erlenmeyer flasks containing nutrient broth (nutrient agar). Besides, a series of flask containing different concentrations of the mixture of mercury, lead and chromium salts, in order to inhibit microbial growth in controlled ranges were set. According to this, in Table 13.1, the concentrations of heavy metals used for getting the different bacterial concentrations are indicated. These experiments were based on the methodology proposed by Amábilis-Sosa *et al.* (2015a,b). Once applied the different heavy metals concentrations as microbial growth inhibiting substances to the bacterial consortia for each level (Table X.1), two samples were taken. One was used to determine CFU by the procedure described by

Benson (2002), and the other one was used for reading the absorbance. Absorbances measurements were carried out to a wavelength of 400 nm using a UV-Visible spectrophotometer GBC Cintra 5.

Tab. 13.1: Bacterial concentration as a function of applied heavy metals concentrations (± standard deviation).

Lead, mg L^{-1}	Mercury, mg L^{-1}	Chromium, mg L^{-1}	Bacterial concentration CFU $mL^{-1} \times 10^5$
0	0	0	300 ± 29
0.82	0.003	0.53	270 ± 31
5.5	0.02	3.5	220 ± 21
16.5	0.06	10.5	100 ± 14
33	0.12	21	39 ± 3
66	0.24	42	0.12 ± 0.01
110	0.4	70	0.02 ± 0.01

The results of absorbance and bacterial concentration, in CFU mL^{-1} were related by means of a calibration curve, where each absorbance value corresponded to a bacterial concentration in order to obtain the resulting curve empirical equation.

Additionally, six wastewater real samples from different sources were collected for measuring its absorbances. These absorbances values were used for calculating their theoretical concentration in CFU mL^{-1} through the calibration curve empirical equation previously obtained. In parallel, bacterial concentrations were determined to the same six samples by the indirect CFU method, and the results were statistically contrasted with the theoretical CFU (obtained from the calibration curve). All analyses were conducted with five replicates.

13.3.3 Analytic determinations and statistical analyzes

Conductivity, pH, redox potential, total suspended solids, volatile solids, and organic carbon, expressed as chemical oxygen demand (COD), were measured for each of the six samples of real wastewater. The first three determinations were made using a multi-parameter water quality sonde Hanna model HI 9828. Meanwhile, suspended solids were quantified by the official Mexican act method (DOF, 2001a), and COD values were obtained by the procedure described in the official Mexican act method (DOF, 2001b). These parameters may indicate the applicability of the proposed quantification method through unusual values, and as a whole, they may be able to indicate the type of wastewater and whether this method is applicable.

To evaluate to certainty of the proposed indirect method, all theoretical results of bacterial concentration were analysed using the statistical method r&R (reproducibility and repeatability), which allows to evaluate both accuracy and precision. The first through the observed variation in the process and the second by a study of accurateness, which examines the differences between average data of the observed quantifications (theoretical) and a reference value or "master value", which in this case is equal to the concentrations in CFU mL^{-1}, measured experimentally (Montgomery, 2012).

The certainty evaluation for the experiment was performed following the ASTM E 691 procedure (1999) and the ISO (1995), where r&R values <10 are reliable values between laboratories or between methods (as for this case). Values higher to 10 and lower to 30 are unreliable, but they can be processed by some analytical or statistical adjustment for acceptance. Finally, the values r&R that had too much variation and uncertainty were not be taken into account.

A covariance analysis was also applied to compare the values of the physicochemical determinations for the different wastewaters studied, and the results of the r&R tests, in order to identify which type of wastewaters were more prone to the applicability of the proposed method. For the analysis all data gathered, the Minitab 15 for Windows statistical package was used.

13.4 Results and discussion

13.4.1 Morphological characterization

In Figure 13.1, morphological characteristics for each one of strains used in this study are shown. Likewise, Gram classification and its genetic identification of the gender of each strain according to the database were published in GenBank (simultaneously available in EMBL, Europe and DNA Data Bank, Japan) under the reference of Salgado *et al.* (2012).

The description in Figure 13.1 indicated that all bacterial strains were of the genus Bacillus and are Gram positive. Result might indicate that its tolerance to heavy metals is derived from the cell wall characteristics, a finding that will guide to future research, as well as by having added protection to its cytoplasm, where surface adsorption phenomena (physical sorption and van der Waals forces) are normally performed, and that seemed to be mainly composed of peptidoglycan (Abou-Shanab *et al.*, 2007; Rathnayake *et al.*, 2010).

In addition to this, the Gram positive bacteria are those that can generate the presence of endospores for its protection, which is another mechanism that may allow its subsistence in the presence of heavy metals (Moynahan *et al.*, 2002; Srinath *et al.*, 2002).

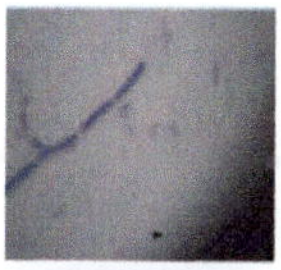

Strain ID: TAN 117
Observation Date: January 20,2014
Target: 100x
Straining and sorting: Gram positive
Description: Bacilli long chain
Genres: *Bacillus subtilis, Bacillus mojavensis and bacillus vallsmortis*

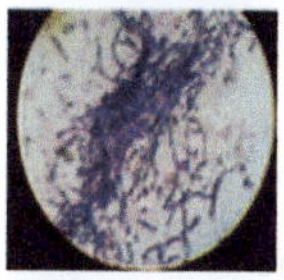

Strain ID: TAN 1113
Observation Date: January 20,2014
Target: 100x
Straining and sorting: Gram positive
Description: Bacilli long chain
Genres: *Bacillus thuringiensis and Bacillus mycoides*

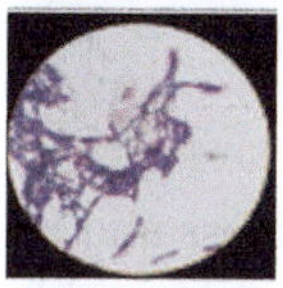

Strain ID: TAN 217
Observation Date: January 20,2014
Target: 100x
Straining and sorting: Gram positive
Description: short and long bacilli no apparent clustering
Genres: *Bacillus sp*

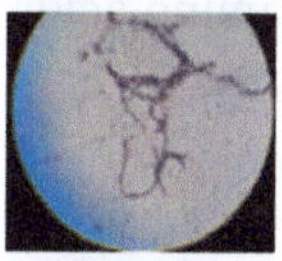

Strain ID: TAN 1115
Observation Date: January 20,2014
Target: 100x
Straining and sorting: Gram positive
Description: short diplobacilos no apparent clustering
Genres: *Bacillus thuringiensis and Bacillus mycoides*

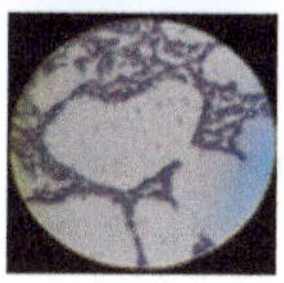

Strain ID: TAN 119
Observation Date: January 20,2014
Target: 100x
Description: short bacilli without grouping
Genres: *Bacillus thuringiensis and Bacillus mycoides and Bacillus weihenstephonensis*

Fig. 13.1: Morphology of each bacterial strain, observed by microscopy.

According to the above mentioned statements, it was observed in the five bacterial strains the presence of endospores when malachite green staining was performed (Hussey & Zayaitz, 2015). Endospores were observed through the green cast (some encased in yellow circles) which were the result of selective staining (Fig. 13.2 and Fig. 13.3).

Using the same images, it can be also noted that some correspond to either terminal, sub terminal, and central rates of intracellular location. Spores are very resistant cells, and thanks to its design they can resist the most aggressive physical and chemical conditions, such as heat, heavy metals, corrosive elements or compounds, and other physical and chemical agents. These resistance properties are due to the highly

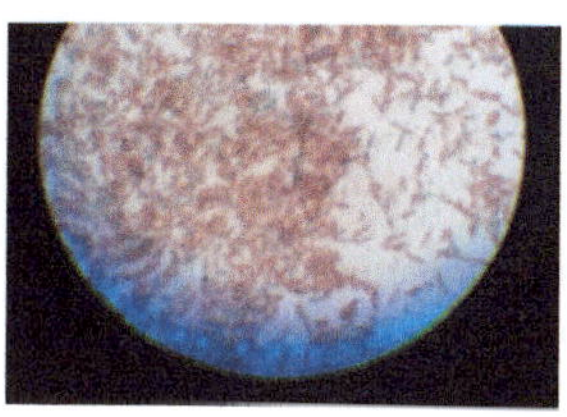

Fig. 13.2: Terminal position of the endospore corresponding to TAN 115.

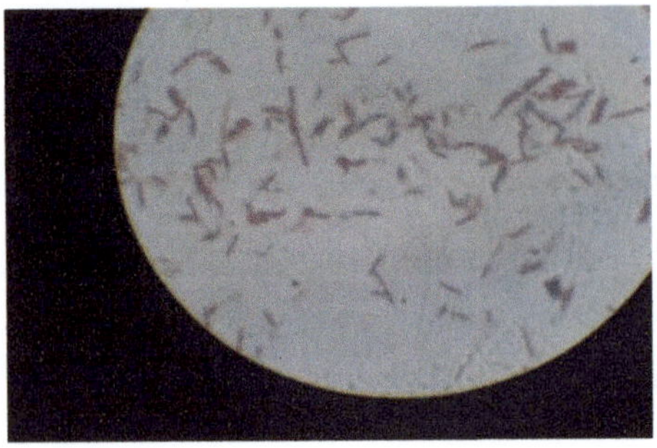

Fig. 13.3: Subterminal position of the endospore corresponding to TAN 117.

selective permeability and composition of their covering layers, its thickness and its composition, mostly based on protein-rich compounds of hydrophobic amino acids and abundant disulphide bridges (Moynahan *et al.*, 2002; Madigan *et al.*, 2009).

13.4.2 Calibration curve

Figure 13.4 shows the samples to produce the calibration curve. Each one contained different concentrations quantified both in CFU as absorbance. At the start of the experiment, all were inoculated in the same manner and proportions, using the bacterial consortium (same nutrient broth, with the same physicochemical characteristics). That is, the different bacterial concentrations were produced by the different concentrations of heavy metals, applied previously to incubation of the microorganisms.

Figure 13.5 refers to the calibration curve, which was constructed based on the results of their respective absorbances and concentrations (CFU mL^{-1}) of the samples considered in Figure 4. According to the generated equation (Fig. 13.5), a correlation factor of 99.35 % was obtained (equivalent to the certainty of using the equation of this curve).

Besides, it was observed that the correlation corresponds to a first order model, indicating as a first approach that the indirect method is efficient. As the bacterial concentration is diminished, although the asymptotic point appears at about

Fig. 13.4: Samples that completed the data used for the bacterial calibration curve. From left to right the density of the consortia is reduced.

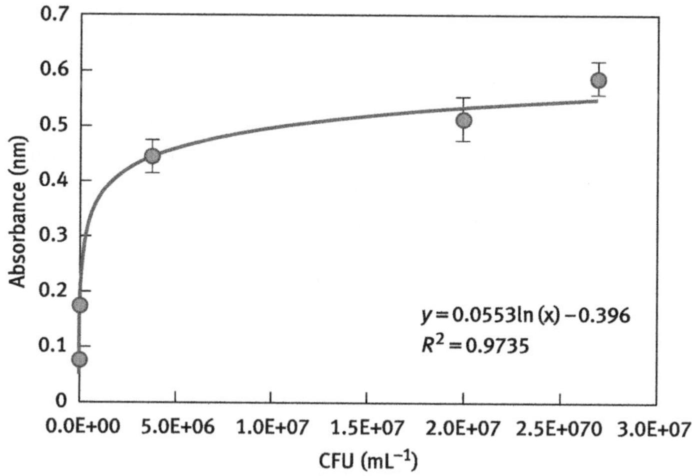

Fig. 13.5: Calibration curve for the correlation between CFU mL^{-1} of bacterial consortia used and their respective absorbance measurements.

20×10^6 CFU mL^{-1}, a value worth over up to four times to that one observed for municipal wastewater (Crites & Tchobanoglous, 2000).

Derived of the tolerance of the consortium used to heavy metals, it was possible to draw a curve with a wide range of bacterial concentrations. Results gathered are supported by those described by Hassen *et al.* (1998) who report that the mixture of heavy metals, due to it are inorganic compounds, oxidize the organic compounds dissolved in the nutrient broth, allowing testing absorbances with microbial consortium.

This procedure has advantages when compared to other studies on inhibitors of microbial growth, especially when the microorganisms used are precisely tolerant to heavy metals (Amor *et al.*, 2001).

13.4.3 Characterization of wastewater samples for application of calibration curve

Table 13.2 shows the description and characterization of wastewater samples, intentionally chosen from different sources (domestic sewage, municipal effluents, and industrial wastewaters) to determine the application degree of the proposed indirect method. That is, to know what kind of samples, the precision and accuracy of the indirect method is certain, according to the experimental method of CFU.

In the Table 13.2, it can be seen that the samples from the mining process exhibit high COD values despite of the fact that samples were colourless and with oxidizing conditions, which clearly denotes industrial processes in which there are complex substances with a high organic and inorganic load (Masters & Ela, 2008). The physicochemical values of the samples of wastewater from domestic and municipal origins, coincide with the characteristic values extensively reported in the literature (Crites & Tchobanoglous, 2000; Davis & Cornwell, 2008; Masters & Ela, 2008).

As a particular case, the wastewater from the treatment plant at the National Autonomous University of Mexico (UNAM in Spanish) that has been treated by an activated sludge system followed by a secondary settling vessel (influent of a chlorination step) have higher values of redox potential and lower values of VSS, in comparison with other samples (Tab. 13.2), which is characteristic of wastewater that have been effectively treated by secondary settling (Crites & Tchobanoglous, 2000; Davis & Cornwell, 2008; Masters & Ela, 2008).

Tab. 13.2: Results of characterization of wastewater samples used for evaluating the proposed indirect method.

Wastewater source	Qualitative description	Absorbance Λ = 400 nm	Characterization parameters				
			COD, mg L^{-1}	VSS, mg L^{-1}	OPR, mV	pH	Conductivity, mS cm^{-1}
UNAM treatment plant (influent)	Brown suspended particles, unpleasant odour	0.6055	462.5	141.3	−15	8.4	1.27
UNAM treatment plant influent to chlorination step (affluent)	Colourless and odourless	0.0159	297	18.5	80	7.5	0.997
Water Tizapa mine effluent flotation process 1	Slight turbidity and suspended particles, odourless	0.0734	687	27.3	28.1	7.1	3.2
Water Tizapa mine effluent flotation process 2	Slight turbidity and suspended particles, odourless	0.0542	687	22.6	50	6.7	3.5
Suspended biomass reactor industrial effluent at laboratory scale 1	Slight turbidity and suspended particles, odourless	0.0065	62.5	7.2	80	2.5	2.12
Suspended biomass reactor industrial effluent at laboratory scale 2	Slight turbidity and suspended particles, odourless	0.0084	27.5	23.1	83	1.9	2.02

13.4.4 Validation of indirect method and influence of physicochemical parameters

Table 13.3 shows the results in CFU mL^{-1} of the wastewater samples calculated by the indirect method and by the technique of CFU analytically validated. Derived from what it is mentioned in the methodology, a repeatability and reproducibility analysis (r&R) was performed, which also included a variance analysis. According to this, in the last column of Table 13.3 the final result of the test of certainty is indicated.

Tab. 13.3: Colonies former-unit for the direct method and indirect method. ± Standard deviation in percent.

Origin of wastewater	CFU mL^{-1} theoretical (calibration curve)	CFU mL^{-1} quantified	Precision (reproducibility indirect method), % r	Accuracy (precision indirect method), % R	Ratio r&R, %
UNAM treatment plant (influent)	1.71×10^6	6.3×10^6±8.4 %	8.2	3.5	22.31
UNAM treatment plant influent to chlorination step (effluent)	7.33×10^1	1.2×10^1±7.6 %	9.3	4.3	8.42
Water Tizapa mine effluent flotation process 1	4.85×10^3	1.15×10^3±8.6 %	8.7	3.8	8.56
Water Tizapa mine effluent flotation process 2	3.43×10^3	3.0×10^3±7.2 %	5.3	3.6	2.3
Suspended biomass reactor industrial effluent at laboratory scale 1	1.44×10^3	1.4×10^1±9.3 %	12.8	45.6	40.9
Suspended biomass reactor industrial effluent at laboratory scale 2	1.49×10^3	1.6×10^1±6.3	11.3	39.7	45.3

Of the six samples tested, the influent to chlorination step from the University wastewater treatment plant (WWTP) and the two in the Tizapa mine effluents have a r&R ratio below 10, indicating that the applied indirect method coincides with precision and accuracy with the analytical method. It is worth mentioning that the effluents of the Tizapa mine are an industrial type of process effluents, which is denoted by the characterization showed in Table 13.2.

The sample of the influent water to the WWTP CU despite being a conventional municipal type, exhibited a r&R ratio above 10 but less than 30 (Tab. 13.3) indicating that it is possible to use the indirect method, as long as the ranges of measurements be analytical and statistically adjusted (ASTM E 691, 1999).

Samples of effluent from the suspended biomass reactor showed r&R ratios above 30, indicating that the indirect method was not applicable to these effluents. It is important to mention that the wastewaters treated by these reactors come from an oil refinery and contain many toxic compounds such as phenols, which when degraded, result in intermediate compounds which modify the characteristics of the wastewater, including the pH which was extremely acid (Tab. 13.2) which may affect the viability of the bacterial cells (Nava-Urrego *et al.*, 2014).

Finally, with regard to analysis of multiple covariance between the parameters of Table 13.3 and the relationships r&R, a negative covariance with respect to volatile suspended solids, VSS, was found ($P<0.05$), considering that this variable is a measurement of an engineering quantifying method of the microorganisms present in the wastewater. The fact that the covariance is negative (a larger number of VSS, less certainty of r&R) might precisely indicate that the principle that quantification by turbidimetry is more efficient at lower concentrations of microorganisms (Madigan *et al.*, 2009), which was already been evidenced by the equation in Figure 13.5, with a first-order behaviour.

Likewise, an analysis of covariance also indicated a sensitivity response with pH, which statistically means that the changes in values of r&R are varying to a greater degree when those of pH do. However, the covariance was not found as such. Here again it should be highlighted that the effect caused by the acidity (pH 1.9–2.5) of the samples from the effluent from suspended biomass reactor might be an issue.

13.5 Conclusions

Based on the obtained results, it can be conclude that the indirect method of rapid bacterial quantification is suitable for domestic and municipal wastewater. It could also be used for some industrial wastewater, provided that they possess pH values within conventional ranges as it was the case for process waters from a flotation step in mining procedures. To corroborate this, it is necessary to conduct more studies focusing on identifying the effect of pH and other physicochemical parameters.

Likewise, it is concluded that the bacterial consortia used in this experiment resulted appropriate to build the calibration curve, which was the basis of calculation for all intercalibration results.

13.6 Acknowledgements

Authors acknowledge the partial financial support from Mexico's UNAM, DGAPA, PAIME Programs "Apoyo a la enseñanza experimental de los laboratorios terminales de las carreras que se imparten en la Facultad de Química de la UNAM", "Apoyo a la enseñanza experimental de las asignaturas terminales de las carreras que se imparten en la Facultad de Química de la UNAM", and "Desarrollo de material didáctico para las asignaturas ingeniería ambiental y estancia académica de la carrera de ingeniería química con base en estudios de caso" (in Spanish) EN103704, PE101709, and PE-100514, respectively. Also, a partial support was received from the Programa de Apoyo a la Investigación y el Posgrado, PAIP, Facultad de Química, UNAM (in Spanish) 50009067.

13.7 References

[1] Abou-Shanab, R.A.I., van Berkum, P. & Angle, J.S. (2007). Heavy metal resistance and genotypic analysis of metal resistance genes in Gram-positive and Gram-negative bacteria present in Ni-rich serpentine soil and in the rhizosphere of Alyssum murale. Chemosphere, 68(2), 360–367.

[2] Amábilis-Sosa, L.E. (2015). Remoción de metales pesados en humedales artificiales inoculados con cepas tolerantes. Doctoral thesis. UNAM, Mexico D.F., Mexico. http://132.248.9.195/ptd2015/mayo/509007094/Index.html

[3] Amabilis-Sosa, L.E., Siebe, Ch., Moeller, G. & Durán-Domínguez-de-Bazúa, M.C. (2015a). Accumulation and distribution of lead and chromium in laboratory-scale constructed wetlands inoculated with metal-tolerant bacteria. International Journal of Phytoremediation, DOI: 10.1080/15226514.2015.1021953

[4] Amábilis-Sosa, L.E., Siebe, Ch., Moeller-Chávez, G. & Durán-Domínguez-de-Bazúa, M.C. (2015b). Mercury, chromium and lead removal by constructed wetlands inoculated with tolerant strains (*Remoción de mercurio, cromo y plomo por humedales artificiales inoculados con cepas tolerantes*). Tecnol. Ciencias Agua (in Spanish), 6(1), 21–34.

[5] Amor, L., Keenes, C. & Veiga, M.C. (2001). Kinetics of inhibition in the biodegradation of monoaromatic hydrocarbons in presence of heavy metals. Biores. Technol., 78(2), 181–185.

[6] ASTM E 691. (1999). Standard practice for conducting an interlaboratory study to determine the precision of a test method. American Society for Testing and Materials. West Conshohocken, PA, US.

[7] Benson HJ. (2002). Microbiological applications: Laboratory Manual in General Microbiology, short version, McGraw Hill, Boston, MA, US.

[8] Bergmann, C. (1847). Über die Verhältnisse der Wärmeökonomie der Tiere zu ihrer Grösse. Göttinger Studien, 3 (1), 595–708.

[9] Bernal-González, M., Toscano-Pérez, D., Falcón, L.I., Alcántara-Hernández, R.J., Salgado-Bernal, I., Solís-Fuentes, J.A. & Durán-Domínguez-de-Bazúa, M.C. (2015). Archaea, a useful group for energy unconventional production: Methane production from sugarcane secondary distillation effluents using thermotolerant strains. In Microbial Bioresources. V.K. Gupta, G.D. Sharma, R. Gaur, M.G. Tuohy, eds. Chapter 23, U.K.: CABI Books.

[10] Cornell University. 2015. https://micro.cornell.edu/research/epulopiscium/bacterial-endospores

[11] Crites, R. & Tchobanoglous, G. (2000). Tratamiento de aguas residuales en pequeñas poblaciones. McGraw-Hill, Bogotá, Colombia.
[12] Davis, M.T. & Cornwell, D.A. (2008). Introduction to Environmental Engineering. Mc GrawHill, 4th Edition, NY, US.
[13] DOF. (2001a). Norma Mexicana NMX-AA-034-SCFI-2001. Análisis de Agua-Determinación de sólidos y sales disueltas en aguas naturales, residuales y residuales tratadas. Método de Prueba. Diario Oficial de la Federación, Gobierno de los Estados Unidos Mexicanos. Mexico DF, Mexico.
[14] DOF. (2001b). Norma Mexicana NMX-AA-030-SCFI-2001. Análisis de Agua-Determinación de la demanda química de oxígeno en aguas naturales, residuales y residuales tratadas. Método de Prueba. Diario Oficial de la Federación, Gobierno de los Estados Unidos Mexicanos. Mexico DF, Mexico.
[15] Hassen, A., Saidi, N., Cherif, M. & Boudabous, A. (1998). Resistance of environmental bacteria to heavy metals. Biores. Technol., 64(1), 7–15.
[16] Hussey, M.A. & Zayaitz, A. (2015). Endospore Stail Protocol. American Society for Microbiology. MicrobeLibrary. http://www.microbelibrary.org/component/resource/laboratory-test/3112-endospore-stain-protocol,2015.
[17] ISO. (1995). Guide to the expression of uncertainty in measurement. BIPM, IEC, ISO, OIML.
[18] Korneli, C., David, F., Biedendieck, R. & Jahn, D. (2013). Getting the big beast to work-systems biotechnology of *Bacillus megaterium* for novel high-value proteins. J. Biotechnol., 163(2), 87–96.
[19] Lüderitz, V. (2001). Towards sustainable water resources management: A case study from Saxony-Anhalt, Germany. Manag. Environ. Qual. Int. J., 15(1), 17–24.
[20] Lui, J., Dong, Y., Xu, H., Wang, D. & Xu, J. (2007). Accumulation of Cd, Pb, and Zn by 19 wetland plant species in constructed wetlands. J. Hazard. Mat., 147(3), 947–653.
[21] Madigan, T.M., Martinko, M.J., Dunlap, V.P. & Clark, P.D. (2009). Brock Biologia de los microorganismos. Pearson Educación, SA, Madrid, Spain.
[22] Masters, G. & Ela, W. (2008). Introduction to Environmental Engineering. Prentice Hall, NY, US.
[23] Metcalf-Eddy. (1972). Wastewater engineering: Collection, treatment, disposal. McGraw-Hill. New York, NY, US.
[24] Miethe, S., Meyer, T., Wöhl-Bruhn, S., Frenzel, A., Schirrmann, T., Dübel, S. & Hust, M. (2013). Production of single chain fragment variable (scFv) antibodies in *Escherichia coli* using LEX bioreactor. J. Biotechnol., 163, 105–111.
[25] Moynahan, O.S., Zabinski, C.A. & Gannon, J.E. (2002). Microbial community structure and carbon-utilization diversity in a mine tailings revegetation study. Restor. Ecol., 10(1), 77–87.
[26] Montgomery, D. (2012). Design and analysis of experiments. Limusa, Wiley, Mexico D.F., Mexico.
[27] Nava-Urrego, L.M., Gasperín-Sánchez, R. & Duran, A. (2014). Comparación de un reactor de biomasa suspendida y un reactor de biomasa adherida para la biodegradación de compuestos tóxicos presentes en aguas residuales de refinerías de petróleo. Rev. Int. Contam. Ambie. (in Spanish), 30(1), 101–112.
[28] Ramírez-Gama, R.M., Luna, B., Velázquez, O., Vierna, L., Mejía, A.G., Tsuzuki, M.G., Hernández, L., Camacho, A., Urzúa, M.C. (2012). Manual de prácticas de microbiología general. UNAM, Facultad de Química, Mexico, DF, Mexico. http://amyd.quimica.unam.mx/pluginfile.php/3447/mod_resource/content/1/Protocolosdepracticas2012_18791.pdf
[29] Rathnayake, I.V.N., Megharaj, M., Bolan, N. & Naidu, R. (2010). Tolerance of heavy metals by Gram positive soil bacteria. Int. J. Civil Environ. Eng., 2(4), 191–195.
[30] Salgado, I., Durán, M.C., Cruz, M., Carballo, M. & Martínez, A. (2012). Bacterias rizosféricas con potencialidad fisiológicas para eliminar materia orgánica de aguas residuales. Rev. Int. Contam. Ambie., 28(1), 17–26.

[31] Sarkis, J. & Cordeiro, J.J. (2001). An empirical evaluation of environmental efficiencies and firm performance: pollution prevention versus end-of-pipe practice. Europ. J. Operation Res., 135(1), 102–113.
[32] Schmidt-Nielson, K. (1984). Scaling: Why is Animal Size so Important? Cambridge University Press, New York, NY, US.
[33] Srinath, T., Verma, T., Ramteke, P.W. & Garg, S.K. (2002). Chromium (VI) biosorption and bioaccumulation by chromate resistant bacteria. Chemosphere, 48(4), 427–435.
[34] WHO. (2001). Upgrading water treatment plants. Wagner, E.G.& Pinheiro, R.G. ISBN 0-419-26050-1 (pbk) 0-419-26040-4. Published on behalf of the World Health Organization. Spon Press London, UK. New York, US. http://www.bvsde.paho.org/bvsacd/who/pinup.pdf

Abdollah Ghasemian, and Zahra Moradpour

14 Cyanobacteria: biotechnological and environmental applications

14.1 Introduction

Cyanobacteria or blue-green algae are prokaryotic microalgae classified as a phylum of eubacteria (Castenholz *et al.*, 2001). The blue-green appearance of these bacteria is originated by the two pigments, chlorophyll *a* (green) and phycocyanin (blue). Cyanobacteria can survive in wide variety of environments. They are photoautotrophic bacteria that mean they can harvest energy from the LIGHT source with their chlorophyll *a* pigment. Also, they consume CO_2 to produce organic compounds. They have a substantial task in the recycling of CO_2 via their photosynthesis, which is similar to photosynthetic plants in oxygen generation. However, some species of cyanobacteria survive under mixotrophic and heterotrophic conditions and as consortia with other microorganisms (Subashchandrabose *et al.*, 2011, 2013). Using water as an electron donor during photosynthesis resulting in the release of oxygen is another important characteristic of these tiny organisms. Furthermore, some cyanobacteria have the ability to fix nitrogen.

According to *Bergey's Manual of Systematic Bacteriology* (Castenholz, *et al.*, 2001), the phylum of cyanobacteria includes 5 orders.

- Order I: Chroococcales are unicellular or nonfilamentous colonial aggregate of cells that reproduce by binary fission or by budding. The most important families are Chroococcaceae (genera *Chroococcus* and *Dactylococcopsis*) and Microcystaceae (genera *Microcystis*, *Anacystis*, *Chondrocystis*, and *Polycystis*).
- Order II: Pleurocapsales are unicellular or nonfilamentous colonial aggregates of cells that reproduce by internal multiple fissions. The most important genus in this order is *Cyanocystis*.
- Order III: Oscillatoriales are filamentous cells that reproduce by binary division in one plane. The families are Spirulinaceae (genus *Spirulina*), Oscillatoriaceae (genera *Lyngbya*, *Oscillatoria* and *Limnoraphis*) and Microcoleaceae (genus *Microcoleus*).
- Order IV: Nostocales are filamentous cells that reproduce by binary division in one plane. The families include Microchaetaceae, Nostocaceae (genera *Anabaena*, *Nostoc* and *Nodularia*), Rivulariaceae (genera *Rivularia* and *Mastigonema*) and Scytonemataceae (genera *Synechococcus*, *Neosynechococcus* and *Thermosynechococcus*).
- Order V: Stigonematales are filamentous cells that reproduce by binary division frequently in more than one plane and produce true branching. The most important genera in this order are *Fischerella* and *Stigonema*.

Reproduction in cyanobacteria occurs principally by binary fission or budding. The doubling time of cyanobacteria is usually about 24 h or more except for some unicellular and oscillatorian strain that is less than 6 h (Carr & Whitton, 1982). Cyanobacteria have a Gram-negative cell wall with a thicker peptidoglycan layer than that of Gram-negative bacteria such as *Proteobacteria*. In several unicellular, colonial, and filamentous cyanobacteria, an envelope (outer sheath) covers this cell wall (Hoiczyk & Baumeister, 1995). The photosynthetic apparatus of cyanobacteria called thylakoid which is formed by endogination of the cytoplasmic membrane (Peschek & Sleytr, 1983). Phycobiliproteins on thylakoids absorb light for the photosystems. These proteins contribute to the blue-green appearance of cyanobacteria (Schluchter *et al.*, 2010). The cytoplasm of cyanobacterial vegetative cells contains several granule components. Polyglucose (glycogen) (Gomez Casati *et al.*, 1999), cyanophycin granules (polymers of arginine and aspartic acid) (Simon, 1971), polyhedral bodies (carboxysomes) which are composed of ribulose bisphosphate carboxylase/oxygenase (RUBISCO) (Stewart & Codd, 1975) and volutin (polyphosphate) granules (Allen, 1984) are among the most important and frequent of these components (Figure 14. 1).

The genome of cyanobacteria is located in nucleoid region called centroplasm. The size of genomes in cyanobacteria (128 strains that are representatives of all main orders) ranges from 1.6 to 8.6×10^6 kD (Herdman *et al.*, 1979). Cyanobacteria from the orders Nostocales and Stigonematales produce some unique differentiated cells. They have a specialized cell in their termini or between vegetative cells of cyanobacteria called heterocyst for fixation of nitrogen in anaerobic condition. These cells are formed when cyanobacteria experience nitrogen deficit in their environments. Heterocysts provide the anaerobic environment required for nitrogen fixation and separate photosynthetic apparatus in vegetative cells from nitrogen-fixing machinery (Adams, 2000). Several of the heterocyst-forming cyanobacteria produce spores or akinetes.

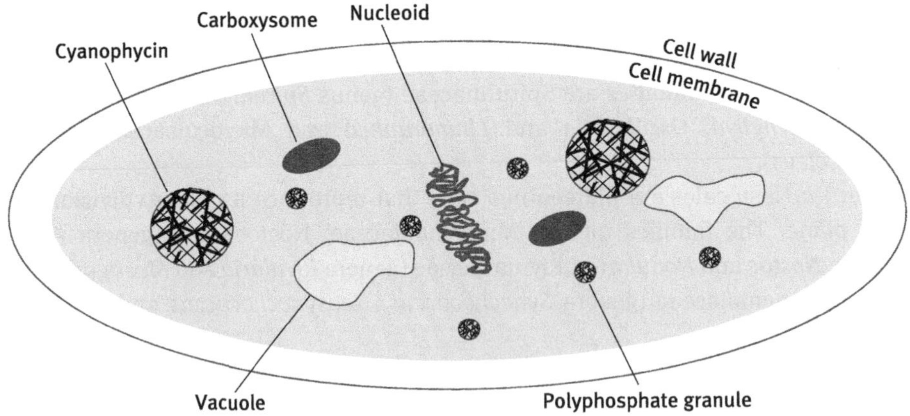

Fig. 14.1: Schematic morphology of a cyanobacterium.

They are resting cells that have thicker walls than vegetative cells. Akinetes are produced under nutrient or light insufficiency. Akinetes contain all vegetative granule components, except polyphosphates granules (Adams & Duggan, 1999).

14.2 Biotechnological applications

Cyanobacteria are comparable in their chemical prolificity (in both aspects of productivity and abundance) with microorganisms such as *myxobacteria* and *streptomycetes* (Nunnery *et al.*, 2010). Potential applications of cyanobacteria in biotechnology have achieved a lot much interest in recent years. Owing to their abilities to synthesize peptide with non-ribosomal system (Polyketide Synthases) and having diverse series of methylases, oxidases, tailoring and other modification enzymes, cyanobacteria are proficient in producing diverse natural products with different chemical structure such as linear peptides (Simmons *et al.*, 2006), cyclic peptides (Sisay *et al.*, 2009), linear lipopeptides (Nogle *et al.*, 2001), depsipeptides (Han *et al.*, 2005), cyclic depsipeptides (Soria-Mercado *et al.*, 2009), fatty acid amides (Chang *et al.*, 2011), swinholides (Andrianasolo *et al.*, 2005), glycomacrolides (Teruya *et al.*, 2009) or macrolactones (Salvador *et al.*, 2010). Moreover, cyanobacteria also possess cytochrome P450 system, a superfamily of oxygenases containing a heme in their structure that give them the capacity for synthesizing and biotransformation of bioactive molecules such as steroids, terpenes and alkaloids, and biodegradation of xenobiotics and contaminants including pesticides, dyes and hydrocarbons (Ke *et al.*, 2005; Kühnel *et al.*, 2008; Alder *et al.*, 2009).

The biotechnological applications of cyanobacteria in diverse areas have been well-documented (Namikoshi & Rinehart, 1996; Singh *et al.*, 2005; Sivonen & Börner, 2008). Cyanobacteria are sources of biologically active molecules such as biofertilizers, biopolymers, biofuels, and food. They are used for biotransformation of different chemicals and have presently been under exploration in medicine and medical diagnostics. Considering the diverse potential uses of cyanobacteria in biotechnology, an overview of these topics is presented here.

14.2.1 Cyanobacteria as biofertilizers

Reduced soil fertility is one of the most important causes of declining food production world-wide. Biofertilizers is a kind of natural fertilizer which includes living organisms. These organisms use natural ingredients such as light, nitrogen, and water to make soil fertile by increasing the supply of essential nutrients to plants. Cyanobacteria are one of the most promising biofertilizers (Mishra & Pabbi, 2004). The use of cyanobacteria as biofertilizers (algalization) can improve growth of plants and harvest yield (Vaishampayan *et al.*, 2001). Field trials confirmed that cyanofertilizer may improve the rice yield by 12–19 % (Mishra & Pabbi, 2004). The cyanobacteria from orders nostocales and stigonematales such as *Anabaena, Nostoc* and *Oscillatoria* absorb the light through photosynthesis process to fix N_2 from the atmosphere and convert it into the form of nitrogen that is usable for plants (Obreht *et al.*, 1993).

The major problem to widespread utilization of cyanobacterial biofertilizer is to find a method to produce an economical and effective substitute for chemical fertilizer (Hashem, 2001). Another challenge is the adaptation of cyanobacteria species with domestic soil. Harnessing of local-isolated cyanobacteria and harvesting them in ponds beside to fields is one of the proposed solutions to overcome these challenges. Therefore, the farmers are able to produce inexpensively their biofertilizers on-site to be employed in their own fields.

14.2.2 Cyanobacteria as a resource for bioenergy

Due to rapid rises in the utilization of energy, constant growing concerns of ecological change and the restricted nature of fossil fuels, the need to develop renewable and sustainable resources is of significant importance (Dincer, 2000).

Biohydrogen is a renewable, commercially viable and eco-friendly source of fuel that does not release CO_2 during production and combustion (Tamagnini *et al.*, 2002). Filamentous cyanobacteria with heterocyst produce biohydrogen more competently than those without (Dutta *et al.*, 2005; Pinzon-Gamez *et al.*, 2005). Cyanobacteria produce hydrogen with two mechanisms. Heterocystous cyanobacteria with nitrogenase activity produce hydrogen during fixation of nitrogen into ammonia accompanied by the reduction of a proton (H^+) to hydrogen. These filamentous cyanobacteria produce hydrogen when they are cultivated under nitrogen restrictive situation (Das & Veziroğlu, 2001). The other mechanism for hydrogen production is through bidirectional hydrogenase enzymes that catalyze both utilization and generation of molecular hydrogen and are found in the in both N_2- and non-N_2-fixing strains of blue-green algae (Tamagnini, *et al.*, 2002).

Production of biohydrogen was reported in several cyanobacterial species from genera including *Anabaena, Calothrix, Oscillatoria, Synechocystis, Cyanothece, Gloeocapsa, Nostoc, Synechococcus, Microcystis, Gloeobacter, Aphanocapsa, Chroococcidiopsis* and *Microcoleus* in different condition of cultivation (Smith *et al.*, 1992; Dutta, *et al.*, 2005; Hallenbeck, 2012). A number of factors affect biohydrogen production in cyanobacteria with different mechanisms. The most important factors are environmental conditions (light, temperature, molecular nitrogen, nitrogen source, micronutrients, and salinity), and intrinsic factors (presence of uptake hydrogenase, sensitivity of hydrogenase and nitrogenase to molecular oxygen, and having heterocyst) (Hallenbeck & Benemann, 2002; Tsygankov *et al.*, 2002). Production of biohydrogen in large scale with different species of *Anabaena* and *Spirulina* has been reported. The production was performed in three different types of bioreactors such as vertical column, tubular and flat panel (Dutta, *et al.*, 2005). Enhancement of hydrogen production was attained by metabolic engineering (Vignais *et al.*, 2006; Mathews & Wang, 2009). However, current research is focused on recognizing novel species of cyanobacteria with higher production yield, the design of better bioreactors and engineering the metabolic and genetic capacity of cyanobacteria for improving biohydrogen production (Srirangan *et al.*, 2011; Sarsekeyeva *et al.*, 2015).

Bioethanol is a different renewable energy and the major biofuel used for transportation worldwide (Gray *et al.*, 2006). Bioethanol can be employed in existing motor engines by combining with fossil fuels (Gray, *et al.*, 2006). Nowadays, bioethanol is commonly produced by the agricultural crops and their lignocellulosic residues (Kim & Dale, 2004). Finding new resources is inevitable, due to concerns such as large-scale production, cost effectiveness and competition with food sources, farmland and water for production of bioethanol from crop and lignocellulosic waste.

Production of bioethanol by cyanobacteria is an attractive research area. Cyanobacteria can be used as producer or as substrate for bioethanol production (Parmar *et al.*, 2011). Cyanobacteria produce minor level of ethanol during anaerobic incubation in the dark as a product of fermentation (Van der Oost *et al.*, 1989), but they have great potential for creating a perfect producer of bioethanol by genetic and metabolic engineering (Deng & Coleman, 1999; Dexter & Fu, 2009). Cyanobacteria are extremely amenable to modification by genetic engineering. To create a bioethanol producer, transgenic *Synechococcus* sp. PCC 7942 and *Synchocystis* sp. PCC 6803 were constructed by introduction of decarboxylase and alcohol dehydrogenase II genes from *Zymomonas mobilis*, a constitutive ethanol producer. These genetically-modified cyanobacteria produced and accumulated bioethanol in their culture medium (Dexter & Fu, 2009; Zhou & Gibbons, 2012). Cyanobacteria were also used indirectly for producing bioethanol. Biomass of some cyanobacteria has up to 60 % carbohydrate in dry weight; hence it is an ideal feedstock for bioethanol production. Biomass of *Synechococcus* sp. PCC 7002 and *Phormidium* sp. were used as starting materials for bioethanol production. *Synechococcus* sp. PCC 7002 cells were hydrolyzed by lysozyme and alpha-glucanases and then fermented with *Saccharomyces cerevisiae*. Enzymatic-hydrolyzed biomass of this cyanobacterium had some nutrient for enhancing fermentation (Karatay, 2014; Möllers *et al.*, 2014). It is more cost effective than yeast extract that is generally used for *Saccharomyces* fermentations. As a different approach, noncrystaline cellulose was produced in a recombinant *Synechococcus leopoliensis*. Because of its straightforward hydrolysis, amorphous cellulose is an excellent starting material for bioethanol production. Cellulose synthesis genes from *Gluconobacter xylinus* were cloned in *Synechococcus leopoliensis* and successfully expressed to produce extracellular amorphous cellulose (Nobles Jr & Brown Jr, 2008).

14.2.3 Cyanobacterial biopolymers

Microorganisms usually accumulate various nutrients in their cytoplasms for upcoming utilization. Cyanobacteria reserve various materials in storage granules such as glycogen, cyanophycin, and volutin. Cyanobacteria assimilate these molecules when the materials are in excess and vegetative cells need not them immediately. Polyhydroxyalkanoates (PHA) are bioplastic polymers that can be easily degraded by microorganisms in nature. Some cyanobacterial species assimilate excess carbon substances into polyhydroxyalkanoates (PHAs), predominantly polyhydroxybutyrate (PHB) and then accumulate in water insoluble granules in their cytoplasms, principally in media with limited nitrogen source (Stal, 1992). PHA producing by cyanobacteria is ideal proposal because they need minimal nutrient necessities for growth. Under nitrogen limiting conditions, cyanobacteria have the potential of assimilating carbon source for instance CO_2 from the atmosphere and transforming it into PHA (Miyake *et al.*, 2000). Because carbon and energy sources for production of PHA by cyanobacteria are inexpensively and readily accessible, making this option commercially noteworthy.

PHA has been isolated from several cyanobacteria, including members of the genera *Chloroglea fritschii* (Carr, 1966), *Spirulina platensis* (Campbell *et al.*, 1982), *Synechocystis* spp. (Hein *et al.*, 1998), *Anabaena cylindrica* (Lama *et al.*, 1996), *Oscillatoria limosa* (Stal *et al.*, 1990), *Aphanothece* sp. (Capon *et al.*, 1983). The most common PHA produced by cyanobacteria is polyhydroxylbutyrate (PHB). The production of PHB is done by β-ketothiolase and acetoacetyl-coenzyme A (CoA) reductase in *Synechocystis* sp. PCC6803 (Taroncher-Oldenburg *et al.*, 2000). Synthesis can be adjusted by changing the carbon source, and concentrations of acetate, nitrogen and phosphorus in the culture medium. Cultivation of *Spirulina* under mixotrophic conditions in the presence of acetate increased PHB synthesis to about 3 % of CDW (cell dry weight) in comparison to the photoautotrophic situation (0.3 wt % of CDW) (Vincenzini *et al.*, 1990). Wastewaters and carbon dioxide can be used for the production of PHA in cyanobacteria which result in more economic production in conjunction with bioremediation of environmental pollutants and carbon dioxide mitigation (Akaraonye *et al.*, 2010).

14.2.4 Cyanobacterial bioactive compounds

Cyanobacteria are a prosperous resource of novel categories of biologically active molecules performing activities such as antibiotic, anticancer, UV-protecting, vitamin or even toxin. These compounds may chemically be carbohydrates, lipids peptides, alkaloids, polyketides and terpenoids with linear or cyclic structure.

Cyanobacteria from orders Oscillatoriales (49 %), Nostocales (26 %), Chroococcales (16 %), Pleurocapsales (6 %) and Stigonematales (4 %) are rich source of secondary metabolites, respectively (Gerwick *et al.*, 2008). Species such as *Anabaena, Nostoc, Microcystis, Lyngbya, Oscillatoria, Phormidium* and *Spirulina* are the most promising for producing variety of high value compounds such as carotenoids, fatty acids, lipopeptides, polysaccharides and other bioactive compounds (Singh, *et al.*, 2005; Abed *et al.*, 2009; Chakdar *et al.*, 2012; Priyadarshani & Rath, 2012; Raja *et al.*, 2015).

14.2.4.1 Cyanotoxins

Several algal bloom-forming cyanobacteria produce toxins (fatal for different animal tissues) including hepatotoxins, neurotoxins, dermatotoxins and cytotoxins (Codd, 1995). Hepatotoxins derived from cyanobacteria are microcystins, nodularins (protein phosphatase inhibitor) and cylindrospermopsin (protein synthesis blocker) (MacKintosh *et al.*, 1990; Froscio *et al.*, 2001). Microcystins are cyclic peptides that show hepatotoxicity to mammals and other animals. Microcystins have been isolated from different cyanobacteria e.g. *Microcystis, Anabaena, Nostoc* and *Planktothrix* (Rantala *et al.*, 2004). Nodularins are another hepatotoxic group isolated from *Nodularia spumigena*. Among the ten variants identified until now, Nodularin-R is the most abundant (Sivonen, 1996). Anatoxins A, homoanatoxin A and saxitoxins are alkaloid neurotoxins that are found in cyanobacteria species like *Anabaena, Aphanizomenon, Cylindrospermum* and *Microcystis* (Rapala *et al.*, 1993; Codd, 2000; Lyra *et al.*, 2001). Toxicity and mortality of these substances are reported worldwide commonly because of drinking of water in the cyanobacterial bloom-forming area. Depending on the bloom density, toxin quantity and the amount of consumed water, severity of toxicity is different (Sivonen *et al.*, 1990; Griffiths & Saker, 2003; Gugger *et al.*, 2005). Moreover, cyanotoxins have favorable use as algaecides, herbicides and insecticides (Berry *et al.*, 2008). They are biodegradable and have short half-life in environments; consequently, they are more eco-friendly than synthetic biocides. Investigations on therapeutic effects of cyanotoxins are in progress and explain in anticancer section. Nevertheless, further studies need to clarify the details concerning the beneficial use of cyanobacterial toxins.

14.2.4.2 Antibiotics

Because of the appearance of multidrug resistance in widespread bacteria, the rapid emergence of new infectious agents, and the possible exploitation of multidrug-resistant agents in bioweapons, the urgent need for new antibiotics is escalating (Penesyan *et al.*, 2015; Schmidt-Dannert, 2015). Cyanobacteria emerge a prospective source of innovative antimicrobial agents to tackle antibiotic resistance (Pandey, 2015). Endeavor to develop new antibiotics from cyanobacteria by researchers has led to many novel compounds with antibacterial activity. Antibacterial compounds are probably the most common metabolites that have been found out in cyanobacteria

(Table 1). These compounds are effective against Gram-positive and/or Gram-negative bacteria.

Major cyanobacterial sources of antibacterial agents are *Fischerell ambigua* (Raveh & Carmeli, 2007; Mo *et al.*, 2009), *Hapalosiphon fontinalis* (Moore *et al.*, 1987), *Tolypothrix nodosa*, *Nostoc* spp. (Jaki *et al.*, 1999; Bui *et al.*, 2007), *Scytonema mirabile*, *Anabaena* spp., *Microcystis aeruginosa*, *Phormidium* sp., *Microcoleous lacustris*.

14.2.4.3 Antivirals

Some molecules with antiviral activity have been determined in blue-green algae. These metabolites are able to block absorption or penetration of viruses into cells or to inhibit replication viruses.

Cyanovirin-N (CVN) is a cyanobacterial lectin isolated from *Nostoc ellipsosporum* (Boyd *et al.*, 1997). This molecule shows affinity to binding with high-mannose oligosaccharides on virus surface and prevents fusion of enveloped viruses with animal cells. CVN is active against enveloped viruses including HIV, influenza, Ebola, hepatitis, and herpesvirus 6 for prophylactic and therapeutic treatment.

Other antiviral compounds such as Spirulan (Hayashi *et al.*, 1996), Nostoflan (Kanekiyo *et al.*, 2005), Scytovirin N (Bokesch *et al.*, 2003), Sulfoglycolipid (Loya *et al.*, 1998) are derived from *Spirulina* sp., *Nostoc flagilliforme*, *Scytonema varium*, *Scytonema* sp., respectively (Table 1).

14.2.4.4 Food and vitamins

In several countries including Chile, Mexico, Peru, the Philippines and some African countries, cyanobacteria such as *Spirulina*, *Nostoc* and *Anabaena* species has been traditionally used as human food and request for cyanobacteria as food and feed supplement has been increased in recent decades (Cannell, 1990). *Spirulina* is an edible cyanobacterium that consists mainly of protein (60 to 71 % of its dry weight), while cow meat provides value about 43 %. *Synechococcus* sp. consists of 63 % protein, 15 % carbohydrate and 11 % lipid (Christaki *et al.*, 2011). These blue-green algae can be used as single-cell protein source. They have unique benefits in comparison with other sources of single-cell protein. In addition to high quantity and quality of its protein, *Spirulina* are rich source of β-carotene, thiamine, riboflavin, absorbable iron, and other minerals, phenolic compounds, anti-inflammatory, antioxidants and essential fatty acids (Annapurna *et al.*, 1991; Sânchez *et al.*, 2003; Deng & Chow, 2010). Supplements of *Spirulina platensis* is now sold in the different forms of powder, tablet and capsule. *Nostoc* and *Aphanizomenon* strains can be utilized as a source of healthy food, because of the high content of their fibers and proteins (Abed, *et al.*, 2009). *Crocosphaera watsonii* is a marine cyanobacterium that can be used as a rich source of vitamin B12 (Bonnet *et al.*, 2010). Various marine cyanobacteria like *Phormidium valderianum* have also been examined as an aquaculture feed source for fish. Increase in the growth rate of

Tilapia fish was reported when they nourished with marine cyanobacteria in indoor and outdoor cultures (Abed, *et al.*, 2009).

14.2.4.5 Anticancer agents

Some cyanobacterial toxins and other secondary metabolites from blue-green algae have activity against malignant cells (Moore, 1982; Tan, 2010; Costa *et al.*, 2012). Cyanotoxins such as microcystins and nodularins are inhibitors of protein phosphatase and interrupt signaling pathways in the cell and disrupt integration of cytoskeleton; thereby inducing cell death (apoptosis) in the tumor. Also, a number of other unique compounds in cyanobacteria show evidence of cytotoxic effect in lab research or/ and clinical trials (Table 14.1). Curacin A from *Lyngbya majuscula* and dolastatin 10 from *Symploca* sp. are two compounds with antimitotic and antiproliferative activity in clinical trials for potential use in neoplasias (Newman & Cragg, 2004). These molecules target tubulin or actin filaments in animal cells and provide lead structure for the synthesis of novel analogues with better pharmacodynamic and pharmacokinetic features (Wipf *et al.*, 2002). Dolastatin derivatives such as soblidotin and tasidotin are in clinical trial steps for use as oncolytic drugs (Garg *et al.*, 2007; Butler, 2008). Cryptophycins is a group of depsipeptides with cytotoxic activity produced by *Nostoc* spp. (Trimurtulu *et al.*, 1994). They are more potent than paclitaxel or vinca alkaloids against neoplastic cells *in vitro* and even on malignant cell lines with multiple drug resistance (MDR). While, clinical trial on cryptophycin-52 in phase II was discontinued because of lacking efficacy *in vivo* and dose-limiting adverse effects, development of analogues of cryptophycins based on structure-activity relationship (SAR) design has been currently undergone and Sanofi-Aventis have two patents on cryptophycin conjugates with immunoglobulins (Bouchard *et al.*, 2012).

Tab. 14.1: Some of antibiotic, antiviral and anticancer compounds derived from cyanobacteria.

Biological activity	Molecule	Source	References
Antibiotic			
	Ambiguines, Hapalindole	*Fischerella* spp.	(Raveh & Carmeli, 2007), (Asthana *et al.*, 2006)
	Nostocine A, Noscomin, Muscoride	*Nostoc* spp.	(Nagatsu *et al.*, 1995; Jaki *et al.*, 1999; Hirata *et al.*, 2003)
	Didehydromirabazole	*Scytonema mirabile*	(Carmeli *et al.*, 1991)
	Malyngolide	*Lyngbya majuscula*	(Cardellina *et al.*, 1979)

Tab. 14.1: (continued)

Biological activity	Molecule	Source	References
	Norharmane	*Nostoc insulare*	(Volk & Furkert, 2006)
	Kawaguchipeptin B	*Microcystis aeruginosa*	(Ishida *et al.*, 1997)
Antiviral			
	Spirulan	*Spirulina* sp.	(Hayashi, *et al.*, 1996)
	Nostoflan	*Nostoc flagilli-forme*	(Kanekiyo, *et al.*, 2005)
	Scytovirin N	*Scytonema varium*	(Bokesch, *et al.*, 2003)
	Sulfoglycolipid	*Scytonema* sp.	(Loya, *et al.*, 1998)
Anticancer			
	Curacin A	*Lyngbya maju-scula*	(White *et al.*, 1995)
	Dolastatin 10	*Symploca* sp.	(Luesch *et al.*, 2001)
	Cryptophycins	*Nostoc* spp.	(Smith *et al.*, 1994)
	Calothrixin A and B	*Calothrix* spp.	(Rickards *et al.*, 1999)
	Apratoxin A-G	*Lyngbya* sp.	(Burja *et al.*, 2001)
	Aurilide B and C	*Lyngbya majuscula*	(Han *et al.*, 2006)
	Caylobolide B	*Phormidium* spp.	(Salvador *et al.*, 2010)
	Kalkitoxin	*Lyngbya majuscula*	(White *et al.*, 2004)
	Majusculamide C and D Deoxymajusculamide D	*Lyngbya majuscula*	(Carter *et al.*, 1984; Moore & Entzeroth, 1988)
	Pitiprolamides	*Lyngbya majuscula*	(Montaser *et al.*, 2010)
	Tasipeptins A and B	*Symploca* sp.	(Williams *et al.*, 2003)
	Nakienones AC	*Synechocystis sp.*	(Nagle & Gerwick, 1995)
	Lagunamide C	*Lyngbya majuscula*	(Tripathi *et al.*, 2011)

14.2.4.6 Pigments and sunscreens

Blue-green algae are photosynthetic bacteria and produce high amounts of chlorophyll *a*, carotenoids and phycobiliproteins. The phycobiliproteins can only be found in the cyanobacteria, red algae and cryptophytes (Sidler, 2004). These pigments are frequently utilized as natural colorants or antioxidants in food, cosmetic and pharmaceutical industries (Sidler, 2004). They also possess other pharmacological activities including antiproliferative, antilipid and neuroprotective (Gonzalez *et al.*, 1999; Romay *et al.*, 2003). Phycocyanin is a phycobiliprotein that is produced by all cyanobacteria. This water insoluble pigment is derived from *Spirulina* and utilized in cosmetics. Phycocyanin is a free radical scavenger and selective inhibitor of cytochrome oxidase; therefore it can be used as anti-inflammatory and hepatoprotective medicines (Reddy *et al.*, 2000). Furthermore, this pigment has favorable fluorescent characteristic for use in fluorescence-based diagnostic tests (Campanella *et al.*, 1999).

Sun protection metabolites are produced by some cyanobacterial species under UV exposure. Scytonemin is a radical-scavenging pigment that is produced by *Nostoc punctiforme* (Soule *et al.*, 2007), *Calothrix* sp., (Dillon & Castenholz, 2003) *Lyngbya aestuarii* (Rath *et al.*, 2012), and shinorine (a specific mycosporine-like amino acid) is produced by *Anabaena variabilis* ATCC 29413 (Balskus & Walsh, 2010).

14.2.5 Cyanobacteria for biotransformation purposes

One of the most interesting applications of cyanobacteria is in the biotransformation of the industrially important materials (Ghasemi *et al.*, 2011). Steroids are important drugs used in medicine as anti-inflammatory, immunosuppressive, antifungal, diuretic, cardiotonic, sex hormones, digestant, antibiotic, neuromuscular blocker, and vitamin D precursor (Foye *et al.*, 2008). Because of the complexity of steroid structure, chemical synthesis of steroids involves multiple steps (37 steps for synthesis of cortisone from deoxycholic acid) and is very expensive and ineffective process; it is more reasonable to produce them by biotransformation of other naturally available steroids (Hirschmann, 1992).

Androst-4-en-3,17-dione (AD) is the primary starting material for partial synthesis of some steroids. The biotransformation of AD by cyanobacteria was reported in several studies. Abul-Hajj and Qian (1986) described the biotransformation of AD to testosterone by different cyanobacterial strain including *Anabaena flos-aquae, Anabaena cylindrica, Osrillatoria prolijeva* and *Phormidium nauturn*. This alteration is performed through reduction of the 17-carbonyl group to hydroxyl by an oxidoreductase enzyme. In some cases, cyanobacteria degraded AD without formation of product suggesting these cyanobacteria could consume AD as carbon and/or energy source. Also, it was found that AD bioconverted to testosterone by cyanobacteria *Nostoc ellipsusporum* and *Nostoc muscorum* (Faramarzi *et al.*, 2006; Moradpour *et al.*, 2006).

Hydrocortisone is another steroid that is metabolized by cyanobacteria. *Nostoc muscorum, Nostoc ellipsosporum* and *Fischerella ambigua* bioconverted hydrocortisone to 11,17,20,21-tetrahydroxypregn-4-en-3-one (major product) and 11-hydroxyandrost-4-en-3,17-dione (Yazdi *et al.*, 2004; Yazdi *et al.*, 2005; Ghasemi *et al.*, 2006).

Terpenes are the biggest and most various class of natural products. They are natural hydrocarbons produced in biological systems as five-carbon (isoprene) units. Most of terpenes are produced in plants, but most structurally-complicated ones found in animals. As well as significant roles in medicine as oncolytic, antioxidants, and anti-infective, they are also used in perfumes, cosmetics, air fresheners and insect repellents (Simonsen *et al.*, 1957).

Biotransformation of two monoterpenes, limonene and limonene epoxide, to corresponding alcohol was performed by *Synechococcus* sp. PCC 7942 (Hamada *et al.*, 2003). This cyanobacterium converted (−)-Limonene to cis- and trans-carveols in 11 % and 9 % yields, respectively, while (+)-limonene was left unaltered. Stereo- and regioselective hydroxylation was carried out at the allylic position of the endocyclic unsaturated carbon-carbon bond. Also, (+)-limonene epoxide biotransformed to (1S,2S,4R)-limonene-1,2-diol and (1S,4R)-limonene-1-ol-2-one in 32 % and 16 % yields, respectively, while (−)-limonene epoxide was left unchanged. This reaction showed the potential of *Synechococcus sp.* PCC 7942 for the enantio- and stereoselective fission of the epoxide ring (Hamada, *et al.*, 2003).

Furthermore, biotransformation of some alkaloids by cyanobacteria was described. Rao *et al.* (1999) found that *Spirulina platensis* has the ability of bioconversion of codeine to morphine with an O-demethylation reaction by the free and immobilized culture of cyanobacterium. Also, codeine can be converted to morphine by *Nostoc muscorum*. But, in this bioconversion, some opiate derivatives like 6-acetylcodeine, oxycodone and norcodeine may be generated as by-products (Niknam *et al.*, 2010).

14.3 Environmental applications

Urbanization, industrialization, development of agriculture and intensive use of chemical substances such as petroleum, hydrocarbons, pesticides and heavy metals produces a huge amount of by-products and contaminants that contribute to environmental pollution. Environmental pollution is currently one of the global challenges and has impacts on the life cycle of all kind of organisms from microorganisms to plants and animals that are ultimately affecting human health as well as the balance of the ecosystem.

There are different methods for pollutant compounds removal from different environments, such as physical techniques (incineration, electroremediation, carbon adsorption, and etc.), chemical methods (oxidation, photocatalysis, and etc.) and biological remediation (microbial and phytoremediation) methods. These approaches can be applied alone or in combination for remediation of environmental pollutants. Bioremediation (biological remediation) is a process that uses organisms (bacteria, fungi and plants) and their enzymes, for detoxification or elimination of pollutant substances (Tong & Yuan, 2012). Bioremediation is the most effective, economic and ecologically sound technique uses natural biological processes for mitigation of pollutant compounds. Also, bioremediation could remove hazardous materials permanently and can be used in combination with physical or chemical methods (Vidali, 2001). Widespread diversity and potential in their metabolic mechanisms together with their ability to survive and being active even in extreme environments- salt lakes, hot springs, highly acidic sites, and the ocean depths- make the microorganisms the best candidates for removal of pollutants from unfavorable sites (Stanier & Bazine, 1977). Microorganisms remediate the pollutant substances, maybe because of the requirement for nutrients and need to eliminate the toxic effects of these materials, or even by chance since some pollutant substances have a similar structure with the natural metabolites of microorganisms.

Searching for more competent microbial bioremediators and developing more powerful strategies is still ongoing. Cyanobacteria are one of the most promising microorganisms that remediate pollutants from the environment. Due to the ubiquitous characteristics of cyanobacteria, cyanoremediation (removal of pollutant by cyanobacteria) is one of the most cost-effective approaches for mitigation of pollution.

14.3.1 Cyanoremediation of metal ions

Metals ions are present as innate components of the earth crust, and their existence in the surrounding media has currently elevated owing to human being actions such as metal mining, metallurgy, energy and fuel production, electroplating, wastewater sludge treatment, and agriculture. They are persistent environmental contaminants since they cannot be degraded or destroyed. They penetrate into the biological systems via food and water and concentrate during their lifespan. Contamination of agricultural fields with toxic metal ions is a critical concern in food production that may affect human health in all around the world.

Some metal ions have a healthy effect on biological systems, but acceptance limits in drinking and potable waters have been established for these metals (like iron, zinc, calcium, cobalt, copper, selenium, vanadium, chromium and magnesium). Nevertheless, some others metal ions such as nickel, cadmium, lead, uranium, silver, beryllium, arsenic and mercury have no desired effect on biological systems and utilization of these metal ions even at very low concentrations can be noxious (cytotoxic, mutagenic and carcinogenic) (Hughes & Poole, 1989; Snow, 1992; Hartwig, 1995).

The traditional approach for metal ions remediation is based on physical (adsorption, ion exchange membrane filtration, electrodialysis and reverse osmosis) and chemical (chemical precipitation, coagulation/flocculation, flotation, evaporation and electrochemical treatment) methods (Fu & Wang, 2011). However, these techniques often suffer from the low elimination yield, creation of toxic byproducts, and poor specificity of procedure and operation scale up particularly for *in situ* practice. Moreover, these processes are not cost effective, mainly in situations of low metal concentration. These disadvantages often restrict the use of physicochemical methods for removing of metal ions.

14.3.1.1 Mechanisms of metal ions bioremediation

There are two approaches in all living systems for interaction with metal ions. Organisms use essential metal ions in a favorable level and prohibit those that are toxic to the balance of their metabolic activities (Cobbett & Goldsbrough, 2002). Elevated concentrations of metal ions above tolerance limit have the potential of causing damage to living systems. These damages may affect metabolic activities, development and differentiation stages of organisms. Cyanobacteria respond to the high concentration of metal ions by diverse mechanisms that can be harnessed for bioremediation applications (Monteiro *et al.*, 2012). The first mechanism is shielding or changing of functional groups of bioactive compounds such as metabolic enzymes or metal ion transferring proteins. In some cases, substitution of metal ions from biologically active substances or transportation of spare metal ions to specific compartments of cytoplasm (e.g., vacuoles or polyphosphate granules) is the response of cyanobacteria and the last (but not the least), increased production of reactive oxygen species that may result in oxidation of vital proteins, lipids and nucleic acids (Kaplan, 2013; Huertas *et al.*, 2014). However, cyanobacteria have developed resistance mechanisms to alleviate metal ions damages. Avoidance and tolerance are the most important mechanisms of resistance. Avoidance mechanism result from passive adsorption of metal ions by functional groups of the cell wall and membrane or through secretion of ligands for complexation with metal ions. Another mechanism for resistance to metal ions is tolerance, which in contrast to avoidance, is an energy dependent mechanism. In this process, metal ions are actively transported in by cyanobacterial cells and accumulated in the cytoplasm. This active assimilation is mainly carried out by transporter system for micronutrient absorption. Understanding the mechanisms of metal-cyanobacteria interactions can provide approaches for bioremediation of these pollutants from the environment (Kaplan, 2013).

Cyanoremediation of metal ions is performed by nonviable and viable cells (wild type, immobilized or engineered) (Gardea-Torresdeya *et al.*, 1998; Rangsayatorn *et al.*, 2004). Most research has on living cyanobacteria, but nonliving cyanobacterial cells, thanks to unique characteristics of their cell surface have a great capacity for removal of metal ions. However, as shown in Figure 14.2, cyanobacterial cells remediate metal ions by a variety of mechanisms depending on the species and the types of pollutant including bioadsorption, bioabsorption, bioflocculation and bioreduction (Gothalwal & Chillara, 2012).

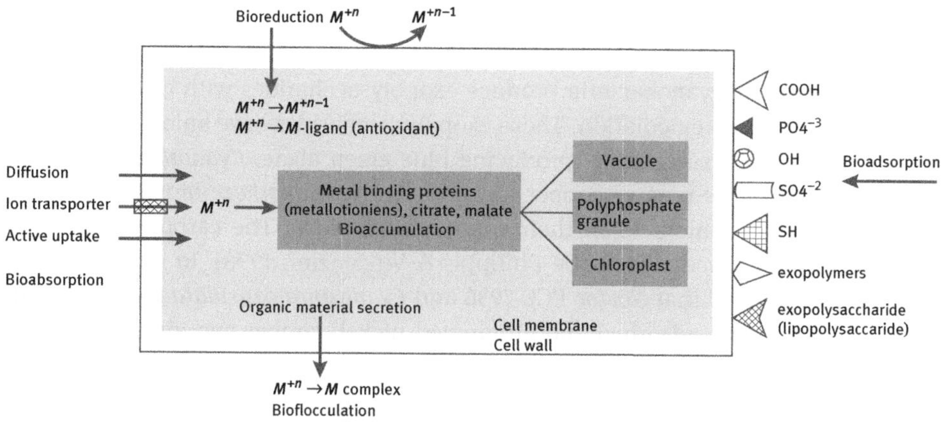

Fig. 14.2: Mechanisms of cyanoremediation of metal ions. For further description refer to text.

Bioadsorption is the most important mechanism for bioremediation of metal ions and carried out via passive binding to various surface functional groups and is a nonmetabolic dependent process which requires no energy. The metal ions bind to cyanobacterial cell surface physically or chemically. Bioadsorption, based on cyanobacterial strains and the environmental situation may involve ion-exchange, complexation or chelating, electrostatic linkage, and surface precipitation (Gavrilescu, 2004). Lipopolysaccharides and peptidoglycans are the main sites of adsorption of metal ions in cyanobacteria (De Philippis & Micheletti, 2009). This passive mechanism is involved in metal ion removal by viable and nonviable cyanobacteria. Temperature, pH of the environment and availability of surface functional groups greatly affect affected the adsorption of metal by cyanobacteria (Kaplan, 2013).

Several cyanobacterial species possess the potential for elimination of different metals ions through passive adsorption on their surface functional groups. It has been revealed that viable cells of *Spirulina platensis* that commonly exist in contaminated area passively adsorbed metal ions such as Cu, Pb, Cr, Ni and Cd via binding of metal ions to surface functional groups and then absorbed them into their cytoplasms (Chojnacka *et al.*, 2005; Şeker *et al.*, 2008).

Nonviable cyanobacterial cells are more potent for metal ion adsorption than viable cells and pretreatment of nonviable cells with calcium chloride and sodium hydroxide increases the yield of adsorption (Mehta & Gaur, 2005; Nagase *et al.*, 2005). The mechanism of bioadsorption is ion exchange of metal ions with proton on the surface of cyanobacterial cells. It has been found that nonviable cells of *Spirulina maxima* adsorbed Pb more efficiently (92 % at pH 5.5) when treated with calcium chloride (Nagase, *et al.*, 2005). Enhancement of Cd removal capacity was also achieved by phosphorylation of *Lyngbya taylori* biomass (Klimmek *et al.*, 2001). Moreover, specific adsorption of cadmium by *Tolypothrix tenuis*, *Anabaena variabilis* and *Microcystis aeruginosa* was explored and discovered that pretreatment with alkaline enhanced the bioadsorption capacity of these cyanobacterial strain for metal ions (Nagase, *et al.*, 2005). Many cyanobacteria produce exopolysaccharides with a unique affinity for metal ions bioremediation. These exopolysaccharides have anionic functional groups. The exopolysaccharide- producing blue-green algae, *Cyanothece* CE 4 and *Cyanospira capsulata* remove copper ions efficiently. *Cyanothece* biomass contains high percentage uronic acid in their exopolysaccharides. The carboxyl groups of uronic acid bind to metal ions (De Philippis & Vincenzini, 1998). In another study, it has been indicated that *Nostoc* PCC 7936 and *Cyanospira capsulata* produced exopolysaccharides that adsorbed and eliminated nickel, copper and zinc in separate and mixed metal contaminated media (Paperi *et al.*, 2006). The biomasses of *Nostoc muscorum* and *Nostoc spongiaeforme* have the potential to be exploited as green and cost-effective biosorbent agents for the elimination of chromium and zinc, respectively (Singh *et al.*, 2003; Gupta & Rastogi, 2008).

Metal ions removal activity of cyanobacteria can be coupled with production of value-added materials such as proteins, carotenoids and pigments. In a study, *Anabaena subcylindrica* and *Nostoc muscorum* were used for bioremediation of copper, cobalt, manganese and lead from wastewater. Because of the competition for resources in mixed cultures, single cultures of cyanobacteria were more efficient than the co-cultures in heavy metal elimination. However, when these cyanobacteria are cultured in sterilized sewage wastewater, the chlorophyll *a*, carotenoids and protein content of *Anabaena subcylindrica* and *Nostoc muscorum* were higher than those cultured in the standard medium (El-Sheekh *et al.*, 2005).

Bioabsorption is a two step process involves the adsorption of metal ions in the first step. The next step is a sustained and active process that occurs along with adsorption of metal ions. The metal ions are entered by micronutrient transporters or ion channels and are followed by binding to ligands such as citrate, malate and metallotioneins, and then the accumulation of bounded metal ions in the cytoplasm or cytoplasmic components (vacuoles, polyphosphate granules and/or chloroplasts) (Kaplan, 2013). Bioabsorption rarely occurs by passive diffusion of free metal ions through cell membrane. Factors such as metal ions concentration, biomass and metabolic condition of cyanobacteria, the presence of metabolic inhibitors, the coexistence of other metal ions, illumination, pH and temperature of the environment affect the metal ions bioabsorption by blue-green algae (Mehta & Gaur, 2005; Kaplan, 2013). Bioremediation of hexavalent chromium was explored by *Oscillatoria laetevirens* and *Oscillatoria trichoides* and shown that these blue-green algae had a high capacity for bioabsorption of this ion metal (Mehta & Gaur, 2005). Bioabsorption and accumulation of metal ions including cadmium, zinc, lead, nickel and copper by *Phormidium* sp. were reported in several experiments (Shanab *et al.*, 2012; Rana *et al.*, 2013). Bioabsorption of metal ions is a pH dependent process in *Phormidium laminosum* and increased in elevated pH (Sampedro *et al.*, 1995). Bioaccumulation of cobalt, copper, and zinc by cyanobacterium *Spirulina platensis* was also described in several studies (Li *et al.*, 2006; Arunakumara *et al.*, 2008). It has been found that *Spirulina plantensis* exchanged chromium, cadmium, copper, lead and nickel by proton ions (Chojnacka, *et al.*, 2005). Other cyanobacterial strains may remove metal ions by the mechanism of ion exchange. It has revealed that *Oscillatoria anguistissima* absorbed zinc and copper by exchange with magnesium (Ahuja *et al.*, 1999). Consortia of cyanobacteria with other microorganisms have a great capacity for removal of metal ions. Consortia of *Spirulina platensis* and sulfate-reducing bacteria remediated copper (79.2 %), zinc (88.0 %) and iron (100 %) from tannery effluent with high removal efficiency (Rose *et al.*, 1998). Mixed culture of cyanobacteria (*Chroococcus* sp., *Pseudoanbaena* sp., *Leptolyngbya* sp.), with microalgae (*Scenedesmus* sp., *Tetraedron* sp., *Chlorella* sp., *Chlorococcus* sp.), diatoms (*Navicula* sp., *Nitzschia* sp., *Cyclotella* sp.) and bacteria produced a biofilter in continuous flowthrough column that removed copper and cadmium, almost completely in 5 min from artificial metal solutions containing cadmium and copper (Loutseti *et al.*, 2009). In such consortia, co-cultured microorganisms are closely related to others genetically and physiologically (Boivin *et al.*, 2007).

Bioflocculation is another mechanism for metal remediation that harnessed by some blue-green algae (Doukyu *et al.*, 2007). Flocculating of metal ions is carried out by cyanobacteria through secretion of exopolymers such as polyhydroxybutyrate, lipopolysaccharide and proteins. These biomolecules have wide range flocculating activity; hence, cyanobacteria with the ability to produce bioflocculating agents can potentially be used for removal of metal ions from aquatic media. *Anabaena* produces a thermostable flocculating factor with glycoprotein structure consisting of neutral sugar, uronic acid and protein (Choi *et al.*, 1998). *Phormidium* secretes a sulfated heteropolysaccharide bioflocculating factor (Fattom & Shilo, 1984) and is used for clarification of waters contaminated with clay particles (Bar-Or & Shilo, 1988).

Bioreduction is involved in detoxification of some metal ions and is often occurred following bioabsorption of metal ions into the cytoplasm of cyanobacteria. Some enzymes and ligands (–sh containing molecule) transform the valence of the metal ions and detoxify them. Bioconversion of hexavalent chromium cation to trivalent chromium cation was carried out by cyanobacteria *Synechocystis* sp. PUPCCC 62 in the presence of sulfate (Khattar *et al.*, 2014). A summary of cyanobacterial strains with metal ion remediation capacity are collected in Table 14.2.

Tab. 14.2: Some cyanobacterial strains with metal ion remediation capacity.

Mechanism of removal	cyanobacteria	Metal ions	References
Bioadsorption	*Spirulina plantensis*	Hg, Cu and Pb	(Johnson & Shubert, 1986; Chojnacka *et al.*, 2004)
	Spirulina maxima	Pb	(Gong *et al.*, 2005)
	Cyanospira capsulate, Cyanothec CE4	Ni, Zn and Cu	(De Philippis *et al.*, 2011)
	Tolipotherix tenuis, Anabaena variabilis, Microcystis aeroginosa	Cu, Zn	(Nagase *et al.*, 1997; Nagase, *et al.*, 2005)
	Nostoc muscorum, Nostoc spongiaforme	Cr, Cu	(Singh, *et al.*, 2003; El-Sheekh, *et al.*, 2005)
	Nostoc PCC 7936, Cyanospira capsulate	Ni, Cu, Zn	(De Philippis *et al.*, 2007)
Bioabsorption			
	Spirulina platensis		(Chojnacka, *et al.*, 2005)
	Oscillatoria quadripunctulata Oscillatoria tenius	Cu , Co, Pb, Zn	(Wilde & Benemann, 1993)
	Oscillatoria laetevirens, Oscillatoria trichoides	Cr (VI)	(Kumar *et al.*, 2011; Miranda *et al.*, 2012)
	Phormidium laminosom		(Sampedro, *et al.*, 1995)
	Synechococcus sp. PCC 7942	Cu, Cd, Pb , Cr (III), and Cr (VI)	(Gardea-Torresdeya, *et al.*, 1998)
	Consortia of cyanobacteria with microalgae and diatoms	Cd, Cu	(Loutseti, *et al.*, 2009)
Bioflocculation			
	Phormidium	Any metal ions	(Fattom & Shilo, 1984)
	Anabaena	Any metal ions	(Choi, *et al.*, 1998)
Bioreduction			
	Synechocystis sp. PUPCCC 62		(Khattar, *et al.*, 2014)

14.3.1.2 Immobilized and genetically engineered cyanobacteria for bioremediation of metal ions

Since immobilization of cyanobacteria provides several advantages over free cyanobacterial cells for removal of metal ions, immobilized cyanobacterial cells on agar, gelatin, polyacrylamide have been used for bioremediation in several experiments (Mallick, 2002). Biosorbents based on immobilized cyanobacterial cells were also designed for the removal of metal ions. These biosorbants can be used for successive cycles of adsorption and desorption, while the capacity of the cyanobacterial cells for adsorption of metal ions remain unchanged (Prasad & Pandey, 2000). Metal ions can be recovered from biosorbents by changing of pH or eluted with a salt or metal chelator solution. Naturally-immobilized cyanobacteria in consortia with other microorganisms are potential biosorbent for metal removal.

Potential of immobilized *Synechococcus* sp. PCC 7942 to remediate metal ions such as copper, lead, nickel, cadmium, chromium (III), and chromium (VI) from solution were studied by Gardea-Torresdey *et al.* (1998) and shown the mechanism of cyanoremediation was bioabsorption and intracellular binding of metal ions to metallothionein. In this study, it was also shown that other cyanobacteria including *Spirulina*, *Prochlorothriz*, *Nostoc*, *Anabaena* and *Fischerella* species had genes for metallothioneins (Gardea-Torresdeya, *et al.*, 1998). *Anacystis nidulans* cells were also immobilized on the alginate and used for bioremediation of metal ions. The immobilized cells adsorbed cadmium and nickel with high efficiency (Awasthi & Rai, 2004).

Bloom-forming cyanobacteria are immobilized microorganisms that are found in nature. They have an outer calyx or capsule around their cell walls. The capsule of cyanobacteria is anionic in nature, owing to the presence of polysaccharides such as uronic acids (De Philippis and Vincenzini, 1998). This outer layer helps to immobilize cyanobacterial cells in their environment. Because of its anionic character, cyanobacterial capsule usually shows very high affinity to adsorb metal ions. *Microcystis* is a bloom-forming cyanobacterium that naturally immobilized in its location and has a high capability for metal ions removal.

As mentioned above, the non-viable biomass of these cyanobacteria is capable for metal bioremediation as viable cells or even better (Mehta & Gaur, 2005). The AlgaSORB is a biological sorbent of non-viable cyanobacterial biomass immobilized in a column. Such packed ion-exchanger columns can be used for bioremediation of metal ions. A biosorbent column based on immobilized cyanobacterial cells (Alga SORB-scy) was designed for the biosorption of trivalent arsenic from water samples (Prasad *et al.*, 2006). Another AlgaSORB column was also fabricated using *Nostoc muscorum* biofilm and found that the column selectively removed copper ions from multi-elemental sample (Prasad & Pandey, 2000).

Genetic engineering of cyanobacteria for improved metal ions tolerance was achieved in several studies. In an experiment, a highly metal sensitive strain of *Synechococcus* was converted to a metal resistant strain by the introduction of a gene that codes a metal binding protein. The recombinant strain harbored an expression vector with metallothionein gene and could grow even in the concentration of cadmium that the wild-type strain did not grow (Sode *et al.*, 1998).

Other metal resistant genetically engineered cyanobacteria were also designed including *Synechococcus elongatus* PCC 7942 (Cu resistant) (McCormick *et al.*, 1999),

Synechococcus elongatus PCC 7942 (Cd resistant) (Erbe *et al.*, 1996), *Synechococcus* sp. NKBG 15041 (Cd resistant) (Sode *et al.*, 1998), *Anabaena* (*Nostoc*) sp. PCC 7120 (Cd resistant) (Ren *et al.*, 1998) and *Anabaena* (*Nostoc*) sp. PCC 7120 (Cd and Zn resistant) (Guo *et al.*, 1999).

14.3.2 Cyanoremediation of hydrocarbons

Hydrocarbon bioremediation is an intricate process that depends on the structure and the quantity of the contaminant. Hydrocarbon pollution can be divided into a different group such as linear, cyclic and aromatic. Linear alkanes are the most susceptible hydrocarbons for biodegradation. The aromatic and cyclic hydrocarbons are more resistant to biodegradation, but the most difficult hydrocarbons for bioremediation are polycyclic aromatic hydrocarbons (PAHs). Most hydrocarbon pollution is originated from petroleum that entered into the environment because of widespread utilization of petroleum as an energy resource and petrochemical agents in the industries. Pollution happens during discovery, production, refinement, transportation and storage of oil and oil products. Hydrocarbon pollutants are toxic for all kinds of life (Sikkema *et al.*, 1995) and cause dangerous ecological contamination that destroys the soil and water recourses (Megharaj *et al.*, 2000). Furthermore, due to human health problems related to hydrocarbon pollution, considerable efforts are directed toward finding an appropriate solution to this problem.

Remediation of hydrocarbon contamination is performed by thermal (incineration, thermal desorption, and heat-enhanced soil vapor extraction), chemical (chemical oxidation, photolysis degradation, and electroremediation), physical (solvent extraction and supercritical fluid extraction) and biological methods (Riser-Roberts, 2010). Remediation with biological methods offers a better solution to reduce hydrocarbon pollutants due to its high efficiency and economic nature. The most important challenge for bioremediation of soil hydrocarbons is the bioavailability of these pollutants. Hydrocarbon contaminants bind to inorganic or organic substances of soil through chemical or/and physical process (complex formation, adsorption and precipitation). The tightness of these binding determines the bioavailability of hydrocarbon pollutants for biodegradation; hence influencing the process efficiency of bioremediation. However, a wide range of variation in water solubility, volatility or reactivity of hydrocarbons may affect their bioavailability in water and soils (Megharaj *et al.*, 2011).

In addition to the potential for surviving in oil-contaminated environments and consuming hydrocarbons as their carbon source, some cyanobacteria associated with other heterotrophic bacteria that have a special enzymatic system for hydrocarbon biodegradation. Therefore, cyanobacteria have a vital role in hydrocarbon bioremediation, either directly or indirectly.

14.3.2.1 Potential of cyanobacteria for hydrocarbon bioremediation

The ability of cyanobacteria to degrade hydrocarbon was confirmed in various investigations. Crude oil contains hydrocarbons in different forms. *Oscillatoria* sp. strain JCM has been found to degrade naphthalene. Naphthalene is the simplest polycyclic aromatic hydrocarbon that metabolized by this blue-green alga (Cerniglia *et al.*, 1980a; Narro *et al.*, 1992a). Also, *Oscillatoria* sp. strain JCM metabolized biphenyl to form 4-hydroxybiphenyl by an oxidation reaction under the photoautotrophic condition (Cerniglia *et al.*, 1980b). *Agmenellum quadruplicatum* mainly bioconverted phenanthrene to trans-9,10-dihydroxy-9,10-dihydrophenanthrene (phenanthrene trans-9,10-dihydrodiol) and 1-methoxyphenanthrene in the photoautotrophic growth situation (Narro *et al.*, 1992b). Alkane hydrocarbons were directly absorbed from crude oil and oxidized to fatty acids by *Microcoleus chthonoplastes* and *Phormidium corium* that revealed both species consumed oil hydrocarbons (Al Hasan *et al.*, 1994; Al-Hasan *et al.*, 1998). The marine blue-green algae *Oscillatoria salina*, *Plectonema terebrans* and *Aphanocapsa* sp. created cyanobacterial mats that remediated about 40 % of the crude oil (Raghukumar *et al.*, 2001). All three cyanobacteria were shown ability for degradation of total crude oil, aromatic (anthracene and phenanthrene) and aliphatic (hexadecane) hydrocarbons. TNT (2,4,6-trinitrotoluene) is a nitroaromatic hydrocarbon were found to remediate by *Anabaena* sp. in batch culture condition at 10 mg/l concentration. However, growth inhibitory effects of TNT or its metabolites on cyanobacterium were reported (Pavlostathis & Jackson, 1999). Efficient elimination of TNT, without growth inhibition, was achieved in a continuous culture of *Anabaena* sp. even at higher (58 mg/l) concentration (Pavlostathis & Jackson, 2002).

14.3.2.2 Indirect role of cyanobacteria in hydrocarbon bioremediation

Although, several studies indicated the direct role of cyanobacteria in biodegradation of hydrocarbons, some researchers believe that cyanobacterial-associated heterotrophic bacteria are responsible for the bioremediation and cyanobacteria have an indirect function and only support the growing of these bacteria (Sorkhoh *et al.*, 1995; Barth, 2003). Cyanobacteria supply oxygen and nitrogen for heterotrophic bacteria. Oxygen is usually involved in the early stages of the aerobic degradation and nitrogen is regularly inadequate in hydrocarbon polluted area. Radwan and Al Hasan (2002) suggested that cyanobacteria could fix the atmospheric CO_2 as carbon source; they would prefer to acquire carbon from their photosynthesis instead of obtaining from biodegradation of hydrocarbons. Furthermore, Abed and Koster (2005) failed to isolate axenic culture of cyanobacteria from oil degrading consortia. However, the importance of the indirect role of cyanobacteria in hydrocarbon biodegradation cannot be ignored.

γ-proteobacteria associated with *Oscillatoria* was found to degrade phenanthrene, dibenzothiophene, pristine and n-octadecane. Although, the heterotrophic bacteria consumed hydrocarbons, cyanobacterium supplied the nitrogen, oxygen and simple organics for degradation activity and survival of γ-*proteobacteria* (Abed & Köster, 2005). Remediation of hydrocarbon-rich wastewater with the consortia of *Burkholderia cepacia* (could consume hydrocarbons as a sole source of carbon and energy) and phototrophic cyanobacteria incuding *Phormidium*, *Oscillatoria* and *Chroococcus* strains were performed using rotating biological contactor (RBC) (Chavan & Mukherji, 2008). This experiment revealed that the consortium of cyanobacteria with the organotrophic bacteria improved the total oil hydrocarbon bioremediation yield by supplying phosphorus and simple organic compounds for *Burkholderia cepacia* (Tang *et al.*, 2010). The cooperation of cyanobacteria with heterotrophic bacteria for degradation of hydrocarbons was also proved in other studies. Phenanthrene was removed by a consortium of *Synechocystis* sp. with *Pseudomonas*-related strain (Abed, 2010) and phenol was efficiently degraded by mixed culture of *Pseudoanabaena* cyanobacterium and heterotrophic bacteria *Pseudomonas* sp. P1 (Kirkwood *et al.*, 2005). The effects of cyanobacterial secreted substances on the growth and biodegradation activity of heterotrophic bacteria was explored and shown that secreted substances encouraged the growth of *Pseudomonas* P1, *Ancylobacter aquaticus* and *Ralstonia eutropha*. These heterotrophic bacteria could consume phenol, dichloroacetate (DCA), and 2,4-dichlorophenoxyacetic acid (2,4-D), respectively. In most cases, cyanobacterial exudates improved the biodegradability of bacteria (Kirkwood *et al.*, 2006).

14.3.2.3 Cyanoremediation of phenolic compounds

Phenolic substances are a group of contaminants that release into the environment because of the industrial activities. They are poisonous for almost all organisms. Conversely, these compounds are used as carbon and energy resource by various microorganisms including bacteria, fungi, microalgae, cyanobacteria and consortia of them (Semple *et al.*, 1999; Wurster *et al.*, 2003). The sensitivity of cyanobacteria to phenolics compounds depends on the number, arrangement and polarity of substituent group(s) on the aromatic ring. Biodegradation of phenol by cyanobacteria was firstly described by Ellis (1997). In this study, *Anabaena cylindrica* and *Phormidium foveolarum* degraded phenol and catechol (Ellis, 1977). Then, biodegradation of naphthalene were examined by nine different cyanobacteria including *Oscillatoria* sp. strain JCM, *Oscillatoria* sp. strain MEV, *Microcoleus chthonoplastes* strain BA-1, *Anabaena* sp. strain CA, *Anabaena* sp. strain 1F, *Agmenellum guadruplicatum* strain PR-6, *Coccochloris elabms* strain 17a, *Nostoc* sp. strain MAC, *Aphanocapsa* sp. strain 6714 (Cerniglia *et al.*, 1980c). All these cyanobacteria metabolized naphthalene and converted naphthalene to at least six metabolites that at least one of the metabolites was 1-naphthol. Some cyanobacteria have polyphenol oxidase and laccase induced in the presence of phenol. In an effort to degrade phenol was reported that *Phormidium valderianum* BDU 30501 was tolerated and efficiently remediated this pollutant in a week via enzymatic activities of polyphenol oxidase and laccase (Shashirekha *et al.*,

1997). Phenol was also found to be effectively degraded by the cyanobacterium *Synechococcus* PCC 7002 under the nonphotosynthetic conditions in dark (Wurster, *et al.*, 2003). *Ortho*-aromatic ring cleavage of phenols was reported as the mechanism of phenol degradation in this case that is similar to the pathway of phenol degradation in heterotrophic bacteria and yeasts. This ring cleavage, which resulted in the production of cis,cis-muconic acid was possibly performed by an extracellular protein that prevented by heat, proteases and metal ions (Wurster, *et al.*, 2003). Other phenolic compounds such as phenantherene and anthracene are also degraded by cyanobacteria. These aromatic hydrocarbons found in oil products were metabolized by *Agmenellum quadruplicatum* and *Phormidium tenue* (Narro, *et al.*, 1992b; Kumar *et al.*, 2009).

14.3.3 Wastewater treatment by cyanoremediation

Agriculture procedures, industrial effluents and the human being activities have generated various contaminants that may change the water and food cycle causing a worldwide concern associated with their ultimate impact on human health. The practice of eliminating pollutants from wastewater is called wastewater treatment. Several methods including physical, chemical and biological treatment are used. The biological techniques and specifically cyanoremediation has an important role in mitigation of wastewater (Henze, 2002).

Several cyanobacterial species were used in wastewater treatment, but the majority of these procedures are under research conditions. Due to the ability to fix atmospheric carbon dioxide and nitrogen, cyanobacteria have the prospective capacity to remediate pollutants from wastewater. Oxygen production in their photosynthesis is very important in wastewater treatment by promoting aerobic biotransformation processes through mixotrophic fashion or consortia with other microorganisms (Subashchandrabose, *et al.*, 2011, 2013). Some of them have the capacity to fix nitrogen that gives them an extra potential over other photosynthetic bacteria for bioremediation of wastewater. Flocculation and sedimentation properties of some cyanobacteria species circumvent the steps for harvesting or physical recovery of the cyanobacterial cells and reduce the cost of bioremediation procedure (Laliberte *et al.*, 1997). Some cyanobacteria also produce high protein biomass can be used in the production of bioenergy, and as animals and fish feed, when they used for wastewater treatment (McGinn *et al.*, 2011). Biomass of cyanobacteria contains high polysaccharides with the ability of contaminant adsorption. Furthermore, some blue-green algae live in extreme environments such as alkaline condition and high phosphorus and nitrogen concentration that can be used for the treatment of wastewater in extreme condition.

14.3.3.1 Cyanoremediation of nitrogen- and phosphorus-containing substances

The elimination of nitrogen and phosphorus from wastewater by cyanobacteria is typically very efficient because these microorganisms need nitrogen and phosphorus for their biochemical activities. Many experiments have established the successful use of the cyanobacteria to remove nitrogenous and phosphorus substances from wastewater.

Factors controlling nitrate absorption by two *Synechococcus* species and *Synechocystis minima* for bioremediation of wastewater were explored by Hu *et al.* (2000). The most efficiency was shown by *Synechococcus* sp. PCC7942. *Synechococcus* sp. PCC6803 and *Synechocystis minima* were also effective in the elimination of nitrate, but not as well as *Synechococcus* sp. PCC7942. Factors such as cell density, inoculums size, fixed-nitrogen concentration, and aeration rate influenced the nitrate removal from wastewater (Hu *et al.*, 2000). Potential of cyanobacterium *Calothrix* sp. for removal of nitrogen and phosphorus ions from wastewater has been discovered. In this study, *Calothrix* sp. sequestered not only nutrient but also improved water quality, electrical conductivity and dissolved oxygen in wastewater (Renuka *et al.*, 2013). The ability of *Phormidium bohneri* for removal of nitrogen and phosphorus from wastewater was also revealed in another experiment. A total depletion of NH_3 and PO_4^{-3} was achieved after 3 and 5 days with *Phormidium bohneri*, respectively. *Phormidium bohneri* has the features of self-aggregation and sedimentation capacity that prevent the steps for harvesting or physical recovery of the cyanobacterial cells; therefore, lowers the cost of bioremediation practice (Dumas *et al.*, 1998). To apply in the treatment of wastewater in a cold area where biological activity is suppressed, a polar growing cyanobacterium *Phormidium tenue* was optimized for remediating nutrient in moderately cold area with temperatures around 15 °C, that are normal throughout fall and spring seasons in the subarctic region (Chevalier *et al.*, 2000).

A technique for improving removal performance of nutrients from wastewater is immobilization of cyanobacterial cell on a support. *Anabaena* CH3 was immobilized in calcium alginate and evaluated for removal of nitrogen in batch and semicontinuous cultures. Elimination rates of ammonium nitrogen were increased about three times in semicontinuous (immobilized) condition (Lee *et al.*, 1995). Also, *Phormidium laminosum* is a thermotolerant cyanobacterium that was immobilized on cellulose hollow fibers. The immobilized cells were used in continuous photobioreactor for bioremediation of nitrate and phosphate at high temperature (above 40 °C). Due to the thermophilic nature of these blue-green algae, it can be used for bioremediation in equatorial waters (Sawayama *et al.*, 1998).

14.3.3.2 Cyanoremediation of pesticides

About 5.2 billion pounds of pesticides including herbicides (40 %), insecticides (17 %) and fungicides (10 %) were used annually in 2006 and 2007 worldwide (Grube *et al.*, 2011). The release of these toxic compounds in the environment leads to various ecological concerns and human health effects. Pesticide biotransformation occurs in three steps. In the first step, the structure of the pesticides is modified to a more water soluble and less harmful compound by a chemical reaction such as oxidation, reduction or/and hydrolysis. Oxidation of pesticides is a common reaction in the biodegradation of these molecules that perform by oxidative enzymes. In the second step, conjugation with a saccharide, glutathione or amino acid will change the product of the first step to a harmless compound. Finally, the product of the second step may undergo further conjugation and produce safe molecule for the microorganism. Enzymes that accomplish biotransformation of pesticides are induced when microorganisms adjust with their contaminated environments (Van Eerd *et al.*, 2009).

Cyanobacteria remediate pesticides through different processes including biotransformation, bioaccumulation and biosorption. Some of these photosynthetic microorganisms utilize pesticides as sole carbon (in heterotrophic fashion) or phosphorus source. Several cyanobacterial species, e.g., *Anacystics nidulans, Microcystes aeroginosa, Oscillatoria animalis, Synechococcus elongatus, Phormidium tenue, Phormidium foveolarum, Nostoc linckia*, and *Nostoc muscorum* have been reported to biotransform different organophosphorus insecticides such as monocrotophos and quinalphos, methyl parathion, malathion, dichlorovos and phosphomidon (Megharaj *et al.*, 1994; Cáceres *et al.*, 2008; Subashchandrabose, *et al.*, 2013). Other cyanobacteria like *Anabaena* sp. and *Aulosira fertilissima* has been reported that remediated organochlorine insecticides. These microorganisms biotransformed DDT (dichlorodiphenyltrichloroethane) to DDD (dichlorodiphenyldichloroethane) and DDE (dichlorodiphenyldichloroethylene). Also, *Aulosira fertilissima* degraded fenitrothion, chlorpyrifos and DDT with a concentration of 6651, 3971 and 1429 ppm, respectively (Lal *et al.*, 1987). *Anabaena* sp., *Nostoc muscorum, Nostoc* sp. MM1, *Nostoc* sp. MM2 and *Nostoc* sp. MM3 were reported that metabolized organophosphorus insecticide fenamiphos by oxidation and hydrolysis. Oxidized metabolite (fenamiphos sulfoxide) is toxic, whereas hydrolyzed products are non-toxic for aquatic organisms (Idi *et al.*, 2014). Dechlorination of lindane was accomplished by *Anabaena* sp. PCC 7120 and *Nostoc ellipsosporum* strain B1453–7 in the presence of nitrate. As ammonia and darkness have an inhibitory effect on dechlorination reaction (Kuritz, 1998), researchers proved by mutational analysis that nitrate reduction system of cyanobacteria was responsible for the reactions (Kuritz *et al.*, 1997). Bioconversion of 2, 4-Dinitrophenol (2, 4-DNP), a precursor for the production of some pesticides, to 2-amino-4-nitrophenol (2-ANP) by *Anabaena variabilis* in the concentration ranges of 5–150 μM and accumulation of 2-amino-4-nitrophenol was also reported (Hirooka *et al.*, 2006).

14.3.3.3 Cyanoremediation of dye compounds

A huge amount of diverse types of dye compounds is liberated by different industries into wastewater that gives rise to severe ecological concerns. These substances may be very dangerous and life-threatening for aquatic biosystems. Along with their direct toxic effects, they decrease the photosynthesis of organisms in deep layers of aquatic biosystems by absorption of different light wavelenghs and prevention of the transmission of its beams in water. Some cyanobacteria have the capacity for bioremediation of dye compounds in aquatic systems. Cyanobacteria remediate dye compounds by three different mechanisms; adsorption, enzymatic degradation and oxygen species generation (that may result in the alkalinization of the environments). While adsorption is a rapid and passive physical process, other mechanisms are slower and active processes that remove dye compounds chemically. In some cases, enzymatic degradation may result in the complete mineralization of dye compounds (Singh & Patel, 2012). *Phormidium valderianum* was found to be efficient in bioremediation of dye compounds. This cyanobacterium immediately eliminated above 90 % of dye compounds (Shah *et al.*, 2001). Rapid elimination of dye compounds is shown that the mechanism of removal is probably physical absorption (Shah, *et al.*, 2001). *Oscillatoria formosa* NTDM02 was examined for bioremediation of dye and was found to decolorize textile dyes very efficiently (Ali *et al.*, 2011). Production of reductase enzymes and alkalinization of the reaction media was proposed as the mechanism of the decolorization reaction (Ali, *et al.*, 2011). Both free and immobilized cells of two dominantly-existed cyanobacteria, *Oscillatoria brevis* and *Westiellopsis* in the dye-polluted area were successfully used for bioremediation of industrial dye effluent (Subramaniyan, 2012). Along with the decolorization of dyes, these cyanobacteria eliminated the mineral and organic materials. Moreover, depletion in biological oxygen demand (BOD) and chemical oxygen demand (COD) up to 95 %, improvement of dissolved oxygen (DO) and full exclusion of nutrient such as nitrates, phosphate and ammonia were reported (Subramaniyan, 2012). The potential of hypersaline-tolerant cyanobacterium *Phormidium tenue* for decolorization of paper mill waste was described. Improvement of physiochemical parameters of effluent were also reported after 20 days of incubation. It was proposed that the mechanism of decolorization might be adsorption of dye compounds (Nagasathya & Thajuddin, 2008).

Melanoidin is the product of Millard reaction between amino and carbonyl groups of amino acids and sugars (Chandra *et al.*, 2008). Melanoidin is discharged mainly by food processing industries into wastewater. Melanoidin was remediated by *Oscillatoria boryana* BDU through production of oxygen species during photosynthetic activity of the microorganism. These oxygen species produced by cyanobacterium decolorized pure and crude dye by approximately 75 % and 65 % in 30 days, respectively (Kalavathi *et al.*, 2001).

14.3.3.4 Consortia of cyanobacteria and other bacteria for wastewater treatment

Consortia of cyanobacteria with other microorganisms are very useful in wastewater treatment. Cyanobacteria provide molecular oxygen for heterotrophic microorganisms by photosynthesis that is utilized as an electron acceptor by heterotrophic microorganisms to decompose organic substances in wastewater. Concurrently, heterotrophic microorganisms provide carbon dioxide for photosynthetic activity of cyanobacteria. This symbiotic relationship along with competition for carbon and nitrogen resources is highly favorable for wastewater remediation by microorganisms (Subashchandrabose, *et al.*, 2011). Consortium of *Halobacterium* US, a halotolerant bacterium with *Oscillatoria* sp. BDU and *Aphanocapsa* sp. BDU has been examined for wastewater bioremediation. This co-culture effectively eliminated calcium and chloride from industrial runoff that did not adversely affect the growing of fish (Subramanian & Uma, 1997).

As mentioned above *Anabaena variabilis* biotransform 2,4-dinitrophenol (2,4-DNP) to 2-amino-4-nitrophenol (2-ANP). However, 2-ANP is a potent mutagenic agent that accumulates in the culture medium. When the mixed culture of *Anabaena variabilis* and *Anabaena cylindrica* was used for remediation of 2,4-DNP from industrial waste, the biodegradation product did not accumulate in the medium (Hirooka, *et al.*, 2006).

14.3.3.5 Mixotrophic cyanobacteria for wastewater treatment

Mixotrophy is the potential for utilization of organic and inorganic (CO_2) carbon for growth. Some cyanobacteria have the ability of grow in the inherent photoautotrophic condition (in the light and consumption of CO_2) in conjunction with heterotrophic (on organic carbon) and mixotrophic (simultaneous utilization of organic carbon and CO_2) conditions (Subashchandrabose, *et al.*, 2013). Soil isolated blue-green algae *Nostoc linckia, Nostoc muscorum, Oscillatoria animalis* and *Phormidium foveolarum* were explored for their ability for degradation of an organophosphorus insecticide in mixotrophic approach. All these cyanobacteria degraded methyl parathion and utilized it as a source of nutrients (Megharaj, *et al.*, 1994). Methyl parathion was also transformed by *Anabaena* sp. PCC7120 to dimethyl *p*-aminophenyl thiophosphate with a reduction reaction on nitro group under aerobic condition in light (Barton *et al.*, 2004). Moreover, different heterocyst-forming cyanobacteria were grown on media with monocrotophos and malathion as a sole phosphorus source. *Aulosira fertilissima* and *Nostoc muscorum* biodegraded these organophosphorus pesticides and used them as phosphorus source (Subramanian *et al.*, 1994).

14.3.3.6 Genetic engineering of cyanobacteria for enhanced biodegradation of wastewater

Engineering of cyanobacteria by the introduction of well-distinguished genes and operons that are responsible for biodegradation activity of potent microorganisms is an approach for improving bioremediation efficacy of cyanobacteria.

A recombinant *Anabaena* sp. was constructed that degraded lindane (γ-pentachlorocyclohexene) to 1,2,4-trichlorobenzene. This cyanobacterium was modified by introduction of the gene linA, confered the ability of lindane degradation to *Pseudomonas paucimobilis* (Kuritz & Wolk, 1995). Also, two transgenic strains of *Anabaena* PCC 7120 and *Nostoc Ellipsosporum* harboring a degradation operon (*fcbABC*) from *Arthrobacter globiformis* were designed. This operon is responsible for dehalogenation of halobenzoates. These genetically-modified cyanobacteria were used for biodegradation of 4-chlorobenzoate (Kuritz, *et al.*, 1997).

14.4 Current challenges and perspectives of cyanoremediation

Cyanobacteria contribute considerably to the primary energy and food chains of various bionetwork, especially aquatic ecosystems, and play a significant role in the cycling of carbon, oxygen and nitrogen. The biotechnological applications of cyanobacteria in diverse areas, such as the production of food, biofertilizer, bioenergies, and bioactive medicinal compounds have been well established.

Human activities along with population expansion and increasing the utilization of fossil fuels and agricultural pesticides result in the release of diverse environmental contaminants that threaten the health of all kinds of life. Removal of these environmental pollutants is performed by physical, chemical and biological techniques. Biological remediation is an eco-friendly process offer a better choice than physicochemical methods for treating contaminants. As it discussed throughout this chapter, among the microorganisms used in the bioremediation of environmental contaminants, cyanobacteria are prospective eco-friendly and green bioremediators. Regardless of much researches have been carried out and many techniques developed, cyanoremediation is still in its infancy. Much more efforts are required to commercialize cyanoremediation. Cyanoremediation has many advantages over other bioremediation techniques due to the use of photosynthetic activity of cyanobacteria to harvest the light energy and convert atmospheric CO_2 into the beneficial product; therefor, they reduce the environmental impacts of greenhouse gas and global warming. Although, commercialization of cyanoremediation is problematic, it is not unrealistic. With ongoing research and development in academia and industry, it is expected that large-scale in situ application of cyanoremediation could become economically reasonable in the future when technological and economical obstacles would be conquered.

Main role players in cyanoremediation process are pollutant, environment and cyanobacteria. Large-scale culture of cyanobacteria, slow growth rate and different sensitivity to light (intensity or duration of light and dark periods), low degradability or removal activity of some strain are major challenges of the cyanobacterial side should be tackled. One prerequisite for large-scale cultivation of cyanobacteria is providing a uniform energy (sunlight) source. Uniform distribution of light is a difficult task due to attenuation when it penetrates through culture media. Consequently, the design of better photobioreactors with high surface to volume is an inevitable requirement for commercializing biotechnological application of cyanobacteria. Exploring contaminated area for adapted fast growing species with better potential for remediation of pollutants and enhancing inherent remediation activity of natural cyanobacteria are important factors affecting the efficiency of cyanoremediation. On the side of pollutant, biodegradability and bioavailability of contaminants are very crucial. Depends on the nature of the contaminant, the soil chemistry, and the environment, the bioavailibility of pollutants in field bioremediation can be improved by some techniques such as surfactant addition. Some cyanobacterial strains produce surfactants that improve the bioavailability of organic pollutant, thereby enhancing the growth of cyanobacteria and other heterotrophic degrading microorganisms and increasing the rate of bioremediation. On the side of the environment, lack of nutrients (simple organic carbon, nitrogen and phosphorus), and the inhibitory effect of the environmental conditions (temperature, salinity, pH, UV, light intensity and soil or aquatic chemistry) are important. It needs to keep in mind that, contaminants are homogeneously used in laboratory scale, but in the environment pollutants are heterogeneously presented. Other points to take into account are the interaction between contaminants and cyanobacteria, the effects of different pollutants on physiological and metabolic condition of cyanobacteria and the impacts of cyanoremediation on the ecosystem. To simulate *in situ* bioremediation in the laboratory scale, using indigenous cyanobacteria or the cyanobacterial consortia is a great and crucial starting point. However, the process is a complex event in the environment. Involving different strains of cyanobacteria and different contaminants along with geophysical and hydrological complexities render this complicated phenomenon.

The key to successful large-scale *in situ* cyanoremediation is to understand the genetic and catabolic potential of cyanobacteria or cyanobacterial consortia with other heterotrophic bacteria to employ for removal of environmental contaminants. Application of genetic and metabolic engineering to improve natural ability of cyanobacteria for abatement of pollution has become extremely promising and several recombinant and transgenic cyanobacteria were developed. Upcoming studies are supposed to focus on genetic and metabolic engineering of currently used species to expand their bioremediation capabilities (with more potent activity toward different contaminants). In recent years, several genetically engineered cyanobacteria were designed but did not examine for *in situ* bioremediation due to regulatory issues that is another problem.

The genome of the capable cyanobacteria should be sequenced to locate novel genes that play key roles in removal or transformation of different contaminants. Post-genomic studies including bioinformatics, functional genomics, structural genomics, transcriptomics, proteomics and metabolomics are also needed to allow for a system-wide approach to cyanobacterial research. Cyanobacterial system biology provides great prospective for enhancing efforts to optimize and validate biotechnological and environmental applications of cyanobacteria and cyanoremediation process and identify the impact of cyanoremediation on the aquatic and soil environment. Using molecular and system biology data, cyanoremediation will be more reliable and safer procedure.

Other techniques such as immobilization on a support or application of naturally immobilized cyanobacterial mats or consortia and encapsulation of sensitive strain improve bioremediation efficiency and help the recovery of some pollutants (metal ions). Coupling cyanoremediation with the production of some useful compounds by cyanobacteria reduced the cost of bioremediation and makes it economical and more eco-friendly (green) process.

14.5 References

[1] Abed, R. M. (2010). Interaction between cyanobacteria and aerobic heterotrophic bacteria in the degradation of hydrocarbons. International Biodeterioration & Biodegradation, 64(1), 58–64.

[2] Abed, R., Dobretsov, S., & Sudesh, K. (2009). Applications of cyanobacteria in biotechnology. Journal of Applied Microbiology, 106(1), 1–12.

[3] Abed, R. M., & Köster, J. (2005). The direct role of aerobic heterotrophic bacteria associated with cyanobacteria in the degradation of oil compounds. International Biodeterioration & Biodegradation, 55(1), 29–37.

[4] Abul-Hajj, Y. J., & Qian, X. (1986). Transformation of steroids by algae. Journal of Natural Products, 49(2), 244–248.

[5] Adams, D. G. (2000). Heterocyst formation in cyanobacteria. Current Opinion in Microbiology, 3(6), 618–624.

[6] Adams, D. G., & Duggan, P. S. (1999). Tansley Review No. 107. Heterocyst and akinete differentiation in cyanobacteria. New Phytologist, 144(1), 3–33.

[7] Ahuja, P., Gupta, R., & Saxena, R. (1999). Zn2+ biosorption by Oscillatoria anguistissima. Process Biochemistry, 34(1), 77–85.

[8] Akaraonye, E., Keshavarz, T., & Roy, I. (2010). Production of polyhydroxyalkanoates: the future green materials of choice. Journal of Chemical Technology & Biotechnology, 85(6), 732–743.

[9] Al-Hasan, R., Al-Bader, D., Sorkhoh, N., & Radwan, S. (1998). Evidence for n-alkane consumption and oxidation by filamentous cyanobacteria from oil-contaminated coasts of the Persian Gulf. Marine Biology, 130(3), 521–527.

[10] Al Hasan, R., Sorkhoh, N., Al Bader, D., & Radwan, S. (1994). Utilization of hydrocarbons by cyanobacteria from microbial mats on oily coasts of the Persian Gulf. Applied Microbiology & Biotechnology, 41(5), 615–619.

[11] Alder, A., Bigler, P., Werck-Reichhart, D., & Al-Babili, S. (2009). In vitro characterization of Synechocystis CYP120A1 revealed the first nonanimal retinoic acid hydroxylase. FEBS Journal, 276(19), 5416–5431.

[12] Ali, D. M., Suresh, A., Praveen Kumar, R., Gunasekaran, M., & Thajuddin, N. (2011). Efficiency of textile dye decolorization by marine cyanobacterium, Oscillatoria formosa NTDM02. African Journal of Basic & Appllied Sciences, 3(1), 9–13.

[13] Allen, M. M. (1984). Cyanobacterial cell inclusions. Annual Reviews in Microbiology, 38(1), 1–25.

[14] Andrianasolo, E. H., Gross, H., Goeger, D., Musafija-Girt, M., McPhail, K., Leal, R. M., Mooberry, S. L., & Gerwick, W. H. (2005). Isolation of swinholide A and related glycosylated derivatives from two field collections of marine cyanobacteria. Organic Letters, 7(7), 1375–1378.

[15] Annapurna, V., Deosthale, Y., & Bamji, M. S. (1991). Spirulina as a source of vitamin A. Plant Foods for Human Nutrition, 41(2), 125–134.
[16] Arunakumara, K., Zhang, X., & Song, X. (2008). Bioaccumulation of Pb2+ and its effects on growth, morphology and pigment contents of Spirulina (Arthrospira) platensis. Journal of Ocean University of China, 7(4), 397–403.
[17] Asthana, R., Srivastava, A., Singh, A., Deepali, Singh, S., Nath, G., Srivastava, R., & Srivastava, B. (2006). Identification of an antimicrobial entity from the cyanobacterium Fischerella sp. isolated from bark of Azadirachta indica (Neem) tree. Journal of Applied Phycology, 18(1), 33–39.
[18] Awasthi, M., & Rai, L. (2004). Adsorption of nickel, zinc and cadmium by immobilized green algae and cyanobacteria: a comparative study. Annals of Microbiology, 54, 257–268.
[19] Balskus, E. P., & Walsh, C. T. (2010). The genetic and molecular basis for sunscreen biosynthesis in cyanobacteria. Science, 329(5999), 1653–1656.
[20] Bar-Or, Y., & Shilo, M. (1988). The role of cell-bound flocculants in coflocculation of benthic cyanobacteria with clay particles. FEMS Microbiology Ecology, 4(3–4), 169–174.
[21] Barth, H.-J. (2003). The influence of cyanobacteria on oil polluted intertidal soils at the Persian Gulf shores. Marine Pollution Bulletin, 46(10), 1245–1252.
[22] Barton, J., Kuritz, T., O'Connor, L., Ma, C., Maskarinec, M., & Davison, B. (2004). Reductive transformation of methyl parathion by the cyanobacterium Anabaena sp. strain PCC7120. Applied Microbiology & Biotechnology, 65(3), 330–335.
[23] Berry, J. P., Gantar, M., Perez, M. H., Berry, G., & Noriega, F. G. (2008). Cyanobacterial toxins as allelochemicals with potential applications as algaecides, herbicides and insecticides. Marine Drugs, 6(2), 117–146.
[24] Boivin, M., Greve, G., Garcia-Meza, J., Massieux, B., Sprenger, W., Kraak, M., Breure, A., Rutgers, M., & Admiraal, W. (2007). Algal-bacterial interactions in metal contaminated floodplain sediments. Environmental Pollution, 145(3), 884–894.
[25] Bokesch, H. R., O'Keefe, B. R., McKee, T. C., Pannell, L. K., Patterson, G. M., Gardella, R. S., Sowder, R. C., Turpin, J., Watson, K., & Buckheit, R. W. (2003). A potent novel anti-HIV protein from the cultured cyanobacterium Scytonema varium. Biochemistry, 42(9), 2578–2584.
[26] Bonnet, S., Webb, E. A., Panzeca, C., Karl, D. M., Capone, D. G., & Wilhelmy, S. A. S. (2010). Vitamin B12 excretion by cultures of the marine cyanobacteria Crocosphaera and Synechococcus. Limnology & Oceanography, 55(5), 1959–1964.
[27] Bouchard, H., Brun, M.-P., Commercon, A., & Zhang, J. (2012). Novel conjugates, preparation thereof, and therapeutic use thereof. US20120225089.
[28] Boyd, M. R., Gustafson, K. R., McMahon, J. B., Shoemaker, R. H., O'Keefe, B. R., Mori, T., Gulakowski, R. J., Wu, L., Rivera, M. I., & Laurencot, C. M. (1997). Discovery of cyanovirin-N, a novel human immunodeficiency virus-inactivating protein that binds viral surface envelope glycoprotein gp120: potential applications to microbicide development. Antimicrobial Agents & Chemotherapy, 41(7), 1521–1530.
[29] Bui, H. T., Jansen, R., Pham, H. T., & Mundt, S. (2007). Carbamidocyclophanes A-E, chlorinated paracyclophanes with cytotoxic and antibiotic activity from the Vietnamese cyanobacterium Nostoc sp. Journal of Natural Products, 70(4), 499–503.
[30] Burja, A. M., Banaigs, B., Abou-Mansour, E., Burgess, J. G., & Wright, P. C. (2001). Marine cyanobacteria– a prolific source of natural products. Tetrahedron, 57(46), 9347–9377.
[31] Butler, M. S. (2008). Natural products to drugs: natural product-derived compounds in clinical trials. Natural Product Reports, 25(3), 475–516.
[32] Cáceres, T. P., Megharaj, M., & Naidu, R. (2008). Biodegradation of the pesticide fenamiphos by ten different species of green algae and cyanobacteria. Current Microbiology, 57(6), 643–646.

[33] Campanella, L., Crescentini, G., & Avino, P. (1999). Chemical composition and nutritional evaluation of some natural and commercial food products based on Spirulina. Analusis, 27(6), 533–540.

[34] Campbell, J., Stevens, S. E., & Balkwill, D. (1982). Accumulation of poly-beta-hydroxybutyrate in Spirulina platensis. Journal of Bacteriology, 149(1), 361–363.

[35] Cannell, R. (1990). Algal biotechnology. Applied Biochemistry & Biotechnology, 26(1), 85–105.

[36] Capon, R. J., Dunlop, R. W., Ghisalberti, E. L., & Jefferies, P. R. (1983). Poly-3-hydroxyalkanoates from marine and freshwater cyanobacteria. Phytochemistry, 22(5), 1181–1184.

[37] Cardellina, J. H., Moore, R. E., Arnold, E. V., & Clardy, J. (1979). Structure and absolute configuration of malyngolide, an antibiotic from the marine blue-green alga Lyngbya majuscula Gomont. The Journal of Organic Chemistry, 44(23), 4039–4042.

[38] Carmeli, S., Moore, R. E., & Patterson, G. L. (1991). Mirabazoles, minor tantazole-related cytotoxins from the terrestrial blue-green alga Scytonema mirabile. Tetrahedron Letters, 32(23), 2593–2596.

[39] Carr, N. (1966). The occurrence of poly-β-hydroxybutyrate in the blue-green alga, Chlorogloea fritschii. Biochimica & Biophysica Acta (BBA)-Biophysics including Photosynthesis, 120(2), 308–310.

[40] Carr, N. G., & Whitton, B. A. (1982). The biology of cyanobacteria: University of California Press.

[41] Carter, D. C., Moore, R. E., Mynderse, J. S., Niemczura, W. P., & Todd, J. S. (1984). Structure of majusculamide C, a cyclic depsipeptide from Lyngbya majuscula. The Journal of Organic Chemistry, 49(2), 236–241.

[42] Castenholz, R. W., Wilmotte, A., Herdman, M., Rippka, R., Waterbury, J. B., Iteman, I., & Hoffmann, L. (2001). Bergey's Manual® of Systematic Bacteriology. Phylum BX. Cyanobacteria; (pp. 473–599): Springer; New York, NY, USA.

[43] Cerniglia, C. E., Gibson, D. T., & Van Baalen, C. (1980a). Oxidation of naphthalene by cyanobacteria and microalgae. Journal of General Microbiology, 116(2), 495–500.

[44] Cerniglia, C. E., Van Baalen, C., & Gibson, D. T. (1980b). Oxidation of biphenyl by the cyanobacterium, Oscillatoria sp., strain JCM. Archives of Microbiology, 125(3), 203–207.

[45] Cerniglia, C. E., Van Baalen, C., & Gibson, D. T. (1980c). Metabolism of naphthalene by the cyanobacterium Oscillatoria sp., strain JCM. Journal of General Microbiology, 116(2), 485–494.

[46] Chakdar, H., Jadhav, S. D., Dhar, D. W., & Pabbi, S. (2012). Potential applications of blue green algae. Journal of Scientific & Industrial Research, 71(1), 13–20.

[47] Chandra, R., Bharagava, R. N., & Rai, V. (2008). Melanoidins as major colourant in sugarcane molasses based distillery effluent and its degradation. Bioresource Technology, 99(11), 4648–4660.

[48] Chang, T. T., More, S. V., Lu, I. H., Hsu, J. C., Chen, T. J., Jen, Y. C., Lu, C. K., & Li, W. S. (2011). Isomalyngamide A, A-1 and their analogs suppress cancer cell migration in vitro. European Journal of Medicinal Chemistry, 46(9), 3810–3819.

[49] Chavan, A., & Mukherji, S. (2008). Treatment of hydrocarbon-rich wastewater using oil degrading bacteria and phototrophic microorganisms in rotating biological contactor: effect of N: P ratio. Journal of Hazardous Materials, 154(1), 63–72.

[50] Chevalier, P., Proulx, D., Lessard, P., Vincent, W., & De la Noüe, J. (2000). Nitrogen and phosphorus removal by high latitude mat-forming cyanobacteria for potential use in tertiary wastewater treatment. Journal of Applied Phycology, 12(2), 105–112.

[51] Choi, C. W., Yoo, S.-A., Oh, I.-H., & Park, S. H. (1998). Characterization of an extracellular flocculating substance produced by a planktonic cyanobacterium, Anabaena sp. Biotechnology Letters, 20(7), 643–646.

[52] Chojnacka, K., Chojnacki, A., & Gorecka, H. (2005). Biosorption of Cr3+, Cd2+ and Cu2+ ions by blue–green algae Spirulina sp.: kinetics, equilibrium and the mechanism of the process. Chemosphere, 59(1), 75–84.
[53] Chojnacka, K., Chojnacki, A., & Górecka, H. (2004). Trace element removal by Spirulina sp. from copper smelter and refinery effluents. Hydrometallurgy, 73(1), 147–153.
[54] Christaki, E., Florou-Paneri, P., & Bonos, E. (2011). Microalgae: a novel ingredient in nutrition. International Journal of Food Sciences & Nutrition, 62(8), 794–799.
[55] Cobbett, C., & Goldsbrough, P. (2002). Phytochelatins and metallothioneins: roles in heavy metal detoxification and homeostasis. Annual Review of Plant Biology, 53(1), 159–182.
[56] Codd, G. (1995). Cyanobacterial toxins: occurrence, properties and biological significance. Water Science & Technology, 32(4), 149–156.
[57] Codd, G. A. (2000). Cyanobacterial toxins, the perception of water quality, and the prioritisation of eutrophication control. Ecological Engineering, 16(1), 51–60.
[58] Costa, M., Costa-Rodrigues, J., Fernandes, M. H., Barros, P., Vasconcelos, V., & Martins, R. (2012). Marine cyanobacteria compounds with anticancer properties: A review on the implication of apoptosis. Marine Drugs, 10(10), 2181–2207.
[59] Das, D., & Veziroğlu, T. N. (2001). Hydrogen production by biological processes: a survey of literature. International Journal of Hydrogen Energy, 26(1), 13–28.
[60] De Philippis, R., Colica, G., & Micheletti, E. (2011). Exopolysaccharide-producing cyanobacteria in heavy metal removal from water: molecular basis and practical applicability of the biosorption process. Applied Microbiology & Biotechnology, 92(4), 697–708.
[61] De Philippis, R., & Micheletti, E. (2009). Heavy metal removal with exopolysaccharide-producing cyanobacteria. Heavy Metals in the Environment, 89–122.
[62] De Philippis, R., Paperi, R., & Sili, C. (2007). Heavy metal sorption by released polysaccharides and whole cultures of two exopolysaccharide-producing cyanobacteria. Biodegradation, 18(2), 181–187.
[63] De Philippis, R., & Vincenzini, M. (1998). Exocellular polysaccharides from cyanobacteria and their possible applications. FEMS Microbiology Reviews, 22(3), 151–175.
[64] Deng, M.-D., & Coleman, J. R. (1999). Ethanol synthesis by genetic engineering in cyanobacteria. Applied & Environmental Microbiology, 65(2), 523–528.
[65] Deng, R., & Chow, T. J. (2010). Hypolipidemic, antioxidant, and antiinflammatory activities of microalgae Spirulina. Cardiovascular Therapeutics, 28(4), e33–e45.
[66] Dexter, J., & Fu, P. (2009). Metabolic engineering of cyanobacteria for ethanol production. Energy & Environmental Science, 2(8), 857–864.
[67] Dillon, J. G., & Castenholz, R. W. (2003). The synthesis of the UV-screening pigment, scytonemin, and photosynthetic performance in isolates from closely related natural populations of cyanobacteria (Calothrix sp.). Environmental Microbiology, 5(6), 484–491.
[68] Dincer, I. (2000). Renewable energy and sustainable development: a crucial review. Renewable & Sustainable Energy Reviews, 4(2), 157–175.
[69] Doukyu, N., Yamagishi, W., Kuwahara, H., Ogino, H., & Furuki, N. (2007). Purification and characterization of a maltooligosaccharide-forming amylase that improves product selectivity in water-miscible organic solvents, from dimethylsulfoxide-tolerant Brachybacterium sp. strain LB25. Extremophiles, 11(6), 781–788.
[70] Dumas, A., Laliberte, G., Lessard, P., & De La Noüe, J. (1998). Biotreatment of fish farm effluents using the cyanobacterium Phormidium bohneri. Aquacultural Engineering, 17(1), 57–68.
[71] Dutta, D., De, D., Chaudhuri, S., & Bhattacharya, S. K. (2005). Hydrogen production by Cyanobacteria. Microbial Cell Factories, 4(1), 36.

[72] El-Sheekh, M. M., El-Shouny, W. A., Osman, M. E., & El-Gammal, E. W. (2005). Growth and heavy metals removal efficiency of Nostoc muscorum and Anabaena subcylindrica in sewage and industrial wastewater effluents. Environmental Toxicology & Pharmacology, 19(2), 357–365.
[73] Ellis, B. (1977). Degradation of phenolic compounds by fresh-water algae. Plant Science Letters, 8(3), 213–216.
[74] Erbe, J., Taylor, K., & Hall, L. (1996). Expression of mouse metallothionein in the cyanobacterium Synechococcus PCC7942. Journal of Industrial Microbiology, 17(1), 41–46.
[75] Faramarzi, M. A., Yazdi, M. T., Ghostinroudi, H., Amini, M., Ghasemi, Y., Jahandar, H., & Arabi, H. (2006). Nostoc muscorum: a regioselective biocatalyst for 17-carbonyl reduction of androst-4-en-3, 17-dione and androst-1, 4-dien-3, 17-dione. Annals of Microbiology, 56(3), 253–256.
[76] Fattom, A., & Shilo, M. (1984). Phormidium J-1 bioflocculant: production and activity. Archives of Microbiology, 139(4), 421–426.
[77] Foye, W. O., Lemke, T. L., & Williams, D. A. (2008). Foye's principles of medicinal chemistry: Lippincott Williams & Wilkins.
[78] Froscio, S. M., Humpage, A. R., Burcham, P. C., & Falconer, I. R. (2001). Cell-free protein synthesis inhibition assay for the cyanobacterial toxin cylindrospermopsin. Environmental Toxicology, 16(5), 408–412.
[79] Fu, F., & Wang, Q. (2011). Removal of heavy metal ions from wastewaters: A review. Journal of Environmental Management, 92(3), 407–418.
[80] Gardea-Torresdeya, J., Arenasb, J., Franciscob, N., Tiemanna, K., & Webbb, R. (1998). Ability of immobilized cyanobacteria to remove metal ions from solution and demonstration of the presence of metallothionein genes in various strains. Journal of Hazardous Substance Research, 1, 2–1.
[81] Garg, V., Zhang, W., Gidwani, P., Kim, M., & Kolb, E. A. (2007). Preclinical analysis of tasidotin HCl in Ewing's sarcoma, rhabdomyosarcoma, synovial sarcoma, and osteosarcoma. Clinical Cancer Research, 13(18), 5446–5454.
[82] Gavrilescu, M. (2004). Removal of heavy metals from the environment by biosorption. Engineering in Life Sciences, 4(3), 219–232.
[83] Gerwick, W. H., Cameron Coates, R., Engene, N., Gerwick, L., Grindberg, R. V., Jones, A. C., & Sorrels, C. M. (2008). Giant marine cyanobacteria produce exciting potential pharmaceuticals. Microbe, 3(6), 277–284.
[84] Ghasemi, Y., Rasoul-Amini, S., & Fotooh-Abadi, E. (2011). The biotransformation, biodegradation, and bioremediation of organic compounds by microalgae. Journal of Phycology, 47(5), 969–980.
[85] Ghasemi, Y., Yazdi, M. T., Dehshahri, A., Niknahad, H., Shokravi, S., Amini, M., Ghasemian, A., & Faramarzi, M. (2006). Algal transformation of hydrocortisone by the cyanobacterium Nostoc ellipsosporum. Chemistry of Natural Compounds, 42(6), 702–705.
[86] Gomez Casati, D. F., Aon, M. A., & Iglesias, A. A. (1999). Ultrasensitive glycogen synthesis in Cyanobacteria. FEBS Letters, 446(1), 117–121.
[87] Gong, R., Ding, Y., Liu, H., Chen, Q., & Liu, Z. (2005). Lead biosorption and desorption by intact and pretreated Spirulina maxima biomass. Chemosphere, 58(1), 125–130.
[88] Gonzalez, R., Rodriguez, S., Romay, C., González, A., Armesto, J., Remirez, D., & Merino, N. (1999). Anti-inflammatory activity of phycocyanin extract in acetic acid-induced colitis in rats. Pharmacological Research, 39(1), 55–59.
[89] Gothalwal, R., & Chillara, S. (2012). Cyanoremediation: A green clean technology. Microorganisms in Environmental Management. (pp767–786): Springer.
[90] Gray, K. A., Zhao, L., & Emptage, M. (2006). Bioethanol. Current Opinion in Chemical Biology, 10(2), 141–146.

[91] Griffiths, D. J., & Saker, M. L. (2003). The Palm Island mystery disease 20 years on: a review of research on the cyanotoxin cylindrospermopsin. Environmental Toxicology, 18(2), 78–93.

[92] Grube, A., Donaldson, D., Kiely, T., & Wu, L. (2011). Pesticides industry sales and usage. US EPA, Washington, DC.

[93] Gugger, M., Lenoir, S., Berger, C., Ledreux, A., Druart, J.-C., Humbert, J.-F., Guette, C., & Bernard, C. (2005). First report in a river in France of the benthic cyanobacterium Phormidium favosum producing anatoxin-a associated with dog neurotoxicosis. Toxicon, 45(7), 919–928.

[94] Guo, X.-X., Shi, D.-J., Xu, X.-D., Ouyang, Y.-X., & Ru, B.-G. (1999). Metal-induced expressing of mammal Metallothionein-I gene in cyanobacteria to promote cadmium-binding preferences. Applied Microbiology & Biotechnology, 52(6), 806–810.

[95] Gupta, V., & Rastogi, A. (2008). Sorption and desorption studies of chromium (VI) from nonviable cyanobacterium Nostoc muscorum biomass. Journal of Hazardous Materials, 154(1), 347–354.

[96] Hallenbeck, P. C. (2012). Hydrogen production by cyanobacteria. Microbial Technologies in Advanced Biofuels Production. (pp. 15–28): Springer.

[97] Hallenbeck, P. C., & Benemann, J. R. (2002). Biological hydrogen production; fundamentals and limiting processes. International Journal of Hydrogen Energy, 27(11), 1185–1193.

[98] Hamada, H., Kondo, Y., Ishihara, K., Nakajima, N., Hamada, H., Kurihara, R., & Hirata, T. (2003). Stereoselective biotransformation of limonene and limonene oxide by cyanobacterium, Synechococcus sp. PCC 7942. Journal of Bioscience & Bioengineering, 96(6), 581–584.

[99] Han, B., Goeger, D., Maier, C. S., & Gerwick, W. H. (2005). The wewakpeptins, cyclic depsipeptides from a Papua New Guinea collection of the marine cyanobacterium Lyngbya semiplena. Journal of Organic Chemistry, 70(8), 3133–3139.

[100] Han, B., Gross, H., Goeger, D. E., Mooberry, S. L., & Gerwick, W. H. (2006). Aurilides B and C, cancer cell toxins from a Papua New Guinea Collection of the marine cyanobacterium Lyngbya m ajuscula. Journal of Natural Products, 69(4), 572–575.

[101] Hartwig, A. (1995). Current aspects in metal genotoxicity. Biometals, 8(1), 3–11.

[102] Hashem, M. A. (2001). Problems and prospects of cyanobacterial biofertilizer for rice cultivation. Functional Plant Biology, 28(9), 881–888.

[103] Hayashi, T., Hayashi, K., Maeda, M., & Kojima, I. (1996). Calcium spirulan, an inhibitor of enveloped virus replication, from a blue-green alga Spirulina platensis. Journal of Natural Products, 59(1), 83–87.

[104] Hein, S., Tran, H., & Steinbüchel, A. (1998). Synechocystis sp. PCC6803 possesses a two-component polyhydroxyalkanoic acid synthase similar to that of anoxygenic purple sulfur bacteria. Archives of Microbiology, 170(3), 162–170.

[105] Henze, M. (2002). Wastewater treatment: biological and chemical processes: Springer Science & Business Media.

[106] Herdman, M., Janvier, M., Rippka, R., & Stanier, R. Y. (1979). Genome size of cyanobacteria. Journal of General Microbiology, 111(1), 73–85.

[107] Hirata, K., Yoshitomi, S., Dwi, S., Iwabe, O., Mahakhant, A., Polchai, J., & Miyamoto, K. (2003). Bioactivities of nostocine A produced by a freshwater cyanobacterium Nostoc spongiaeforme TISTR 8169. Journal of Bioscience & Bioengineering, 95(5), 512–517.

[108] Hirooka, T., Nagase, H., Hirata, K., & Miyamoto, K. (2006). Degradation of 2, 4-dinitrophenol by a mixed culture of photoautotrophic microorganisms. Biochemical Engineering Journal, 29(1), 157–162.

[109] Hirschmann, R. (1992). The cortisone era: aspects of its impact. Some contributions of the Merck Laboratories. Steroids, 57(12), 579–592.

[110] Hoiczyk, E., & Baumeister, W. (1995). Envelope structure of four gliding filamentous cyanobacteria. Journal of Bacteriology, 177(9), 2387–2395.

[111] Hu, Q., Westerhoff, P., & Vermaas, W. (2000). Removal of nitrate from groundwater by cyanobacteria: quantitative assessment of factors influencing nitrate uptake. Applied & Environmental Microbiology, 66(1), 133–139.

[112] Huertas, M. J., López-Maury, L., Giner-Lamia, J., Sánchez-Riego, A. M., & Florencio, F. J. (2014). Metals in cyanobacteria: analysis of the copper, nickel, cobalt and arsenic homeostasis mechanisms. Life, 4(4), 865–886.

[113] Hughes, M. N., & Poole, R. K. (1989). Metals and micro-organisms: Chapman and Hall, London.

[114] Idi, A., Nor, M. H. M., Wahab, M. F. A., & Ibrahim, Z. (2014). Photosynthetic bacteria: an eco-friendly and cheap tool for bioremediation. Reviews in Environmental Science & Bio/ Technology, 1–15.

[115] Ishida, K., Matsuda, H., Murakami, M., & Yamaguchi, K. (1997). Kawaguchipeptin B, an antibacterial cyclic undecapeptide from the cyanobacterium Microcystis aeruginosa. Journal of Natural Products, 60(7), 724–726.

[116] Jaki, B., Orjala, J., & Sticher, O. (1999). A novel extracellular diterpenoid with antibacterial activity from the cyanobacterium Nostoc commune. Journal of Natural Products, 62(3), 502–503.

[117] Johnson, P. E., & Shubert, L. E. (1986). Accumulation of mercury and other elements by Spirulina (Cyanophyceae). Nutrition Reports International, 34(1063), 70.

[118] Kalavathi, D. F., Uma, L., & Subramanian, G. (2001). Degradation and metabolization of the pigment– melanoidin in distillery effluent by the marine cyanobacterium Oscillatoria boryana BDU 92181. Enzyme & Microbial Technology, 29(4), 246–251.

[119] Kanekiyo, K., Lee, J.-B., Hayashi, K., Takenaka, H., Hayakawa, Y., Endo, S., & Hayashi, T. (2005). Isolation of an antiviral polysaccharide, nostoflan, from a terrestrial cyanobacterium, Nostoc f lagelliforme. Journal of Natural Products, 68(7), 1037–1041.

[120] Kaplan, D. (2013). Absorption and adsorption of heavy metals by microalgae. Handbook of Microalgal Culture.(pp. 602–61): John Wiley & Sons, Ltd.

[121] Karatay, S. E. (2014). Usage of thermophilic cyanobacterial biomass for bioethanol production. Environmental Progress & Sustainable Energy, 34(3), 903–907.

[122] Ke, N., Baudry, J., Makris, T. M., Schuler, M. A., & Sligar, S. G. (2005). A retinoic acid binding cytochrome P450: CYP120A1 from Synechocystis sp. PCC 6803. Archives of Biochemistry & Biophysics, 436(1), 110–120.

[123] Khattar, J., Parveen, S., Singh, Y., Singh, D., & Gulati, A. (2015). Intracellular uptake and reduction of hexavalent chromium by the cyanobacterium Synechocystis sp. PUPCCC 62. Journal of Applied Phycology, 27 (2) 827–837.

[124] Kim, S., & Dale, B. E. (2004). Global potential bioethanol production from wasted crops and crop residues. Biomass & Bioenergy, 26(4), 361–375.

[125] Kirkwood, A., Nalewajko, C., & Fulthorpe, R. (2005). The impacts of cyanobacteria on pulp-and-paper wastewater toxicity and biodegradation of wastewater contaminants. Canadian Journal of Microbiology, 51(7), 531–540.

[126] Kirkwood, A., Nalewajko, C., & Fulthorpe, R. (2006). The effects of cyanobacterial exudates on bacterial growth and biodegradation of organic contaminants. Microbial Ecology, 51(1), 4–12.

[127] Klimmek, S., Stan, H.-J., Wilke, A., Bunke, G., & Buchholz, R. (2001). Comparative analysis of the biosorption of cadmium, lead, nickel, and zinc by algae. Environmental Science & Technology, 35(21), 4283–4288.

[128] Kühnel, K., Ke, N., Cryle, M. J., Sligar, S. G., Schuler, M. A., & Schlichting, I. (2008). Crystal structures of substrate-free and retinoic acid-bound cyanobacterial cytochrome P450 CYP120A1. Biochemistry, 47(25), 6552–6559.
[129] Kumar, M. S., Muralitharan, G., & Thajuddin, N. (2009). Screening of a hypersaline cyanobacterium, Phormidium tenue, for the degradation of aromatic hydrocarbons: naphthalene and anthracene. Biotechnology Letters, 31(12), 1863–1866.
[130] Kumar, M. S., Rajeshwari, K., Johnson, S., Thajuddin, N., & Gunasekaran, M. (2011). Removal of Pb (II) by immobilized and free filaments of marine Oscillatoria sp. NTMS01 and Phormidium sp. NTMS02. Bulletin of Environmental Contamination & Toxicology, 87(3), 254–259.
[131] Kuritz, T. (1998). Cyanobacteria as agents for the control of pollution by pesticides and chlorinated organic compounds. Journal of Applied Microbiology, 85(S1), 186S–192S.
[132] Kuritz, T., Bocanera, L. V., & Rivera, N. S. (1997). Dechlorination of lindane by the cyanobacterium Anabaena sp. strain PCC7120 depends on the function of the nir operon. Journal of Bacteriology, 179(10), 3368–3370.
[133] Kuritz, T., & Wolk, C. P. (1995). Use of filamentous cyanobacteria for biodegradation of organic pollutants. Applied & Environmental Microbiology, 61(1), 234–238.
[134] Lal, S., Lal, R., & Saxena, D. (1987). Bioconcentration and metabolism of DDT, fenitrothion and chlorpyrifos by the blue-green algae Anabaena sp. and Aulosira fertilissima. Environmental Pollution, 46(3), 187–196.
[135] Laliberte, G., Lessard, P., De La Noüe, J., & Sylvestre, S. (1997). Effect of phosphorus addition on nutrient removal from wastewater with the cyanobacterium Phormidium bohneri. Bioresource Technology, 59(2), 227–233.
[136] Lama, L., Nicolaus, B., Calandrelli, V., Manca, M. C., Romano, I., & Gambacorta, A. (1996). Effect of growth conditions on endo-and exopolymer biosynthesis in Anabaena cylindrica 10 C. Phytochemistry, 42(3), 655–659.
[137] Lee, C. M., Lu, C., Lu, W. M., & Chen, P.-C. (1995). Removal of nitrogenous compounds from wastewaters using immobilized cyanobacteria Anabaena CH3. Environmental Technology, 16(8), 701–713.
[138] Li, Z.-Y., Guo, S.-Y., & Li, L. (2006). Study on the process, thermodynamical isotherm and mechanism of Cr (III) uptake by Spirulina platensis. Journal of Food Engineering, 75(1), 129–136.
[139] Loutseti, S., Danielidis, D., Economou-Amilli, A., Katsaros, C., Santas, R., & Santas, P. (2009). The application of a micro-algal/bacterial biofilter for the detoxification of copper and cadmium metal wastes. Bioresource Technology, 100(7), 2099–2105.
[140] Loya, S., Reshef, V., Mizrachi, E., Silberstein, C., Rachamim, Y., Carmeli, S., & Hizi, A. (1998). The inhibition of the reverse transcriptase of HIV-1 by the natural sulfoglycolipids from cyanobacteria: contribution of different moieties to their high potency. Journal of Natural Products, 61(7), 891–895.
[141] Luesch, H., Moore, R. E., Paul, V. J., Mooberry, S. L., & Corbett, T. H. (2001). Isolation of dolastatin 10 from the marine cyanobacterium Symploca species VP642 and total stereochemistry and biological evaluation of its analogue symplostatin 1. Journal of Natural Products, 64(7), 907–910.
[142] Lyra, C., Suomalainen, S., Gugger, M., Vezie, C., Sundman, P., Paulin, L., & Sivonen, K. (2001). Molecular characterization of planktic cyanobacteria of Anabaena, Aphanizomenon, Microcystis and Planktothrix genera. International Journal of Systematic & Evolutionary Microbiology, 51(2), 513–526.

[143] MacKintosh, C., Beattie, K. A., Klumpp, S., Cohen, P., & Codd, G. A. (1990). Cyanobacterial microcystin-LR is a potent and specific inhibitor of protein phosphatases 1 and 2A from both mammals and higher plants. FEBS Letters, 264(2), 187–192.

[144] Mallick, N. (2002). Biotechnological potential of immobilized algae for wastewater N, P and metal removal: a review. Biometals, 15(4), 377–390.

[145] Mathews, J., & Wang, G. (2009). Metabolic pathway engineering for enhanced biohydrogen production. International Journal of Hydrogen Energy, 34(17), 7404–7416.

[146] McCormick, P. M., Cannon, G. C., & Heinhorst, S. (1999). Expression of the copper metallothionein CUPI from Saccharomyces cerevisiae in the cyanobacterium Synechococcus R2-PIM8 (smtA). Current Microbiology, 38(3), 155–162.

[147] McGinn, P., Dickinson, K., Bhatti, S., Frigon, J.-C., Guiot, S., & O'Leary, S. B. (2011). Integration of microalgae cultivation with industrial waste remediation for biofuel and bioenergy production: opportunities and limitations. Photosynthesis Research, 109(1–3), 231–247.

[148] Megharaj, M., Madhavi, D., Sreenivasulu, C., Umamaheswari, A., & Venkateswarlu, K. (1994). Biodegradation of methyl parathion by soil isolates of microalgae and cyanobacteria. Bulletin of Environmental Contamination & Toxicology, 53(2), 292–297.

[149] Megharaj, M., Ramakrishnan, B., Venkateswarlu, K., Sethunathan, N., & Naidu, R. (2011). Bioremediation approaches for organic pollutants: a critical perspective. Environment International, 37(8), 1362–1375.

[150] Megharaj, M., Singleton, I., McClure, N., & Naidu, R. (2000). Influence of petroleum hydrocarbon contamination on microalgae and microbial activities in a long-term contaminated soil. Archives of Environmental Contamination & Toxicology, 38(4), 439–445.

[151] Mehta, S., & Gaur, J. (2005). Use of algae for removing heavy metal ions from wastewater: progress and prospects. Critical Reviews in Biotechnology, 25(3), 113–152.

[152] Miranda, J., Krishnakumar, G., & Gonsalves, R. (2012). Cr6+ bioremediation efficiency of Oscillatoria laete-virens (Crouan & Crouan) Gomont and Oscillatoria trichoides Szafer: kinetics and equilibrium study. Journal of Applied Phycology, 24(6), 1439–1454.

[153] Mishra, U., & Pabbi, S. (2004). Cyanobacteria: a potential biofertilizer for rice. Resonance, 9(6), 6–10.

[154] Miyake, M., Takase, K., Narato, M., Khatipov, E., Schnackenberg, J., Shirai, M., Kurane, R., & Asada, Y. (2000). Polyhydroxybutyrate production from carbon dioxide by cyanobacteria. Twenty-First Symposium on Biotechnology for Fuels & Chemicals. (pp. 991–1002): Springer.

[155] Mo, S., Krunic, A., Chlipala, G., & Orjala, J. (2009). Antimicrobial ambiguine isonitriles from the cyanobacterium Fischerella ambigua. Journal of Natural Products, 72(5), 894–899.

[156] Möllers, K. B., Cannella, D., Jørgensen, H., & Frigaard, N.-U. (2014). Cyanobacterial biomass as carbohydrate and nutrient feedstock for bioethanol production by yeast fermentation. Biotechnology for Biofuels, 7, 64.

[157] Montaser, R., Abboud, K. A., Paul, V. J., & Luesch, H. (2010). Pitiprolamide, a proline-rich dolastatin 16 analogue from the marine cyanobacterium Lyngbya majuscula from Guam. Journal of Natural Products, 74(1), 109–112.

[158] Monteiro, C. M., Castro, P. M., & Malcata, F. X. (2012). Metal uptake by microalgae: Underlying mechanisms and practical applications. Biotechnology Progress, 28(2), 299–311.

[159] Moore, R. E. (1982). Toxins, anticancer agents, and tumor promoters from marine prokaryotes. Pure & Applied Chemistry, 54(10), 1919–1934.

[160] Moore, R. E., Cheuk, C., Yang, X. Q. G., Patterson, G. M. L., Bonjouklian, R., Smitka, T. A., Mynderse, J. S., Foster, R. S., & Jones, N. D. (1987). Hapalindoles, antibacterial and antimycotic alkaloids from the cyanophyte Hapalosiphon fontinalis. The Journal of Organic Chemistry, 52(6), 1036–1043.

[161] Moore, R. E., & Entzeroth, M. (1988). Majusculamide D and deoxymajusculamide D, two cytotoxins from Lyngbya majuscula. Phytochemistry, 27(10), 3101–3103.
[162] Moradpour, Z., Torshabi, M., Faramarzi, M. A., Zolfaghary, N., & Zarrini, G. (2006). Microalgal Transformation of Androst-4-en-3, 17-dione. Research Journal of Microbiology, 1(3), 289–293.
[163] Nagasathya, A., & Thajuddin, N. (2008). Decolourization of paper mill effluent using hypersaline cyanobacterium. Research Journal of Environmental Sciences, 2(5), 408–414.
[164] Nagase, H., Inthorn, D., Isaji, Y., Oda, A., Hirata, K., & Miyamoto, K. (1997). Selective cadmium removal from hard water using NaOH-treated cells of the cyanobacterium Tolypothrix tenuis. Journal of Fermentation & Bioengineering, 84(2), 151–154.
[165] Nagase, H., Inthorn, D., Oda, A., Nishimura, J., Kajiwara, Y., Park, M.-o., Hirata, K., & Miyamoto, K. (2005). Improvement of selective removal of heavy metals in cyanobacteria by NaOH treatment. Journal of Bioscience & Bioengineering, 99(4), 372–377.
[166] Nagatsu, A., Kajitani, H., & Sakakibara, J. (1995). Muscoride A: A new oxazole peptide alkaloid from freshwater cyanobacterium Nostoc muscorum. Tetrahedron Letters, 36(23), 4097–4100.
[167] Nagle, D. G., & Gerwick, W. H. (1995). Nakienones AC and nakitriol, new cytotoxic cyclic C 11 metabolites from an Okinawan cyanobacterial (Synechocystis sp.) overgrowth of coral. Tetrahedron Letters, 36(6), 849–852.
[168] Namikoshi, M., & Rinehart, K. (1996). Bioactive compounds produced by cyanobacteria. Journal of Industrial Microbiology, 17(5–6), 373–384.
[169] Narro, M. L., Cerniglia, C. E., Van Baalen, C., & Gibson, D. T. (1992a). Evidence for an NIH shift in oxidation of naphthalene by the marine cyanobacterium Oscillatoria sp. strain JCM. Applied & Environmental Microbiology, 58(4), 1360–1363.
[170] Narro, M. L., Cerniglia, C., Van Baalen, C., & Gibson, D. (1992b). Metabolism of phenanthrene by the marine cyanobacterium Agmenellum quadruplicatum PR-6. Applied & Environmental Microbiology, 58(4), 1351–1359.
[171] Newman, D. J., & Cragg, G. M. (2004). Advanced preclinical and clinical trials of natural products and related compounds from marine sources. Current Medicinal Chemistry, 11(13), 1693–1713.
[172] Niknam, S., Faramarzi, M. A., Abdi, K., Yazdi, M. T., Amini, M., & Rastegar, H. (2010). Bioconversion of codeine to semi-synthetic opiate derivatives by the cyanobacterium Nostoc muscorum. World Journal of Microbiology & Biotechnology, 26(1), 119–123.
[173] Nobles Jr, D. R., & Brown Jr, R. (2008). Transgenic expression of Gluconacetobacter xylinus strain ATCC 53582 cellulose synthase genes in the cyanobacterium Synechococcus leopoliensis strain UTCC 100. Cellulose, 15(5), 691–701.
[174] Nogle, L. M., Okino, T., & Gerwick, W. H. (2001). Antillatoxin B, a neurotoxic lipopeptide from the marine cyanobacterium Lyngbya majuscula. Journal of Natural Products, 64(7), 983–985.
[175] Nunnery, J. K., Mevers, E., & Gerwick, W. H. (2010). Biologically active secondary metabolites from marine cyanobacteria. Current Opinion Biotechnology, 21(6), 787–793.
[176] Obreht, Z., Kerby, N. W., Gantar, M., & Rowell, P. (1993). Effects of root-associated N2-fixing cyanobacteria on the growth and nitrogen content of wheat (Triticum vulgare L.) seedlings. Biology & Fertility Soils, 15(1), 68–72.
[177] Pandey, V. (2015). Cyanobacterial natural products as antimicrobial agents. International Journal Current Microbiology & Applied Sciences, 4(1), 310–317.
[178] Paperi, R., Micheletti, E., & De Philippis, R. (2006). Optimization of copper sorbing–desorbing cycles with confined cultures of the exopolysaccharide-producing cyanobacterium Cyanospira capsulata. Journal Aplied Microbiology, 101(6), 1351–1356.
[179] Parmar, A., Singh, N. K., Pandey, A., Gnansounou, E., & Madamwar, D. (2011). Cyanobacteria and microalgae: a positive prospect for biofuels. Bioresource Technology, 102(22), 10163–10172.

[180] Pavlostathis, S. G., & Jackson, G. H. (1999). Biotransformation of 2, 4, 6-trinitrotoluene in Anabaena sp. cultures. Environmental Toxicology & Chemistry, 18(3), 412–419.

[181] Pavlostathis, S. G., & Jackson, G. H. (2002). Biotransformation of 2, 4, 6-trinitrotoluene in a continuous-flow Anabaena sp. system. Water Research, 36(7), 1699–1706.

[182] Penesyan, A., Gillings, M., & Paulsen, I. T. (2015). Antibiotic discovery: combatting bacterial resistance in cells and in biofilm communities. Molecules, 20(4), 5286–5298.

[183] Peschek, G. A., & Sleytr, U. B. (1983). Thylakoid morphology of the cyanobacteria Anabaena variabilis and Nostoc MAC grown under light and dark conditions. Journal Ultrastructure Research, 82(2), 233–239.

[184] Pinzon-Gamez, N. M., Sundaram, S., & Ju, L.-K. (2005). Heterocyst differentiation and H2 production in N2-fixing cyanobacteria. In AIChE Annual Meeting, Conference Proceedings. Cincinnati, OH. pp. 8949–8951.

[185] Prasad, B., & Pandey, U. (2000). Separation and preconcentration of copper and cadmium ions from multielemental solutions using Nostoc muscorum-based biosorbents. World Journal of Microbiology & Biotechnology, 16(8–9), 819–827.

[186] Prasad, B. B., Banerjee, S., & Lakshmi, D. (2006). An AlgaSORB column for the quantitative sorption of arsenic (III) from water samples. Water Quality Research Journal of Canada, 41(2), 190–197.

[187] Priyadarshani, I., & Rath, B. (2012). Bioactive compounds from microalgae and Cyanobacteria: utility and applications. International Journal of Pharmaceutical Sciences & Research, 3(11), 4123–4130.

[188] Radwan, S. S., & Al-Hasan, R. H. (2002). Oil pollution and cyanobacteria. The Ecology of Cyanobacteria. (pp. 307–319): Springer.

[189] Raghukumar, C., Vipparty, V., David, J., & Chandramohan, D. (2001). Degradation of crude oil by marine cyanobacteria. Applied Microbiology & Biotechnology, 57(3), 433–436.

[190] Raja, R., Hemaiswarya, S., Ganesan, V., & Carvalho, I. S. (2015). Recent developments in therapeutic applications of cyanobacteria. Critical Reviews in Microbiology(0), 1–12.

[191] Rana, L., Chhikara, S., & Dhankhar, R. (2013). Assessment of growth rate of indigenous cyanobacteria in metal enriched culture medium. Asian Journal of Experimental Biology Sciences, 4(3), 465–471.

[192] Rangsayatorn, N., Pokethitiyook, P., Upatham, E., & Lanza, G. (2004). Cadmium biosorption by cells of Spirulina platensis TISTR 8217 immobilized in alginate and silica gel. Environment International, 30(1), 57–63.

[193] Rantala, A., Fewer, D. P., Hisbergues, M., Rouhiainen, L., Vaitomaa, J., Börner, T., & Sivonen, K. (2004). Phylogenetic evidence for the early evolution of microcystin synthesis. Proceedings of the National Academy of Sciences of the United States of America, 101(2), 568–573.

[194] Rao, S. R., Tripathi, U., & Ravishankar, G. (1999). Biotransformation of codeine to morphine in freely suspended cells and immobilized cultures of Spirulina platensis. World Journal of Microbiology & Biotechnology, 15(4), 465–469.

[195] Rapala, J., Sivonen, K., Luukkainen, R., & Niemelä, S. I. (1993). Anatoxin-a concentration in Anabaena and Aphanizomenon under different environmental conditions and comparison of growth by toxic and non-toxic Anabaena strains– a laboratory study. Journal of Applied Phycology, 5(6), 581–591.

[196] Rath, J., Mandal, S., & Adhikary, S. P. (2012). Salinity induced synthesis of UV-screening compound scytonemin in the cyanobacterium Lyngbya aestuarii. Journal of Photochemistry & Photobiology B: Biology, 115, 5–8.

[197] Raveh, A., & Carmeli, S. (2007). Antimicrobial ambiguines from the cyanobacterium Fischerella sp. collected in Israel. Journal of Natural Products, 70(2), 196–201.

[198] Reddy, C. M., Bhat, V. B., Kiranmai, G., Reddy, M. N., Reddanna, P., & Madyastha, K. (2000). Selective inhibition of cyclooxygenase-2 by C-phycocyanin, a biliprotein from Spirulina platensis. Biochemical & Biophysical Research Communications, 277(3), 599–603.

[199] Ren, L., Shi, D., Dai, J., & Ru, B. (1998). Expression of the mouse metallothionein-I gene conferring cadmium resistance in a transgenic cyanobacterium. FEMS Microbiology Letters, 158(1), 127–132.

[200] Renuka, N., Sood, A., Ratha, S. K., Prasanna, R., & Ahluwalia, A. S. (2013). Nutrient sequestration, biomass production by microalgae and phytoremediation of sewage water. International Journal of Phytoremediation, 15(8), 789–800.

[201] Rickards, R. W., Rothschild, J. M., Willis, A. C., de Chazal, N. M., Kirk, J., Kirk, K., Saliba, K. J., & Smith, G. D. (1999). Calothrixins A and B, novel pentacyclic metabolites from Calothrix cyanobacteria with potent activity against malaria parasites and human cancer cells. Tetrahedron, 55(47), 13513–13520.

[202] Riser-Roberts, E. (2010). Remediation of petroleum contaminated soils: biological, physical, and chemical processes: CRC Press.

[203] Romay, C., Gonzalez, R., Ledon, N., Remirez, D., & Rimbau, V. (2003). C-phycocyanin: a biliprotein with antioxidant, anti-inflammatory and neuroprotective effects. Current Protein & Peptide Science, 4(3), 207–216.

[204] Rose, P., Boshoff, G., Van Hille, R., Wallace, L., Dunn, K., & Duncan, J. (1998). An integrated algal sulphate reducing high rate ponding process for the treatment of acid mine drainage wastewaters. Biodegradation, 9(3–4), 247–257.

[205] Salvador, L. A., Paul, V. J., & Luesch, H. (2010a). Caylobolide B, a macrolactone from symplostatin 1-producing marine cyanobacteria Phormidium spp. from Florida. Journal of Natural Products, 73(9), 1606–1609.

[206] Sampedro, M., Blanco, A., Llama, M., & Serra, J. (1995). Sorption of heavy metals to Phormidium laminosum biomass. Biotechnology & Applied Biochemistry, 22.

[207] Sánchez, M., Bernal-Castillo, J., Rozo, C., & Rodríguez, I. (2003). Spirulina (Arthrospira): An edible microorganism: A review. Universitas Scientiarum, 8(1), 7–24.

[208] Sarsekeyeva, F., Zayadan, B. K., Usserbaeva, A., Bedbenov, V. S., Sinetova, M. A., & Los, D. A. (2015). Cyanofuels: biofuels from cyanobacteria. Reality and perspectives. Photosynthesis Research, 1–12.

[209] Sawayama, S., Rao, K., & Hall, D. (1998). Nitrate and phosphate ion removal from water by Phormidium laminosum immobilized on hollow fibres in a photobioreactor. Applied Microbiology & Biotechnology, 49(4), 463–468.

[210] Schluchter, W. M., Shen, G., Alvey, R. M., Biswas, A., Saunee, N. A., Williams, S. R., Mille, C. A., & Bryant, D. A. (2010). Phycobiliprotein biosynthesis in cyanobacteria: structure and function of enzymes involved in post-translational modification. Advances in Experimental Medicine & Biology, 675, 211–228.

[211] Schmidt-Dannert, C. (2015). NextGen microbial natural products discovery. Microbial biotechnology. 8 (1), 26–28.

[212] Şeker, A., Shahwan, T., Eroğlu, A. E., Yılmaz, S., Demirel, Z., & Dalay, M. C. (2008). Equilibrium, thermodynamic and kinetic studies for the biosorption of aqueous lead (II), cadmium (II) and nickel (II) ions on Spirulina platensis. Journal of Hazardous Materials, 154(1), 973–980.

[213] Semple, K. T., Cain, R. B., & Schmidt, S. (1999). Biodegradation of aromatic compounds by microalgae. FEMS Microbiology Letters, 170(2), 291–300.

[214] Shah, V., Garg, N., & Madamwar, D. (2001). An integrated process of textile dye removal and hydrogen evolution using cyanobacterium, Phormidium valderianum. World Journal of Microbiology & Biotechnology, 17(5), 499–504.

[215] Shanab, S., Essa, A., & Shalaby, E. (2012). Bioremoval capacity of three heavy metals by some microalgae species (Egyptian Isolates). Plant Signaling & Behavior, 7(3), 392–399.
[216] Shashirekha, S., Uma, L., & Subramanian, G. (1997). Phenol degradation by the marine cyanobacterium Phormidium valderianum BDU 30501. Journal of Industrial Microbiology & Biotechnology, 19(2), 130–133.
[217] Sidler, W. A. (2004). Phycobilisome and phycobiliprotein structures. The molecular biology of cyanobacteria. (pp. 139–216): Springer.
[218] Sikkema, J., De Bont, J., & Poolman, B. (1995). Mechanisms of membrane toxicity of hydrocarbons. Microbiological Reviews, 59(2), 201–222.
[219] Simmons, T. L., McPhail, K. L., Ortega-Barría, E., Mooberry, S. L., & Gerwick, W. H. (2006). Belamide A, a new antimitotic tetrapeptide from a Panamanian marine cyanobacterium. Tetrahedron Letters, 47(20), 3387–3390.
[220] Simon, R. D. (1971). Cyanophycin Granules from the Blue-Green Alga Anabaena cylindrica: A Reserve material consisting of copolymers of aspartic acid and arginine. Proceedings of the National Academy of Sciences of the United States of America, 68(2), 265–267.
[221] Simonsen, J. L., Ross, W. C. J., & De Mayo, P. (1957). The terpenes: University Press Cambridge.
[222] Singh, N., Asthana, R., & Singh, S. (2003). Characterization of an exopolysaccharide mutant of Nostoc spongiaeforme: Zn2+-sorption and uptake. World Journal of Microbiology & Biotechnology, 19(8), 851–857.
[223] Singh, N. K., & Patel, D. (2012). Microalgae for bioremediation of distillery effluent. Farming for Food & Water Security (pp. 83–109): Springer.
[224] Singh, S., Kate, B. N., & Banerjee, U. (2005). Bioactive compounds from cyanobacteria and microalgae: an overview. Critical Reviews in Biotechnology, 25(3), 73–95.
[225] Sisay, M. T., Hautmann, S., Mehner, C., Konig, G. M., Bajorath, J., & Gutschow, M. (2009). Inhibition of human leukocyte elastase by brunsvicamides A-C: cyanobacterial cyclic peptides. ChemMedChem, 4(9), 1425–1429.
[226] Sivonen, K. (1996). Cyanobacterial toxins and toxin production. Phycologia, 35(6S), 12–24.
[227] Sivonen, K., & Börner, T. (2008). Bioactive compounds produced by cyanobacteria. The cyanobacteria: molecular biology, genomics and evolution, (pp159–197): Caister Academic Press.
[228] Sivonen, K., Niemelä, S., Niemi, R., Lepistö, L., Luoma, T., & Räsänen, L. (1990). Toxic cyanobacteria (blue-green algae) in Finnish fresh and coastal waters. Hydrobiologia, 190(3), 267–275.
[229] Smith, C. D., Zhang, X., Mooberry, S. L., Patterson, G. M., & Moore, R. E. (1994). Cryptophycin: a new antimicrotubule agent active against drug-resistant cells. Cancer Research, 54(14), 3779–3784.
[230] Smith, G., Ewart, G., & Tucker, W. (1992). Hydrogen production by cyanobacteria. International Journal of Hydrogen Energy, 17(9), 695–698.
[231] Snow, E. T. (1992). Metal carcinogenesis: mechanistic implications. Pharmacology & Therapeutics, 53(1), 31–65.
[232] Sode, K., Yamamoto, Y., & Hatano, N. (1998a). Construction of a marine cyanobacterial strain with increased heavy metal ion tolerance by introducing exogenic metallothionein gene. Journal of Marine Biotechnology, 6, 174–177.
[233] Soria-Mercado, I. E., Pereira, A., Cao, Z., Murray, T. F., & Gerwick, W. H. (2009). Alotamide A, a novel neuropharmacological agent from the marine cyanobacterium Lyngbya bouillonii. Organic Letters, 11(20), 4704–4707.
[234] Sorkhoh, N., Al-Hasan, R., Khanafer, M., & Radwan, S. (1995). Establishment of oil-degrading bacteria associated with cyanobacteria in oil-polluted soil. Journal of Applied Bacteriology, 78(2), 194–199.

[235] Soule, T., Stout, V., Swingley, W., Meeks, J., & Garcia-Pichel, F. (2007). Molecular genetics and genomic analysis of scytonemin biosynthesis in Nostoc punctiforme ATCC 29133. Journal of Bacteriology, 189(12), 4465–4472.

[236] Srirangan, K., Pyne, M. E., & Chou, C. P. (2011). Biochemical and genetic engineering strategies to enhance hydrogen production in photosynthetic algae and cyanobacteria. Bioresource Technology, 102(18), 8589–8604.

[237] Stal, L. J. (1992). Poly (hydroxyalkanoate) in cyanobacteria: an overview. FEMS Microbiology Reviews, 9(2–4), 169–180.

[238] Stal, L. J., Heyer, H., & Jacobs, G. (1990). Occurrence and role of poly-hydroxy-alkanoate in the cyanobacterium Oscillatoria limosa. Novel Biodegradable Microbial Polymers (pp. 435–438): Springer.

[239] Stanier, R., & Bazine, G. (1977). Phototrophic prokaryotes: the cyanobacteria. Annual Reviews in Microbiology, 31(1), 225–274.

[240] Stewart, W., & Codd, G. (1975). Polyhedral bodies (carboxysomes) of nitrogen-fixing blue-green algae. British Phycological Journal, 10(3), 273–278.

[241] Subashchandrabose, S. R., Ramakrishnan, B., Megharaj, M., Venkateswarlu, K., & Naidu, R. (2011). Consortia of cyanobacteria/microalgae and bacteria: biotechnological potential. Biotechnology Advances, 29(6), 896–907.

[242] Subashchandrabose, S. R., Ramakrishnan, B., Megharaj, M., Venkateswarlu, K., & Naidu, R. (2013). Mixotrophic cyanobacteria and microalgae as distinctive biological agents for organic pollutant degradation. Environment international, 51, 59–72.

[243] Subramanian, G., Sekar, S., & Sampoornam, S. (1994). Biodegradation and utilization of organophosphorus pesticides by cyanobacteria. International Biodeterioration & Biodegradation, 33(2), 129–143.

[244] Subramanian, G., & Uma, L. (1997). Role of cyanobacteria in pollution abatement. Recent Advances in Ecobiological Research, 1, 435–443.

[245] Subramaniyan, V. (2012). Treatment of Dye Industry Effluent Using Free and Immobilized Cyanobacteria. Journal of Bioremediation & Biodegradation, 3 (10), 165–170.

[246] Tamagnini, P., Axelsson, R., Lindberg, P., Oxelfelt, F., Wünschiers, R., & Lindblad, P. (2002). Hydrogenases and hydrogen metabolism of cyanobacteria. Microbiology & Molecular Biology Reviews, 66(1), 1–20.

[247] Tan, L. T. (2010). Filamentous tropical marine cyanobacteria: a rich source of natural products for anticancer drug discovery. Journal of Applied Phycology, 22(5), 659–676.

[248] Tang, X., He, L., Tao, X., Dang, Z., Guo, C., Lu, G., & Yi, X. (2010). Construction of an artificial microalgal-bacterial consortium that efficiently degrades crude oil. Journal of hazardous materials, 181(1), 1158–1162.

[249] Taroncher-Oldenburg, G., Nishina, K., & Stephanopoulos, G. (2000). Identification and analysis of the polyhydroxyalkanoate-specific β-ketothiolase and acetoacetyl coenzyme A reductase genes in the cyanobacterium Synechocystis sp. strain PCC6803. Applied & Environmental Microbiology, 66(10), 4440–4448.

[250] Teruya, T., Sasaki, H., Kitamura, K., Nakayama, T., & Suenaga, K. (2009). Biselyngbyaside, a macrolide glycoside from the marine cyanobacterium Lyngbya sp. Organic Letters, 11(11), 2421–2424.

[251] Tong, M., & Yuan, S. (2012). Physiochemical technologies for HCB remediation and disposal: A review. Journal of Hazardous Materials, 229, 1–14.

[252] Trimurtulu, G., Ohtani, I., Patterson, G. M., Moore, R. E., Corbett, T. H., Valeriote, F. A., & Demchik, L. (1994). Total structures of cryptophycins, potent antitumor depsipeptides from the blue-green alga Nostoc sp. strain GSV 224. Journal of the American Chemical Society, 116(11), 4729–4737.

[253] Tripathi, A., Puddick, J., Prinsep, M. R., Rottmann, M., Chan, K. P., Chen, D. Y.-K., & Tan, L. T. (2011). Lagunamide C, a cytotoxic cyclodepsipeptide from the marine cyanobacterium Lyngbya majuscula. Phytochemistry, 72(18), 2369–2375.

[254] Tsygankov, A., Fedorov, A., Kosourov, S., & Rao, K. (2002). Hydrogen production by cyanobacteria in an automated outdoor photobioreactor under aerobic conditions. Biotechnology & Bioengineering, 80(7), 777–783.

[255] Vaishampayan, A., Sinha, R., Hader, D.-P., Dey, T., Gupta, A., Bhan, U., & Rao, A. (2001). Cyanobacterial biofertilizers in rice agriculture. The Botanical Review, 67(4), 453–516.

[256] Van der Oost, J., Bulthuis, B., Feitz, S., Krab, K., & Kraayenhof, R. (1989). Fermentation metabolism of the unicellular cyanobacterium Cyanothece PCC 7822. Archives of Microbiology, 152(5), 415–419.

[257] Van Eerd, L. L., Hoagland, R. E., Zablotowicz, R. M., & Hall, J. C. (2009). Pesticide metabolism in plants and microorganisms. Weed Science, 51(4), 472–495.

[258] Vidali, M. (2001). Bioremediation. an overview. Pure & Applied Chemistry, 73(7), 1163–1172.

[259] Vignais, P. M., Magnin, J.-P., & Willison, J. C. (2006). Increasing biohydrogen production by metabolic engineering. International Journal of Hydrogen Energy, 31(11), 1478–1483.

[260] Vincenzini, M., Sili, C., De Philippis, R., Ena, A., & Materassi, R. (1990). Occurrence of poly-beta-hydroxybutyrate in Spirulina species. Journal of Bacteriology, 172(5), 2791–2792.

[261] Volk, R.-B., & Furkert, F. H. (2006). Antialgal, antibacterial and antifungal activity of two metabolites produced and excreted by cyanobacteria during growth. Microbiological Research, 161(2), 180–186.

[262] White, J. D., Kim, T.-S., & Nambu, M. (1995). Synthesis of curacin A: a powerful antimitotic from the cyanobacterium Lyngbya majuscula. Journal of the American Chemical Society, 117(20), 5612–5613.

[263] White, J. D., Xu, Q., Lee, C.-S., & Valeriote, F. A. (2004). Total synthesis and biological evaluation of (+)-kalkitoxin, a cytotoxic metabolite of the cyanobacterium Lyngbya majuscula. Organic & Biomolecular Chemistry, 2(14), 2092–2102.

[264] Wilde, E. W., & Benemann, J. R. (1993). Bioremoval of heavy metals by the use of microalgae. Biotechnology Advances, 11(4), 781–812.

[265] Williams, P. G., Yoshida, W. Y., Moore, R. E., & Paul, V. J. (2003). Tasipeptins A and B: new cytotoxic depsipeptides from the marine cyanobacterium Symploca sp. Journal of Natural Products, 66(5), 620–624.

[266] Wipf, P., Reeves, J. T., Balachandran, R., & Day, B. W. (2002). Synthesis and biological evaluation of structurally highly modified analogues of the antimitotic natural product curacin A. Journal of Medicinal Chemistry, 45(9), 1901–1917.

[267] Wurster, M., Mundt, S., Hammer, E., Schauer, F., & Lindequist, U. (2003). Extracellular degradation of phenol by the cyanobacterium Synechococcus PCC 7002. Journal of Applied Phycology, 15(2–3), 171–176.

[268] Yazdi, M., Ghasemi, Y., Ghasemian, A., Shokravi, S., Niknahad, H., Amini, M., Dehshahri, A., & Faramarzi, M. (2005). Bioconversion of Hydrocortisone by Cyanobacterium Fischerella ambigua PTCC 1635. World Journal of Microbiology & Biotechnology, 21(6–7), 811–814.

[269] Yazdi, M. T., Arabi, H., Faramarzi, M. A., Ghasemi, Y., Amini, M., Shokravi, S., & Mohseni, F. A. (2004). Biotransformation of hydrocortisone by a natural isolate of Nostoc muscorum. Phytochemistry, 65(15), 2205–2209.

[270] Zhou, R., & Gibbons, W. (2012). Genetically engineered cyanobacteria. US20120276637 A1.

Vedavyas R. Niveditha, and Kandikere R. Sridhar

15 Improvement of functional attributes of kernels of wild legume *Canavalia martima* by *Rhizopus oligosporus*

15.1 Introduction

Supply of nutritious, safe and affordable protein sources is the major challenge for food scientists and nutritionists. The incessant raise in the population and poor supply of protein rich foods in developing countries has resulted in protein energy malnutrition (Boye *et al.*, 2010). Research on vegetable proteins has gained importance due to expensive and inadequate supply of animal proteins. Legumes being rich in proteins (18–25 %), minerals, vitamins and slow release carbohydrates constitute affordable alternative food source (Tharanathan and Mahadevamma, 2003). Recently, research on indigenous wild legumes as potential source of food and pharmaceuticals has gained immense importance (Bhat and Karim, 2009). Optimum utilization of legume proteins depends upon the functional properties as they play important role in preparation, processing, formulation, storage, food quality, sensory characteristics and consumer acceptability. Legume proteins serve as food ingredients as well as improve the nutritional and functional properties. The functional properties depend on various intrinsic factors (e.g. protein structure, size and conformation), environmental factors (e.g. temperature, pH and ionic strength) and the presence of additional components. Food industries are in need of suitable functional properties in legumes to develop cost effective and valuable products (Boye *et al.*, 2010). By improvement of functional properties, legume proteins can serve better in manufacturing texturized food products for human consumption (Chel-Guerrero *et al.*, 2002).

Canavalia maritima is an underutilized wild legume distributed throughout the tropical and subtropical coastal sand dunes (CSD) of Southeast Asia, South America and Africa (Vatanparast, 2010). The indigenous land races of *C. maritima* are the potential source of nutrition, bioactive principles and antioxidants (Bressani *et al.*, 1987; Martinez and Psuty, 2004). Besides nutritional value, *C. maritima* serves as cover crop, green manure and fix atmospheric nitrogen with native rhizobia (Arun and Sridhar, 2004; Bhagya and Sridhar, 2009). It also possesses valuable agronomic features like high fertility index, high biomass production, high yield (720–1,500 kg seeds/ha), tolerance to drought and resistance to pests (Akpapunam and Sefa-Dedeh, 1997; Bresseni *et al.*, 1987; Seena and Sridhar, 2006). It is suitable for cultivation in marginal soils of arid to semiarid regions along with cultivated legumes. The coastal dwellers and other ethnic groups of Southwest India, Sri Lanka and China utilize *Canavalia* seeds as source of food after processing (e.g. soaking, cooking and removal of seed coat/testa) (Siddhuraju and Becker, 2001; Seena *et al*, 2007; Bhagya and Sridhar,

2009). The seeds serve as a substitute for mashed potatoes and also roasted seeds are used in preparation of coffee-like beverage in Latin America (Bourdy *et al.*, 1992; Xu *et al.*, 2009; Vadivel *et al.*, 2010). Raw Seeds of *C. maritima* are endowed with appreciable amounts of proteins (25–35 %), carbohydrates (50–60 %), fats (1–3 %), minerals and amino acids (Seena and Sridhar, 2006). Several strategies have been followed to enhance the nutritional value of legume seeds especially by cooking, germination, roasting and fermentation. Solid substrate fermentation (SSF) with food-grade fungus (*Rhizopus oligosporus* or Temph mold) has been reported to improve the nutritional quality in cowpea, soybean and mung bean (Prinyawiwatkul et al., 1997; McCue and Shetty, 2003; Randhir and Shetty, 2007). Studies on the functional properties of seeds of *Canavalia* spp. are meagre (Sridhar and Seena, 2006). There are no attempts on improvement of functional properties of seeds of *C. maritima* through fermentation for effective utilization or formulation of nutrient-rich food or feed. Therefore, the aim of the present study is to evaluate the cooking properties of *C. maritima* beans and functional properties of cooked and solid-substrate fermented split beans with *Rhizopus oligospurus* for food product formulations.

15.2 Seeds and Processing

The seed samples of *Canavalia maritima* Thouars were sampled from coastal sand dunes of Someshwara, Southwest India (12° 47′ N, 74° 52′ E) during summer season (February-March, 2010). Healthy undamaged seeds were separated from dry pods, sun-dried for 2–3 days followed by evaluation of physical and cooking properties. The seeds were dehusked and beans were subjected to pressure-cooking and solid substrate fermentation by *Rhizopus oligosporus* var. *oligosporus* (Saito) Schipper and Stalpers (MTCC # 556; strain designation # 22959; Institute of Microbial Technology, Chandigarh, India). The beans (25 g) were transferred to conical flask (250 ml), soaked in distilled water (1:3 w/v) for pressure-cooking (6.5 L, Deluxe stainless steel; *TTK* Prestige™, Prestige Ltd., India). The cooked beans were spread on aluminium foil, dried at 42±2 °C, milled (Wiley Mill, 30 mesh) and stored in air-tight glass containers. Another set of beans (25 g) was cooked in conical flasks (250 ml capacity), cooled to laboratory temperature, inoculated with two 5 mm plugs of 3-day-old cultures of *Rhizopus oligosporus* grown on potato-dextrose agar medium and allowed for solid substrate fermentation (SSF) up to one week at 37±1 °C. Cooked beans formed tight cake with mycelial mat on fermentation and it was spread on aluminium foil, dried at 42±2°C, powdered and stored in air-tight glass containers.

15.3 Assessment of Functional Properties

Flours of cooked and fermented and cooked beans of *C. cathartica* were assessed for protein solubility, gelation property, water- and oil-absorption capacities, emulsion properties and foam properties.

15.3.1 Protein Solubility

Protein solubility of cooked and fermented flours was determined according to Were *et al.* (1997). Cooked and fermented seed flours 125 mg each was blended with distilled water (25 ml) in a blender (Philips HL1643, Philips India Ltd., Kolkata, India). The solution was mixed using a magnetic stirrer (1 hr, 20 °C) and centrifuged (12000 g, 20 min, 4 °C). The supernatant was filtered through glass wool and nitrogen was estimated by micro-Kjeldahl method (Humphries, 1956). The soluble protein (%) profile was determined (nitrogen solubility × 6.25). The effect of ionic strength (0.2–1.0M NaCl) and pH (2–10) on protein solubility (PS%) was determined based on the nitrogen content in supernatant (NS) and in flour (NF):

$$\text{Ps}\ (\%) = (\text{Ns} \div \text{Nf}) \times 100 \tag{1}$$

15.3.2 Gelation

The least gelation capacity (LGC) of cooked and fermented flours was evaluated based on the method outlined by Coffman and Garcia (1977). Cooked and fermented seed flour suspensions in distilled water (2–20 %; in 10 ml) was transferred to test tubes and heated in a boiling water bath (1 hr) and cooled to room temperature. The samples in tubes were further cooled (4 °C, 2 hr) and the LGC was detected when the sample from inverted test tube did not slip down. The effect of ionic strength (0.2–1M NaCl) as well as pH (2–10) on LGC was determined.

15.3.3 Water- and Oil-Absorption Capacities

Water-absorption capacity (WAC) and oil-absorption capacity (OAC) of cooked and fermented flours were determined following the method by Beuchat (1977). The flours (0.5 g) were mixed with deionised water (5 ml) or edible oil (Fortune Sunlite Refined Sunflower Oil, Adani Wilmar Ltd., Gujarat, India) in a graduated centrifuge tube and stirred (1 min). The solution was allowed to stand at room temperature (27±2 °C; 30 min), centrifuged (5000 g; 30 min) and the volume of supernatant was measured. Water- or oil-absorption capacity was expressed as ml of water or oil absorbed per gram of flour.

15.3.4 Emulsion Properties

Emulsion capacity and stability of cooked and fermented flours were determined according to Neto *et al.* (2001). The flour was suspended in distilled water (10 mg/ml), aliquots (5 ml) were homogenized (1 min) with edible oil (5 ml) and centrifuged (1100 g; 5 min). Height of emulsified layer (H_e) and height of total contents (H_t) in the tube were determined to calculate the emulsion activity (E_a%):

$$E_a\,(\%) = (H_e \div H_t) \times 100 \tag{2}$$

Emulsion stability was determined by heating the emulsion (80 °C, 30 min) before centrifugation (1100 g; 5 min). The effect of flour concentration (2–10 %), ionic strength (0.1–1.0M NaCl) and pH (2–10) on emulsion activity and stability were determined.

15.3.5 Foam Properties

Foam capacity and foam stability were determined according to Kabirullah and Willis (1982). Flour sample (2 g) was dispersed in distilled water (100 ml) and whipped vigorously (2 min) in blender at speed 1. To determine the foam capacity (F_c%), the volume before (V_b) and after (V_a) whipping was recorded:

$$F_c\,(\%) = (V_a \div V_b) \times 100 \tag{3}$$

The foam stability was determined as the volume of foam that remained after several hours at room temperature (27±2 °C) and expressed as the percentage of initial foam volume. The foam stability (%) was assessed at 30 min intervals up to 6–7 hr. The effect of ionic strength (0.1–1M NaCl) and pH (2–10) on foam capacity and stability were also determined.

15.4 Data Analysis

The *t*-test was employed to assess the difference in functional properties between cooked and fermented bean flours (StatSoft Inc., 2008). The relationship within and between functional properties considering the highest values against proximal characteristics of cooked and fermented beans (data from Niveditha and Sridhar, 2015), the principal component analysis (PCA) was employed (SPSS 16.0: www.spss.com). The PCA score plot was drawn to group different classes of functional properties vs. proximal characteristics of cooked and fermented beans.

15.5 Discussion

15.5.1 Protein Solubility

Protein solubility (PS) is an important property as it influences other functional properties of the seed flours. It is also useful in projecting the lipid-water interactions in food systems (Kinsella, 1976). According to Bernardi *et al.* (1991), solubility of denatured proteins could be increased by bacterial or fungal fermentation. Similarly, fermented beans of *C. maritima* in the present study showed drastic improvement in PS (35.5 vs. 41 %) (Fig. 15.1). Unlike cooked beans, in fermented beans increased molarity from 0.2 to 0.4M resulted in elevation of PS (17.3 vs. 19.1 %), however, cooked as well as fermented beans showed the highest PS at 1M (20.3–22.6 %). The PS was significantly elevated in fermented than cooked beans at 0.2M and 0.6M ($p<0.05$). Decreased PS in cooked flours may be attributed to elevation of surface hydrophobicity, extreme denaturation, coagulation of protein and disulfide formation due to heating. Better PS of *R. oligosporus* fermented beans over cooked and raw beans were also evident in cowpea, sorghum and peanut (Quinn and Beuchat, 1975; Prinyawiwatkul *et al.*, 1997; Elkhalifa *et al.*, 2005).

The pH had a significant impact on the PS profile between cooked and fermented beans of *C. maritima*. Fermentation significantly improved the PS across the pH range compared to cooked beans ($p<0.05$) with exception of pH 10. The high PS of fermented flours at a pH 4 (27.8 %) may be due to the proteolytic activity of the fungus yielding peptides and free amino acids and also due to the inactivation of phytates during fermentation (Beuchat, 1977; Elkhalifa *et al.*, 2005). As depicted in Figure 15.1, minimum (6.2 %, pH 8) and maximum (11.1 %, pH 10) PS was seen at basic pH in cooked beans. The raw bean flours of *C. maritima* and peanut showed maximum PS at pH 10 (Seena and Sridhar 2005; Yu *et al.*, 2007). On the contrary, fermented beans showed minimum (11.5 %) and maximum (27.8 %) PS at acidic pH (pH 2 and pH 4). Elkhalifa *et al.* (2005) reported minimum PS in fermented sorghum flours at pH 6 (34–42 %). The highest PS in fermented beans of *C. maritima* at pH 4 was almost equivalent to that of raw bean flours *C. maritima* and *C. cathartica* (27.8 vs. 27.8–28.1 %) (Seena and Sridhar, 2006). At pH 4, unfermented flours of many legumes (*C. cathartica, C. ensiformis, Mucuna, Sorghum*, faba beans and peas) exhibited least PS (Sosulaski and McCurdy, 1987; Adebowale and Lawal, 2004; Elkhalifa *et al.*, 2005; Seena and Sridhar, 2005). Elevated PS in raw seed flours of *C. cathartica, C. ensiformis, C. maritima* and *Vigna unguiculata* at extreme acidic and alkaline conditions has been reported by several authors (Chel-Guerrero *et al.*, 2002; Ragab *et al.*, 2004; Seena and Sridhar, 2005). Higher PS in fermented flours imparts certain desirable characteristics in food formulations and finds its potential application in the preparation of baby foods, diet drinks, carbonated drinks and desserts (Idouraine *et al.*, 1997).

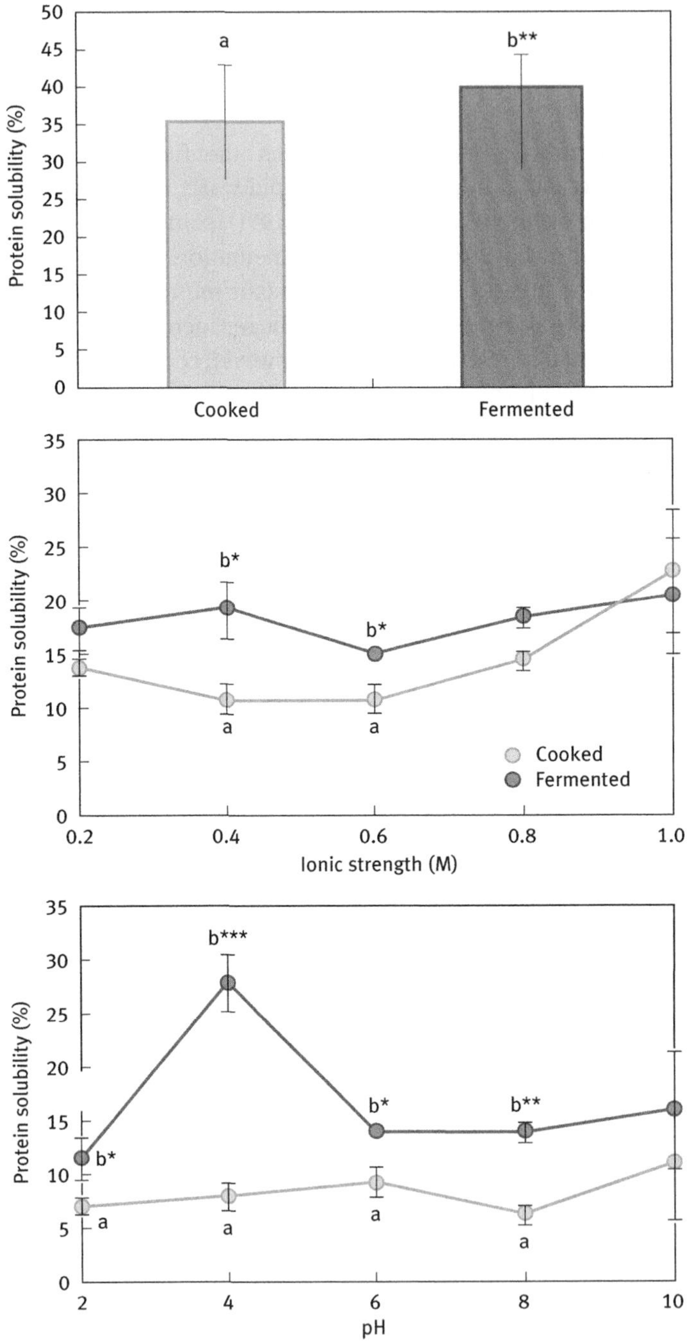

Fig. 15.1: Profile of protein solubility (PS) of cooked and fermented bean flours of *C. maritima* in distilled water, ionic concentration and pH (n=5, mean± SD) (different alphabets between cooked and fermented flours significantly differed: *, $p<0.05$; **, $p<0.01$; ***$p<0.001$).

15.5.2 Water-Absorption Capacity

The water-absorption capacity (WAC) indicates the ability of a product to associate with water and it gives an indication of the amount of water available for gelatinization. Carbohydrate content of seed flours is one of the major factors influencing the WAC (Paredes-López *et al.*, 1991). Fermented beans resulted in the significant decrease in WAC ($p<0.001$) (2.8 vs. 2.4 ml/g) (Fig. 15.2). It was comparable with raw seeds of *C. maritima*, *Mucuna pruriens* and protein isolates of *C. ensiformis* (2.2–2.7 ml/g) (Chel-Guerrero *et al.*, 2002; Seena and Sridhar 2005; Bhat *et al.*, 2008) and higher than raw flours of *C. cathartica*, cow pea, green gram, mung bean and pea (Mesallam and Hamza, 1987; Chau *et al.*, 1997; Prinyawiwatkul *et al.*, 1997; Yu *et al.*, 2007).

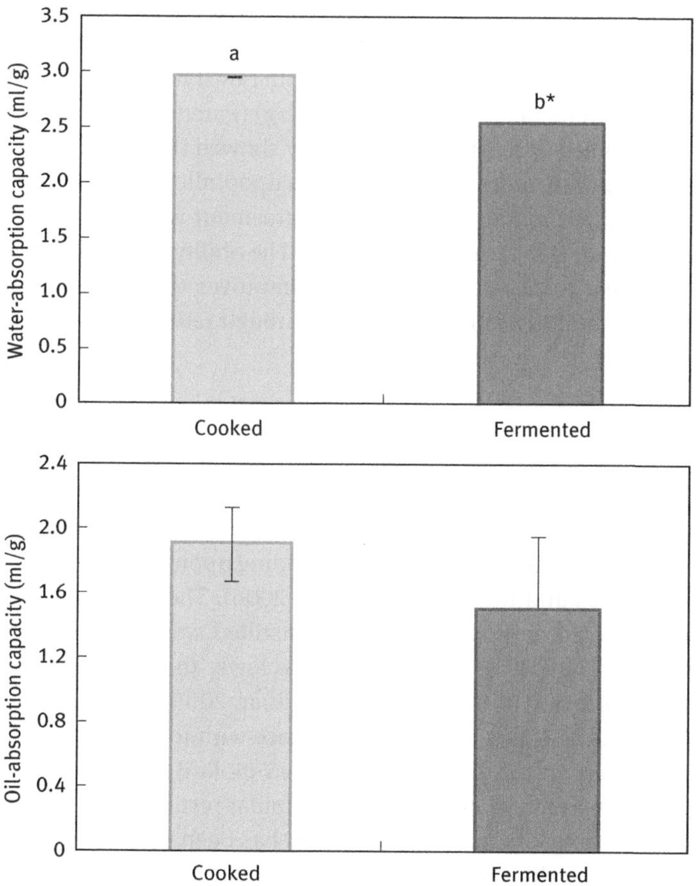

Fig. 15.2: Water- and oil-absorption capacities (WAC and OAC) of cooked and fermented bean flours of *C. maritima* (n=5, mean ± SD) (different alphabets between cooked and fermented flours significantly differed: *, $p<0.001$).

Fermentation significantly decreased the WAC of sorghum flour by about 7 %, which is similar to the present study. Fermentation with *R. oligosporus* drastically decreased the carbohydrate content (63.3 vs. 51.3 %) (Niveditha and Sridhar, 2015), which has influenced WAC in the present study. The difference in the WAC between cooked and fermented seed flours can also be attributed to the protein fractions (albumins and globulins) as well as external factors like stirring velocity, pH and protein concentration (Chel-Guerrero *et al.*, 2002).

15.5.3 Oil-Absorption Capacity

The oil absorption capacity (OAC) of cooked beans was higher compared to fermented beans (1.9 vs. 1.5 ml/g) (Fig. 15.2), and well above those reported for *Vigna ungiculata* and *Phaseolus calcaratus* (0.7–1.3 ml/g) (Prinyawiwatkul *et al.*, 1997; Chau and Cheung, 1998). The OAC of fermented flour is comparable with raw flours of *C. cathartica*, *C. maritima* and protein isolate of chick pea (1.4–1.7 ml/g) (Paredes-López *et al.*, 1991; Seena and Sridhar, 2005). Results of the present study showed that cooked and fermented flours of *C. maritima* were more hydrophilic than lipophilic despite of processing methods (cooking and fermentation). This is in agreement with the results reported by Prinyawiwatkul *et al.* (1997) on cowpea flours. The ability of cooked and fermented flours of *C. maritima* to absorb water and oil improves the palatability, flavor retention and enhance shelf life in meat products through reduction of moisture and fat loss.

15.5.4 Gelation

Gelation is an important functional property of globular proteins commonly used in food industry. It differs among the legume seed flours depending upon the interaction between proteins, carbohydrates and lipids (Moure *et al.*, 2006). The least gelation concentration (LGC) for fermented and cooked flours in distilled water was 14 and 20 %, respectively (Fig. 15.3). The LGC of cooked flour was lower than raw flour of *C. cathartica* (16 %) and *C. maritima* (18 %) (Seena and Sridhar, 2005). The LGC was minimum (14 %) at 0.6M in cooked flours and tend to increase with the higher molar concentration. A linear increase in the LGC with increased molarity in fermented flours may be attributed to decrease in protein unfolding. Similar results on the effect of ionic strength on gelling capacity of *Cajanus cajan* flours has been reported (Akintayo *et al.*, 1999). The LGC was in the range of 12–14 % for cooked and 18–22 % for fermented flours between pH 2 and 10. Gelation and WAC of seed flours are interrelated (Fernandez-Quintela *et al.*, 1997; Prinyawiwatkul *et al.*, 1997). The low WAC in fermented than cooked flours of *C. maritima* could explain the poor gelation capacity. In addition to ionic strength and pH, other factors like polysaccharide concentration,

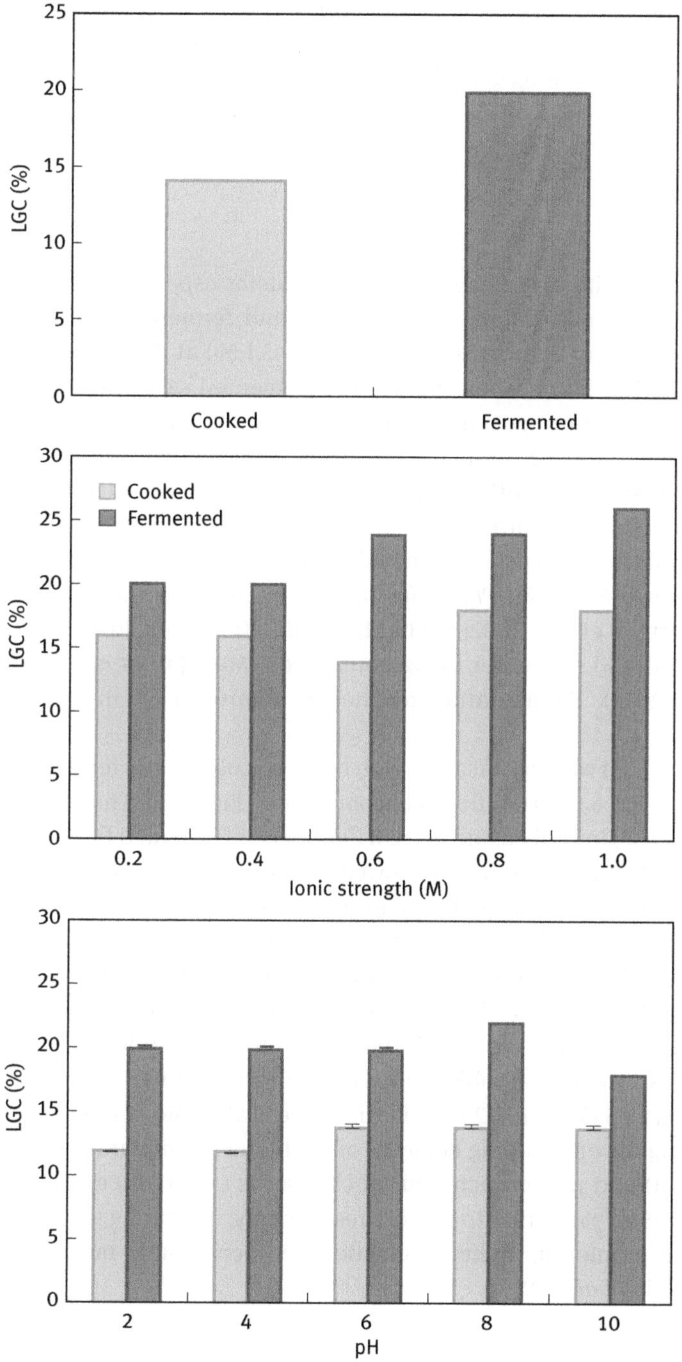

Fig. 15.3: Least gelation concentration (LGC) of cooked and fermented bean flours of *C. maritima* in distilled water, ionic concentration and pH (average of five independent tirals).

extent of proteolysis and presence of phenolic compounds influence the gelling properties. High amount of total phenolics in fermented flours of *C. maritima* over cooked flours might have influenced the gelling capacity of fermented flours (Niveditha *et al.*, 2014).

15.5.5 Emulsion Properties

Emulsion formation and stability play a key role in food systems especially in salad dressing, frozen desserts and mayonnaise. Both cooked and fermented flours of *C. maritima* exhibited good emulsion capacity (EC) (42.6–45.1 %) at different flour concentrations (Fig. 15.4). The EC between cooked and fermented samples did not vary significantly at different flour concentrations except for 2 % ($p<0.01$). The EC was highest at 4 % (44.6 %) and 10 % (45.1 %) for fermented and cooked flours, respectively. It was not substantially differed between cooked and fermented flours at different ionic strengths (0.2–1M) although it was highest at 0.6M ($p>0.05$), while it differed significantly only at pH 8 ($p<0.05$). The EC values at different pH are comparable with *C. ensiformis* (48–55 %), *Phaseolus lunatus* (41.8 %), sorghum flour (41.2–49.4 %) (Chel-Guerrero *et al.*, 2002; Elkhalifa *et al.*, 2005), but lower than chia (51–56 %), sesame (81 %) and other oleaginous seeds (Mahajan *et al.*, 1999; Vásquez-Ovando *et al.*, 2012). Fermentation resulted in improved EC in peanut flours (Yu *et al.*, 2007).

The emulsion stability (ES) was significantly high between cooked and fermented flours at 8 % ($p<0.01$) and 10 % ($p<0.05$) flour concentrations (Fig. 15.5). The ES was minimum at 0.4 and 1M (97 %) in cooked and at 0.2M (94.3 %) for fermented flour with highest ES (100 %) at 0.6M. Fermented flours showed better ES than cooked flours, reaching 100 % at pH 4 and 6, which is higher than the values reported for soybean protein isolate (98 %) and fermented sorghum flour (52.1 %) (Chau *et al.*, 1997; Elkhalifa *et al.*, 2005). The higher EC and ES of fermented flour might be due to the action of proteases produced by *R. oligosporus* during fermentation (Nowak and Szebiotko, 1992; Varzakas, 1994). Proteases degrade the polypeptides into oligopeptides, consequently increase PS and expose hydrophobic groups to water and oil interface, which results in enhanced EC and ES (Yu *et al.*, 2007). Particle size is also one of the important factors in determining the emulsifying behavior of the flours (Damodaran, 1997). Corn flour gluten (<53 µm) and protein-rich flour of chia seeds (~140 µm) exhibited emulsifying activity (46.2–56.2 %) similar to that of present study, indicating the influence of microparticles in promoting higher interaction between protein molecules (Wu, 2001; Vásquez-Ovando *et al.*, 2012).

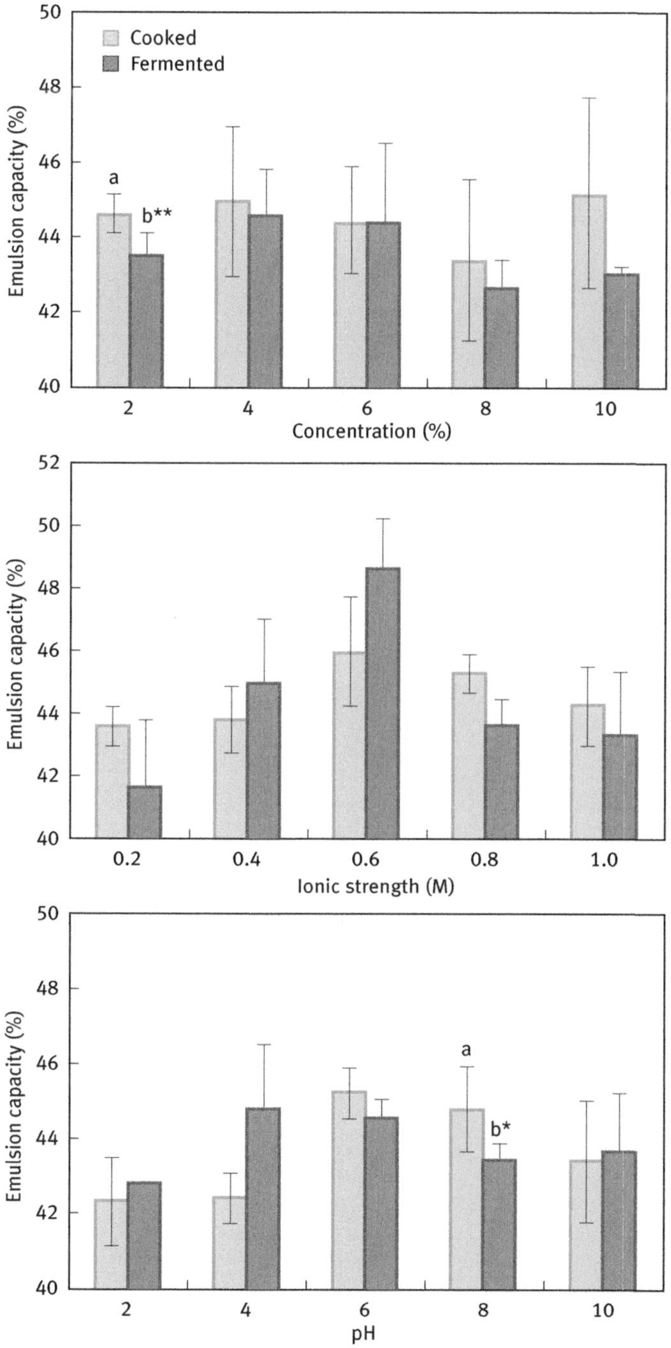

Fig. 15.4: Emulsion capacity (EC) of cooked and fermented bean flours of *C. maritima* in distilled water, ionic concentration and pH (n=5, mean ± SD) (different alphabets between cooked and fermented flours significantly differed: *, p<0.05; **, p<0.01).

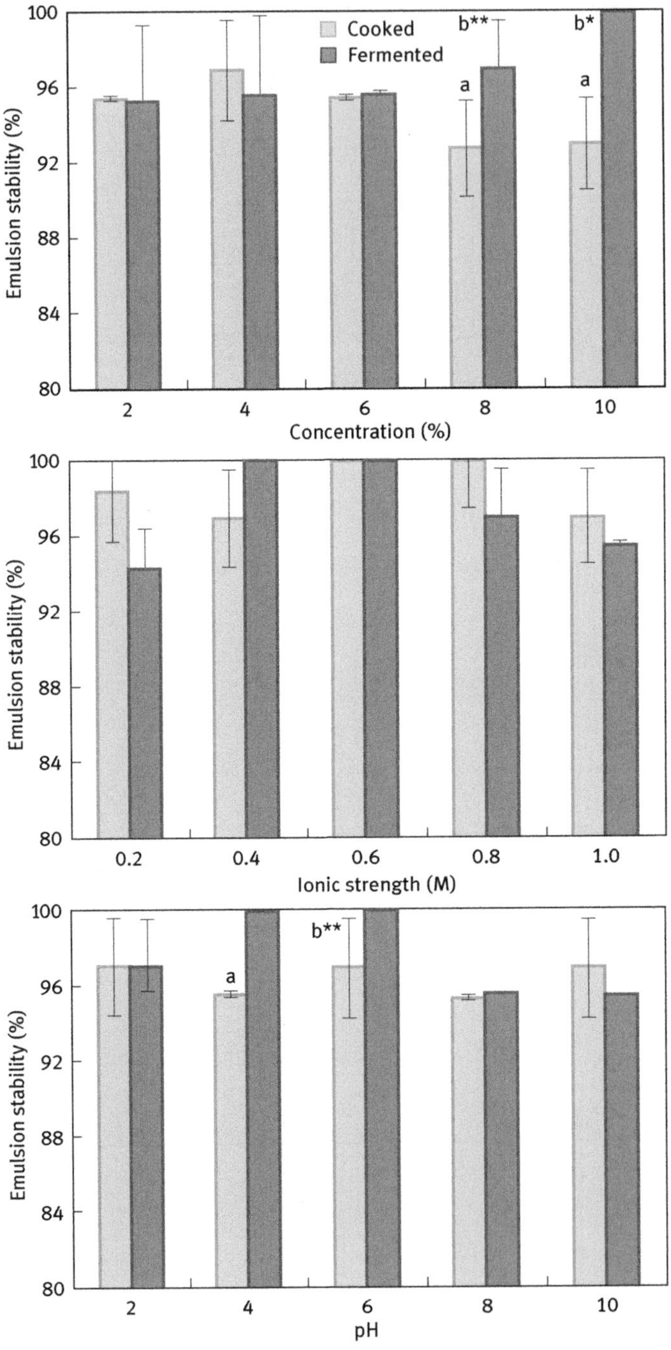

Fig. 15.5: Emulsion stability (ES) of cooked and fermented bean flours of *C. maritima* in distilled water, ionic concentration and pH (n=5, mean ± SD) (different alphabets between cooked and fermented flours significantly differed: *, $p<0.05$; **, $p<0.001$).

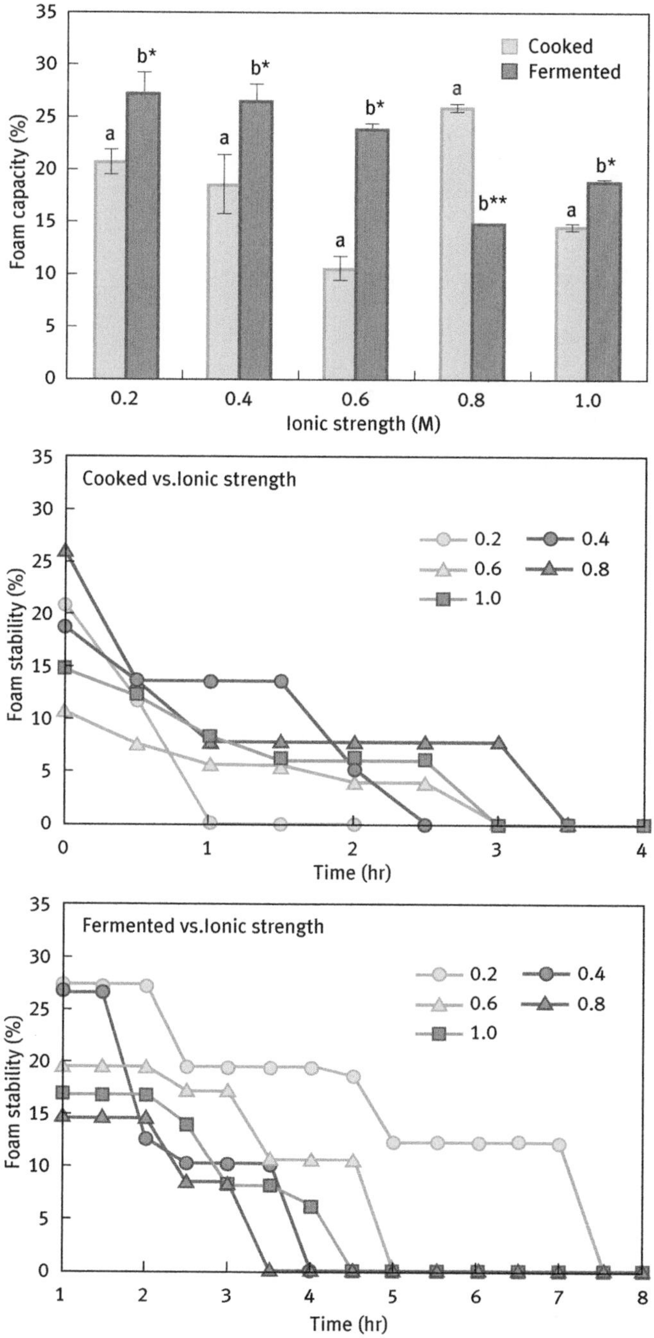

Fig. 15.6: Foam capacity (FC) and foam stability (FS) of cooked and fermented bean flours of *C. maritima* in different ionic concentration (n=5, mean ± SD) (different alphabets between cooked and fermented flours significantly differed: *, p<0.01; **, p<0.001).

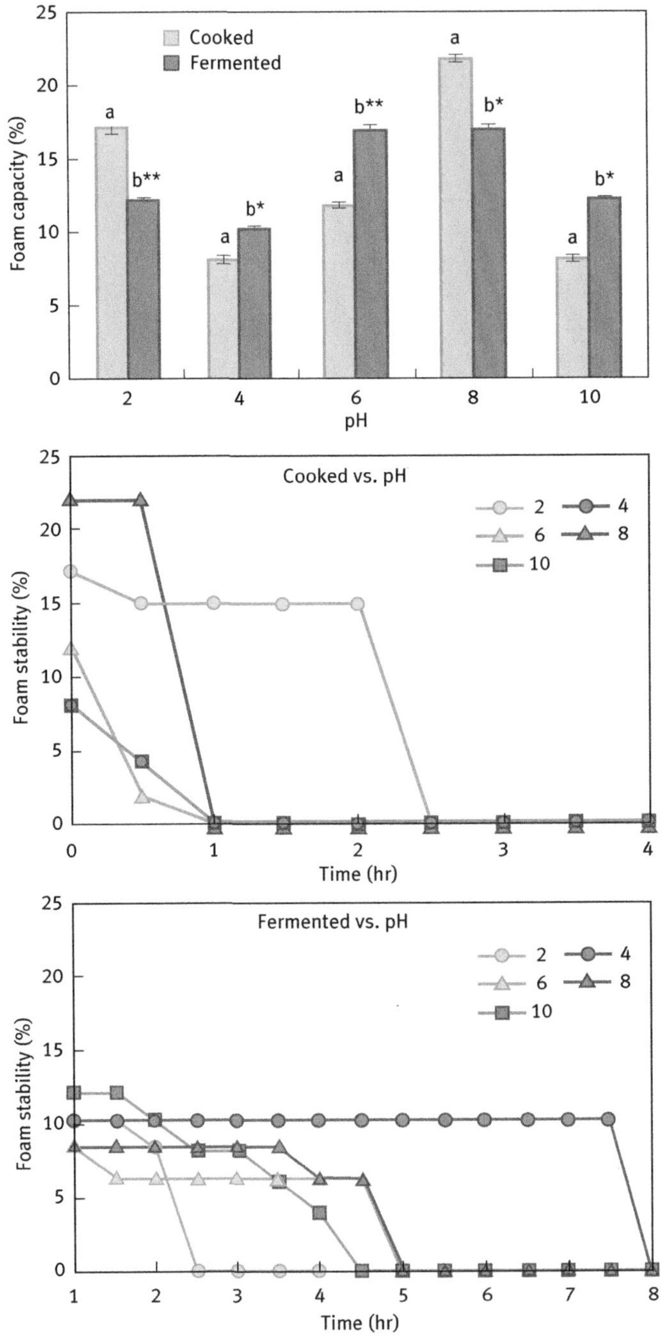

Fig. 15.7: Foam capacity (FC) and foam stability (FS) of cooked and fermented bean flours of *C. maritima* in different pH (n=5, mean ± SD) (different alphabets between cooked and fermented flours significantly differed: *, p<0.01; **, p<0.001).

15.5.6 Foam Properties

The foam capacity (FC) and foam stability (FS) were low in cooked flours at most of the ionic strength tested (Fig. 15.6). The FC was highest at 0.8M for cooked (25.9 %) and at 0.2M for fermented (27.08 %) flours. Linear decrease in FC was seen with increasing ionic strength in fermented flours with an exception of 1M (0.2>0.4>0.6>0.8M: 27.1>26.6>23.9>14.8 %). Similarly, cooked flours showed poor FC at increasing ionic strength except for 0.8M. The FS was prolonged in fermented than cooked flours. Variation in FC was pH-dependent in cooked as well as fermented flours with lowest at pH 4 (8.2 %, 10.2 %, respectively) and highest at pH 8 (21.9 %, 17 %) (Fig. 15.7). This result is closely in agreement with raw flours of *C. cathartica* and *C. maritima* (Seena and Sridhar, 2005). Both the samples showed improved FC at acidic as well as alkaline pH and the FS at different pH was better in fermented over cooked flours, however, linear increase in FC and FS was seen in raw flours (Seena and Sridhar, 2006). The overall increase in the FC in fermented flour may be attributed to the high protein content (Niveditha and Sridhar, 2015) and its solubility (Fig. 15.1). As albumin fraction is known to play a key role in foam properties (Damodaran, 1997), low albumin content in cooked compared to fermented beans (10.3 vs. 14.5 %) (Niveditha and Sridhar, 2015) might have negative effect on FC and FS of cooked *C. maritima*. Fermented flours of *C. maritima* can be used as foaming agent in frozen desserts, cakes, whipped toppings and ice-cream mixes, where high foam capacity and stability are essential.

15.6 Principal Component Analysis

The PCA for cooked beans consists of two components (Eigen value, <1) with 100 % total variation (PC1, 70.24 %; PC2, 29.76 %) (Fig. 15.8). Out of three distinct clusters, one situated towards the right of PCA score plot and the rest two towards left. The right cluster embodied with crude protein and carbohydrates vs. protein solubility (NaCl), emulsion capacity/stability (pH and NaCl) and foam capacity (pH). Towards left of the plot, crude fiber clustered with emulsion capacity/stability (flour concentration) and water-absorption capacity. Total lipid clustered with protein solubility (water) and foam capacity (NaCl). Oil-absorption capacity and protein solubility (pH) remained alone towards right and left of the plot, respectively.

Two principal components (Eigen value, <1) in fermented beans showed 100 % total variation (PC1, 57.42 %; PC2, 42.58 %) (Fig. 15.9). Among the four distinct clusters, two clustered towards right and the rest two towards left of the plot. In the first right cluster, crude protein and total lipids combined with protein solubility (pH), emulsion stability (flour concentration, NaCl and pH). In second right cluster, protein solubility (NaCl) combined with foam capacity (pH). Among two clusters towards left, first cluster showed combination of oil-absorption capacity with emulsion capacity

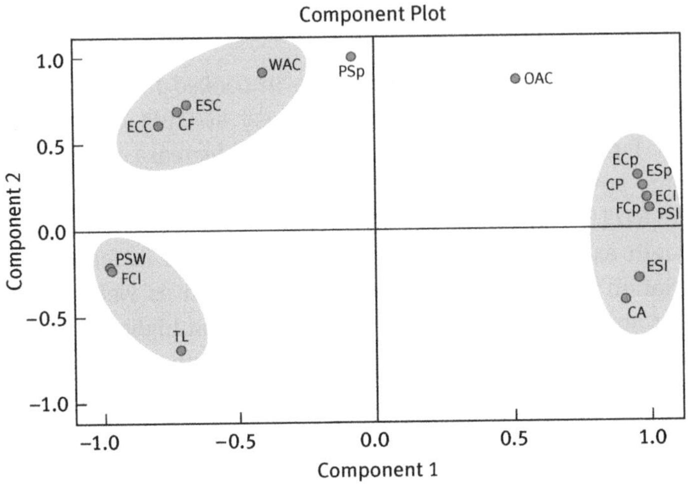

Fig. 15.8: Principal component analysis (PCA) score plot (PC1, 70.24 %; PC2, 29.76 %) for proximal and functional properties of cooked bean flours of *C. maritima* (Proximal properties: CP, crude protein; TL, total lipids; CF, crude fiber; CA, carbohydrates; Functional properties: PSW, protein solubility in water; PSC, PS in ionic concentration; PSp, PS in pH; WAC, water-absorption capacity; OAC, oil-absorption capacity; ECC, emulsion capacity in flour concentration; ECI, EC in ionic concentration; ECp, EC in pH; ESC, emulsion stability in flour concentration; ESI, ES in ionic concentration; ESp, ES in pH; FCI, foam capacity in ionic concentration; FCp, FC in pH).

(NaCl and pH), while in the second cluster, carbohydrates combined with water-absorption capacity and foam capacity (NaCl). The rest three parameters, crude fiber, protein solubility (water) (towards left) and emulsion capacity (flour concentration) (towards right) stood alone.

Comparison of cooked and fermented beans based on PCA score plots reveals that in cooked beans, crude protein and carbohydrates mainly responsible for foam properties, while in fermented beans, it was due to crude proteins and total lipids. However, high emulsion capacity (flour concentration) and water-absorption capacity in cooked beans were under the influence of crude fiber corroborating the observations by Betancur *et al.* (2004) in *C. ensiformis*. On the contrary, lowered crude fiber and carbohydrate contents in fermented beans were responsible for low water-absorption capacity. Even though there is a drastic difference in the functional properties between raw and cooked beans due to heat-denaturation of proteins (Akpapunam and Sefa-Dedeh, 1997), fermented beans showed improved proximal properties along with selectively-improved functional properties.

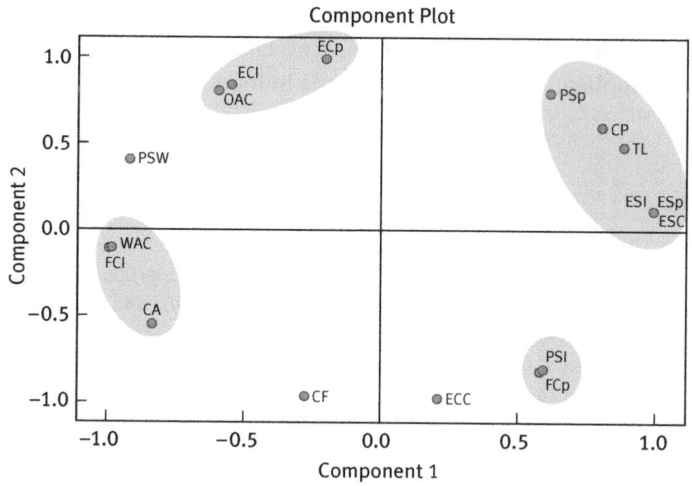

Fig. 15.9: Principal component analysis (PCA) score plot (PC1, 57.42 %; PC2, 42.58 %) for proximal and functional properties of fermented bean flours of *C. maritima* (Proximal properties: CP, crude protein; TL, total lipids; CF, crude fiber; CA, carbohydrates; Functional properties: PSW, protein solubility in water; PSC, PS in ionic concentration; PSp, PS in pH; WAC, water-absorption capacity; OAC, oil-absorption capacity; ECC, emulsion capacity in flour concentration; ECI, EC in ionic concentration; ECp, EC in pH; ESC, emulsion stability in flour concentration; ESI, ES in ionic concentration; ESp, ES in pH; FCI, foam capacity in ionic concentration; FCp, FC in pH).

15.7 Summary

Cooked and *Rhizopus oligosporus* solid substrate fermented beans of *Canavalia maritima* of coastal sand dunes of Southwest of India with considerable amount of crude proteins (29–41 %), carbohydrates (51–63 %), low amount of total lipids (1–1.7 %) and crude fiber (0.4–1.6 %) is useful in production of valuable food or pharmaceutical products. Improved protein solubility was seen in fermented than in cooked beans at various ionic strengths and pH, but the gelation property, water-/oil-absorption capacities were not improved on fermentation. The emulsion capacity and stability were selectively improved in different flour/ionic concentrations and pH. The foam capacity was also behaved similar to emulsion capacity, while the foam stability was consistently prolonged in fermented beans. Adequate hydrophobic and hydrophilic balance might be responsible for good emulsion properties than foam properties. Based on the improved proximal and functional properties, it is possible to utilize another coastal sand dune/mangrove landrace *Canavalia cathartica* as composite flour for improved nutraceutical formulations. Further assessment of functional properties of protein isolates of fermented beans of *C. maritima* might facilitate production of value-added indigenous nutraceutical product. For instance, improvement in protein functionality through phosphorylation, acetylation and succinylation might be effective in wider application of *Canavalia* beans in food and pharmaceutical industries.

15.8 References

[1] Adebowale, K.O. & Lawal, O.S. (2004). Comparative study of the functional properties of bambara groundnut (*Voandzeia subterranean*), jack bean (*Canavalia ensiformis*) and mucuna bean (*Mucuna pruriens*) flours. Food Research International, 37, 355–365.

[2] Akintayo, E.T., Oshodi, A.A. & Suoso, K.O. (1999). Effect of ionic strength and pH on the foaming and gelation of pigeon pea (*Cajanus cajan*) protein concentrates. Food Chemistry, 66, 51–56.

[3] Akpapunam, M.A. & Sefa-Dedeh, S. (1997). Some physiological properties and antinutritional factors of raw, cooked and germinated jack bean (*Canavalia ensiformis*). Food Chemistry, 59, 121–125.

[4] Arun, A.B. & Sridhar, K.R. (2004). Symbiotic performance of fast-growing rhizobia isolated from the coastal sand dune legumes of west coast of India. Biology and Fertility of Soils, 40, 435–439.

[5] Bernardi L.S., Pilosof, A.M.R. & Bartholomal, G.B. (1991). Enzymatic modification of soy protein concentrates by fungal and bacterial proteases. Journal of American Oil Chemists' Society, 68, 102–105.

[6] Betancur, A.D., Peraza, M.G., Moguel, O.Y. & Fuertes, B.S. (2004). Physicochemical characterization of lima bean (*Phaseolus lunatus*) and Jack bean (*Canavalia ensiformis*) fibrous residues. Food Chemistry, 84, 287–295.

[7] Beuchat, L.R. (1977). Functional and electrophoretic characteristics of succinylated peanut flour protein. Journal of Agricultural and Food chemistry, 25, 258–261.

[8] Bhagya, B. & Sridhar, K.R. (2009). Ethnobiology of coastal sand dune legumes of southwest India. Indian Journal of Traditional Knowledge, 9, 611–620.

[9] Bhat, R. & Karim, A.A. (2009). Exploring the nutritional potential of wild and underutilized legumes. Comprehensive Review of Food Science and Food Safety, 8, 305–331.

[10] Bhat, R., Sridhar, K.R., Young, C.-C., Arun, A.B. & Ganesh, S. (2008). Composition and functional properties of raw and electron beam irradiated *Mucuna pruriens* seeds. International Journal of Food Science and Technology, 43, 1338–1351.

[11] Bourdy, G., Cabalion, P., Amade, P. & Laurent, D. (1992). Traditional remedies used in the Western Pacific for the treatment of ciguatera poisoning. Journal of Ethanopharmacology, 36, 163–174.

[12] Boye, J., Zare, F. & Pletch, P. (2010). Pulse proteins: Processing, characterization, functional properties and applications in food and feed. Food Research International, 43, 414–431.

[13] Bressani, R., Brenes, R.S., Gracia, A. & Elias, L.G. (1987). Chemical composition, amino acid content and protein quality of *Canavalia* spp. seeds. Journal of Science of Food and Agriculture, 40, 17–23.

[14] Chau C.F., Cheung, K & Wong. Y.S. (1997). Functional properties of protein isolates from three Chinese indigenous legume seeds. Journal of Agricultural and Food Chemistry, 45, 2500–2503.

[15] Chau, C.F & Cheung, P.C.K. (1998). Functional properties of flours prepared from three Chinese indigenous legume seeds. Food Chemistry, 61, 429–433.

[16] Chel-Guerrero, L., Pearez-Flores, V., Betancur-Ancona, D. & Daavila-Ortiz, G. (2002). Functional properties of flours and protein isolates from *Phaseolus lunatus* and *Canavalia ensiformis* seeds. Journal of Agricultural and Food Chemistry, 50, 584–591.

[17] Chel-Guerrero, L., Pearez-Flores, V., Betancur-Ancona, D. & Daavila-Ortiz, G. (2002). Functional properties of flours and protein isolates from *Phaseolus lunatus* and *Canavalia ensiformis* seeds. Journal of Agricultural and Food chemistry, 50, 584–591.

[18] Coffman, C.W. & Garcia, V.V. (1977). Functional properties and amino acid content of protein isolate from mung bean flour. Journal of Food Technology, 12, 473–484.

[19] Damodaran, S. (1997). Food Proteins: An overview. In S. Damodaran & A. Paraf (Eds.), Food Proteins and their Applications (pp. 1–30). Marcel Dekker: New York.

[20] Elkhalifa, A.E.O., Schiffler, B. & Bernhardt, R. (2005). Effect of fermentation on the functional properties of sorghum flour. Food Chemistry, 92, 1–5.

[21] Fernandez-Quintela, A., Macarulla, M.T., Del Barrio A.S. & Martinez, J.A. (1997). Composition and functional properties of protein isolates obtained from commercial legumes grown in northern Spain. Plant Foods for Human Nutrition, 51, 331–342.

[22] Humphries, E.C. (1956). Mineral composition and ash analysis. In K. Peach & M.V. Tracey (Eds.), Modern Methods of Plant Analysis, Volume # 1 (pp. 468–502). Berlin: Springer.

[23] Idouraine, A., Yensen, S. & Weber, C. (1997). Tepary Bean Flour, Albumin and Globulin Fractions Functional Properties Compared with Soy Protein Isolate. Journal of Food Science, 56, 1316–1318.

[24] Kabirullah, M. & Willis, R.B.H. (1982). Functional properties of acetylated and succinylated sunflower protein isolate. Food Technology, 17, 235–249.

[25] Kinsella, J.E. (1976). Functional properties of proteins in Foods – A survey. Critical Reviews in Food Science and Nutrition, 7, 219–232.

[26] Mahajan, A., Bharadwaj, S. & Dua, S. (1999). Traditional processing treatments as a promising approach to enhance the functional properties of rape seed (*Brassica campestris* var. *toria*) and sesame seed (*Sesamum indicum*) meals. Journal of Agricultural and Food Chemistry, 47, 3093–3098.

[27] Martinez, M.L. & Psuty, N.P. (2004). Coastal Dunes: Ecology and Conservation. Berlin and Heidelberg: Springer-Verlag.

[28] McCue, P. & Shetty, K. (2003). Role of carbohyddrate-cleaving enzymes in phenolic antioxidant mobilizatin from whole soybean fermented with *Rhizopus oligosporus*. Food Biotechnology, 17, 27–37.

[29] Mesallam, A.S. & Hamza, M.A. (1987). Studies on green gram (*Phaseolus aureus*) protein concentrate and flour. Plant Foods for Human Nutrition, 37, 17–27.

[30] Moure, A., Sineiro, J., Dominguez, H. & Parajo, J.C. (2006). Functionality of oilseed protein products: A Review. Food Research International, 39, 945–963.

[31] Neto, V.Q., Narain, N., Silvia, J.B. & Bora, P.S. (2001). Functional properties of raw and heat-processed cashew nut (*Anacardium occidentale* L.) kernel protein isolate. Nahrung, 45, 258–262.

[32] Niveditha, V.R. & Sridhar, K.R. (2015). Nutritional qualities of fermented beans of coastal sand dune wild legume *Canavalia maritima*. In R.R. Watson, J.A. Tabor, J.E. Ehiri, & V.R. Preedy (Eds.), Handbook of Public Health in Natural Disasters (pp. 441–462). The Netherlands: Wageningen Academic Publishers.

[33] Niveditha, V.R., Sridhar, K.R. & Tomita-Yokotani, K. (2014). Improvement of bioactive potential of *Canavalia* beans of coastal sand dunes by solid-substrate fermentation using *Rhizopus oligosporus*. Current Nutrition & Food Science, 10, 308–313.

[34] Nowak, J. & Szebiotko, K. (1992). Some biochemical changes during soybean and pea tempeh fermentation. Food Microbiology, 9, 37–43.

[35] Paredes-López, O, Ordorica, F. & Olivares, M. (1991). Chickpea protein Isolates: Physico-chemical, functional and nutritional characterization. Journal of Food Science, 56, 726–729.

[36] Prinyawiwatkul, W., Beuchat, L.R., McWatters, K.H. & Phillips, R.D. (1997). Functional properties of cowpea (*Vigna unguiculata*) flour as affected by soaking, boiling and fungal fermentation. Journal of Agricultural and Food Chemistry, 45, 480–486.

[37] Quinn, M.R. & Beuchat, L.R. (1975). Functional property changes resulting from fungal fermentation of peanut flour. Journal of Food Science, 40, 475–478.

[38] Ragab, D.M., Babiker, E.E. & Eltinay, A.H. (2004). Fractionation, solubility and functional properties of cowpea (*Vigna unguiculata*) proteins as affected by pH and/or salt concentration. Food Chemistry, 84, 207–212.

[39] Randhir, R. & Shetty, K. (2007). Mung beans processed by solid-state bioconversion improves phenolic content and functionality relevant for diabetes and ulcer management. Innovative Food Science and Emerging Technology, 8, 197–204.

[40] Seena, S. & Sridhar, K.R. (2005). Physicochemical, functional and cooking properties of under explored legumes, *Canavalia* of the southwest coast of India. Food Research International, 38, 803–814.

[41] Seena, S. & Sridhar, K.R. (2006). Nutritional and microbiological features of little known legumes, *Canavalia cathartica* Thouars and *C. maritima* Thouars of the southwest coast of India. Current Science, 90, 1638–1650.

[42] Seena, S., Sridhar, K.R. & Arun, A.B. (2007). *Canavalia cathartica* of southwest coast of India – A neglected wild legume. Plant Genetic Resources Newsletter # 150, 16–20.

[43] Siddhuraju, P. & Becker, K. (2001). Species/variety differences in biochemical composition and nutritional value of Indian tribal legumes of the genus *Canavalia*. Nahrung, 45, 224–233.

[44] Sosulski, F.W. & A.R. McCurdy. (1987). Functionality of flours, protein fractions and isolates from field peas and faba bean. Journal of Food Science, 52, 1010–1014.

[45] Sridhar, K.R. & Seena, S. (2006). Nutritional and antinutritional significance of four unconventional legumes of the genus *Canavalia* – A comparative study. Food Chemistry, 99, 267–288.

[46] StatSoft Inc. (2008). Statistica Version 8. Tulsa, Oklahoma, USA: StatSoft.

[47] Tharanathan, N. & Mahadevamma, S. (2003). Gain legumes – A boon to human nutrition. Trends in Food Science and Technology, 14, 507–518.

[48] Vadivel, V., Doss, A. & Pugalenthi, M. (2010). Evaluation of nutritional value and Protein quality of raw and differentially processed sword bean (*Canavalia gladiata* (Jacq.) DC.) seeds. African Journal of Food Agriculture Nutrition and Development, 10, 2850–2865.

[49] Varzakas, T. (1994). *Rhizopus oligosporus* mycelia penetration and enzyme diffusion in soya bean tempeh. Process Biochemistry, 33, 741–747.

[50] Vásquez-Ovando A., Betancur-Ancona, D. & Chel-Guerrero, L. (2012). Physicochemical and functional properties of a protein-rich fraction produced by dry fractionation of chia seeds (*Salvia hispanica* L.). CyTA Journal of Food, DOI: 10.1080/19476337.2012.734163.

[51] Vatanparast, M. (2010). Phylogeography of a pantropical plant with sea-drifted seeds: *Canavalia rosea* (Sw.) DC. (Fabaceae), Dissertation. Graduate School of Science, Japan: Chiba University.

[52] Were, L., Hettiarachchy, L. & Kalapathy, U. (1997). Modified soy proteins with improved foaming and water hydration proteins. Journal of Food Science, 62, 821–824.

[53] Wu, Y. (2001). Emulsifying activity and emulsion stability of corn gluten meal. Journal of Science of Food and Agriculture, 81, 1223–1227.

[54] Xu, M.-J., Huang, X.-P., Li, M., Sun, W., Cui, J.-R. & Lin, W.-H. (2009). Cytotoxic and proapoptotic activities of medicarpin from *Canavalia maritima* (Aubl.) via the suppression of NFκB activation in HeLa cells. Journal of Chinese Pharmaceutical Sciences, 18, 331–336.

[55] Yu, J., Ahmedna, M. & Goktepe, I. (2007). Peanut protein concentrate: Production and functional properties as affected by processing. Food Chemistry, 103, 121–129.

Printed and bound by CPI Group (UK) Ltd, Croydon, CR0 4YY

13/09/2026

14960851-0005